ENVIRONMENTAL LIFE CYCLE ASSESSMENT

MEASURING THE ENVIRONMENTAL PERFORMANCE OF PRODUCTS

RITA SCHENCK AND PHILIP WHITE, EDITORS

ISBN: 978-0-9882145-5-2

Printed in the United States of America

Preface

The field of Life Cycle Assessment has flourished and broadened immensely since the Coca-Cola Company performed the first LCA in 1969. Given the burgeoning application of LCA, and the continuing evolution of the methodology, a single person can no longer be an expert in all aspects of LCA. Just as the fields of biology and engineering are supported by many sub-specialties, experts in many branches of environmental science, engineering, and computer science increasingly support the practice of LCA.

LCA continues to advance in achieving its promise of being a comprehensive measure of the environmental performance of products. This allows science-based and reliable comparisons between products that enable decision-makers to make informed choices that reduce the overall environmental impacts of products and services. Because nearly all human-sourced environmental impacts derive from the things we buy and sell, this is surely a good thing for the health of the biosphere.

At this juncture we take pleasure in publishing this textbook that is intended to assist a college-level student in becoming an entry-level LCA practitioner. It results from the work of many minds, each expert in their sub-discipline. Without their support and hard work, this book could not exist. We hope this text to be the instructive and thorough support to LCA educators that it was designed to be. Moreover, we hope that students understand the implications of their decisions as producers and consumers moving forward, and that their choices help to preserve a vibrant ecology for all people and all living things, in current and future generations.

RITA SCHENCK & PHILIP WHITE, EDITORS

CONTRIBUTORS

Buyung Agusdinata

Carina Alles

Catherine Benoit

Mikhail Chester

Andreas Ciroth

Matthew Eckelman

Matthias Finkbeiner

Bill Flanagan

Roland Geyer

Tom Gloria

Troy Hawkins

Rich Helling

Andrew Henderson

Shawn Hunter

Wes Ingwersen

Greg Keoleian

Christoff Koffler

Chris Meinrenken

Ivo Mersiowsky

Anna Nicholson

Hanna-Leena Pesonen

Tom Redick

Matt Pietrowski

Bev Sauer

Rita Schenck

Bengt Steen

Susanne Veith

Thorsten Volz

Chris Weber

Philip White

Ron Wroczynski

Fu Zhao

Table of Contents

FRAMEWORK OF LIFE CYCLE ASSESSMENT

PHILIP WHITE AND MIKHAIL CHESTER

1.1 WHAT IS LIFE CYCLE ASSESSMENT?

The vast majority of objects that people use are created and transformed from their natural state by other people. From a miniscule rivet to the immense networks of urban sprawl, from the foods we consume to the information systems on which we rely, the quantity of artifacts on our planet grows, as does the environmental damage these artifacts cause. As the human population encroaches upon the finite limits of the Earth's biosphere, we increasingly need to understand the environmental impacts of the systems we design and use. How can we evaluate, in a consistent and rational way, the total environmental impacts of human-made objects, services, and systems?

Environmental life cycle assessment (LCA) is the analytical framework for quantifying the resources used and the impact to the environment and human health by a product, service or system over its entire life cycle (ISO 1997). For most product systems, this means the period when raw materials are extracted from nature (the biosphere) to the period when these materials are processed, as well as the, manufacture of the product system, the distribution to the user, the use and potential upgrade of the product, and the product's eventual land-filling, incineration or recycling.

A wide range of disciplines, including system engineers, product designers, environmental chemists, policy makers, architectural planners, material scientists, and project managers, use LCA because it supports the process of identifying and improving environmental performance. LCA strives to deliver superior environmental outcomes by providing metrics based on an entire product system. The knowledge it provides helps to avoid shifting environmental impacts between life cycle stages and among different types of impacts (see page 2).

As new environmental problems and regulations emerge, there is increased interest in applying the LCA framework for better overall environmental outcomes. As cities and regions develop greenhouse gas reduction goals, LCA emerges as a common approach used to assess mitigation strategies and policy goals. As our knowledge of environmental impacts improves and goals are set to establish better balance between human and natural systems, the LCA framework is likely to expand its presence in product, process, service, and activity environmental assessment.

The comprehensive approach of LCA promotes the complete documentation and evaluation of material as it flows upstream and downstream from an actor (such as a manufacturer) in the product life cycle. A manufactured product is typically supported by a large network of suppliers and sub-suppliers. Likewise, the downstream information includes potentially diverse purchasers with disparate modes of use and end-of-life waste treatment. Wielding this complex information gives the LCA practitioner substantial power.

CENTRAL CONCEPTS OF LCA

LCA embodies several core principles that support the process of quantifying, comparing, and integrating detailed environmental and human health information of a product, service, activity, or physical system. A novice to LCA may find these attributes to be somewhat abstract, but with practice and experience their value becomes clearly evident.

LCA STRIVES TO DELIVER SUPERIOR ENVIRONMENTAL OUTCOMES

LCA helps product system teams develop products and services with few, if any, negative effects on the environment and human health.

> Environmental LCA can be expressed in several synonyms: eco-balance, life-cycle analysis, and cradle-to-grave analysis. All the terms allude to the collection of life information over the entire life cycle of a product system.

LCA CONSIDERS THE ENTIRE LIFE CYCLE OF A PRODUCT SYSTEM FROM CRADLE-TO-GRAVE

LCA typically evaluates environmental impacts in stages, identifying the point where they occur. Potential life cycle stages include extraction of raw materials from nature, refining of raw materials into technical materials, processing of technical materials into components, assembling components into finished products, using and maintaining the product system, upgrading of the product system (if applicable), and end-of-life treatment. Inclusion of multiple life cycle stages is known as vertical integration.

Not all LCA studies cover all life cycle stages. The selection of applicable life cycle stages to be modeled depends on the characteristics of the product system in question. Products such as automobiles or refrigerators that require energy in use often have their greatest impacts in the use phase. The LCA should include the use phase and evaluate the impacts of electricity or fuel consumption. In contrast, for products that consume no energy, such as newspaper newsprint, the use stage creates a minimal impact; often, such studies ignore the use phase. Impacts from consumable products, such as beer, often exclude the impacts of human waste (the use stage usually includes refrigeration), although the packaging of beer is often modeled at the end-of-life stage (Figure 1.1). Likewise, the common distribution stage from the perspective of a manufacturer is usually from the factory to the purchaser; however, the distribution of materials or products also occurs within and among all stages in the life cycle.

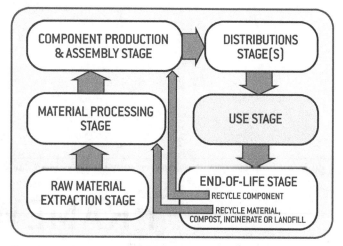

Figure 1.1. Common Life Cycle Stages of a Product System

The comprehensive approach of LCA promotes the complete documentation and evaluation of material as it flows upstream and downstream from an actor (such as a manufacturer) in the product life cycle. A manufactured product is typically supported by a large network of suppliers and sub-suppliers. Likewise, the downstream information includes potentially diverse purchasers with disparate modes of use and end-of-life waste treatment. Wielding this complex information gives the LCA practitioner substantial power.

LCA INTEGRATES ALL ECOSYSTEM HEALTH, HUMAN HEALTH, AND RESOURCE DAMAGES

An impact category exhibits damage to ecological health, human health or resource availability. Impacts assessed on a global or regional basis can consist of impact categories for ecological health (such as climate change, aquatic eutrophication, and ecotoxicity), human health (such as human respiratory health and human toxicity), and resource depletion (such as freshwater depletion or fossil fuel depletion). Resource depletion typically refers only to substances, such as refined metals, that are of economic value to humans. An impact category can create effects in more than one general type of impact. For instance climate change, ozone layer depletion, and photochemical smog all affect both human health and ecosystem health; however, one type of impact is more dominant in each case. Figure 1.2 visualizes some potential impact categories, organized according to the dominant type of impact. Inclusion of multiple impact categories is known as horizontal integration. Environmental impacts are explored in more detail in Chapter 9: Natural Science, Chapter 10: Impact Assessment and Modeling, and Chapter 11: LCIA Methods. LCA is holistic, integrating over space and time

LCA integrates relevant physical and environmental health

HUMAN HEALTH	ECOSYSTEM HEALTH	RESOURCE USE
HUMAN TOXICITY	CLIMATE CHANGE	FOSSIL FUELS
RESPIRATORY EFFECTS	AQUATIC ECOTOXICITY	MINERALS
IONIZING RADIATION	TERRESTRIAL ECOTOXICITY	FRESH WATER
OZONE LAYER DEPLETION	AQUATIC ACIDIFICATION	TOPSOIL
PHOTOCHEMICAL SMOG	AQUATIC EUTROPHICATION	
CARCINOGENICITY	TERRESTRIAL ACIDIFICATION	
	LAND TRANSFORMATION	

Figure 1.2. Selected Environmental Impact Categories Used in LCA

phenomena through a consistently applied quantitative framework. The phenomena are captured in all life cycle phases, with impacts modeled in multiple impact categories, according to the defined temporal and geographical scope of the assessment. Each LCA describes the specific timeframe and geographical scope. LCA usually models global and regional impacts, and not local ones (such as noise). The methodology allows for the modeling of changes in human elements in the product system and their respective environmental impacts over time, while the models used for characterizing impacts employ steady state conditions.

WHAT IS THE RISK OF EXCLUDING A RELEVANT LIFE CYCLE STAGE IN AN LCA?

The marketing department of Zipco, an electric car manufacturer, created a promotional campaign for its urban electric vehicle. "We've got it!" said the marketing director, "Our new slogan is 'Drive guilt-free with Zipco's Zero Emission City Car'." After months of expensive television and print advertisements, the company received heavy criticism that Zipco City Cars were not zero-emission vehicles, because the electricity generating plants produced significant amounts of emissions to charge the cars' batteries. After dropping sales, Zipco changed its marketing message for the City Car to reflect the reality of the entire system life cycle of the electric vehicle, noting that the electric vehicles produced fewer life cycle emissions than gasoline-powered automobiles.

WHAT IS THE RISK OF EXCLUDING A CRUCIAL IMPACT CATEGORY IN AN LCA?

An LCA practitioner was hired to assess several types of beer production, including the process of growing wheat, hops, and barley in addition to the brewing processes. The template of impact categories in the assessment included common categories (such as climate change, water eutrophication, acidification ozone depletion, human toxicity, ecotoxicity, and fossil fuel depletion). After many months of difficult work, when the assessment was complete, she presented the results to the beer brewer who sponsored the work. After reviewing the LCA results, the brewer told the LCA practitioner, "Beer is made from agricultural products, so why did you not include water use and land use, which are essential aspects of agriculture, in the study? We seriously question the strength of the conclusions of the study when these impact categories are missing from the study." Indeed, the exclusion of these pertinent impact categories seriously undermined the veracity of the assessment's results.

THE PRACTICE OF LCA IS STANDARDIZED

LCA practice is standardized through a series of international standards known as the ISO 14040 series. Conformity to these standards assures that all the steps for an LCA are performed in a transparent and reproducible fashion. This standardization does not limit the creativity or scope of an LCA study, but sometimes LCA studies do not follow the standard. This is especially common in academic studies.

LCA ALLOWS COMPARISON OF RADICALLY DIFFERENT PRODUCT SYSTEMS DELIVERING SIMILAR SERVICES

The functional unit in LCA enables comparison of any products or service systems that deliver the same service, regardless of dissimilar physical configurations and product system requirements. For instance, the functional unit "mile of human transport" allows for the comparison of the life cycle impacts of an automobile and a bicycle measured per mile of transport.

LCA METHODOLOGY FOLLOWS FOUR CLEARLY DEFINED PHASES:

1. *Goal and Scope Definition Phase:* Specifying the objectives, contents, and pertinent choices of the study

2. *Inventory Analysis Phase:* Compiling an inventory of relevant energy and material inputs, and environmental releases

3. *Impact Analysis Phase:* Calculating the potential impacts created by the identified inputs and emissions

4. *Interpretation Phase:* Interpreting the results throughout each stage to help make a more informed decision

Figure 1.3 visualizes the iterative flow of information in the LCA process. LCA is usually applied iteratively, whereby the results of the study are modified and improved with more

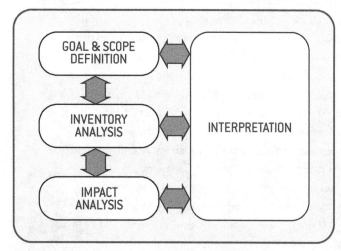

Figure 1.3. Methodological Framework for LCA (ISO 14040 and 14044)

accurate inventory data. An assessment that conforms to ISO contains specific kinds of information in each phase of the LCA report. ISO requirements are explored in Chapter 2: ISO 14040 and 14044, and reporting is explored in Chapter 18: Writing an LCA Report.

LCA PROVIDES LAYERED DATA

LCA combines data from several sources (inventory data, characterization data, weighting, and normalization data) that deliver complex, multi-layered assessment results. Results can be separated and expressed according to impact category, life cycle stage, system subset, or product constituent.

Life cycle inventory (LCI) data quantify resource extractions, land use, and substance emissions. LCI data inform us about the substance inputs and outputs of the system being assessed; however, a proper LCA must also include an impact assessment that maps each of these inventory inputs and outputs to an actual impact. LCI is explored in more detail in Chapter 5: Life Cycle Inventory and Chapter 6: Data Quality.

Life cycle impact analysis (LCIA) calculates discrete impacts in separate impact categories, as reported in different impact indicator units. LCIA provides the most purely scientific results from an LCA. LCIA results can also report the specific substances emitted from the product system that create the largest impacts. LCIA is explored in more detail in Chapter 10: Impact Assessment and Modeling, and Chapter 11: LCIA Methods.

Normalization contextualizes impact results. Likewise, weighted results reflect impacts that have been modified by culturally defined priorities among the impact categories. Both normalization and weighting are optional steps according to ISO; they can assist in the interpretation of LCIA results. Damage assessment is impact assessment at the endpoint, at an aggregated level. Normalization, weighting, and damage assessment are explored in Chapter 12: Decision Support Calculations.

1.2 ENVIRONMENTAL ASSESSMENT METHOD OVERVIEW

A broad range of analytical and risk assessment tools can assist in quantitatively modeling and understanding the environmental effects of physical systems. Table 1-1 provides an overview of LCA among common environmental analysis methods. Some approaches measure economic costs; some approaches measure physical phenomena, and other approaches measure both.

USING ECONOMIC METRICS	USING PHYSICAL METRICS	USING COMBINED METRICS
Cost Benefit Analysis	**Procedural Tools**	
Life Cycle Costing	Environmental Health & Safety Protocols (EHS)	Eco-efficiency
Input-Output Analysis	Environmental Management Systems (EMS)	LC Sustainability Assessment (LCSA)
	Environmental Impact Assessment (EIA) and	Economic Input-Output Life Cycle Assessment (LCA)
	Permits	Hybrid (EIO + Process-based) LCA
	Toxicity Screening	
	Environmental Procurement	
	Risk Analysis Tools	
	Risk Analysis (RA)	
	Environmental Risk Assessment (ERA)	
	Product System Tools	
	Checklists	
	Matrices	
	Ecological Footprint	
	Material Input per Unit of Service (MIPS)	
	Cumulative Energy Demand (CED)	
	Life Cycle Assessment (LCA)	

Table 1.1. Overview of Environmental Analysis Tools

MIDPOINT VERSUS ENDPOINT IMPACTS

Midpoint impacts are measured in specific impact category units. Land occupation can be measured in square meter of land used, water eutrophication can be measured in gram of phosphorus equivalent, and terrestrial eco-toxicity can be measured in pound of triethylene glycol (TEG) equivalents. Midpoint impact characterization usually has lower uncertainty than endpoint characterization.

End-point impacts are measured in generic units that combine specific impact categories. Land use, water eutrophication, and terrestrial ecotoxicity can all be measured as ecosystem damage in Potentially Disappeared Fraction of species per square foot per year (PDF*m2*-year). Endpoint characterization usually has higher uncertainty than midpoint characterization.

TOOLS USING ECONOMIC METRICS

COST-BENEFIT ANALYSIS

Cost benefit analysis (CBA) addresses a single activity in which all impacts are expressed in monetary terms. For example, a CBA could be used to assist a government office in determining whether an alternative program or investment plan will deliver the largest economic return over a defined period of time. A CBA could also be used to compare the cost of infrastructure projects or waste management. Environmental scientists have criticized CBA for its under-representation of ecological damage and its propensity to convert natural capital to financial capital.

LIFE CYCLE COSTING OR TOTAL COST ACCOUNTING

Life cycle costing (LCC), or total cost accounting (TCA), measures the costs that are directly accrued by actors along the life cycle of a product system. LCC can be considered to be analogous to LCA, except that it measures internalized and externalized economic costs (Hunkeler et al. 2008). LCC can focus on the costs of the manufacturer or user, or can expand to include the life cycle costs to the environment or society.

INPUT-OUTPUT ANALYSIS

Input-output analysis (IOA) models the consequences of investment or policy changes in an economic sector at a national level. IOA was created in the 1930s by Wassily Leontief, who posthumously received a Nobel Prize in Economics. A typical application of EIO LCA includes prioritizing environmental policies to support specific economic sectors (such as agricultural commodities).

TOOLS USING PHYSICAL METRICS

ENVIRONMENTAL HEALTH AND SAFETY PROTOCOLS

Manufacturing companies use Environmental Health and

Safety (EHS) protocols to insure compliance with environmental regulations and company policies. These quality-management procedures regularly monitor conditions on the site and report on performance results.

ENVIRONMENTAL MANAGEMENT SYSTEMS

Environmental Management Systems (EMS) are voluntary programs that define environmental goals for a site or an entire organization, measure annual performance, and report on progress towards the goal. For instance, ISO 14001 comprises an environmental management system that is analogous to ISO 9000 quality management, while the Eco-Management and Audit Scheme (EMAS) is a voluntary European program. EMS and EMAS involve environmental audits of the applicable sites, which may be planned or impromptu.

ENVIRONMENTAL IMPACT ASSESSMENT PERMITS

National or local laws may require an Environmental Impact Assessment (EIA) to assess the potential environmental impacts of a proposed activity, building, or infrastructural installation on a specific location. Other laws may require that detailed conditions be met for an enterprise to qualify for an environmental permit or license for a particular construction, manufacture, or emission site.

TOXICITY SCREENING

A producer can adopt a toxicity-screening protocol that identifies the toxic and hazardous substances contained in, or emitted by, a material or component. If the product meets the protocol requirements, it may receive certification to be used for marketing purposes (Heine 2008).

ENVIRONMENTAL PROCUREMENT

Enterprises can establish procurement protocols that specify environmental requirements for purchased materials, components, and services. These can take the form of voluntary agreements between stakeholders, or be part of purchasing negotiations with vendors in a supply chain.

RISK ANALYSIS

Risk analysis (RA), also known as probabilistic risk assessment, uses a stochastic approach to characterize the impacts of an activity or action. The overall risk is the product of the probability of the action occurring multiplied by the seriousness of the effect of its occurrence. RA typically focuses on human health in a local or regional context, and is often used for evaluating large industrial installations. RA is different than risk perception, which is the degree to which a potential activity is perceived to be dangerous to the public.

ENVIRONMENTAL RISK ASSESSMENT

Environmental risk assessment (ERA) scientifically examines the risks resulting from natural events, technologies, processes, chemical agents, biological agents, radiological agents, and industrial activities that pose potential threats to ecosystems and people. Human and environmental risk assessment (HERA) focuses on risks to human health. ERA often estimates the predicted environmental concentration (PEC) of a hazardous substance in the air, soil or water, and if it exceeds the predicted no-effect concentration (PNEC) of the substance, an environmental problem is probable (IPCS 2000).

CHECKLISTS

An enterprise can compile a list of environmental characteristics that must be met for a particular product or process. Successful completion of the list may be required for all products or processes in the enterprise.

MATRICES

An enterprise may employ a matrix (also referred to as matrix LCA) of environmental impacts and life cycle stages, and enlist environmental analysts in using the matrix to evaluate a product or process (Graedel 1996). The analysts estimate the impact categories in which significant impacts will occur during each life-cycle phase. The matrices may also employ weighting. The speed at which such matrices can be completed is beneficial. However, the validity of their results depends completely on the accuracy of the analyst who applies the matrices.

MATERIAL INPUT PER UNIT OF SERVICE

Material Input per Unit of Service (MIPS) divides the material inputs of a product or service by the quantity of service that the system delivers over its entire life cycle (Schmidt-Bleek 1993). MIPS tabulates the input side of LCA; it does not account for emissions or environmental impacts.

CUMULATIVE ENERGY DEMAND

Cumulative Energy Demand (CED) calculates the total amount of energy needed to produce a product or service (Frischknecht et al. 2003). The total requirements may include many types of energy (such as fossil fuels, nuclear, biomass, solar, wind, geothermal energy, and hydropower).

ECOLOGICAL FOOTPRINT

Ecological footprint provides an accounting system for bio-capacity, measuring the total environmental pressure per unit area (Wackernagel & Rees 1996). Like LCA, it is a steady state model focusing on the global scale. Unlike LCA, it accounts for energy demand on a biological basis and converts all impact categories to the unit area metric. Although its rigor for assessing overall environmental performance is debatable (Fiala 2008), Ecological Footprint strives to provide robust perspectives on biotic resource use.

LIFE CYCLE ASSESSMENT

Life cycle assessment (LCA) calculates the many types of environmental, human health, and resource use impacts of the entire life cycle of a product. By including relevant data to support a complete model of all environmental impacts, LCA strives to avoid shifting burdens between impact categories and life cycle stages. LCA includes the unique concept of the functional unit, which enables comparisons of radically different product systems that deliver the same service.

TOOLS COMBINING ECONOMIC AND ENVIRONMENTAL METRICS

ECO-EFFICIENCY

LCC can be integrated with LCA in eco-efficiency analyses that measure environmental impact per unit currency, or cost per unit environmental impact.

ECONOMIC INPUT OUTPUT LIFE CYCLE ASSESSMENT

Economic input output life cycle assessment (EIO LCA) joins monetary flows among economic sectors with the emissions associated with economic transactions (Hendrickson, Lave, and Matthews 2006). In contrast to process-based LCA, which defines a system boundary (usually excluding labor, capital investments, and large scale infrastructure) beyond which data are not accounted, EIO LCA uses the national economy as the system boundary and strives to include all externalized data within the national boundary.

HYBRID LCA

Hybrid LCA combines process and EIO approaches to employ the most powerful attributes of each method. IOA and Hybrid Process/EIO LCA are discussed in more detail in Chapter 7: EIO LCA.

LIFE CYCLE SUSTAINABILITY ASSESSMENT

LCC can be used in conjunction with LCA and social LCA in life cycle sustainability assessments (LCSA). LCC and LCSA are more fully described in Chapter 16: Parallel Life Cycle Methods.

APPLIED ENVIRONMENTAL ANALYSIS EXAMPLE

A manufacturer plans a new manufacturing facility to produce an innovative product line. The environmental managers and members of the marketing and product development departments discuss the methods that will be used for these activities. In this situation, they may decide to apply the following tools (other companies and enterprises may select other tools):

Environmental health and safety officers: EIA and ERA (new site evaluation), and checklists (general compliance)

Business planning, product development and marketing: Checklists, LCA, and chemical screening

Purchasing and supply chain management: Environmental purchasing

1.3 LCA METHODOLOGICAL CHARACTERISTICS

LCA IS AN ENVIRONMENTAL RISK APPROACH, NOT PREVENTION APPROACH

LCA is a guiding framework for measuring resource, human health, and environmental perturbations, and connecting those perturbations to their impact potential. It focuses on environmental performance rather than environmental hazards.

LCA IS GLOBAL AND REGIONAL, NOT LOCAL

Because product value chains are distributed globally, LCA models impacts at large scales and not at local scales. The results of a LCA often transcend geo-political boundaries and practitioners must consider this when using LCA to make recommendations at a local scale.

LCA IS STEADY STATE, NOT DYNAMIC

We can consider characterization models to work under the assumption that background conditions in the environment are not changing significantly over time. Although certainly some conditions may vary over time, it is unnecessary to make dynamic models that accommodate those fluctuations.

LCA IS QUANTITATIVE; IT DOES NOT USE PASS/FAIL CRITERIA

This means that LCA provides quantitative results that are expressed relative to the function of a product. These relative LCA results must be compared to another product system to determine their significance. The methodology does not identify good performance versus bad performance, but leaves that to the reader.

LCA CAN TAKE AN ATTRIBUTIONAL OR CONSEQUENTIAL APPROACH

It can model the impacts across a large system and allocate

these effects to a particular process (attributional LCA) or can model changes to a system or the resulting impacts from a policy or decision (consequential LCA). Consequential LCA is explored in Chapter 8: Advanced Modeling.

LCA HAS LIMITATIONS

Over time, practitioners of LCA learn the constraints of the methodology. For instance, methodological inconsistencies between different LCA's can lead to quite different LCA results. System boundaries may not be identical on two LCA studies, even if they evaluate the same product system. Likewise, if a material or input process in an LCA creates multiple products, allocation of the environmental impacts to the different products is required. The accuracy of a study depends on the accuracy of the inventory data and LCIA methods, and a wide range of decisions that can increase uncertainty of the LCA results. These challenging aspects of LCA underscore the necessity to clearly document the system boundary, the allocation method and the assessment uncertainty when reporting LCA results. Allocation methods are discussed in Chapter 8: Advanced Modeling and uncertainty analysis is covered in Chapter 15: Bias and Uncertainty.

Confusion about attribution versus change oriented (consequential) LCA can create methodological challenges in organizing and conducting the study. Further, heterogeneous characterization mechanisms, including complex processes like toxicity, pose a significant obstacle to developing empirically rigorous LCIA methods. Differences of opinion about the robustness of midpoint versus endpoint characterization methods are common (midpoint and endpoint characterization are covered in Chapters 11: LCIA Methods and 12: Decision Support Calculations). Also, the need for more regionalized LCI data and regionalized characterization methods continues to grow.

1.4 UBIQUITY OF ENERGY IN LCA

Energy is a fundamental stock and flow in any LCA that enables the production of desirable and undesirable system outputs. LCA builds upon mass and energy balance principles, and requires that inputs into a system be equal to accumulation plus outputs. The LCA framework can be used to track energy in its many forms or to assess energy systems. There are many different characterizations of energy, including primary versus secondary, fossil versus non-fossil, renewable versus non-renewable, and electrical versus fuel. Depending on the goal and interest of the LCA practitioner, energy may be tracked using one or more characterization factors. From cradle-to-grave, energy may change form, perform useful

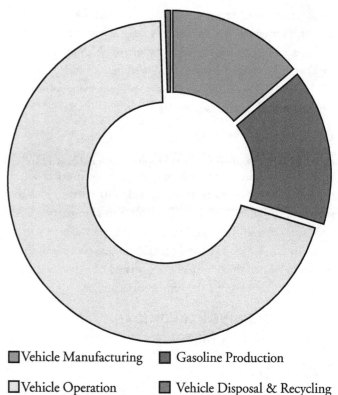

■ Vehicle Manufacturing   ■ Gasoline Production

☐ Vehicle Operation   ■ Vehicle Disposal & Recycling

Figure 1.4. Cradle-to-Grave Energy Use of a Typical Automobile in Megajoules (GREET 2012)

work, become waste, or embed itself in materials, but it cannot be destroyed. Because LCA is built on mass and energy balance principles, it is intrinsically positioned to capture energy inputs, energy transformation, embedded energy in materials, and energy output from processes.

LCA has been widely used for energy systems analysis to assess both energy flows and environmental impacts. From fuel production to electricity generation, LCA has emerged as the preeminent framework for assessing the environmental effects of energy feedstocks and system configurations. Energy systems analysis has become increasingly used in the assessment of emerging non-fossil forms. The analysis of biofuels has come to heavily rely on LCA. Early biofuel LCA studies focused on determining the ratio of biofuel output to energy input of different feedstocks (OECD 1991). Recent efforts have focused on determining the greenhouse gas footprint of feedstocks once upstream processes are accounted for, co-products are assessed, and land use effects are determined (Searchinger et al. 2008). Similarly, emerging renewable electricity generation technologies have used LCA (Pehnt 2006). LCA is also used to assess the effectiveness of carbon capture and storage technologies at electricity generation facilities (Sathre et al. 2012).

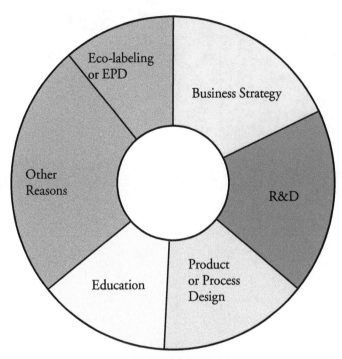

Figure 1.5. Motivations for Conducting LCAs (Cooper and Fava 2006)

tenance, or end-of-life phases, or iii) determine the effects of decisions. LCA emerged simultaneously in the US and Europe in the 1970s and 1980s (Franklin and Hunt 1972) after a decade of tumultuous adolescence and with support from a variety of sources, including the US Environmental Protection Agency and the Society for Toxicology and Chemistry, as a framework for better assessing chemical toxicity. Many early LCAs focused on improving the performance of products or removing high impact inputs (Lave et al. 1995). By the early 2000s, the framework was being used to assess more complex processes, services, and activities (Horvath and Hendrickson 1998). LCA is built on principles that can be generalized for assessing any system capable of being expressed with cradle-to-grave processes, so it has been widely used by individuals in industry, government, academia, and non-government organizations (NGOs). It has also begun making its way into policy regulations (CA 2008) and eco-labeling (Houe and Grabot 2009).

LCA INFORMS PRODUCT DESIGN, IMPROVEMENT, AND DEVELOPMENT

Product development is one of the most frequent applications of LCA. An industry can use LCA software internally or hire an external consultant to perform assessments that complement the product creation process. The LCAs will identify environmentally crucial characteristics of a given product type and signal where the greatest opportunities for improvement lie. For example, an LCA of lighting systems usually indicates the design team should focus on efficient energy use.

LCA ENABLES ENVIRONMENTAL PRODUCT DECLARATIONS

LCA supports the development of eco-labels and environmental product declarations (EPDs). The systematic structure of LCA makes it possible to compare similar products and services. If an LCA conforms to the ISO 14040-series, those who work in the industry, or with the government, or with a NGO can use the LCA results to make a product comparison. Public disclosure of LCA results requires an authorized procedure and rigorous peer review, especially for "comparative assertions disclosed to the public" (ISO 2006). Eco-labels are explored in more detail in Chapter 19: LCA-based Product Claims.

LCA SUPPORTS DIALOGUE AMONG STAKEHOLDERS

The comprehensive scope of LCA provides many opportunities for dialogue among the diverse actors along the life cycle of a product or service. The discussion could address the scope and methods used within the assessment, or use results of one or more LCAs to influence a wide range of decisions. Stake-

Because energy systems are core infrastructure services that tie into every facet of the built environment, LCA has become a valuable tool for understanding the broad effects and environmental externalities that are inherent to the systems. Figure 1.4 shows the importance of using LCA in the analysis of energy in a complex system, which in this case is a typical North American automobile. Both vehicle manufacturing and gasoline production require significant quantities of energy, even when compared to the energy consumed to move the vehicle (GREET 2012). The end-of-life phase of the vehicle includes recycling, which results in the offset of energy use in the production primary (virgin) materials. This avoided energy use can therefore be credited to the automobile life cycle.

1.5 USES FOR LCA

The uses of LCA ultimately depend on the needs of the parties who organize the studies. For instance, a survey of LCA practitioners (Figure 1.5) indicated that LCA is mostly conducted for research and development (18%), to support business strategy (18%), as input to product or process design (15%), for education (13%), and for product labeling or environmental product declarations (11%).

The LCA framework can be used to assess products, processes, services, and activities, or the complex systems in which they reside. LCAs can be structured to i) compare alternatives, ii) improve design, development, use, main-

holder discussions could embrace the producer, members of the supply chain, members of government, NGOs, users of the product or service, purchasers, and/or competitors (Schenck 2010).

LCA SUPPORTS POLICY DEVELOPMENT

The broad and inclusive scope of LCA, including many impact categories and many life cycle stages, makes a robust method for supporting decisions about governmental policies regulations. Governments may convene meetings with multiple stakeholders, as mentioned above, to discuss possible policies on a wide range of topics, including waste management, packaging, product policy, energy policy, and green building.

As we have improved our understanding of the environmental impact and resource constraint externalities produced by the large-scale application of industrial technologies, LCA has transitioned from being helpful to being necessary. LCA applications evolve. Tough environmental questions will require a framework that not only assesses the comprehensive footprint of complex systems, but does so in a way that allows decision makers to dynamically understand the outcomes of their choices. For instance, new technologies like biofuels and electric vehicles make it apparent that a life cycle approach is the only way to understand and mitigate impacts because traditional methods of only evaluating one life cycle phase or one type of environmental impact are insufficient.

The expanding use of LCA to support EPDs, which are defined as "quantified environmental data for a product with pre-set categories of parameters based on the ISO 14040 series of standards, but not excluding additional environmental information" (Hiejungs, et al. 2008). Third-party certified LCA-based labels provide an increasingly important basis for assessing the relative environmental merits of competing products. Third-party certification plays a major role in today's industry. Independent certification can show a company's dedication to safe, environmentally friendly products to customers and NGOs (US EPA 2010).

The use of LCA is rapidly increasing, as practitioners and academics continually find ways of improving the framework, and apply it to novel problems, in order to generate rich information for policy and decision makers. For many problems, LCA is becoming the preferred approach for understanding the environmental and human health impacts of products, processes, services, and activities. LCA offers a rich and rewarding field of study with tremendous capacity to reduce the burdens on human health, depleting natural resources, and our stressed biosphere.

PROBLEM EXERCISES

1. Consider the life cycle of a T-shirt from five different perspectives:

    A a clothing user

    B a clothing manufacturer

    C a washing/drying machine manufacturer

    D a detergent manufacturer

    E a municipality supplying water, sewage water processing, and solid waste disposal

From the perceptive of each of these participants, list the life cycle stages with material or energy flows that are relevant to the participant. Comment on the reasons each participant should (or should not) be allowed to exclude life cycle stages that they have no control over from an LCA of the T-shirt. For instance, should a washing machine manufacturer be responsible to model the impacts of the formulation in a laundry detergent?

2. An LCA can measure the environmental performance of competing physical systems that deliver the same service. For example, human transport can be provided by a gasoline powered automobile, a bicycle, or a jumbo jet airliner. Using your creative capacities, identify at least two services and describe at least three significantly different ways of delivering each service.

3. Different kinds of environmental impact phenomena can be measured with different methods, depending on the needs of the client and the legal requirements that may apply to the activity in question. You advise a company about the appropriate methods for their situation. They will build a factory that will manufacture asphalt roof tiles and sell them throughout North America. What methods do you recommend that they use? Explain in one sentence why you think each method should be used.

4. As a student of life cycle assessment (LCA), you will learn to identify questions that can only be answered with LCA. Describe three such questions and then answer the following: i) What is unique about this problem such that LCA is the suitable approach? ii) How might the answers to these questions differ if LCA was not used? For example, the answer to the question may focus strictly on a single phase (manufacturing, use) in the life cycle.

5. A chemical producer has decided that a life cycle impact assessment is needed so that it can better understand how its products might impact human health and the options for re-engineering its processes in order to reduce these impacts. They hire you to develop the assessment, which identifies two major hotspots in the production process that should partic-ularly concern the chemical producer. First, the basic input to chemical production is natural gas which the chemical company has been purchasing from the Oil and Gas Company. The Oil and Gas Company operates wells in Ohio and Pennsylvania, and uses toxic chemicals pumped in the ground to reach and release the gas. Second, the chemical company also uses toxic chemicals to turn the natural gas into the final products that are released to the market. With this knowledge, what strategies could the chemical company consider to reduce life-cycle impacts in the production process? What are the differences in options when they are targeting their own operations versus those of their suppliers?

6. The regional transportation planning sustainability manager is considering whether to promote investment in new bus or rail service in the county. Part of the manager's recommendation will be based on the greenhouse gas and other air emissions changes that might occur in the region from the addition of new bus or electric rail service as residents shift out of their cars. The manager sees that the implementation of new bus service could utilize the existing roadways, while the new rail service would require the construction of a new track and associated infrastructure. Ultimately, the manager wants to know the per passenger kilometer traveled gas air emissions differences between car, bus, and rail. In at least four sentences, explain what you would include and important requirements for this LCA. Limit yourself to considering propulsion (the burning of fuel for cars, buses, and electricity generation for the train) and infrastructure construction (in the case of the train).

International Standards for Life Cycle Assessment

Matthias Finkbeiner

2.1 Introduction

The basic concept of life cycle assessment (LCA), which is to analyze all environmental burdens in all life cycle stages from cradle to grave, was convincing at its inception. However, the substantial implementation of LCA practice in real-world decision-making and as a tangible contribution to reducing environmental stress on our biosphere, did not happen immediately. After significant progress in recent decades, LCA is now well established in private and public organizations.

The establishment of international standards significantly supported the development of LCA practice. The international standards of life cycle assessment (International Organization for Standardisation (ISO) 14040 series) were of utmost importance for the broad acceptance of LCA worldwide. The current ISO standards of LCA (ISO 14040 and ISO 14044, both in 2006) are the only relevant international standards that are broadly referenced by users and other standardization processes.

Background and Relevance

In the early days of LCA, the vested interests of the study commissioners often skewed the results of the studies. While the general concept of LCA appealed to many stakeholders, the credibility of the method was severely damaged by such biased studies. Because of the lack of commonly accepted procedures and methods, governments were reluctant to apply LCA for their policy developments. Most companies had a risk-aversive strategy towards LCA because they were either afraid of market distortions due to unjustified claims by competitors, or of barriers to trade, or of mandatory reporting requirements by public policy.

This attitude was impaired by the tendency of some LCA practitioners to oversell the tool. For some practitioners, LCA was no longer a tool, but rather a religion of sorts to help determine what was good and what was evil. During the nineties, LCA practitioners from the academic and consulting worlds typically belonged to different schools of thought and fought about the right way to implement LCA practices, the right allocation approach, the right impact assessment method, the best LCA software, and so on.

The setting urgently called for harmonization of the method. A first success was achieved by the *LCA Code of Practice*, created by the Society of Environmental Toxicology and Chemistry (SETAC 1993). SETAC still serves as a major scientific organization that deals with LCA. The *LCA Code of Practice* was an important seed document for the standardization process summarized in Section 2.2.

The successful implementation of LCA international standardization achieved a clearer perspective of what LCA can do. More importantly, it defined what it cannot do. It established a common language of terms and key methodological requirements, but it did not standardize the particular method to perform LCA. By giving the users and providers of LCA an equally important voice, standards achieved consensus on the basic rules and framework of LCA. In addition, the standards made the limitations of LCA transparent and provided strict requirements for the most contentious application of LCA, the comparative assertions intended to be disclosed to the public (the claim of overall environmental equivalence or superiority of products). As such, the establishment of international standards was of utmost importance for broad acceptance of LCA by all stakeholders around the world.

ISO and its standardization process provided the structure that supported the standard's development. It has a membership of over 160 national standards institutes from countries large and small, as well as industrialized and developing countries from all regions of the world. An ISO standard is developed by a panel of experts within a technical committee. Once the need for a standard has been established, experts meet in a working group established for this purpose to discuss and negotiate a draft standard. As soon as a draft has been developed, it is shared with ISO members who comment and vote on it. If a consensus is reached, the draft becomes an ISO standard. If consensus is not achieved, the draft returns to the technical committee for further edits (ISO 2012a).

International Organization for Standardization

ISO does not decide when to develop a new standard (2012b). Instead, ISO responds to a request from industry members or other stakeholders such as consumer groups. Typically, an industry sector or group communicates the need for a standard to its national member who then contacts ISO. ISO standards are developed by groups of experts from all over the world that are part of larger groups called technical committees. These technical committees negotiate all aspects of the standard, including its scope, key definitions, and content. The technical committees are made up of experts from the relevant industry, as well as experts from consumer associations, academia, NGOs and government. Finally, developing ISO standards is a consensus-based approach and comments from stakeholders are taken into account.

While the ISO standards for LCA are sometimes criticized for not sufficiently addressing certain issues, the standards represent the global consensus on those methodological features for which such a consensus exists. Some stakeholders desire more specific stipulations (e.g., allocation procedures or a default set of impact categories) or a particular impact assessment method, but no global stakeholder consensus exists for these

options. It makes no sense to blame the standards for this state of affairs, as it is the natural result of the democratic procedure to develop an ISO standard.

2.1.2 History of ISO and Standards of LCA

The early days of the standardization process were challenging because of the lack of consensus on many methodological issues. ISO technical committee ISO/TC 207 'Environmental management' is responsible for developing and maintaining the ISO 14000 family of standards. ISO/TC 207 was established in 1993 (Finkbeiner 2012). The membership of ISO/TC 207 is among the highest of any ISO technical committee and is both broad and diverse in representation. National delegations of environmental experts from over 100 countries participate in ISO/TC 207, including over 25 developing countries (ISO 2010). Within ISO/TC207/SC5 the early standardization work was organized in five working groups (Marsmann et al 1997). They produced the first set of international standards of LCA, which include the following:

- ISO 14040 – Environmental Management – Life Cycle Assessment – Principles and Guidelines (ISO 14040 1997)
- ISO 14041 – Environmental Management – Life Cycle Assessment– Goal and Scope Definition and Inventory Analysis (ISO 14041 1998)
- ISO 14042 – Environmental Management – Life Cycle Assessment – Life Cycle Impact Assessment (ISO 14042 2000) and
- ISO 14043 – Environmental Management – Life Cycle Assessment – Life Cycle Interpretation (ISO 14043 2000).

The publication of these first international standards of LCA comprised an essential step to consolidate procedures and methods. However, the parallel development of documents in different working groups led to some inconsistencies between the first generation of standards that have been corrected in the first revision completed in 2006.

The revision process resulted in ISO 14044 (Finkbeiner et al. 2006). The main targets of this revision were improving readability, applicability, and merging the previous standards. This was fulfilled by two new standards:

- a revised ISO 14040 standard ('Environmental Management – Life Cycle Assessment– Principles and Framework') (ISO 14040 2006) and
- a new standard 14044 containing all requirements ('Environmental Management – Life Cycle Assessment – Requirements and Guidelines') (ISO 14044 2006).

The working group dealing with the revision arrived at consensus to reduce the number of standards was and to reduce the number of annexes. Technical modifications were made as well, but the modified technical content was in line with the previous requirements and served mainly as a clarification, and as a correction, of errors and inconsistencies. They included the addition of several definitions (product, process, etc.), the addition of principles for LCA, clarifications concerning LCA intended for use in comparative assertions for public disclosure, system boundaries, the critical review panel, and the addition of an annex about applications.

The unanimously approved revisions represented a consensus of all countries and stakeholders. The versions of ISO 14040 and 14044 developed in 2006 are still valid today. As part of the systematic review procedure of ISO standards, there was an inquiry on the need for revision to all member bodies in 2009. The result of the inquiry was an almost unanimous confirmation of the existing standards.

In recent years, we observed additional momentum generated by the development of new approaches built on the basis of classical LCA. This includes:

- single-issue LCAs like carbon footprinting or water footprinting
- beyond-environment LCAs like life cycle costing, social LCAs
- eco-efficiency assessments or life cycle sustainability assessments
- beyond-product-LCAs like scope 3 type LCAs of organizations or sector-based IO-LCAs
- beyond-quantification LCAs like type III environmental product declarations or other types of environmental labels and claims.

While some of these additional standards are part of the ISO/TC207 family, additional public and private standardization bodies entered the market with their own standards. In particular, the carbon footprint discussions led to a proliferation of different guidelines and standards (Finkbeiner 2009). Despite these developments, ISO 14040 and ISO 14044 remain by far the most relevant standards in the field. Some of their main contents and key features are introduced in the following sections.

2.2 Terms and Definitions

Every standard establishes a common understanding of key definitions. Both ISO 14040 and 14044 contain the same set of 46 terms and definitions. Some of the main LCA terms are given in this section. Complete coverage of all definitions goes beyond the scope of this section and readers are advised to go back to the original documents (ISO 14040 2006 and ISO 14044 2006).

As a fundamental term, *LCA* is defined as "compilation and evaluation of the inputs, outputs and the potential environmental impacts of a product system throughout its life cycle" (ISO 14040 2006, 3.2). The complementary definition of the *life cycle* itself is "consecutive and interlinked stages of a product system, from raw material acquisition or generation from natural resources to final disposal" (ISO 14040 2006, 3.1).

Another set of definitions addresses the individual life cycle phases. As a starting point, the *life cycle inventory analysis (LCIA)* is defined as "phase of life cycle assessment involving the compilation and quantification of inputs and outputs for a product throughout its life cycle (ISO 14040 2006, 3.3). The term *product* comprises "any good or service" (ISO 14040 2006, 3.9). In the LCIA phase, the LCA practitioner defines the *product system* for a particular LCA study, which is a "collection of unit processes with elementary and product flows, performing one or more defined functions, and which models the life cycle of a product" (ISO 14040 1006, 3.28). Defining the boundaries of the product system to be studied can be a difficult task.

One methodological element to support the proper and consistent definition of the system boundaries are *cut-off criteria* as "specification of the amount of material or energy flow or the level of environmental significance associated with unit processes or product system to be excluded from a study" (ISO 14040 2006, 3.18). After the system is established, the data collection is usually a time-consuming part of the LCI phase. Data gaps and data quality have to be documented for the later interpretation phase. The ISO standards have several paragraphs addressing the different data quality requirements that serve as input for assessing the *data quality* of a study as "characteristics of data that relate to their ability to satisfy stated requirements" (ISO 14040 2006, 3.19).

The final goal of the LCI phase is to identify the elementary flows associated with the product system studied. Elementary flows cross the system boundary between the product system and the environment, and therefore serve as input and interface to the following impact assessment phase. *Elementary flows* are defined as "material or energy entering the system being studied that has been drawn from the environment without previous human transformation, or material or energy leaving the system being studied that is released into the environment without subsequent human transformation" (ISO 14040 2006, 3.12).

After the collection of the process data, the product system is related to the *functional unit,* which is the "quantified performance of a product system for use as a reference unit" (ISO 1404 2006, 3.20). If different products are compared, the functional unit must be the same. However, if the performance characteristics of the compared products are not equal, they have to be compared based on different *reference flows* as "measure of the outputs from processes in a given product system required to fulfil the function expressed by the functional unit (ISO 14040 2006, 3.29)."

If the product system studied contains multi-functional processes (i.e., production processes that have more than one valuable product or waste treatment processes that treat more than one type of waste), and the functional unit requires only one of them, the LCA practitioner has to make choices about how to assign the flows of a process to its different products. The standard offers some opportunities to avoid allocation, but in practice it is often necessary to perform allocation or "partitioning the input or output flows of a process or a product system between the product system under study and one or more other product systems" (ISO 14040 2006, 3.17).

Life cycle impact assessment (LCIA) represents the "phase of life cycle assessment aimed at understanding and evaluating the magnitude and significance of the potential environmental impacts for a product system throughout the life cycle of the product" (ISO 14040 2006, 3.4).

The LCIA phase consists of several steps. A key element is the concept of the *impact category* as a "class representing environmental issues of concern to which life cycle inventory analysis results may be assigned" (ISO 14040 2006, 3.39). The impact categories represent relevant environmental issues such as global warming, ozone depletion, toxicity, or resource depletion. They are quantified by indicators that are related to the category endpoint, which is the "attribute or aspect of natural environment, human health, or resources, identifying an environmental issue giving cause for concern" (ISO 14040 2006, 3.36). The category endpoint represents the ultimate stage in the environmental mechanism in which the potential damage occurs, for example human health. LCIA indicators can be defined anywhere along the cause-effect-chain as long as they are relevant for the category endpoint.

The final phase of the LCA is *life cycle interpretation.* It is defined as a "phase of life cycle assessment in which the findings of either the inventory analysis or the impact assessment, or both, are evaluated in relation to the defined goal and scope in order to reach conclusions and recommendations" (ISO 14040 2006, 3.5). The main target of the interpretation is to derive robust results and conclusions based on a self-evaluation of the study with regard to completeness, sensitivity, consistency and limitations. This is particularly important for a *comparative assertion* or "environmental claim regarding the superiority or equivalence of one product versus a competing product that performs the same function" (ISO 14040 2006, 3.6).

If we intend to communicate such a comparative assertions to the public, ISO 14044 requires a conformity assessment by an independent panel of at least three experts. For other applications of LCA, this is just a recommendation and can be performed by an individual expert, (and not necessarily by a review panel). The conformity assessment in ISO 14040 and 14044 is called *critical review* and defined as "process intended to ensure consistency between a life cycle assessment and the principles and requirements of the International Standards on life cycle assessment" (ISO 14040 2006, 3.45).

2.3 PRINCIPLES, KEY FEATURES AND LIMITATIONS OF LCA

The first edition of ISO 14040 from 1997 was titled "Principles and Guidelines" (ISO 14040 1997). However, this standard did not contain any principles. This inconsistency was removed during the revision by adding seven principles of LCA (ISO 14040 2006). These principles represent fundamental guidance for the LCA practitioner for planning and conducting an LCA.

The first principle is called *life cycle perspective* and contains the obvious notion that "LCA considers the entire life cycle of a product, from raw material extraction and acquisition, through energy and material production and manufacturing, to use and end of life treatment and final disposal. Through such a systematic overview and perspective, the shifting of a potential environmental burden between life cycle stages or individual processes can be identified and possibly avoided" (ISO 14040 2006, 4.1.2).

The second principle, *environmental focus,* clarifies that LCA in its pure form is restricted to the environmental dimension of sustainability and that "economic and social aspects and impacts are, typically, outside the scope of the LCA" (ISO 14040 2006, 4.1.3).

The third principle, *relative approach and functional unit,* is more technical as it describes the functional unit as a key concept of LCA. The principle states that "LCA is a relative approach, which is structured around a functional unit. This functional unit defines what is being studied. All subsequent analyses are then relative to that functional unit, as all inputs and outputs in the LCI and consequently the LCIA profile are related to the functional unit" (ISO 14040 2006, 4.1.4).

The fourth principle addresses LCA as an *iterative approach*, which describes "the individual phases of an LCA use results of the other phases. The *iterative approach* within and between the phases contributes to the comprehensiveness and consistency of the study and the reported results" (ISO 14040 2006, 4.1.5).

The fifth principle gives *transparency* a strong role to deal with the trade-off between "the inherent complexity in LCA…[and]…proper interpretation of the results" (ISO 14040 2006, 4.1.6).

The sixth principle is *comprehensiveness* in the sense that "LCA considers all attributes or aspects of natural environment, human health and resources. By considering all attributes and aspects within one study in a cross-media perspective, potential trade-offs can be identified and assessed" (ISO 14040 2006, 4.1.7).

Last, but not least, the seventh principle establishes the *priority of scientific approach*. It states that "decisions within an LCA are preferably based on natural science. If this is not possible, other scientific approaches (e.g. from social and economic sciences) may be used or international conventions may be referred to. If neither a scientific basis exists nor a justification based on other scientific approaches or international conventions is possible, then, as appropriate, decisions may be based on value choices" (ISO 14040 2006, 4.1.8).

Next to the principles, the standard defines fifteen key features of an LCA. One key feature is that there is no "one-size-fits-all" approach to LCA, because "there is no single method for conducting LCA. Organizations have the flexibility to implement LCA…in accordance with the intended application and the requirements of the organization" (ISO 1404 2006). This flexibility stresses the importance of the goal and scope definition phase of LCA, because "the depth of detail and time frame of an LCA may vary to a large extent, depending on the goal and scope definition…[and with]…respect [to] confidentiality and proprietary matters" (ISO 14040 2006).

Several LCIA features try to avoid the exaggeration of the value of LCIA results by clarifying that LCA addresses potential environmental impacts. LCA does not predict absolute or precise environmental impacts due to the following factors:

- the relative expression of potential environmental impacts to a reference unit
- the integration of environmental data over space and time
- the inherent uncertainty in modelling of environmental impacts

- the fact that some possible environmental impacts are clearly future impacts

In addition, "there is no scientific basis for reducing LCA results to a single overall score or number, since weighting requires value choices" (ISO 14040 2006).

These two key features address implicitly some of the inherent limitations of LCA according to the standard. To make the benefits and the useful application of LCA credible, it is necessary to be transparent and explicit about its limitations as well. ISO 14040 and ISO 14044 clearly acknowledge that any LCA study has its limitations. Therefore, the limitations of every study have to be documented in the goal and scope definition (ISO 14040 2006, 5.2.1.2). Moreover, there is a specific chapter on limitations (5.4.3 Limitations of LCIA) of the LCIA phase, which be found in ISO 14040 2006.

The LCIA addresses only the environmental issues that are specified in the goal and scope. Therefore, LCIA is not a complete assessment of all environmental issues of the product system under study. LCIA cannot always demonstrate significant differences between impact categories and the related indicator results of alternative product systems. This may be due to:

- limited development of the characterization models, sensitivity analysis and uncertainty analysis for the LCIA phase,
- limitations of the LCI phase, such as setting the system boundary, that do not encompass all possible unit processes for a product system or do not include all inputs and outputs of every unit process, since there are cut-offs and data gaps,
- limitations of the LCI phase, such as inadequate LCI data quality which may, for instance, be caused by uncertainties or differences in allocation and aggregation procedures, and
- limitations in the collection of inventory data appropriate and representative for each impact category.

ISO also cautions that "the lack of spatial and temporal dimensions in the LCI results introduces uncertainty in the LCIA results. The uncertainty varies with the spatial and temporal characteristics of each impact category. There are no generally accepted methodologies for consistently and accurately associating inventory data with specific potential environmental impacts. Models for impact categories are in different stages of development" (ISO 14040 2006).

Furthermore, ISO 14044 2006, 4.4.5 adds that "an LCIA shall not provide the sole basis of comparative assertion intended to be disclosed to the public of overall environmental superiority or equivalence, as additional information will be necessary to overcome some of the inherent limitations in the LCIA. Value choices, exclusion of spatial and temporal, threshold and dose-response information, relative approach, and the variation in precision among impact categories are examples of such limitations. LCIA results do not predict impacts on category endpoints, exceeding thresholds, safety margins or risks."

All the limitations have to be addressed by making them transparent in the reporting of an LCA study (ISO 14040 2006, 6.), especially when drawing conclusions: "report the results and conclusions of the LCA in an adequate form to the intended audience, addressing the data, methods and assumptions applied in the study, and the limitations thereof."

In summary, the international standards of LCA explicitly recognize the limitations of the method. They explain the potential benefits and uses of LCA, and they are also particularly cautious about the need to draw conclusions that explicitly consider the limitations of each study.

2.4 KEY REQUIREMENTS ACCORDING TO ISO 14044

Requirements, or in standard language "shalls," have to be fulfilled for compliance, whereas recommendations, or "shoulds," are just guidance, but not mandatory. In the first version of the ISO 14040 series, all four standards (ISO 14040-43, see 1.2) contained requirements. This required the LCA practitioner to consult four documents in parallel. Therefore, all technical requirements were transferred to ISO 14044, making it the only reference document for the LCA practitioners. The new ISO 14040 aims to provide a description of LCA principles and framework that is readable and accessible not only for LCA practitioners, but also for a broader target audience. The revised 14040 contains just a single, formal requirement of compliance with the new ISO 14044 standard (Finkbeiner 2006) in order to avoid ambiguous claims of conformity.

14044 reflects two levels of sophistication when describing requirements. The basic set of requirements applies to all kinds of LCA studies. Additional and more rigorous requirements are defined for studies intended to support comparative assertions intended to be disclosed to the public. According to 14044, "an exact and clear description of this application of LCA requires the two intentions: the intention for the use for a comparative assertion and the intention to disclose it to the public. Because of the importance of this issue for the overall credibility of LCA, unambiguousness, clarity and accuracy were seen as more crucial than conciseness in this case" (Fink-

beiner 2006). As a result of the strict requirements for comparative assertions, studies making this claim are extremely rare.

2.4.1 REQUIREMENTS FOR ALL LCA APPLICATIONS

Coverage of all requirements goes beyond the scope of this section and readers are advised to consult the original document for complete reference (ISO 14044 2006). A fundamental requirement establishes the four phases of LCA by defining that "LCA studies shall include the goal and scope definition, inventory analysis, impact assessment and interpretation of results. LCI studies shall include definition of the goal and scope, inventory analysis and interpretation of results" (ISO 14044 2006).

When defining the goal of the study, the standard requires that the following items must be unambiguously stated:

- the intended application
- the reasons for carrying out the study
- the intended audience, (i.e., to whom the results of the study are intended to be communicated)
- whether the results are intended to be used in comparative assertions intended to be disclosed to the public

The first three points document the background and motivation of the study, whereas the fourth point requires expressing explicitly whether the study is intended for this particular application for which additional requirements apply (see Section 4.2).

As a next step the scope definition addresses the specification of a set of items that basically represent the backbone of every LCA study. According to ISO 14044 (2006), for the scope of an LCA, the following items shall be considered and clearly described:

- the product system to be studied
- the functions of the product system or, in the case of comparative studies, the systems
- the functional unit
- the system boundary
- allocation procedures
- LCIA methodology and types of impacts
- interpretation to be used
- data requirements
- assumptions
- value choices and optional elements
- limitations
- data quality requirements
- type of critical review, if any
- type and format of the report required for the study

For each of these elements, the standard provides specific requirements. A completed goal and scope definition includes the plan for how the following steps of the LCA study are supposed to be performed.

The life cycle inventory (LCI) phase includes steps like data collection, data validation, relation of data to unit processes, relation of data to the functional unit, and data aggregation. Among the basic requirements for this phase, the rules to deal with multi-functional processes will be presented here as an example. Multi-functional processes include multi-output processes (processes that produce more than one product) and multi-input processes (waste treatment processes that treat more than one type of waste). Both cases pose a challenge for the compilation of an inventory of a particular product and results in the so-called allocation problem. ISO 14044 (2006) requires documenting and explaining the chosen allocation procedures and provides a hierarchy of steps to deal with the problem:

STEP 1: Wherever possible, allocation should be avoided by

1 dividing the unit process to be allocated into two or more sub-processes and collecting the input and output data related to these sub-processes, or

2 expanding the product system to include the additional functions related to the co-products.

STEP 2: Where allocation cannot be avoided, the inputs and outputs of the system should be partitioned between its different products or functions in a way that reflects the underlying physical relationships between them; i.e. they should reflect the way in which the inputs and outputs are changed by quantitative changes in the products or functions delivered by the system.

STEP 3: Where physical relationship alone cannot be established or used as the basis for allocation, the inputs should be allocated between the products and functions in a way that reflects other relationships between them. For example, input and output data might be allocated between co-products in proportion to the economic value of the products. Further explanation of allocation can be found in Chapter 8: Advanced Modeling.

In the LCIA phase, the standard does not define a default set of impact categories or particular impact category indicators because there is no global consensus on these. However, it clearly differentiates between the mandatory and optional elements of impact assessment. ISO 14044 (2006) states that the LCIA phase shall include the following mandatory elements:

- selection of impact category indicators and characterization models;

- assignment of LCI results to the selected impact categories, (classification);
- calculation of category indicator results (characterization).

In addition to the mandatory elements of LCIA, there could be optional elements depending on the goal and scope of the LCA. They are described in ISO 14044 (2006) as

- normalization: calculating the magnitude of category indicator results relative to reference information;
- grouping: sorting and possibly ranking of the impact categories;
- weighting: converting and possibly aggregating indicator results across impact categories using numerical factors based on value choices; pre-weighted data should remain available;
- data quality analysis: better understanding the reliability of the collection of indicator results, the LCIA profile.

The standard contains a relatively large set of requirements and recommendations for all these steps of LCIA. As an example, the impact categories, category indicators, and characterization models should be internationally accepted, should avoid double counting, should be scientifically and technically valid, should be based upon a distinct identifiable environmental mechanism or reproducible empirical observation, and should be environmentally relevant.

Deriving information from the data compiled in the study is the key task of the life cycle interpretation phase of an LCA. It is composed of three main elements, according to ISO 14044 (2006):

- identification of the significant issues based on the results of the LCI and LCIA phases of LCA;
- an evaluation that considers completeness, sensitivity and consistency checks;
- conclusions, limitations, and recommendations.

The purpose of the evaluation of the significant parameters is to assess the reliability of the final results and conclusions by determining how they are affected by uncertainties in the data, allocation methods, calculation of category indicator results, or other factors. This includes a self-assessment for consistency and checks whether differences in data quality, along a product life cycle and between different product systems, are consistent with the goal and scope of the study. It also determines whether regional and/or temporal conditions, allocation rules, system boundary and the elements of impact assessment have been consistently applied to all product systems. The evaluation must take the goal of the study and the final intended use of the study results into account.

"The final objective of...the life cycle interpretation is to draw conclusions, identify limitations and make recommendations for the intended audience of the LCA. [It is recommended to] draw preliminary conclusions and check that these are consistent with the requirements of the goal and scope of the study, including, in particular, data quality requirements, predefined assumptions and values, methodological and study limitations, and application-oriented requirements. Recommendations shall be based on the final conclusions of the study, and shall reflect a logical and reasonable consequence of the conclusions" (ISO 14044 2006).

The ISO-standards for LCA contain a separate section on reporting that contains requirements which are generally applicable for all LCA studies, as well as additional sets of requirements for third-party reports and for comparative assertion intended to be disclosed to the public (see Section 4.2). On the general level, ISO 14044 (2006) requires that "the results and conclusions of the LCA shall be completely and accurately reported without bias to the intended audience. The results, data, methods, assumptions and limitations shall be transparent and presented in sufficient detail to allow the reader to comprehend the complexities and trade-offs inherent in the LCA. The report shall also allow the results and interpretation to be used in a manner consistent with the goals of the study." The requirements that apply to the critical review as conformity assessment for ISO 14040 and ISO 14044 will be discussed separately in Section 2.5.

ADDITIONAL REQUIREMENTS FOR COMPARATIVE ASSERTIONS

The application of LCAs to support comparative assertions intended to be disclosed to the public has strong implications on third parties. As a consequence, ISO 14044 (2006) specifies additional requirements to those summarized in Section 4.1. Certain aspects are mandatory for comparative assertions, including:

- "The equivalence of the systems being compared shall be evaluated before interpreting the results. Systems shall be compared using the same functional unit and equivalent methodological considerations such as performance, system boundary, data quality, allocation procedures, decision rules on evaluating inputs, and outputs and impact assessment. Any differences between systems regarding these parameters shall be identified and reported."
- While an LCI study without impact assessment is a feasible choice for any other application, an LCIA is required for comparisons used in comparative assertions to be disclosed to the public.

- "The LCIA shall employ a sufficiently comprehensive set of category indicators. The comparison shall be conducted by category indicator.
- Weighting shall not be used in LCA studies intended to be used in comparative assertions intended to be disclosed to the public."
- Several data quality requirements and sensitivity analyses are required, rather than recommended.
- Finally, more comprehensive reporting requirements apply as described in paragraph 5.3 of ISO 14044.
- In order to decrease the likelihood of misunderstandings or negative effects on external interested parties, a critical review panel of interested parties is mandatory, whereas critical reviews are merely recommended for all the other applications.

The last point addresses the critical review of LCA as a particular form of the conformity assessment with the standards, which is described in more detail in Section 2.5.

2.5 CRITICAL REVIEW AS CONFORMITY ASSESSMENT

According to ISO 14040 (2006) "a critical review is a process to verify whether an LCA has met the requirements for methodology, data, interpretation and reporting and whether it is consistent with the principles. A critical review can neither verify nor validate the goals that are chosen for an LCA by the study commissioner, nor the ways in which the LCA results are used." The content and elements to be checked are detailed in ISO 14044 (2006). The critical review process shall ensure that:

- the methods used to carry out the LCA are consistent with.... [ISO 14044],
- the methods used to carry out the LCA are scientifically and technically valid,
- the data used are appropriate and reasonable in relation to the goal of the study,
- the interpretations reflect the limitations identified and the goal of the study, and
- the study report is transparent and consistent.

The ISO-standards differentiate between the critical review of individual experts who can be either internal or external to the organization performing the LCA and the critical review performed by a panel of interested parties. For the latter case, "an external independent expert should be selected by the original study commissioner to act as chairperson of a review panel of at least three members. Based on the goal and scope of the study, the chairperson should select other independent qualified reviewers. This panel may include other interested parties affected by the conclusions drawn from the LCA, such as government agencies, non-governmental groups, competitors and affected industries."

In general, "the review statement and review panel report, as well as comments of the expert and any responses to recommendations made by the reviewer or by the panel, shall be included in the LCA report" (ISO 14044 2006). The standards do not give detailed guidelines for the critical review procedure, but a common critical review process has emerged in the LCA community that satisfied all stakeholders. For the mandatory case of comparative assertions disclosed to the public, but also in many cases for which a critical review is not mandatory, study commissioners often decide to perform critical reviews to improve their studies and to support their credibility.

One of the LCA standards' key success factors is that it does not operate an accreditation scheme that tries to ensure quality by bureaucracy, but rather ensures quality by making the individual reviewer personally accountable. In order to document this well-established critical review practice in a more formal way, 'ISO 14071 Environmental Management — Life Cycle Assessment — Requirements and Guidelines for Critical Review Processes and Reviewer Competencies' was created.

2.6 SPIN-OFF STANDARDS

While ISO 14040 and ISO 14044 are clearly the core standards of LCA, a growing number of LCA related standards are available. These are often referred to as spin-off standards. One of these spin-off standards, ISO 14025 on type III environmental declarations, is part of the ISO 14020 series of eco-labeling standards (ISO 14025 2006). ISO 14025 provides a standardized reporting format for LCAs.

Several non-normative documents, or technical reports, were developed with examples for their application. Their development began in parallel to the first generation of core LCA standards:

- ISO/TR 14047:2012 Environmental Management – Life Cycle Assessment — Illustrative examples on how to apply ISO 14044 to impact assessment ISO/TR 14049:2012
- Environmental Management – Life Cycle Assessment – Illustrative examples on how to apply ISO 14044 to goal and scope definition and inventory analysis

Finally, the TS Environmental Management – Life Cycle Assessment – Data Documentation Format' (ISO/TS 14048

2002) provides the requirements for a data documentation format to be used for transparent and unambiguous documentation, as well an exchange of LCA and LCI data.

CARBON FOOTPRINT

A process known as carbon footprinting contributed significantly to the growing use of life cycle- based assessment tools and to a proliferation of guides and standards. On the ISO level, there is ISO/TS 14067 Carbon footprint of products – Requirements and Guidelines for Quantification and Communication (ISO/TS 14067 2013). According to the introduction in the TS, "International Technical Specification is based on existing ISO standards, e.g. ISO 14020, ISO 14025, ISO 14040 and ISO 14044 and aims to set more specific requirements for the quantification and communication of carbon footprints of products (CFP). Specifically, using life cycle assessment according to this International Standard with climate change as the single impact category may offer benefits through:

- providing requirements for the methods to be adopted in assessing the CFP;
- facilitating the tracking of performance in reducing GHG emissions;
- assisting in the creation of efficient and consistent procedures to provide CFP information to interested parties;
- providing a better understanding of the CFP such that opportunities for GHG reductions may be identified;
- providing CFP information to encourage changes in consumer behavior which could reduce in GHG emissions through improved purchasing, use and disposal decisions;
- providing correct and consistent communication of CFPs which supports comparability of products in a free and open market;
- enhancing the credibility, consistency and transparency of the quantification, reporting and communication of the CFP;
- facilitating the evaluation of alternative product design and sourcing options, production and manufacturing methods, raw material choices, recycling and other end-of-life stages;
- facilitating the development and implementation of GHG management strategies and plans across product life cycles as well as the detection of additional efficiencies in the supply chain (ISO/TS 14067 2013).

While the TS document now totals approximately fifty pages, most of its content copies the core standards of LCA, while the additional CFP specific requirements for quantification remain relatively few.

WATER FOOTPRINT

The water footprinting impact characterization method has been neglected for many years due to a lack of awareness and a lack of appropriate means for assessing water consumption (Berger and Finkbeiner 2012). The methodological developments in this field are in development and there is not yet a definitive water footprinting method, but rather different approaches to analyzing the water consumption of organizations along product life cycles (Berger and Finkbeiner 2010).

The first version of the water footprint standard will primarily clarify the terminology and key methodological concepts, rather than settle on a preferred method. Decisions about volumetric versus impact-oriented water footprint methods are fundamental. An inventory of water volumes is currently not accepted to be scientifically sufficient to address the issue of water scarcity. To put it in LCA terms: a pure water inventory is not necessarily meaningful and should just serve as a basis for water impact assessment and not as a metric itself. This is acknowledged by the upcoming ISO 14046, which currently defines a water footprint on the impact level as "parameters that quantify the potential environmental impacts related to water" (ISO 14046.CD.1 2012).

According to the current committee draft document (ISO 14046.CD.1 2012), the scope is defined as "specifying principles, requirements and guidelines to assess and report the water footprints of products, processes and organizations based on life cycle assessment (LCA). The standard provides requirements and guidance for calculating and reporting a water footprint as a stand-alone assessment or as part of a more comprehensive environmental assessment. The water footprint is calculated as one impact indicator result or multiple impact indicator results."

ECO-EFFICIENCY ASSESSMENT

While the single-issue footprinting standards discussed in Sections above provide specific requirements for particular impact aspects of LCA, a complementary approach towards broadening the scope of traditional LCA also exists. While the measurement of the environmental dimension of sustainability with LCA is now well established, it is crucial to expand this environmental focus towards life cycle based sustainability assessments (Kloepffer 2008). Recently, complementary approaches were developed for the economic (life cycle costing – LCC) and the social (social LCA – SLCA) dimensions of sustainability. An integrated concept, life cycle sustainability

assessment, or LCSA, has been proposed because the life cycle perspective is inevitable for all sustainability dimensions in order to achieve reliable and robust results (Finkbeiner et al. 2010). Eco-efficiency and resource efficiency are intermediate concepts that combine the environmental and economic dimensions of sustainability, but do not address the social dimension.

As a consequence, the standard ISO 14045:2012:Environmental Management – Eco Efficiency Assessment of Product Systems – Principles, Requirements and Guidelines represents an important step due to the broader focus beyond environmental issues only (ISO 14045 2012). Eco-efficiency assessment is a quantitative management tool which enables the consideration of life cycle environmental impacts of a product system alongside its product system value to a stakeholder.

Within eco-efficiency assessment, environmental impacts are evaluated using LCA as prescribed by other International Standards (ISO 14040, 14044). Consequently, eco-efficiency assessment shares with LCA many important principles such as life cycle perspective, comprehensiveness, functional unit approach, iterative nature, transparency and priority of scientific approach.

The value of the product system may be chosen to reflect, for example, its resource, production, delivery or use efficiency, or a combination of these. The value may be expressed in monetary terms or other value aspects (e.g. functional or aesthetic). The key objectives of this International Standard are to:

- establish clear terminology and a common methodological framework for eco-efficiency assessment;
- enable the practical use of eco-efficiency assessment for a wide range of product (including service) systems;
- provide clear guidance on the interpretation of eco-efficiency assessment results;
- encourage the transparent, accurate and informative reporting of eco-efficiency assessment results (ISO 14045 2012).

2.7 CONCLUSION AND OUTLOOK

ISO 14040 and ISO 14044 are the standards that comprise the constitution of LCA. They comprise the only current and relevant international standard documents on LCA, which are broadly referenced by users and other standardization processes. The standards contributed significantly to the transition of LCA from a misused "greenwashing machine" towards a serious, robust, and professional decision support tool for public and private organizations.

The constitution of LCA should thus be respected and protected by everyone. It is fair to ask for more details and more specific stipulations in future versions, if global consensus evolves on such issues. However, if such a consensus for moving forward does not exist, we should at least protect the current level of agreement.

Companies benefit from standards because they provide a level playing field in the market. They help avoiding misuse and market distortions that can result from unjustified claims. This benefit even happens passively, if these standards have reached a certain financial or political relevance. Additional benefits include the technical guidance the standards provide, as well as their credibility among stakeholders internationally. However, standards do not provide a detailed one-size-fits-all solution. They tell you where the playing field is and set some rules, but do not tell how to win. They also do not always tell you where to go, but rather where you should *not* go. But in systems that support participation of citizens and democracy, a set of commonly accepted rules is a prerequisite for consistent and sustainable development.

Future activities for additional standards aim at specifying particular parts of LCA methodology (e.g., a critical review of LCA on organizations), while other activities expand the environmental focus towards these three sustainability dimensions: resource efficiency, life cycle costing, social LCA, and life cycle sustainability assessment. Hopefully, these developments will support the credible and robust use of LCA for making real-world decisions through life cycle management and life cycle sustainability management (Finkbeiner 2011).

EXERCISE 1. LOCATING COMPLETE ISO 14000-SERIES STANDARDS DOCUMENTS

ISO standards are unique documents because they are carefully protected. To fund ISO and its work, all ISO standards must be purchased. In this out of classroom exercise, you will search for complete copies of the ISO 14000-series standards. At a minimum, you should try to find 14040 and 14044. If you can find the entire suite, that would be even better.

If you are studying at a college or university, you can visit the nearest branch of the school library and speak with a librarian. If you are not part of a college library system, you can go to a public library. You can explain that you are looking for complete copies of these standards for a course. If the library does not have copies on site, you can ask whether they exist within the college library system in another branch (possibly the engineering library). If the standards are not within the immediate library system, you could request if it is possible to borrow a copy through interlibrary loan.

To describe your search process, answer these questions:
a) Where did you search for the documents?
b) Were you successful in finding copies, and if so, which standards did you find?
c) If you were able to see the complete standards, were you allowed to carry them out of the library?
d) If you were able to see the complete standards, what was your impression of them? Were they long and technical, or short and easy to understand? Please explain in three or more sentences.
e) If you were not successful, explain what happened.

EXERCISE 2. ISO 14040 AND 14044 CLAIMS AND STATEMENTS

Determine whether the following claims or statements are likely to be compliant with ISO 1400 and 14044. Answer yes or no and explain why.

A) Our cleaning spray is greener than the others.
B) Our new bottle weighs 30% less than our old one.
C) This paint formulation has 15% lower global warming potential, 30% lower acidification potential, and 20% higher marine eutrophication potential than an average paint used in the U.S.
D) The environmentally-superior performance of this toothpaste was supported in a critically reviewed LCA (details available at ourbettertoothpaste.com).
E) Using our fuel additive can increase your car's fuel economy by 10%.
F) This bag is environmentally friendly.

G) The future compliance of this manufacturing site with EPA and state emissions regulations was assured by conducting a LCA.

EXERCISE 3. FORMING A CRITICAL REVIEW COMMITTEE

You are the owner of LCASrUS, a consultancy focused on LCA and sustainability issues. You have been hired to organize a critical review under the ISO standards for a LCA of a new automobile tire that includes recycled materials, and has a longer life and better fuel economy than conventional tires. This critical review has been created so that advertising material can be developed comparing the new tire with others. You think the product sounds great, so you organize a review committee consisting of yourself (an expert in LCA, particularly of food products and packaging) and two other LCA experts that you have worked with on previous projects. Describe how such a committee may or may not meet the expectations of ISO 14044.

EXERCISE 4. LCA PROJECTS WITHIN CORPORATIONS

You are a recently-hired "sustainability expert" at XYLA corporation. You have been asked to work with an R&D team to conduct an LCA on three product concepts they have to determine which offers the greatest environmental benefits (or the least environmental burdens). The goal of the study is to iodentify the product the R&D team should focus on for further development. What sort of critical review should be planned for this project and who should be involved?

EXERCISE 5. LCA STUDY ATTRIBUTES

You enjoy an afternoon catching up on LCA publications and journal articles. Listed below are attributes of some of the studies you have read. Is each study likely to be compliant with ISO 14040 and 14044 or not, and why?

1. All the results are presented only using ecopoints (from Ecoindicator 99).
2. A statement of the environmental preference of one product compared to others is based on three new, previously unpublished metrics.
3. The study includes clear calculations of category indicators, but no normalization or weighting
4. A product carbon footprint report based on the Greenhouse Gas Protocol or similar non-ISO standards
5. An LCA for comparative assertions that includes calculation of only the global warming potential.
6. An attributional LCA that uses solely economic allocation.

ETHICS

Rita Schenck and Tom Redick

LCA is a tool used to help make decisions about environmental improvement, environmental marketing, and environmental policy. If information in a life cycle study is incorrect, either through deliberate or accidental error, or if the practitioner does not properly explain the limits of the available analysis, then decisions may be made that cause environmental, social, or economic damage rather than improvement. Many people could be hurt by a poor environmental decision based on an erroneous study. As a result, the practice of LCA is a deeply ethical activity.

The ethical responsibility of the practitioner is especially strong given that that the potential environmental damage can be substantial, even if the damage is never detected by the layman. Like the errors in engineering design, real damage can occur, potentially leading to increased climate change, increased ecosystem losses, and many other impacts involving loss of life. Unlike errors in engineering, where design failure becomes apparent over time (when structures fail, for example), errors in LCA are cumulative and subtle. The LCA accumulates estimates of environmental impacts across the entire life cycle of the product and may project impacts far into the future. Only decades later might one see the negative outcome of poor life cycle environmental decisions. However, tracking these decisions back to a poor analysis will, by that point, be nearly impossible.

To avoid these negative environmental outcomes, the practitioner must understand the implications of decisions in study design and follow through with careful analysis. Finally, the practitioner will clearly communicate the results of the study, including limits to the knowledge that can be derived from it. Although some uncertainty in LCA studies always exists, the key is being transparent about that uncertainty. Chapters 6, on data quality and 18, on reporting, give guidance on how to do this.

In addition to the ethical responsibilities related to performance of the LCA itself, LCA practitioners need to understand ethical issues related to employee relations, customer and vendor relations, good business practice and legal requirements. Often, the practitioner is privy to confidential business information and the practitioner must take care to avoid conflicts of interest. This chapter describes the ethical responsibilities of LCA practitioners, and provides tools to help navigate the ethical uncertainties that are part of being a good LCA practitioner.

3.1 Ethical Approaches

The study of ethics is not new. When Aristotle wrote *The Ethics* in the 4th century BC, he described ethics as performing good actions, living virtuously (which he described as being the "golden mean") and showing concern for human well-being. He identified happiness as the greatest good and stated that virtuous action was the way to achieve the greatest happiness for the greatest number of people. However, Aristotle also made it clear that generalizations can take one only so far, as many ethical decisions depend on individual circumstances.

Most of Western ethical philosophy adopted Aristotle as its

basis. Some philosophers, such as Immanuel Kant, described ethics as a form of purely rational decision making based on *a-priori* moral principles. Kant believed that if one had a clear grasp of those principles, then ethical decisions were a question of logical thought. David Hume, on the other hand, thought that moral actions grew out of moral feeling. While those feelings might lead one to develop abstractions that could be described as moral principles, in Hume's view rational thought had little to do with ethics. Kant credited Jean-Jacques Rousseau's writings on innate conscience with guiding his own system of ethics, and Rousseau's writings also incorporate a form of early "Biophilia" (Wilson 1984) with respect to the natural environment's positive influence on human ethics. Rousseau's philosophy, in contrast to the cerebral Kantian system, placed nature in a preeminent position, in what was an early Enlightenment predecessor to the modern eco-consciousness movement. In his writings on purity of spirit in pre-civilization mankind, Rousseau foreshadowed the modern class of extremely environmentally-conscious individuals and organizations that go "off the grid," or households and corporate factories that have "zero waste" status.

Jean-Jacques Rousseau

Modern neuroscience sheds some light on these issues, as does the emerging field of evolutionary biology, which finds the roots of conscientious behavior toward others in genetic tendencies that promote "kin selection"(Hamilton, 1964). Kin selection is believed to reflect the reproductive success of the genes that one carries—the more closely one individual is related to another, the more likely the first individual will provide assistance despite the cost.

Applying environmental ethics to policy decision-making is a relatively recent phenomenon. This concept holds that the natural world has value in and of itself, and that deliberate damage to the environment is wrong. These concepts are also sometimes called deep ecology. In the United States, early proponents of this view include Aldo Leopold (1949) and John Muir (1894).

The concepts of right and wrong, and good and evil, are powerful motivators of human behavior. Such concepts are both the glue that holds society together and a force that has been the basis of human conflicts throughout history, such as when violent acts are condoned when fighting against evil.

Another approach to ethics is the application of the Golden Rule: do unto others as you would have them do unto you. This approach to ethical issues is focused on the decisions of the individual rather than on the decisions of society and organizations. Golden Rule ethical concepts underlie much of the legal framework that has come to be called business ethics. Nearly all LCA work is performed as some part of business activity, and LCA practitioners need to understand the legal requirements related to their work. These requirements vary from country to country, and even from state to state. A few examples are shown below.

3.2 BUSINESS ETHICS

Business ethics are characterized by legal underpinnings that can either be legislated or may be based on case law. Business ethics vary from place to place, but they can be divided into three sets: how an individual or company treats employees, how an individual or company treats customers and how an individual or company treats competitors. Business ethics are critical to a practitioner, even though they indirectly relate to the practice of LCA. The credibility of the organization, and thus the inferred credibility of the organization's LCA studies, is strongly affected by business ethics.

EMPLOYEE CONCERNS

In most countries, labor laws address these issues: the number of hours of work that can be demanded of individuals; the minimum wage that must be paid to employees; and health and safety concerns on the job. In general, the employer holds the responsibility to provide a safe working environment and to keep records of on-the-job accidents. Worker insurance is usually required and the cost of insurance is tied to the number of claims filed. These factors provide the motivation for employers to decrease on-the-job illnesses and injuries in order to reduce insurance costs.

Many countries have anti-discrimination laws related to employment, including hiring processes, the receiving of benefits and promotions, and firing procedures. For example, in most Western countries discriminating against individuals on the basis of gender, age, and disability is illegal. In most countries, discrimination based on sexual orientation is also protected under law, and sexual harassment is generally illegal. Several international conventions address these topics.[1]

1. http://www.ilo.org/global/standards/introduction-to-international-labour-standards/conventions-and-recommendations/lang–en/index.htm

In the US, issues related to intellectual property are controlled by contract. Typically, if an individual is working for a company develops new intellectual property, the property belongs to the company and not the individual. In the context of LCA studies, therefore, new ideas about how to perform LCAs become a part of the intellectual property of the employer. Sometimes employment contracts arrange to share intellectual property between employer and employee.

Some employment contracts require that the employee sign a non-compete agreement, which restricts the employee from working in the same field, or with the same customers, for a specified period of time after leaving the company. Often, a company invests a substantial amount of time and money in developing its customers, and employees who leave a company and take its customers with them to a new employer could potentially be subject to legal claims related to the intellectual property of the first employer.

Regardless of issues related to employment, practitioners have the ethical responsibility to credit the work of others properly. This means, for example, that when one is a co-author of a study, one makes sure to mention the other authors when describing the work regardless of the venue. When one is expanding on the ideas of another, one makes sure to properly reference the work of that other individual or group.

Market oriented issues

In the US and elsewhere (e.g. the United Kingdom, Australia, the Netherlands, and Canada), government regulators have oversight of "green marketing" claims and can prosecute companies whose advertising goes too far, particularly those which make claims that reach into the field of environmental sustainability. LCAs can either help to validate or contradict claims of sustainability, as they are more comprehensive in scope than claims relating to specific impacts or issues (e.g., dolphin-safe tuna[2]). As claims are increasingly made for indirect as well as direct environmental effects, attorneys predict increased attention to both compliance within corporations and enforcement by government agencies (Cox 2006). This section briefly reviews the legal boundaries for such marketing representations, including LCA labeling standards, in the US and the UK.

Companies In a wide range of industries are seeking to use sustainability for part of their marketing claims approaches. In particular, Wal-Mart and Whole Foods Market are both hailed for their sustainability efforts in the book *Green to Gold* (Esty & Winston 2006). In the book, former Wal-Mart

CEO Lee Scott promised that sustainability efforts would help protect the company's "license to grow" and challenges Whole Foods Market for market share among consumers who care about their ecological footprint. Another book on green marketing (Cox 2008) suggests that "the corporate takeover of organic food retailing has been the industrialization of organic agriculture" (pp 65-73) and cites critics who see corporations "exploiting organic agriculture's feel-good image even when selling conventional products" (ibid. p. 411).

Most companies are careful about making claims, including "green" claims, about the environmental attributes of their products or services. But companies can always take LCA results out of context to make broad environmental claims or even sustainability claims. Regulators will scrutinize these claims under applicable green marketing law or regulation. These risks arise from US law, and have counterparts overseas. For example, the United Kingdom Advertising Standards Authority (UK-ASA)[3] challenged the World Wildlife Fund US for its campaign marketing certified "sustainable" palm oil from Malaysia and other sustainability claims made in products sold. The practitioner has the responsibility to caution the client (internal or external) about the limitations of the LCA results.

The Federal Trade Commission (FTC) has made its "Green Guides" available since 1992 to assist marketing departments everywhere in avoiding "deceptive or misleading" environmental claims, otherwise known as "greenwashing." These guidelines were updated in 2012 to provide marketers with critical guidance about making legally valid environmental claims in labeling, advertising, promotional materials, and all other forms of marketing whether asserted directly or by implication, or through words, symbols, emblems, logos, depictions, and product brand names. The guiding principle behind the Green Guides is that environmental claims should not be deceptive to the consumer.

As in the U.K., the FTC finds problems with broad claims of "sustainable" product, and requires that claims be as limited and clear as possible, whether they relate to the product itself, to its packaging, or to some aspects of both. Because an LCA that makes comparative assessments between products may find one product to be less environmentally sustainable than another, would the first product violate green marketing or would the system allow for degrees of sustainability to be

---

2. www.earthisland.org/dolphinSafeTuna/consumer

3. UK-ASA is the FTC's counterpart in the United Kingdom and it has gained notoriety in the past by ruling in against companies including Shell and Lexus over green claims in their advertising, in disputes related to allegedly deceptive advertising

British Advertising Authority Rejects "Sustainability" Claims

On September 10, 2009, the British press reported that UK-ASA ordered withdrawal of a press campaign that made environmental claims about palm oil from Malaysia. Based on the facts before it, UK-ASA found the word "sustainable" to be "misleading" to consumers. UK-ASA ruled that the palm oil company's sustainability certification by the third-party Roundtable on Sustainable Palm Oil (RSPO) proved little, because the "greening" of biofuels was "still the subject of debate" worldwide. US-ASA rejected the claims of the Worldwide Fund for Nature, which argued that the RSPO led to measurable environmental and social progress among Malaysia's palm producers. Among other aspects which reduce environmental degradation from palm oil production, measures such as zero-discharge operations, HCVF (high conservation value forest) surveys throughout Borneo, and limiting new plantings on forest land not already cleared did not satisfy the green marketing requirements of the U.K.

This RSPO decision follows in the wake of a similar decision on US sources of sustainable cotton that question whether "sustainability" can be defined at all, and which

also note that the UK-ASA is still not ready to accept biotech crops as sustainable. In March 2008, UK-ASA ruled that the claim "SOFT, SENSUAL AND SUSTAINABLE" in the US Cotton Council advertising had "misleadingly implied the sustainability of CCI's cotton was universally agreed" upon, a claim which ASA found to be untrue. In light of this decision, the ability of any sustainability claim to pass muster, given the lack of a generally accepted definition, places doubt on proposals to make sustainable labels, at least in the UK.

claimed? This has not been addressed in the implementation of the Green Guides, but the future might hold hazards for those claiming "environmental sustainability" without adequate LCA procedures to defend this claim.

The ISO 14044 standard lays out stringent requirements for comparative assertions (public claims of the equivalence or superiority of one product over another). Among the requirements to be met, the data must be equivalent and the system boundaries the same. Competitors almost never have equivalent data for both their own processes and those of their competitors. More often the case that current data may exist for one company and published data (that is five to ten years old) exists for another. Inasmuch as production methods change rapidly (usually towards lower-impacting methods), the newer data will probably be "cleaner." In fact, a comparative study usually finds that the commissioner's results (based on current data) are better than a competitor's results (based on old data).

Besides the issues related to green claims, businesses have an ethical responsibility to avoid monopolistic business practices. Microsoft encountered this kind of problem when it tried to make its Internet browser an integral part of its Office

Suite programs. This attempt was struck down as illegal in both the US and Europe. Monopolistic practices can only occur when companies have a majority of a given market. At the moment, the field of LCA is growing rapidly, with more and more companies in the business of providing LCA services. However, this need not be the permanent state of affairs. After a period of expansion, there will probably be an aggregation of LCA services, with fewer and fewer companies taking on the role of LCA service providers. If or when such a situation occurs, monopoly considerations may come into play.

Conflicts of Interest

A conflict of Interest occurs when a person in a position of trust has divided loyalties. For example, a conflict of interest can occur if a practitioner performs a comparison between the LCA of two products, one made by his or her employer and one made by another company. Having such a conflict does not mean that the person has done anything wrong, but it does mean that the person may be biased towards his or her employer. For this reason, the study's funder or funders, as well as the affiliations of the LCA professionals who participate in it, should always be disclosed.

LCA AS AN AID TO LIABILITY PREVENTION

Preventing liability is a risk-management exercise. A liability is identified and its intensity (cost) is modified by the likelihood of its occurrence. If measured with reasonable accuracy (taking into consideration lessons learned in tort litigation such as those noted above) these measurements – and the prevention initiatives they generate – can prevent liability for much lower cost than would otherwise occur.

LCA can identify impacts that carry potential liability, such as climate change or the production of toxic nano-scale compounds imbedded in the supply chain. To the extent that industry leaders use LCA as a starting point for risk management, the benefits of LCA analysis will increase over time. Risk managers hoping to demonstrate the benefit of addressing a hazard can use an LCA-Life Cycle Costing study to show the potential long-term liability expenses associated with a particular hazard that has been limited or avoided.

The demand for this kind of liability management is growing. For example, the Carbon Tracker Project assembled over 70 large global investors to request that 45 major oil and gas producers disclose the current and proposed future carbon emissions of their products. The assumption is that these emissions are at high risk of being regulated or even banned. The insurance industry, in particular the re-insurance industry that insures insurers, have a financial stake in climate change, for they pay out claims related to severe weather events. Such claims total billions of dollars annually.

It has been proposed that LCA metrics could be the basis of economic investments for ecosystem and social services for what is called Intergenerational Finance. This system is intended to provide some kind of bonding method (either government or business bonds) in which current environmentally or socially beneficial actions are paid for by future generations who reap the benefits.

CORPORATE SOCIAL RESPONSIBILITY

Corporations are increasingly being held to high ethical standards through the practice of corporate social responsibility (CSR). Industry organizations such as the Business for Social Responsibility and The Future 500 drive the adoption of an internally regulated approach for a company to integrate social issues into its work practices. In addition to several local and national codes, two well-recognized international standards, ISO 26000 and SA 8000, have codified what this means, and many corporations have Corporate Social Responsibility (CSR) officers. Sometimes, these programs include LCA approaches to managing social issues. The primary driver for CSR programs is the recognition that companies need a license to operate, and a polluting company that does not take good care of its employees, customers, communities, and vendors at every link in the value chain may lose the ability to operate as a result of poor public relations.

PROFESSIONAL ISSUES

An LCA practitioner always works in a team, with different members of the team undertaking different tasks. In such an environment, managing relationships with honesty, transparency, and tact is essential. When starting a project, the practitioner must make clear its expected outcomes, and identify members of the team to execute certain parts of the work. This approach manages the expectations of the team members and makes the process flow smoothly. Project management is described in detail in Chapter 17.

Every LCA is an opportunity to learn new things, but this learning is built on previous knowledge. As a member of a team, the LCA practitioner has the responsibility to make clear as soon as possible the limitations he or she may have with project. If the practitioner has no experience in evaluating the impacts of power generation, for example, he or she must make that clear when an LCA on electric grids is planned. Additional talent sets can be brought to bear at the planning step of the study.

The practitioner should always give credit where credit is due: all LCA studies depend on the work of many individuals. This should be acknowledged. Likewise, all references should be properly cited in the final report.

Finally, a practitioner has the responsibility to keep current in the field, and to help others keep current. The field of LCA is a young one, with many changes expected to come about as a result of ongoing research. No one can be said to be an expert in all parts of the field. Many LCA studies are reviewed and the practice of reviewing LCA studies provides an immediate opportunity for LCA professionals to assist their peers in keeping current in the field.

3.3 DUE CARE IN LCA

The field of LCA encompasses a broad range of studies, such as screening LCAs meant to inform further work, as well as LCAs for the following purposes: internal decision making; short academic projects; carbon footprinting; cradle-to-gate studies for common materials that represent a regional or global average; and environmental product declarations for a single product. The level of care applied to each of these studies depends on the study audience and the decisions supported. The larger the audience and the more wide-ranging the implications of the decision, the more care must be taken in completing the study.

For example, if a company seeks to replace its current process with a more environmentally friendly one, the audience is limited to the management team making the decision. The implications of the decision can vary, but typically the impacts would fall on a particular local population: the workers and neighbors of the facility. At the other extreme, one might have an LCA to support a national regulation on biofuels. The LCA result would have an immensely large audience (such as an entire nation) that potentially affect an even larger population (such as the world).

While a practitioner should do a careful job in the first case, in the second case he or she would make doubly sure to understand the implications of every assumption and characterize the uncertainty of every conclusion. Tools such as sensitivity and uncertainty analysis are suited to this task. These tools are presented in later chapters.

The fundamental thing to keep in mind is that LCA is a form of modeling, and models are simplified representations of reality. To the extent that models make accurate predictions about the real world, they are good models. To the extent that the predictions they make are inaccurate, the models are poor. The smaller the system that is being modeled, the simpler the model and the more likely the results will be accurate. With notable exceptions, such as the IPCC's climate change models and the World Meteorological Organization's stratospheric ozone depletion model, few life cycle impact models have ever been tested in the real world. Models such as those implied by most life cycle impact assessments models at global or regional scale are weak predictors of actual environmental damage. Instead, they are simplified tracking models of the resources consumption and emissions of product systems.

The ethical practice of LCA requires that the practitioner understand the limitations of the model being used and com-municate these limitations to the client. Then using the best available science, and careful review of one's own work, one can produce an LCA that can help inform the client.

CONFIDENTIALITY OF DATA AND THIRD-PARTY REVIEWS

Regardless of the LCA model used, the LCA practitioner is likely to have access to confidential data from the hiring company in order to perform the study. Many companies require non-disclosure agreements prior to sharing data, but even in the absence of such agreements the practitioner is ethically responsible to retain the confidentiality of business information. If a third-party review is undertaken, a non-disclosure agreement should be part of the contract retaining the reviewer.

If an LCA is intended to be used for public comparative assertions, the ISO standard requires that the study be reviewed by a third party, who may be a single person or who may be a group of individuals. Typically, a third-party review panel will include an LCA expert, an expert in the industry represented in the LCA, and a third person who may be an expert or who may represent an outside interest (such as an environmental advocate).

Sometimes, a group of companies want to create an average LCI report that aggregates their individual LCI information. In such cases, the group will retain a LCA professional who signs non-disclosure agreements with each company, aggregates the data, and publishes the aggregated results. Such agreements are behind most cradle-to-gate LCI datasets found in public databases, such as the plastic resin data published by the American Chemistry Council (ACC).

3.4 THE ACLCA ETHICS STATEMENT

The American Center for Life Cycle Assessment (ACLCA) published its Code of Ethics in 2004. The Code of Ethics was the first LCA-specific code of ethics that covers the broadest range of ethics issues related discussed in some detail above. It serves as a quick reminder of the ethical issues that should be considered during the practice of LCA.

ACLCA
American Center for
Life Cycle Assessment

The best available science does not necessarily mean latest available science. Our understanding of natural science changes daily. The process of science is one of hypothesis development and testing, as well as publication of the results at each step of the process. Life Cycle Impact Assessment (LCIA) is based on consensus models of processes in nature. For that reason, LCIA is significantly behind the latest information about natural processes. Models used in LCIA should be tested by a diverse set of practitioners before coming into general use.

PREAMBLE

Environmental Life Cycle Assessment (LCA) is a comprehensive yardstick of the environmental performance of goods and services. LCA is a powerful tool to influence human behavior and environmental outcomes through management and engineering decisions, public policy and purchasing decisions. The majority of those using the output of a life cycle assessment are not in a position to verify every element of the LCA. Life Cycle Assessment professionals, those performing, providing life cycle data, and developing life cycle tools and models therefore have great responsibility to provide unbiased, accurate and transparent data and analyses to the greatest extent possible. In doing so they use the work of their hands and minds to build a more sustainable world, and they support the strength and quality of the LCA profession.

REQUIREMENTS

The American Center for Life Cycle Assessment supports adherence to this code of ethics when conducting LCAs. All members of the American Center for Life Cycle Assessment therefore make the following commitments:

1. To make maximum use of national and international standards, particularly the ISO 14040 series standards, and U.S. federal FTC guidelines for claims when conducting an LCA;
2. To avoid real or perceived conflicts of interest whenever possible, and to disclose them to affected parties when they do exist;
3. To be honest and realistic in stating claims or estimates based on available data;
4. To the greatest extent consistent with retaining confidential business information, to disclose data and estimates in a full and transparent manner;
5. To clearly distinguish between professional extrapolations and value judgments when developing, performing and using life cycle assessments;
6. To improve the understanding among LCA professionals, the general public and decision-makers of Life Cycle Assessment, its appropriate application, and potential consequences;
7. To maintain and improve one's technical competence and to undertake tasks for others only if qualified by training or experience, or after full disclosure of pertinent limitations;
8. To seek, accept, and offer honest criticism of technical work, to acknowledge and correct errors, and to credit properly the contributions of others;

9. To treat fairly all persons regardless of such factors as race, religion, gender and sexual orientation, disability, age, or national origin; and
10. To assist colleagues and co-workers in their professional development and to support them in following this code of ethics.

ADDITIONAL READING

Trevino, Linda K., and Katherine A. Nelson. *Managing Business Ethics (5th Ed.)*. John Wiley and Sons, 2011.

Shaw, William H. *Business Ethics: A Textbook with Cases (7th Edition)*. Boston, Wadsworth, 2011.

Stanford Encyclopedia of Philosophy. http://plato.stanford.edu/

PROBLEM EXERCISES

1. When doing a life cycle study, a practitioner can encounter many potential ethical pitfalls. This exercise is intended to help you understand where to find some of those pitfalls.

Scenario: You are retained by a company to conduct an LCA of balloons. It compares rubber (latex) balloons with balloons made of mylar (a form of polyethylene terephthalate [PET]). Both types of balloons contain the same amount of helium gas. Both can be printed with attractive messages. The rubber balloons last about 24 hours before the helium escapes, while the Mylar balloons last for a week before the helium leaks out. The rubber balloon weighs 45 grams, while the Mylar balloon weighs 20 grams. The study concluded that the rubber balloon is more environmentally friendly because it is made of a plant-derived and biodegradable product, while the Mylar is synthesized from petroleum and is not biodegradable.

The study chose a functional unit of "containing 4 liters of helium at standard temperature and pressure". The impact categories evaluated were the following: climate change, acidification, eutrophication, and fossil fuel depletion. It assumes that the rubber is fully degraded at end of life and that the Mylar is all put into a landfill.

To understand more about the systems being compared, you will search the Internet to learn how latex and mylar are each made, as well as the end-of-life alternatives exist for each, including biodegradation, recycling, landfilling, and waste-to-energy. List the potential places where the choices made in the study might have unfairly caused the rubber alternative to be advantaged over the mylar alternative, or vice-versa.

2. You perform a comparative LCA study and have a client who wants to make broad claims of the superior sustainability of their product versus another's. What should you tell the client?

3. What particular aspects of LCA drive the need for high ethical practice?

4. What is deep ecology?

5. You have a client new to the field of environmental LCA, and he or she needs to justify the resources (both time and money) spent on it. What arguments might you give to help justify the expense?

Unit Processes

Beverly Sauer and Greg Keoleian

4.1 Unit Processes

ISO standards on LCA define a unit process as the "smallest element considered in the life cycle inventory analysis for which input and output data are quantified" (ISO 14044: 2006, Section 3.34). ISO does not define "smallest element." In practice, the level of detail that constitutes a unit process can vary depending on the level at which data is available or can be collected (e.g., facility-level process data or data for a specific process within a facility). Depending on the scope and goals of the LCA, aggregated cradle-to-material data sets encompassing many individual process steps can even be treated as unit processes, as described later in this section. In LCA, unit processes for each life cycle stage are linked to construct a complete life cycle model of the system being analyzed. Examples of unit processes for different life cycle stages include the following:

Raw material extraction

- Minerals and fossil fuels: mining, drilling, blasting, and crushing
- Agriculture and forestry: planting, fertilizing, irrigating, and harvesting

Material production

- Metals and minerals: milling, refining, and smelting
- Chemicals: mixing, distillation, crystallization, and evaporation

Material converting and product manufacturing

- Metals: machining, rolling, stamping, drawing, welding, heat treating, and plating
- Plastics: blow molding, injection molding, thermoforming, and sheet and film extrusion

Use

- Reusable container: storing food
- Automobile operation: driving, fuel use, and charge depletion (electric vehicles)

Service/Maintenance

- Reusable container: washing
- Automobile service: washing, parts and fluids replacement, fueling, and charging (electric vehicles)

End of life management

- Landfilling, recycling, reuse, composting, and combustion

Each unit process has input and output flows that link to other related unit processes. For example, a process that uses grid electricity must link to data sets for extraction, processing, transport, and combustion of each of the fuels in the grid mix used to produce electricity for that region, as shown in Figure 4.1. In addition, a comprehensive model would include the construction processes for power plants and renewable energy technologies such as wind turbines.

In practice, data for life cycle processes are available at varying levels of detail and completeness. Often, many unit

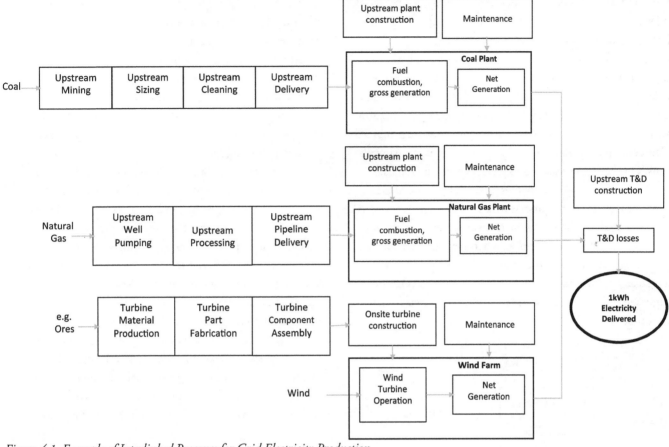

Figure 4.1. Example of Interlinked Processes for Grid Electricity Production
Note: Some steps in the flow diagram, such as Turbine Material Production and Turbine Part Fabrication, are represented as a single box for simplicity, but would actually include multiple unit processes.

processes occur within an industrial facility, necessitating careful analysis to determine the quantities and types of input and output flows that are associated with specific unit processes occurring within the facility.

The depth of analysis for a process should be consistent with the scope and goals of the study. For example, a life cycle study analyzing improvement strategies for steel production requires detail on unit process modules such as iron ore mining, sintering, the blast furnace process to produce pig iron, and the basic oxygen furnace process for steelmaking. On the other hand, if the analyst is constructing a life cycle inventory for a mid-sized automobile that is comprised of many types of materials and components, the sequence of steps for steel production can be represented as a single process for the materials production stage of the vehicle life cycle.

Process-based life cycle inventory

Process-based life cycle inventory (LCI), as its name suggests, is life cycle modeling based on linking individual related unit process data sets to form a complete model of the life cycle of a product system. This is the most detailed approach to life cycle modeling. The unit process approach to life cycle modeling offers many advantages, but it also poses significant challenges.

Benefits of unit process modeling approach

Flexibility is a key advantage of the unit process approach. Updating or conducting sensitivity analyses on a life cycle model that is built on a unit process basis is relatively easy. The unit process approach allows users to change data within contributing unit processes (e.g., the amount of natural gas used for process energy), or change the factors used to link related processes (e.g., the amount of output from one process needed as an input in the next process, such as the weight of resin to mold a two-liter bottle). A unit process approach provides further flexibility by allowing nesting and weighted averaging of processes. For example, the data set for production of an average kilogram of crude oil can be set up using data sets for oil production in individual countries multiplied by the percentage of oil produced in each country. In this way, the composite oil production data set can easily be updated if oil supply percentages change, or if there is a change in the oil production process for any supplying country.

Ideally, for maximum transparency and flexibility in modeling, data sets for processes and materials should be made available on a unit process basis. In practice, however, process data are sometimes published exclusively on an aggregated basis in which the data for two or more linked processes are combined. One could publish data in an aggregated format for many reasons. One common reason is to protect confidential data.

If only one or two companies produce a given material, showing data on a unit-process basis may divulge confidential process details or allow a company to "back out" information about a competitor's operations. When data for a specific unit process are confidential, then the data can only be shown aggregated with data for one or more related unit processes. Figure 4.2 illustrates an aggregated data set for two sequential unit processes.

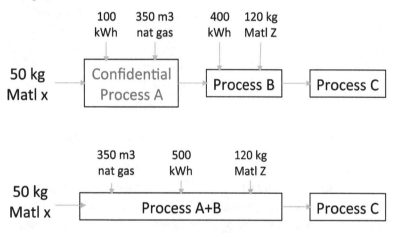

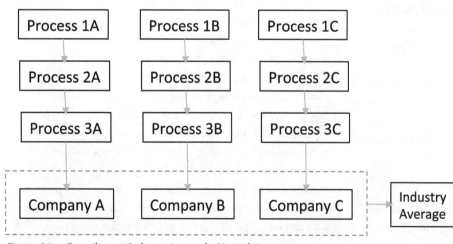

Figure 4.2. Aggregating Sequential Unit Processes (A+B) to Protect Confidential Data

This type of aggregation still allows users to model changes in the individual input and output flows for the combined data set (e.g., substituting a different grid mix kWh to model production taking place in a different geographic region). When working with confidential data, care must be taken to ensure that any data sets published are sufficiently aggregated so that stoichiometry, mass balances, and other physical relationships cannot be used to extract confidential process details on material or energy use.

When aggregating data sets to compile an industry average, the data can be combined using horizontal or vertical averaging. The terminology can be understood by envisioning the sequence of process steps arranged vertically, with data for individual companies horizontally arranged. In the vertical averaging method, the data sets for each company's specific process sequence and supply chain are compiled; a weighted industry average is then calculated based on each company's share of total production or market share of the final output.

In the horizontal averaging method, an average data set is compiled for each unit process; the average unit process data sets are then linked using average input/output factors. The two averaging approaches are illustrated in Figure 4.3a and Figure 4.3b.

Life cycle data can also be published on a fully aggregated basis, known as a "system process." In this form of aggregation, only the inputs from nature and outputs to nature are shown; all information and assumptions on intermediate contributing process details and amounts are "locked in" and cannot be adjusted. This method fully protects any confidential details of the data, but all modeling flexibility and transparency is lost, and the data set can only be used "as is." If changes are made to any contributing process (such as an improvement in energy or material efficiency for a converting process, or a shift to a less carbon-intensive electricity mix used for the process electricity), there is no way to apply this change to a fully aggregated data set. Similarly, if any allocations were applied in the development of the fully aggregated data set, evaluating alternative allocation approaches is impossible.

Unit process data sets are not

Figure 4.3a. Compiling an Industry Average by Vertical Averaging

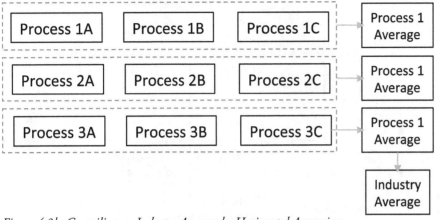

Figure 4.3b. Compiling an Industry Average by Horizontal Averaging

ipate in providing data. In this situation, the practitioner must conduct research to develop a good understanding of the processes so that reasonable estimates can be developed to fill the data gaps. These may include calculations based on stoichiometry and reaction chemistry, or may make use of process modeling software, government statistics on energy use by industry sectors, and published emission factors (e.g., US EPA AP-42).

Knowing when an estimate is "good enough" can be difficult. From a practical standpoint, the analyst may wish to start with a basic estimate, then use contribution analysis and sensitivity analysis to evaluate the influence of the estimate on overall system results and conclusions. If the estimated data has a large influence on the study outcome, then further refinement may be warranted.

always ideal. In cases where unallocated data cannot be shown due to confidentiality issues, published unit process data sets may be based on a specific allocation approach so that alternative allocations are not possible. In addition, depending on the intended use of the data, an industry average system process data set, based on the industry mix of producers and technologies in the region of interest, may be more representative than a unit process data set that covers only one of the technologies used or that represents a different region.

CHALLENGES OF UNIT PROCESS MODELING APPROACH

Because developing a process-based LCI is a data-intensive effort, availability of data can be a major challenge. When gathering process data to construct a life cycle model, these problems are commonly encountered:

- No data available for a specific unit process or material.
- The available data set is incomplete (e.g., material and energy requirements may be available, but no information about process emissions).
- The available data set is not representative of the process technology and/or geography for the current system being modeled.
- The available data set is old enough that improvements in material or energy efficiency may have been made, or process technology has changed.

For less common situations in which multiple data sets are available for a given process or material, the practitioner must review the data sets to determine which, if any, provides the best representation for the current system being modeled in terms of appropriate age, technology, and geographic location.

When data gaps exist for industry processes not directly involved in the life cycle project (e.g., industries located several steps back in the supply chain or industries for competing products), these parties will probably not be willing to partic-

4.2 PROCESS LIFE CYCLE INVENTORY DATA

The variety of data sources that can be used for LCI modeling is discussed in Chapter 5: Life Cycle Inventory. The focus of this section is on collecting primary process data to support LCI. Primary data is data gathered from facilities or individuals directly involved in conducting the unit process. Primary data may be measured directly (e.g., by weighing material outputs and inputs, metering equipment, or sampling emissions) or indirectly (e.g., by making calculations based on plant records of material purchasing, utilities, and plant production).

When data is collected for a unit process, the practitioner will need to take into account the background processes required to produce each of the direct inputs to the process reported by the producer. To identify all the unit processes needed for modeling, the practitioner generally starts with the product or service that is being analyzed, then traces back through all the associated materials and processes to the original flows from nature. An example for polyethylene terephthalate (PET) resin production is provided in Figure 4.4. An analysis of PET resin production would begin with data on the processes to produce PET by polymerization from dimethyltryptamine (DMT) and from terephthalic acid (PTA). The PTA polymerization process requires inputs of PTA and ethylene glycol. Ethylene glycol requires inputs of ethylene oxide. Ethylene oxide requires inputs of ethylene and oxygen, and so on. Each process is traced backward based on the input materials required, thus creating a process tree.

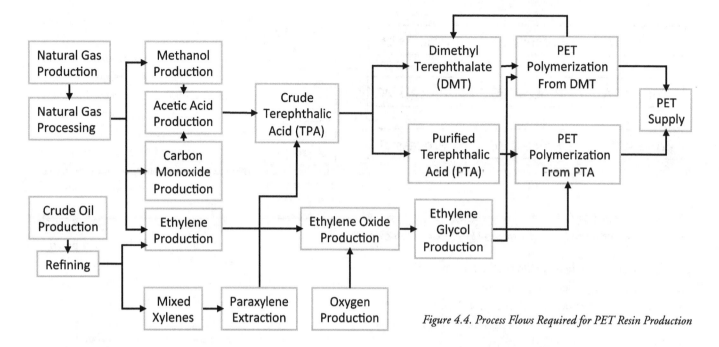

Figure 4.4. Process Flows Required for PET Resin Production

The bill of materials for a unit process is often used as the starting point for tracing back to raw material extraction. In a full LCI, the analyst must also identify all the processes associated with the use and end-of-life management of the unit process outputs being evaluated. This can only be done in the context of a full LCI after the functional unit for the analysis is defined in the scoping phase.

ISO 14044 Annex A provides a simple example of a unit process data collection sheet as exhibit A.4. Essential types of information that must be gathered and documented when developing primary unit process data sets are described in the following sections.

REPORTING BASIS

Understanding and correctly interpreting the process data reported is key. Facilities may report data on the basis of a specific amount of output, such as 1000 kilograms of product or 1000 units of product, or they may report based on a given period of facility operation. If reporting for a given time period, knowing the physical units of output produced during the time period is necessary, as this information is needed to normalize the operating data to an appropriate unit of output for linking with other unit processes (more on this in Section 4.3). The reporting period needs to be sufficiently long that the process data reflect average, steady-state production that is not unduly influenced by project startups, periodic production changeovers, or seasonal variations.

MATERIAL INPUTS TO THE PROCESS

Input materials may include both flows from nature and flows from industrial processes. Material reporting should include information on the types and quantities of material used in the process, including any relevant information on concentrations, if materials are used as solutions, and incoming transport of materials. Materials reported should include not only inputs that become part of the useful output product, but also ancillary materials (i.e., materials that are used or consumed in the process, but do not directly become part of the output product).

Examples of ancillary materials include catalysts that must be periodically replaced or replenished, lubricants and cooling fluids required to maintain the operating equipment, and detergents used for periodic cleaning of equipment. The amounts of ancillary materials should be scaled to the reporting basis. Materials should be specified by chemical name and Chemical Abstracts Service (CAS) registry number to the greatest extent possible. Brand names of materials can be useful, if published information about the makeup of the material is available from the supplier. Material safety data sheets (MSDS) are sometimes useful, but they often do not contain the necessary level of information on composition needed to support unit process modeling.

In addition to ancillary material inputs, capital equipment and facilities also represent material inputs for a process. This capital equipment and the facilities should be amortized (i.e. normalized to the amount of product output over the service life of the equipment). In general, capital equipment and facilities have long service lives and produce large quantities of output over their useful lives. As a result, the capital equipment impacts allocated to a specific output quantity of product are relatively small compared to the impacts associated with inputs of the raw materials, energy, and ancillary materials required to manufacture the product. For example, LCAs of a coal-powered boiler indicate that excluding plant manufacturing affects major system indicators (e.g., energy and global warming potential) by less than 1% (Spath et al. 1999). Capital equipment and infrastructure can, however, represent significant contributions to certain industry sectors. An example would be the photovoltaics and wind turbines used for electricity generation from renewable sources (Frischknecht et al. 2007).

USEFUL OUTPUTS OF THE PROCESS

Useful outputs include the primary output of interest as well as other useful material or energy co-products, including process scrap that is used off-site. Some process scrap or rejected product may be recycled on site in a closed-loop process (e.g., re-pulping paper "broke" at an integrated paper mill, or regrinding clean plastic molding scrap and returning it to the molder feed). Quantities of material recycled on-site in this manner generally do not need to be reported separately, as the material efficiencies should be reflected in the net amounts of purchased input material required to produce a given amount of output.

If a process produces heat or energy as a useful co-product, the user should indicate whether the energy is utilized by the current process (which reduces the need for purchased fuels) or is used by some other process outside the boundaries of the system being evaluated; in the latter case, credit may be given for the exported energy. If any output streams are solutions, the concentration of the output stream should be noted in order to facilitate material balances, particularly in systems where an output stream from a later process is recycled back to an earlier process step at a different concentration.

PROCESS ENERGY REQUIREMENTS

Data must be collected on the types and quantities of process fuels used to produce the amount of output reported on the data form. The practitioner needs a good understanding of the processes and operations that are included in the energy reported by the facility in order to avoid double counting or underreporting. Process energy should include the energy requirements for operating the equipment that is directly used in the process of interest, as well as energy used for supporting services and operations that are necessary for the process, such

as compressed air and vacuums, chillers, emission control systems, wastewater treatments, and process-related space conditioning (e.g., clean room operation).

Process energy requirements may be determined by equipment metering, which are calculations based on equipment operating specifications or facility utility records. If a plant is reporting energy use based on site-utility records, the utility amounts may include energy use that is not directly associated with the unit process (e.g., energy for office equipment and space conditioning, warehouse lighting, or cafeteria operations). If steam is used for process energy, determining the types and quantities of fuel used to generate the steam is important, as is taking into account the steam generating efficiency.

WATER USE

Water use can include water that is directly utilized in facility processes (e.g., for preparing solutions or slurries, or for cleaning processes). It can also include irrigation water, cooling water circulated in closed loop systems or open evaporative systems, water used to produce process steam, or water used to operate turbines. The water use data should identify the source of the water used (river, ocean, or municipal water supply) and the net consumption of water (water that is not returned to the system from which it was withdrawn). If municipal water is used, requirements for pretreatment of the incoming municipal water should be included in the background modeling. Examples of consumptive uses of water include water that becomes part of the product, evaporative losses of water, and water that is withdrawn from surface or groundwater and returned to a different body of water.

SOLID WASTES

Examples of process solid wastes include residual wastes from chemical processes, sludges from pulping operations, trim scrap from converting processes, out-of-spec products, and packaging wastes from incoming shipment of process materials. The end fate of the materials should be noted, so that the appropriate impacts are modeled (e.g., emissions from aerobic or anaerobic decomposition of wastes that are land-applied or landfilled). Process scrap, off-spec products, and packaging wastes that are recycled or reused, either on-site or off-site, should not be reported as waste. If useful energy is recovered from combustion of waste products, it should be noted so that appropriate credit can be given. Associated combustion emissions may need to be calculated for materials that are burned on-site, while combustion emissions for material sent off-site for disposal would generally be reported in the material disposal data set.

ATMOSPHERIC AND WATERBORNE EMISSIONS

Emissions reporting can be one of the most challenging parts of the data collection process. Emissions may include substances released directly from the process, such as emissions from a chemical reaction, or emissions associated with on-site combustion of fuels or waste materials. The process energy reported on the LCI data form is normally linked in the LCI model to data sets for the production and combustion of the corresponding fuels, knowing whether the emissions reported on the form already include combustion emissions in order to adjust the model and avoid double-counting of fuel combustion emissions.

Differentiating between emissions that are directly released to the environment and emissions that are routed to control equipment or treatment is essential. If control systems are used to reduce emissions before they are released, the operating requirements for the control systems (e.g., materials, energy, and solid wastes) should be included in the process modeling. Emission naming conventions, speciation, and concentrations are important emission reporting issues that can affect the life cycle impact assessment stage of the LCA. More discussion on these issues is provided in Section 4.4.

LAND TRANSFORMATION AND OCCUPATION

Information on land use is most often tracked for agricultural processes, but may be tracked for other processes as well. Land use data may include land transformation (the area of land that is converted from one form to another, such as the conversion of an acre of natural forest land to managed crop land) and land occupation (the area of land occupied by a manufacturing plant). The importance of scaling land transformation and land-use data appropriately cannot be underestimated. For example, the land space occupied by a refinery should be allocated over the total output of the refinery during the time it operates on that land. Land transformation and occupation reporting may also include any data available on habitats or species affected by the conversion or use of the land.

OTHER DATA

If the analysis will include life cycle costing, economic information should be gathered for input and output flows. There is also growing interest in social LCA, which addresses issues such as human rights, working conditions, health and safety, and socio-economic effects. A set of guideline for social LCA, including comparisons with environmental LCA methodology, was published by UNEP-SETAC in 2009. Much of the data needed for social LCA may more appropriately be gathered at the enterprise level (e.g., a company's fair labor practices or use of child labor), rather than quantified on a basis of product output at the unit-process level.

METADATA ABOUT THE PROCESS

Users of data should understand the technology, geographic location, and time period represented by the data set. When compiling industry averages, information on a facility's or company's share of total production of the output is also needed. These "data about data" are called "metadata." For data sets that will be made available in LCI databases, practitioners need the information in order to evaluate whether the data set is suitably representative of the specific technology and geographic region for the system they are trying to model. Geographical location is important not only in terms of identifying the country where the process is taking place (which can affect supply chain, electricity grid, and transport modeling), but also in specifying the location of the facility (which should also be reported, as it may be relevant for assessing regional issues such as water scarcity).

Information on the time frame is useful when evaluating whether the data are reasonably current, or whether it is likely that improvements in efficiency or improvements in technology have been made since the data set was first developed. Metadata can also include information on the how the reported data reported was developed (e.g., direct metering of equipment, plant utility records, material purchasing records, stack tests, permit levels, or emission factors), together with descriptions of calculation processes used by facility staff to allocate plant data to specific processes or products within the facility.

MATERIAL BALANCE

The unit-process data set should be reviewed for completeness, including a material balance to check for discrepancies between the weights of process inputs and outputs. Imbalances may indicate a failure to properly account for co-products, process scrap or wastes, or emissions. Adjustments may need to be made to account for changes in mass associated with fluctuations in material moisture content or inputs of materials that are not tracked as purchased materials (e.g., oxygen from air). Any significant discrepancies should be investigated and resolved.

When collecting unit-process data, process reporting should be as complete as possible. Background process data may not be available for modeling materials that are used in small quantities (e.g., catalysts or pigments), but the decision to research production of these materials, or to identify appropriate surrogate data, will be made in the context of the cut-off rules established during the scoping phase of the LCA in which the data sets are being used. The cut-off rules define contribution thresholds for excluding materials from the analysis, and are often expressed based on mass, energy, or toxicity contribution to the overall system being analyzed.

UNIT PROCESS VERSUS SYSTEM PROCESS EXAMPLE

Examples are provided in tables 4.1 and 4.2 of a unit process for the production of general purpose polystyrene resin from ethylbenzene/styrene, and a system process (cradle-to-resin, or CTR, data set for production of general purpose polystyrene resin, to illustrate the differences in the two types of data sets. These examples were downloaded from the U.S. LCI Database. The unit process data set includes both flows directly to and from nature (e.g., emission releases) as well as

flows to or from other processes (e.g., electricity, processed natural gas combusted in boiler, transport by various modes, and ethylbenzene/styrene); meanwhile, the system process data set expresses all inputs and outputs needed to produce polystyrene resin in terms of flows to and from nature (e.g., resources, emissions to air, and emissions to water). For example, where the unit process data set lists inputs of electricity, the system process data set lists the types and amounts of resources used to generate the electricity for the resin production steps, as well as the emission releases associated with extraction and combustion of fuels to generate the electricity.

4.3 UNIT PROCESS OUTPUT BASIS

Unit process data sets must be expressed on a basis that is useful for linking to other unit process data sets. Depending on the process, appropriate bases could be a kilogram of

Name	Location	Category	SubCategory	Unit	Polystyrene, general purpose, at plant
Location					RNA
InfrastructureProcess					0
Unit					kg
Electricity, at grid, US, 2008	RNA			kWh	0.115079365
Natural gas, combusted in industrial boiler	RNA			m3	0.020039067
Transport, barge, diesel powered	RNA			tkm	0.242316516
Transport, barge, residual fuel oil powered	RNA			tkm	0.805702416
Transport, ocean freighter, diesel powered	RNA			tkm	0.0250753
Transport, ocean freighter, residual fuel oil powered	RNA			tkm	0.225677696
Transport, combination truck, diesel powered	RNA			tkm	0.244990363
Transport, train, diesel powered	RNA			tkm	0.331789477
Ethylbenzene styrene, at plant	RNA			kg	0.999
White mineral oil, at plant	RNA			kg	0.00257
Dummy_Disposal, solid waste, unspecified, to municipal incineration	RNA			kg	0.000017
Dummy_Disposal, solid waste, unspecified, to sanitary landfill	RNA			kg	0.00063
Dummy_Disposal, solid waste, unspecified, to waste-to-energy	RNA			kg	0.00154
Water, process, unspecified natural origin/m3		water	unspecified	m3	0.00017928
Carbon monoxide		air	unspecified	kg	0.000019
Methane, difluromonochloro-, HCFC-22		air	unspecified	kg	0.000001
Nitrogen oxides		air	unspecified	kg	0.000043
NMVOC, non-methane volatile organic compounds, unspecified		air	unspecified	kg	0.00012
Organic substances, unspecified		air	unspecified	kg	0.00001
Particulates, unspecified		air	unspecified	kg	0.000024
Sulfur oxides		air	unspecified	kg	0.00000033
Ammonia		water	unspecified	kg	0.0000005
BOD5, Biological Oxygen Demand		water	unspecified	kg	0.0000001
Chromium, unspecified		water	unspecified	kg	0.00000005
Cyanide		water	unspecified	kg	0.000000001
Dissolved solids		water	unspecified	kg	0.001
Hydrocarbons, unspecified		water	unspecified	kg	0.00000001
Iron		water	unspecified	kg	0.00000001
Lead		water	unspecified	kg	0.00000005
Nickel		water	unspecified	kg	0.00000005
Oil and grease, unspecified		water	unspecified	kg	0.0000001
Phenol compounds, unspecified		water	unspecified	kg	0.000000001
Phosphates		water	unspecified	kg	0.000001
Suspended solids, unspecified		water	unspecified	kg	0.0000005
Zinc		water	unspecified	kg	0.00000005
Polystyrene, general purpose, at plant	RNA			kg	1

Table 4.1. Unit Process Data Set for Production of General Polystyrene Resin for Ethylbenzene/Styrene

Name	Location	Category	SubCategory	Unit	Polystyrene, general purpose, at plant, CTR
Location					RNA
InfrastructureProcess					0
Unit					kg
Coal, lignite, in ground		resource	fossil-	kg	0.018432788
Coal, bituminous, in ground		resource	fossil-	kg	0.2049328
Energy, from biomass		resource	unspecified	MJ	0.034036583
Energy, from hydro power		resource	unspecified	MJ	0.17948824
Energy, geothermal		resource	unspecified	MJ	0.009470487
Energy, kinetic (in wind), converted		resource	unspecified	MJ	0.009379965
Energy, recovered		resource	unspecified	MJ	0.010008351
Energy, solar		resource	unspecified	MJ	4.13017E-06
Gas, natural, 49.8 MJ per kg, in ground (EMR)		resource	fossil-	kg	0.41211165
Gas, natural, 36.7 MJ per m3, in ground		resource	fossil-	m3	0.85488256
Oil, crude, 43.7 MJ per kg, in ground (EMR)		resource	fossil-	kg	0.72575106
Oil, crude, 43.7 MJ per kg, in ground		resource	fossil-	kg	0.1501221
Uranium oxide, 332 GJ per kg, in ore		resource	ground-	kg	4.69884E-06
Water, consumptive use, unspecified origin/m3		resource	unspecified	m3	0.011427306
2-Chloroacetophenone		air	unspecified	kg	3.07071E-13
5-methyl Chrysene		air	unspecified	kg	1.97772E-12
Acenaphthene		air	unspecified	kg	4.58461E-11
Acenaphthylene		air	unspecified	kg	2.24742E-11
Acetaldehyde		air	unspecified	kg	5.05729E-08
Acetophenone		air	unspecified	kg	6.58008E-13
Acrolein		air	unspecified	kg	3.21601E-08
Aldehydes, unspecified		air	unspecified	kg	8.62763E-08
Ammonia		air	unspecified	kg	1.75292E-05
Ammonium chloride		air	unspecified	kg	2.49397E-07
Anthracene		air	unspecified	kg	1.88784E-11
Antimony		air	unspecified	kg	3.31397E-09
Arsenic		air	unspecified	kg	4.86214E-08
Benzene		air	unspecified	kg	9.7022E-05
Benzene, chloro-		air	unspecified	kg	9.65079E-13
Benzene, ethyl-		air	unspecified	kg	1.19875E-05
(additional rows of emissions to air and water in full data set are omitted here to keep example table to a single page)					
Styrene		water	unspecified	kg	9.99555E-07
Sulfate		water	unspecified	kg	0.000127271
Sulfide		water	unspecified	kg	9.40162E-07
Sulfur		water	unspecified	kg	1.09623E-06
Surfactants		water	unspecified	kg	1.02397E-07
Surfactants, unspecified		water	unspecified	kg	1.24367E-07
Suspended solids, unspecified		water	unspecified	kg	0.004590822
Tetradecane		water	unspecified	kg	1.03483E-08
Thallium		water	unspecified	kg	5.41006E-09
Tin		water	unspecified	kg	1.01838E-07
Titanium, ion		water	unspecified	kg	3.94599E-07
TOC, Total Organic Carbon		water	unspecified	kg	5.62572E-07
Toluene		water	unspecified	kg	5.01043E-07
Vanadium		water	unspecified	kg	5.65562E-08
Xylene		water	unspecified	kg	2.1135E-07
Yttrium		water	unspecified	kg	2.97293E-09
Zinc		water	unspecified	kg	1.06655E-06
Polystyrene, general purpose, at plant, CTR	RNA			kg	1
Recovered energy	RNA			MJ	0.0097

Table 4.2. Cradle-to-Resin System Data Set for Production of General Purpose Polystyrene Resin (Condensed)

output, a liter of output, an energy unit of output (such as MJ), or a finished unit of product output. There may be more than one basis that can be used; for example, the outputs of a petroleum refinery can be expressed in terms of physical units (kilograms or liters) of the different output streams, or expressed on the basis of energy units, taking into account both the physical amounts and energy content of the different output streams.

Ideally, the goal is to construct unit processes that specify only input and output flows relating directly to the product output of interest. However, if multiple useful outputs result from a process or from a facility, as is often the case, the practitioner must determine the shares of the various process inputs and outputs (e.g., material inputs, energy, water, wastes, and emissions) that should be assigned to each of the output co-products. By examining the sequence of processes within a facility in more detail, it may be possible to track the inputs and outputs to specific product flows so that allocation can be avoided. In other cases, such as a chemical reaction that produces two or more useful co-products, some form of allo-

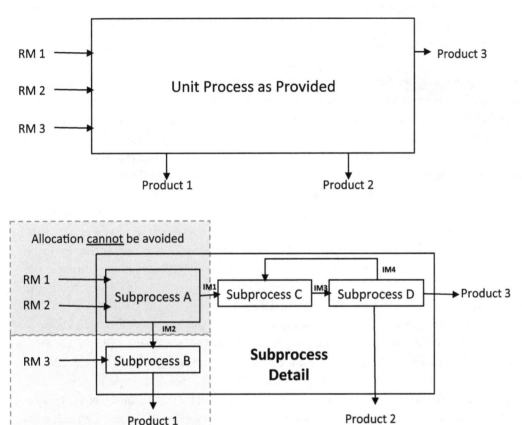

Figure 4.5. Using Subprocess Detail to Avoid Allocation where Possible. (RM = raw material, IM = intermediate material)

cation is required. Figure 4.5 illustrates sub-processes that are included within a unit process data set. The figure shows that allocation of Raw Material (RM) 3 can be avoided, because it is only used for Product 3, while allocation is required for Raw Materials 1 and 2 because they are both used to produce both Intermediate Materials (IM) 1 and 2.

When allocation cannot be avoided, careful consideration should be given to selecting the most appropriate method, as the choice of allocation method can significantly influence the results. For example, in the production of methylene diphenyl diisocyanate (MDI), hydrogen chloride is produced as a co-product. If mass allocation is used (i.e., process inputs and emissions allocated equally over the total mass of outputs produced), the greenhouse gas emissions per kg of MDI are 2.39 kg CO_2 eq. If a price-based allocation is used

(i.e., process inputs and emissions allocated over the total amounts of outputs based on their relative economic values), a larger share of the emissions is allocated to the more valuable MDI, and the CO2 eq. per kg of MDI are 3.69 kg (Polyurethane Eco-Profile, Plastics Europe 2012). Allocation issues and ISO preferred hierarchy for allocation are discussed in more detail in Chapter 8: Advanced System Modeling.

Once the output basis (e.g., mass, energy, and units of finished product) and the method for any necessary allocations is decided upon, the process data reported by the facility can then be allocated appropriately to the process outputs and expressed in terms of a useful reporting basis for linking to other unit processes that use the outputs. For example, a plant's annual output of a chemical is not a useful output basis for linking to a process that requires only a few kilograms of that chemical; therefore, unit process data sets are usually expressed on the basis of a single unit of output for convenience in linking to other processes.

4.4 MATHEMATICAL DESCRIPTION OF UNIT PROCESSES

FLOW TYPE, DESCRIPTION, QUANTITY, AND UNIT

Unit processes can be batch operations or flow operations that are continuous. They can run at a steady state, but often are dynamic and vary on different time scales. For example, agricultural processes are generally seasonal, while animal based systems can be extremely complex to characterize, as detailed milk production models must account for calf and cow population dynamics and related changes in feed rations over time.

Many chemical operations operate at a steady state with periodic shutdowns for maintenance, product changeovers, or when production quotas are met. Startup and shutdown unit operations can result in transient behavior and emissions that differ from steady state operation. Life cycle models generally do not characterize the system dynamics, but rather represent average conditions over a specified time period (e.g., annual basis).

Unit processes are most commonly provided in the form of a data set reporting types and quantities of input and output flows. However, unit processes can be represented mathematically in many forms, including chemical reaction equations, mass balance equations, and energy balance equations. These equations are required in life cycle modeling to bring primary and secondary data that has been collected in relation to the functional unit of analysis. In addition, mathematical models can be useful for identifying sources of error in data collection, can serve as a basis for allocation, and can be used for estimating input or output flows where data are not reported. Model verification and validation are also important for ensuring accuracy of model results. Verification procedures are used to determine whether the computational model is accurately constructed and correctly calculating. Validation determines how well the model represents real-world processes, which can be more difficult to establish.

A variety of qualitative and quantitative information and data are necessary to characterize unit processes, including the naming and description of processes, naming and description of material and energy input and output flows, and the specific data indicating the magnitude of the input and output flows, including the proper units, with respect to the reference flow for the system. Typical mass units are kilograms (pounds in the US), and energy units are expressed in megaJoules (British thermal units in the US).

MARGINAL VERSUS AVERAGE DATA

The concepts of average data (which are used for attributional modeling) and marginal data (which are required for consequential modeling) are often a source of confusion. The example of power plants and the electrical grid can be used to highlight some of the nuances in describing the unit operations for electricity generation. Generating plants are characterized as baseload, intermediate, or peaking, depending on their mode of operation. Baseload plants, such as coal and nuclear plants, typically generate electricity at 80- 90% of their nameplate capacity over the course of a year. Peaking plants are dispatched for short periods to meet fluctuations in

demand. Emission factors for the grid can be reported following many different protocols. Average emission factors can be calculated for all plants in a grid by dividing total emissions by total generation. These emissions would be based on the mix of fuels used by the continuously operating baseload plants in the grid, plus the intermediate and peaking plants. This is considered attributional LCA, as the total emissions from the generation mix are assigned a distribution to all the electricity produced. Marginal emission factors can be reported to account for emissions to meet marginal demand by the grid. For example, if one wanted to determine the emissions from additional electricity demand (e.g., the use of air conditioners on a hot summer day), she or he could estimate the emission factors for the peaking plants of a grid alone. This is considered consequential LCA, as it evaluates the consequences of a change from the baseline. If one were interested in studying longer term impacts from electricity demand over a ten year period, this model would need to account for changes to the grid fuel mix over time. Attributional and consequential modeling are discussed in more detail in Chapter 8: Advanced System Modeling.

Clearly, the ability to conduct consequential modeling is limited by the level of unit process detail available. For example, if the current average supply of a chemical is based on a mix of several technologies, then a consequential analysis of meeting increased demand for that chemical requires data on the individual technology (or mix of technologies) that would provide the marginal increase in production. Consequential analysis cannot be conducted if the only data set available is the compiled average of the current mix of technologies with no detail on the individual contributing technologies.

Reporting key statistics other than average values for characterizing unit processes is valuable, whenever possible. Key statistics include upper and lower bounds on flows, standard deviations, and probability distributions, if possible. These statistics can be useful in conducting sensitivity analyses and uncertainty analyses, including Monte Carlo simulations. More discussion on statistics in LCA can be found in Chapter 14: Statistics in LCA.

4.5 ELEMENTARY (ECOSPHERE) AND INTERMEDIATE (TECHNOSPHERE) FLOWS

Distinguishing the origin of process flows is useful for characterizing the impact on natural systems. Elementary flows are (1) material and energy resources that are extracted from the natural environment (ecosphere) before their use and transformation in the economy (technosphere) and (2) material and energy flows released back **to** the ecosphere without subsequent human intervention. Examples of elementary flows from the environment include ores extracted by mining processes, or crude oil or natural gas obtained by drilling. In the case of mineral ores, the metals that are refined from these ores are classified as virgin (primary) metals, in contrast with metals derived or recycled from secondary sources that already exist in the economy (e.g., secondary aluminum recovered from recycling of postconsumer beverage containers).

Examples of elementary flows **to** the environment include air pollutant emissions that are directly released from vehicle tailpipes or emissions from the stack of a power plant after air pollutant scrubbing processes. Intermediate flows represent material and energy flows between technosphere (technological) processes. Technosphere flows typically are accompanied by financial transactions. Examples of intermediate flows include the flow of electricity to and from the electric grid, the flow of ethylene from a hydrocracker to a polyethylene manufacturer, the flow of polyethylene from a resin manufacturer to a bottle molder, or the flow of particulate matter to a dust collector. Distinguishing between flows of product materials and process or ancillary materials is also useful. Figure 4.6 shows a simple example highlighting the distinction between elemental and intermediate flows for primary aluminum production.

NAMING CONVENTIONS AND ADDITIONAL DESCRIPTORS

In characterizing unit processes, standard naming conventions and additional descriptors are necessary to facilitate data exchange and accurate impact assessment modeling. The *ILCD Handbook: Nomenclature and other conventions* (European Commission 2010) provides useful guidance on defining naming schemes of flows and processes. Because several different names can often be used for the same chemical, Chemical Abstract Service (CAS) registry names should be used whenever possible for consistent identification of substances. For example, the substance with CAS number 71-55-6 is referred to by International Union of Pure and Applied Chemistry (IUPAC) as 1,1,1-trichloroethane, but the same substance can also be called methyl chloroform or chlorothene. If different names for the same substance are used in different unit process data sets, it can become quite difficult to ensure that the substance is properly captured in impact assessment methods.

For materials, specifying alloys, rather than only the names of the base metals, can be essential. Examples of specific descriptors for materials taken from the ILCD Handbook include "Stainless steel hot rolled coil; annealed and pickled, grade 304, austenitic, electric arc furnace route; production mix, at plant; 18% chromium, 10% nickel," "Corrugated board boxes; consumption mix; 16.6% primary fibre, 83.4% recycled fibre," and "Polyethylene terephthalate (PET) granulate; bottle grade; production mix, at plant."

Speciation is also important for more accurate impact assessment modeling. For example, impacts cannot be accurately assigned to emissions reported simply as "metal ions" or "non-methane VOCs." These broadly defined categories can include a range of different substances, each with different potential impacts. For example, iron and mercury are both metal ions, but a kilogram of mercury emissions has different impacts from a kilogram of iron emissions. Different forms of the same metal can also have different impacts, as in the case of trivalent and

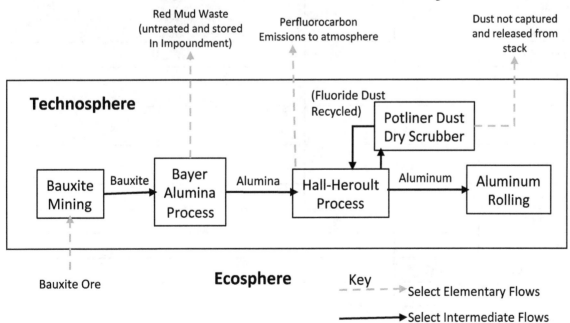

Figure 4.6. Select Elementary Flows from the Ecosphere to the Technosphere and Intermediate Flows between Key Aluminum Production Processes within the Technosphere.

hexavalent chromium. Similar issues apply to flows and emissions reported using general terms such as "particulates" or "sodium hydroxide solution." Information on the specific chemical composition of emissions, as well as the size of particulates and the concentration of solutions, are needed to properly classify and characterize potential impacts.

For emissions, the receiving environmental compartment should be indicated, if known. ISO 14044 names "emissions to air, water and soil" as top-level classification, but LCIA methods can require further specification of compartments in which distinctions can be made between releases to freshwater and releases to sea water. Similarly, atmospheric emissions can be further distinguished between releases to regions with high population density and releases to areas of low population density. Reporting by sub-compartments helps improve impact assessment modeling. For example, human exposures (and associated human health impacts) for a given amount of a specific

emission contributing to respiratory impacts would be higher for emissions released in areas with higher population density.

In summary, a detailed, thorough, and consistent approach to developing unit process data is essential to support transparent process-based LCAs and more accurate impact assessment modeling. The process flow diagram is important for communicating the scope and system boundaries for analysis, and provides a valuable blueprint for organizing and conducting the life cycle inventory analysis and impact assessment. Detailed descriptions and documentation of the unit process data sources are also critical for a high-quality study.

PROBLEM EXERCISES

1. Identify which processes are elementary flows (directly to/from nature) and which are intermediate flows (to/from another technosphere process). If classified as intermediate, briefly explain why.

	Elementary	Intermediate	If Intermediate, why?
Bauxite ore, in ground, at mine site			
Iron ore, at blast furnace			
Trees, standing, in forest			
Corn plant, in field			
Wood chips, at sawmill			
Corn grain, at wet mill			
Limestone, crushed, at mine site			
Coal, in ground			
Crushed coal, at utility plant			
Electricity, from coal, at generating facility			
Grid electricity, at manufacturing plant			
Tap water input to washing operation			
River water, at plant inlet			
Stack emissions to scrubber			
Vehicle tailpipe emissions			
Fugitive methane from landfill			
Methane emissions to flare			
CO2 from methane flare			
Spent plating solution to treatment plant			

2. Two confidential process data sets are shown in the figures below. Process A uses input of raw materials (RM) X and Y to produce Product A, while Process B uses inputs of RM X, RM Z, and Product A to produce Product B. Use the data sets for the two proprietary processes to prepare a combined data set for 100 kg output of Product B that protects the details of the individual processes.

3. Identify a relatively simple metal containing product system that consists of no more than a half dozen materials. Develop a process flow diagram for this product system tracing major life cycle processes that encompass material production, manufacturing, use and service, and end-of-life management stages of each material. Label the key unit operations (e.g., mining, pig iron production in blast furnace, steelmaking in basic oxygen furnace, and stamping) and material flows across the life cycle (iron ore, steel, stamped steel, and scrap steel).

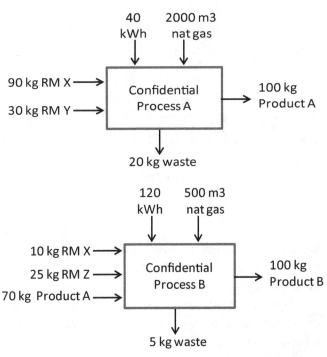

	per 100 kg Product A	per 100 kg Product B
Inputs (RM = raw material)		
kg RM X	90	10
kg RM Y	30	
kg RM Z		25
kg Product A		70
kWh	40	120
cu meters natural gas	2000	500
check total kg inputs	**120**	**105**
Outputs		
kg Product A	100	
kg Product B		100
kg solid waste to landfill	20	5
check total kg outputs	**120**	**105**

Life Cycle Inventory

Christoph Koffler, Roland Geyer, and Thorsten Volz

5.1 System Function, Functional Unit, and Reference Flows

One of the most essential steps during the goal and scope definition phase of a life cycle analysis (LCA) is the proper definition of the system function and the resulting functional unit. The functional unit is used to scale each product system under study so that it produces the product outputs necessary to satisfy the same requirements. This procedure is also known as the relative approach, which distinguishes LCA from other environmental management tools such as environmental performance evaluation, environmental impact assessment, and risk assessment, all of which are based on absolute rather than relative environmental loads (ISO 14044:2006a, Section 4.4.1).

The proper definition of the functional unit is a necessary prerequisite for a fair comparison between different product systems. The unit of comparison is defined in terms of the service provided to the user by the product in order to render meaningful and defensible results. Comparative LCAs communicated to the public are subject to scrutiny by various stakeholders; the credibility of the results, therefore, starts with the proper definition of the functional unit.

System function

Under ISO 14040 and ISO 14044 (ISO 2006a, 2006b), the system function describes the performance of a product system. This means that it is not the physical manifestation of the product itself that is captured here, but rather the service

provided by that product. Questions like "What does the product do for me?" or "What is it that one gets in return for the money spent on the product?" are helpful in terms of coming up with a list of possible system functions. These descriptions should be purely qualitative at first; quantification is the next step.

Here are some examples of rather simple and clear cut system functions:

- paperweight: the system function is 'hold loose sheets of paper in place'.
- pen: the system function is "apply ink to paper".
- beverage cans: the system function is "enclose a beverage'.

Many products provide multiple functions. For example, while one of beer's historical main functions certainly was to provide its consumers with safe and affordable nutrition and hydration, one may also argue that the alcohol content of today's beers, along with the temporary improvement of one's self-esteem, changed overall mood, and reduced perception of risk, is the actual system function of the product if consumed responsibly and in moderate amounts.

Another example of multi-functional products are automotive parts, which may have to satisfy a list of requirements, including mechanical properties, crash behavior, vibration characteristics, surface properties, and corrosion resistance. Information and communication technology (ICT) is another good example for multi-functional product systems (Hischier and Reichart 2003). A smart phone can provide virtually thousands of functions when the appropriate software is installed; as a result, the ability to make phone calls became one more function among others.

Whenever multiple system functions are possible, one should aim to identify the main product function (i.e., the product function that is of the highest relevance with regard to the goal of the LCA at hand). If multiple functions appear reasonable, it may be necessary to conduct a scenario analysis to determine the effect of different functional units for the outcome of the study (Mathey et al. 2007).

FUNCTIONAL UNIT

The functional unit (FU) of a product system is defined as the "quantified performance of a product system for use as a reference unit" (ISO 2006a, Section 3.20). Once the proper system function has been identified and described, it needs to be specified with regard to the following aspects according to (Smith Cooper 2003):

- its unit
- its magnitude
- its duration
- its level of quality

Going back to the beer example, if moving people into the desired state of mild euphoria is accepted as the core system function of the product, then its specification would mean answering the following questions:

- How many people are supposed to get mildly euphoric?
- What does "mildly euphoric" mean exactly? How do I measure it?
- Over what timeframe should this condition be maintained?
- What other requirements need to be fulfilled?

Clearly, no precise answers based purely on natural science exist to the above questions. The quantification is therefore mostly a convention. The only requirement is that it stays within the boundaries of reason so that it can be easily communicated to the intended audience of the study.

For example, a given study may require that one hundred people consume beer until they reach a blood alcohol content (BAC) of 0.05 and that they should maintain this level of mild intoxication over a duration of four hours. The level of quality could then be related to the gender distribution within the group of consumers, as well as to requirements about flavor, serving size, or serving format of the beer (draft, keg, bottles, or cans).

The functional unit for this example may therefore be defined as follows:

Establishing and maintaining a BAC of 0.05 in fifty women and fifty men over a duration of four hours using 12 fl. oz. servings of beer.

This functional unit deliberately excludes the alcohol content of the beer and its drinking temperature in order to facilitate comparisons between different styles of beer that may have different alcohol contents or different recommended drinking temperature (which in turn would lead to a different energy demand for cooling). These differences will subsequently be accounted for when scaling each involved product system to provide the above functional unit, a process which is called the calculation of the reference flows (see below).

Also, the desired duration does not have to coincide with the expected lifetime of any of the products under study. While the concept of "lifetime," in terms of durability, does not apply to the above example of a food and beverage product beyond its shelf life, for technical product systems the ratio of desired duration and expected lifetime in years (or hours) of operation needs to be used to establish how many products are necessary to fulfill the functional unit for each product system. That way, differences in durability between products can be properly accounted for. These ratios do not have to be integers; quantities that fall between whole units of product, such as 0.7 or 15.2 pieces of product, are required to ensure an apples-to-apples comparison of products.

REFERENCE FLOWS

ISO 14044:2006, Section 3.29, defines the reference flows (RFs) as "measure of the outputs from processes in a given product system required to fulfill the function expressed by the functional unit" (ISO 2006a). They are therefore the physical manifestations of the respective product quantities and other relevant flows needed to provide the functional unit. Proceeding with the beer example, additional information is needed in order to properly calculate the RFs for different beers. First, you need to know the average body weights of the individuals involved. Using average US values from the 2008 National Health Statistics Report (McDowell et al. 2008), the statistical body weight is 88.3 kg (194.7 lb.) for men and 74.7 kg (164.7 lb.) for women over twenty years of age, which conveniently coincides with the general US legal drinking age.

Using an online blood alcohol content calculator (www.globalrph.com/bac.cgi), a man of above-average weight would need roughly two and a half 12 oz. beers with an alcohol content of 4.17 % (which equals a Standard Drink Equivalent, or SDE, containing 0.5 fluid oz. of pure alcohol each) to reach a BAC of 0.05, while a woman would only need about 1.8 standard drinks. Table 5.1 shows how many liters of beer it would therefore take to establish the desired BAC of 0.05 within the group for different types of beers. Naturally, the higher the alcohol content of the beer, the lower the necessary volume.

At this point, one still needs to account for the required duration of four hours. Assuming that the body metabolizes 0.01 BAC per forty minutes, and disregarding the time it takes to initially establish the target BAC, 0.06 additional BAC would have to be provided to make up for the losses occurring over a four-hour time period. This results in a factor 2.2 with regard to the values displayed in Table 5.1. Further assuming a beer density of about 1.05 kg/liter, the resulting beer reference flows would therefore result in the following calculations:

* Amstel light™ 92 l * 2.2 * 1.05 kg/l = ~213 kg
* Samuel Adams™ Boston Lager 66 l * 2.2 * 1.05 kg/l = ~152 kg
* Bud Light™ Platinum 54 l * 2.2 * 1.05 kg/l = ~125 kg

If the chosen BAC or the size of the group would have been twice as high, these reference flows would double as well. This linear scalability of product systems is an essential characteristic of LCA models (see Section 6.3Chapter 6: Data Quality in LCA).

For the beer product system, one would also have to account for primary packaging (beer bottles). Due to the lack of exact values, it should therefore be assumed that an average 12 oz. glass beer bottle weighs about 200 grams and an average tinplate bottle cap about 2 grams. Based on the values displayed above, one could therefore calculate the RFs of bottles and caps needed in each product system to provide the functional unit. One would further need to consider secondary and tertiary packaging for shipping and retail (e.g., cardboard, pallets, and shrink wrap), as well as for the electricity needed to cool the beer and maintain an appropriate drinking temperature.

The above calculations reveal some important aspects of the FU definition and RF calculation:

* While there is only one functional unit in comparative LCA studies, RFs can (and in most cases will) be different for each product system under study.
* LCA is essentially based on mass and energy flows; therefore, the RFs should likewise be expressed on a mass or energy basis.
* Product systems in LCA are linearly scalable. The RF to satisfy a certain functional unit is exactly half of the RF that it would take to satisfy double that FU.
* User behavior is difficult to capture in the functional unit. For example, if everybody would drink their beers faster than their metabolization rates require (which is probably not uncommon), the blood alcohol would significantly exceed the functional unit of 0.05 BAC and the group would run out of beer well before four hours are up. Likewise, a car's fuel efficiency will heavily depend on the driver's individual driving style, and a cell phone's battery life will depend just as heavily on individual usage patterns.
* The assumptions made about the metabolization rate, the beer density, and the bottle and cap weights are uncertain. The effect of uncertain assumptions on the overall results should therefore be tested in sensitivity and uncertainty analyses in the interpretation phase of the study (see Chapter 13).

	Alcohol content (vol.%)	Standard-Drink-Equivalents (SDE) per 12 fluid ounces	Volume required for group to reach a 0.05 BAC*
Amstel light™	3.5%	0.84	~92 liters
Samuel Adams™ Boston Lager	4.9%	1.18	~66 liters
Bud Light™ Platinum	6.0%	1.44	~54 liters

Table 5.1. Alcohol Contents of Different Beers, Standard-Drink Equivalents, and Required Volumes
* Calculated as (1.8*50+2.5*50)/SDE * 12 fl. oz. * 0.03 l/fl. oz.

THE CASE OF ENVIRONMENTAL PRODUCT DECLARATIONS

In non-comparative LCA studies, such as those conducted to create Environmental Product Declarations (EPD) according to ISO 14025:2006c, RF and FU are frequently synonymous, with a certain quantity of product output representing both. This is sometimes referred to as the *declared unit* to make clear that it is not based on product function (CEN 2012). For example, communicating the environmental profile of 12 oz. of beer is legitimate without further quantification of the product functions. The above example shows that the same volume of different types of beer can have quite different functions; this is true in terms of alcohol content as well as nutritional values. A practitioner should at least provide his or her audience with enough information about the product so that the reader can scale the results of LCA studies in order to allow a fair comparison between different products.

In summary, FU and RFs can be defined following the below work flow:

- Start by qualitatively describing different possible system functions. Use active verbs to capture the service (or services) provided by the product system(s) in question.
- Choose the most appropriate system function for the goal of the study. If more than one system function appears reasonable, include additional functions in a scenario analysis.
- Specify the system function in terms of unit, magnitude, duration, and level of quality.
- Scale each product system under question by calculating the magnitude of RFs necessary to fulfill the functional unit, taking into account each product's expected lifetime.
- Identify and document uncertain assumptions that need to be analyzed in a subsequent sensitivity and/or uncertainty analysis.

5.2 LIFE CYCLE INVENTORY MODELING

The aim of a life cycle inventory (LCI) is to account for and quantify all relevant interactions between a product system and the natural environment. These interactions are quantified as elementary flows, which are defined in Section 5.5 and ISO 14044:2006a as material and energy flows from the environment into the product system without prior human transformation, or from the product system to the environment without any subsequent human transformation. Typical elementary inputs are fossil fuels and metal ores. Typical elementary outputs are air emissions, treated wastewater, and solid wastes. Elementary flows do not have to be material or energy flows. Examples of such flows are noise, land occupation, and land transformation. The process of developing an LCI from a functional unit is called inventory modeling. Inventory modeling consists of the following steps:

1. Define the FU.
2. Define the RFs of all alternative product systems to be studied.
3. For each product system, identify all unit processes that are to be included (i.e., that are within the system boundary of the product system).
4. Collect suitable process inventories PI_i for all identified unit processes $i \in \{1,2,3,lln\}$.
5. For each unit process i, determine the activity level λ_i that is required so that, collectively, the unit processes of a product system generate the RF and thus provide the FU. Scale each unit process with its activity level, $\overline{PI_i} = \lambda_i \cdot PI_i$.
6. Eliminate any intermediate flows that cross the system boundary of the inventory model and are not part of the RF through process subdivision, system expansion, or allocation.
7. Aggregate the inputs and outputs of all scaled unit processes of a product system to one vector of inputs and outputs $LCI = \sum_{i=1}^{n} \overline{PI_i}$. This vector is called the LCI of the product system.

Section 5.1 covered the definition of the FU and the resulting RFs, the first two steps of LCI modeling. The next step is to delineate the product system(s) that provide the RFs. A product system is defined in ISO 14044:2006 as a collection of unit processes that models the life cycle of a product. Depending on the definition of the FU and the scope of the LCA, a product system may cover the life cycle of several products rather than just one, or only parts or fractions of a product. In some cases, a product system may not cover the entire life cycle of a product (e.g., in cradle-to-gate studies that exclude use and end-of-life management of the product).

Regardless of the FU and RFs, in practical terms a product system is always modeled as a set of unit processes. Unit processes are defined in Section 5.1 of ISO 14044:2006 as the smallest element in LCI analysis for which input and output data are quantified. The scope of a unit process can vary widely. It can be as small as modeling the operation of a plastic injection molding machine or as large as describing all processes involved in cradle-to-gate aluminum sheet production. The latter includes many processes, including bauxite mining, alumina production through the Bayer process, aluminum production through the Hall-Heroult process, and aluminum casting and rolling, as well as production and use of all energy and additional material inputs, such as electricity, cryolite and carbon anodes. These aggregated unit processes are sometimes called system processes, to distinguish them from single processes.

The scope and level of detail that are required for each unit process depend on the goal and scope of the LCA. Practitioners commonly divide the product system into a foreground system, which is typically modeled in greater detail, and a background system, which is frequently modeled with less detail, (e.g., using aggregated cradle-to-gate processes). At the same time, the possible level of detail may be significantly limited by the availability of necessary data. Lack of data availability may therefore make it necessary to modify the goal and scope of the LCA.

PROCESS FLOW DIAGRAMS

A process flow diagram visualizes the processes that make up a product system. It visually subdivides the product system into a set of mutually exclusive, and collectively exhaustive, processes and shows their linkages. Figure 5.1 shows an example of a process flow diagram for a polyethylene teraphthalate (PET) bottle with a paper label and a polypropylene (PP) bottle cap. It shows core processes in the life cycle of a PET bottle and how they are connected through intermediate flows.

Intermediate flows are defined in Section 5.5 of ISO 14044:2006 as occurring between the processes of a product system. Intermediate flows are also called technosphere flows. They serve a different role than elementary flows. Elementary flows (also called ecosphere flows) are the flows that are responsible for the environmental impacts caused by the product system. They are not shown on a process flow diagram. Examples of elementary flows are CO_2 emissions causing climate change impacts, NO_2 emissions causing acidification impacts, O_2 flows causing reduced oxygen tension, and lead emissions causing toxicity impacts. Intermediate flows do not cause any environmental impacts. The electricity delivered to various production processes and the fuel delivered to transportation processes in Figure 5.1 do not cause any direct environmental impacts. The CO_2 and NO_2 emissions from fuel production and combustion, and electricity production, cause climate change and acidification impacts. A gasoline spill that leads to soil contamination is not an intermediate but an elementary flow.

Intermediate flows have a different role in LCA. They help us identify all processes that are required in order for the product system to deliver the RF and thus provide the FU. They also enable us to quantify how much of each process output is required. If the FU of the product system in Figure 5.1 is the packaging of 1 liter of beverage, the RF is the amount of PET bottle required to contain 1 liter of beverage

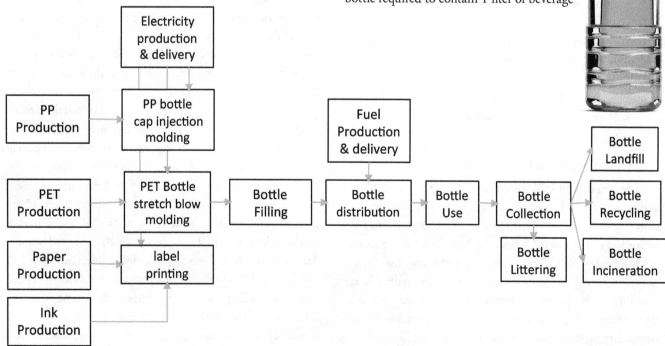

Figure 5.1.A. High-level process flow diagram of a polyethylene teraphthalate (PET) bottle with a paper label and a polypropylene (PP) bottle cap. Only intermediate flows are shown.

(e.g., around 1.7 20 oz. bottles). The intermediate inputs and outputs of the PET stretch blow molding process would then inform us that 36 grams of PET and 1.8 kWh of electricity are required to produce enough PET bottles to contain 1 liter of beverage (Kuczenski and Geyer 2013).

these machines is probably insignificant compared to the life cycle impacts of the product system shown in Figure 5.1. The task of LCI modeling, therefore, is not to construct the most comprehensive model of a product system, but rather to identify those processes and flows that significantly contribute

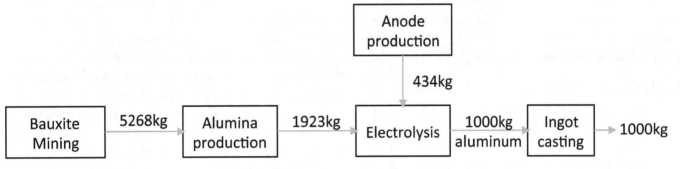

Figure 5.2. High-level process flow diagram of global primary aluminum production (IAI 2007)

Figure 5.2 provides another example. The process flow diagram tells us that for 1,000kg of cast aluminum, 1,000kg of liquid aluminum are required. Also, 1,000kg of liquid primary aluminum require (on global average) 1,923kg of alumina and 435kg of carbon anode. On global average, alumina production (i.e., "the Bayer process") requires 5,268kg of bauxite to produce 1,923kg of alumina. Sometimes Sankey diagrams are used to depict the product system, where the width of the arrows is shown proportionally to the intermediate flow quantities.

SYSTEM BOUNDARY SELECTION AND CUT-OFF CRITERIA

One major issue in LCI modeling is determining which processes should be included and which should be excluded (Step 3 of the LCI modeling procedure). While LCI modeling aims to construct comprehensive models of product systems, including every process that is involved in generating an RF is not possible. Consider the following dilemma: The product system in Figure 5.1 involves stretch blow molding of a PET bottle.

The unit process of stretch blow molding requires PET pellets and electricity as inputs, but it also requires the use of the stretch blow-molding machine. If the total output of such a machine over its lifetime is one million PET bottles, one PET bottle requires one millionth of a stretch blow-molding machine. So strictly speaking, the product system in Figure 5.1 should include the life cycle of the stretch blow-molding machine. But what about the machines required to manufacture the stretch blow-molding machines? If those machines make one million stretch blow-molding machines over their lifetimes, then 1 PET bottle requires one trillionth of those machines.

The environmental impacts of making one trillionth of

to the total impact of providing the FU and excluding all others. This task is called system boundary selection. The iterative nature of LCA is particularly evident in the process of system boundary selection. Initially included processes and flows may be omitted once their insignificance becomes evident.

Conversely, initially omitted processes and flows may have to be included once their significance becomes evident. System boundary selection is typically supported by so-called cut-off rules or criteria. According to paragraph 4.2.3.3.3 in ISO 14044:2006, cut-off criteria can be based on how processes or flows contribute to total mass, energy, or environmental impact of the product system (ISO 2006a). Section 5.6 in ISO/TR 14049 (2000) gives three examples of such cut-off criteria:

1. The sum of all materials included shall be more than 99% of the total mass inflow to the system.
2. The sum of all processes included shall be more than 99% of the total energy requirement of the system.
3. The cumulative contribution of the included processes to an environmental impact category is 90% or more of the initially calculated total impact in that category.

The three examples above reveal a basic dilemma of system boundary selection and cut-off criteria: How can we determine that certain processes and flows can be excluded without already having established their contribution to the total mass inflow, energy input, or environmental impact of the product system? A resolution to this paradox is to create initial estimates based on an inclusive system boundary, make exclusions based on those initial estimates and well-defined cut-off criteria, and then continue with detailed inventory modeling using the narrowed system boundary.

Findings in literature can also be helpful for making boundary choices. For instance, a considerable body of literature analyzes the contribution of capital equipment, such as the blow molding machine in the PET bottle example, to total environmental impact. It turns out that capital equipment can be excluded in many product systems, but contributes significantly, and thus needs to be included, in some systems (e.g., agricultural machinery in agricultural product systems or infrastructure in transportation systems).

5.3 Life Cycle Inventory Calculations

The main purpose of intermediate flows, which is scaling every unit process to the RF, will be demonstrated in more detail shortly. First, we look at how the mass balance principle can be applied to input and output flows for validation and plausibility checks.

Using mass balancing for validation and plausibility checks

The mass balance principle potentially allows us to validate process inventories or to infer missing input and output flows. Applied to inventory modeling, it simply states that everything that enters a process must also leave the process. Unless nuclear reactions are involved, conservation of mass holds at the elemental level (for every chemical element, the amount of input and output must be the same). We use the example from Figure 5.2 to demonstrate this. The basic reaction equation of reducing alumina into aluminum through electrolysis is as follows:

$$2Al_2O_3 + 3C \rightarrow 4Al + 3CO_2 \quad (1)$$

We can use this equation to derive a mass balance in atomic mass units u:

$$2 \cdot (2 \cdot 27 + 3 \cdot 16)u + 3 \cdot 12u \rightarrow 4 \cdot 27u + 3 \cdot (12 + 2 \cdot 16)u$$
$$(2)$$

$$204u + 36u \rightarrow 108u + 132u \quad (3)$$

This means that, stoichiometrically, 1,000 kg of pure aluminum requires of $\frac{204}{108}1,000kg = 1,889kg$ alumina, which is close to the value reported by IAI (2007). However, stoichiometrically, electrolysis of of aluminum requires only $\frac{36}{108}1,000kg = 333kg$ of carbon. This is much less than the 435kg of carbon anode consumption reported by IAI (2007), all of which is petrol coke or pitch (i.e., mostly carbon). This seeming discrepancy can be explained with the carbon from the anode reacting with oxygen in the ambient air (Bayliss 2013). IAI reports 1,557 kg of process CO2 emissions from

electrolysis of 1,000 kg of aluminum (not shown in Figure 5.3), which is stoichiometrically consistent with a carbon consumption of around 435kg because $\frac{44}{12} \cdot 435kg = 1,595kg$. The average chemical composition of bauxite is roughly 38% Al_2O_3, 20% H_2O, and 42% oxides of Fe, Si, and Ti (IAI 2007, USGS 1986). On average, 5,268 of bauxite therefore contain around 0.38 kg bauxite = 2,002 kg of alumina, which is again consistent with the intermediate flows shown in Figure 5.3.

Applying the mass balance principle to the intermediate flow exchanges between unit processes is a powerful way to assess the validity, or at least the plausibility, of individual process inventories and of the way they are connected within the whole product system. However, it should not be expected that every material or elemental balance can be closed using mass balancing, especially in the case of complex unit processes containing large numbers of physical and chemical transformations.

Determining the activity levels of all unit processes

The main purpose of intermediate flows is the scaling of every unit process to the RF. Every unit process can be thought of as a long list of well-defined inputs and outputs, all scaled to a certain quantity of main process function, typically a unit amount of its main intermediate output. Each input and output is characterized by a name, a unit, and a quantity. An example would be 5,268kg of bauxite. By convention, input quantities are negative and output quantities are positive. In math, a list of numbers is also called a vector. A simple example of a unit process is given in Figure 5.3.

$$PI_{al\_el} = \begin{pmatrix} -1923kg \ alumina \\ -435kg \ anode \\ -15,289kWh \ electricity \\ 1,000kg \ aluminum \\ 1,557kg \ CO_2 \\ 14.9kg \ NO_X \\ 0.13kg \ CF_4 \\ 0.013kg \ C_2F_6 \end{pmatrix}$$

Figure 5.3. Select process inventory data for aluminum electrolysis (IAI 2007)

Figure 5.3 shows select global average inventory data of electrolysis of alumina into liquid aluminum as reported by IAI (2007). That ability to scale process inventory linearly is an important standard assumption in LCA. For example, if the FU of packaging 1,000 liters of beverage was met with aluminum cans, the RF would be around 40.5 kg of 12 oz.

aluminum cans (Kuczenski and Geyer 2013). If the yield of converting aluminum ingots into cans was 75%, the RF would therefore require $\frac{40.5 kg}{0.75} = 54 kg$ of liquid aluminum.

$$\overline{PI}_{al\_el} = 0.054 \cdot \begin{pmatrix} -1923kg\ alumina \\ -435kg\ anode \\ -15,289kWh\ electricity \\ 1,000kg\ aluminum \\ 1,557kg\ CO_2 \\ 14.9kg\ NO_X \\ 0.13kg\ CF_4 \\ 0.013kg\ C_2F_6 \end{pmatrix} = \begin{pmatrix} -104kg\ alumina \\ -23kg\ anode \\ -826kWh\ electricity \\ 54kg\ aluminum \\ 84kg\ CO_2 \\ 0.8kg\ NO_X \\ 7g\ CF_4 \\ 0.7g\ C_2F_6 \end{pmatrix}$$

Figure 5.4. Process inventory from Figure 6.4 scaled to activity level 0.054

The linearity assumption means that the resulting process inventory for electrolysis can be derived by multiplying every flow in Figure 5.3 by 0.054. The scaled process inventory is shown in Figure 5.4. If the FU would be packaging 10,000 liters of beverage instead, the RF would be 405 kg of 12 oz. aluminum cans and the associated activity level for electrolysis would be 0.54. The figures 0.53 and 0.54 are called activity levels, for which we will use the symbol λ. The scaled process inventory $\overline{PI}_{al\_el} = \lambda \cdot PI_{al\_el}$ tells us how much air emissions are generated by the amount of aluminum electrolysis required to generate the reference flow. It also tells us how much alumina, anode, and electricity are needed.

Using the linearity assumption, we could thus derive activity levels for the process inventories of alumina, anode, and electricity production, and scale each of the production processes to the right amount of output. If the process inventories of alumina, anode, and electricity production have intermediate inputs, the scaling procedure would be repeated for the process inventories of their production processes. Scaling is applied to all unit processes upstream of the reference flow (e.g., 40.5 kg of 12 oz. aluminum cans) until we arrive at unit processes that have only elementary input flows, which is when the beginning of the product life cycle has been reached (see left side of Figure 5.1). The same is done going downstream until either the system boundary has been reached (e.g., in a cradle-to-gate LCA), or results in processes that have only elementary outputs, when the end of the

product life cycle has been reached (see right side of Figure 5.1). Once the activity levels λ_i of all unit processes PI_i have been determined that way, and all unit processes have been scaled to the RF according to $\overline{PI}_i = \lambda_i \cdot PI_i$, step 5 of the LCI modeling procedure is complete.

5.4 COMPUTATIONAL STRUCTURE OF LIFE CYCLE INVENTORY CALCULATIONS

Simple product systems only consist of chain-like or tree-like intermediate flow structures from cradle-to-grave (or left to right in Figure 5.1). The procedure outlined in Section 5.3 is sufficient to scale the entire product system. However, when an intermediate output of a process is used upstream of the process, a loop is created in the product system and scaling all unit processes is no longer as simple as described above. This section explains the generic way to scale all unit processes of a product system with an instructional example of a product system with four unit processes, four intermediate flow types, two elementary flow types, and the RF. The approach applies equally to all product systems with any number of unit processes and flows.

$$PI_1 = \begin{pmatrix} a_{11} \\ a_{21} \\ a_{31} \\ a_{41} \\ b_{11} \\ b_{21} \end{pmatrix}, PI_2 = \begin{pmatrix} a_{12} \\ a_{22} \\ a_{32} \\ a_{42} \\ b_{12} \\ b_{22} \end{pmatrix}, PI_3 = \begin{pmatrix} a_{13} \\ a_{23} \\ a_{33} \\ a_{43} \\ b_{13} \\ b_{23} \end{pmatrix}, PI_4 = \begin{pmatrix} a_{14} \\ a_{24} \\ a_{34} \\ a_{44} \\ b_{14} \\ b_{24} \end{pmatrix}, RF = \begin{pmatrix} q_1 \\ q_2 \\ q_3 \\ q_4 \end{pmatrix}$$

Figure 5.5. Formalism for four unit processes PI_i with four intermediate flow types a_j, two elementary flow types b_k, and RF

LCI CALCULATIONS AS SOLVING A LINEAR EQUATION SYSTEM

Figure 5.5 shows the generic inventory data of the four unit processes. The product system is balanced by scaling each process PI_i with an unknown activity level λ_i. This results in the following linear equation system between the activity levels λ_i and the intermediate flows a_{ji}:

$$\lambda_1 \cdot a_{11} + \lambda_2 \cdot a_{12} + \lambda_3 \cdot a_{13} + \lambda_4 \cdot a_{14} = q_1$$
$$\lambda_1 \cdot a_{21} + \lambda_2 \cdot a_{22} + \lambda_3 \cdot a_{23} + \lambda_4 \cdot a_{24} = q_2 \quad (4)$$
$$\lambda_1 \cdot a_{31} + \lambda_2 \cdot a_{32} + \lambda_3 \cdot a_{33} + \lambda_4 \cdot a_{34} = q_3$$
$$\lambda_1 \cdot a_{41} + \lambda_2 \cdot a_{42} + \lambda_3 \cdot a_{43} + \lambda_4 \cdot a_{44} = q_4$$

In actual inventory models, many of the intermediate flows a_{ji} are zero because not every process exchanges flows with every other process. Figure 5.6 shows a hypothetical example of the generic inventory model in Figure 5.5. Remember that the first four numbers of each process inventory PI_i are intermediate flows and the remaining two are elementary flows. Only the intermediate flows are used to calculate the unknown activity levels λ_i.

$$PI_1 = \begin{pmatrix} 1 \\ -0.05 \\ 0 \\ 0 \\ 1 \\ 0.2 \end{pmatrix}, PI_2 = \begin{pmatrix} -7 \\ 1 \\ 0 \\ 0 \\ 0.8 \\ 0.1 \end{pmatrix}, PI_3 = \begin{pmatrix} -4 \\ -1 \\ 1 \\ 0 \\ 0.5 \\ 0.1 \end{pmatrix}, PI_4 = \begin{pmatrix} -2 \\ 0 \\ -1 \\ 10 \\ 0.2 \\ 0.05 \end{pmatrix}, RF = \begin{pmatrix} 0 \\ 0 \\ 0 \\ 10 \end{pmatrix}$$

Figure 5.6. Hypothetical example of the generic formalism from Figure 5.5

Equation (5) shows the linear equation system (4) for the example in Figure 5.6:

$$
\begin{aligned}
1 \cdot \lambda_1 - 7 \cdot \lambda_2 - 4 \cdot \lambda_3 - 2 \cdot \lambda_4 &= 0 \\
-0.05 \cdot \lambda_1 + 1 \cdot \lambda_2 - 1 \cdot \lambda_3 &= 0 \\
1 \cdot \lambda_3 - 1 \cdot \lambda_4 &= 0 \\
10 \cdot \lambda_4 &= 10
\end{aligned}
\tag{5}
$$

Solving (5) is straightforward and yields the activity levels $\lambda_1 = 20, \lambda_2 = 2, \lambda_3 = 1, \lambda_4 = 1$. Equation (6) shows the balanced inventory model of the example in Figure 5.6.

$$20 \cdot \begin{pmatrix} 1 \\ -0.05 \\ 0 \\ 0 \\ 1 \\ 0.2 \end{pmatrix} + 2 \cdot \begin{pmatrix} -7 \\ 1 \\ 0 \\ 0 \\ 0.8 \\ 0.1 \end{pmatrix} + 1 \cdot \begin{pmatrix} -4 \\ -1 \\ 1 \\ 0 \\ 0.5 \\ 0.1 \end{pmatrix} + 1 \cdot \begin{pmatrix} -2 \\ 0 \\ -1 \\ 10 \\ 0.2 \\ 0.05 \end{pmatrix} = \begin{pmatrix} 0 \\ 0 \\ 0 \\ 10 \\ 22.3 \\ 4.35 \end{pmatrix} \tag{6}$$

The right-hand side of equation (6) is the life cycle inventory (LCI) of the example in Figure 5.6. The balanced inventory model also allowed us to calculate the total amount of elementary flows, $b_1 = 22.3$ and $b_2 = 4.35$ and $b_2 = 4.35$. These elementary flow results will be used to conduct impact assessment, which is explained in Chapters 10 through 12.

SOLVABILITY OF LINEAR EQUATION SYSTEMS

The previous sections described how calculating the activity levels of all unit processes involves solving a system of linear equations. Linear equation systems in general can have no solution, exactly one solution, or an infinite number of solutions. An LCI model, on the other hand, is expected to have exactly one solution. To have exactly one solution a linear equation system needs to have as many equations as it has unknown variables. In LCI modeling, the unknown variables are the activity levels λ_i. There are as many activity levels as there are unit processes PI_i. The number of equations is equal to the number of different intermediate flows sa_j.

A system is called underdetermined if the system is solvable but has fewer equations than unknown variables. An underdetermined system will have an infinite number of solutions.

An example of an LCI model with fewer equations (i.e., intermediate flows with unknown variables or unit processes) would be a product system with a refinery that receives crude oil from many different drilling operations (see Figure 5.7). Under-determination is solved by simply adding relevant information. In Figure 5.7, this could be the fraction of crude oil input coming from drilling operation A.

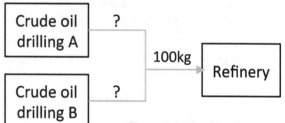

Figure 5.7. Two Unit Processes Producing the Same Type of Intermediate Flow

If more independent equations exist than unknown variables, the system is called overdetermined. An overdetermined system has no solution unless equations are eliminated. A classic example of a product system with more intermediate output types than unit processes is a crude oil refinery (see Figure 5.8). In LCA, such a situation is called co-production or multi-functionality and leads to a so-called allocation issue. Allocation is covered in Chapter 8: Advanced System Modeling. So mathematically speaking, solving an allocation issue means eliminating equations from an overdetermined linear equation system.

LCI CALCULATIONS USING MATRIX ALGEBRA

Inventory models of complex product systems can contain dozens, or even hundreds, of unit processes. While this does not change the computational structure of the inventory calculations, it requires a suitable mathematical formalism such

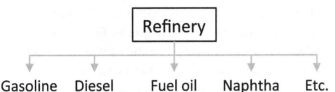

Figure 5.8. A Unit Processes Producing Different Types of Intermediate Flows (co-production)

as matrix algebra. Matrix notation is uniquely suited to handle and solve large linear equation systems like the ones generated by complex product systems. Many introductory textbooks on matrix algebra exist, so this section is limited to outlining its application to LCI modeling.

The first step is to repeat the LCI example from Figure 5.6 in matrix notation and then introduce the generic matrix notation of LCI commonly used in the literature. Equation (7) shows the linear equation system from equation (5) in matrix notation. Equation (8) shows the matrix notation for equation (4), while equation (9) shows the matrix notation of a generic life cycle inventory model. In matrix A, the number of columns equals the number of unit processes, and the number of rows equals the number of different intermediate flows. In matrix notation, solving the linear equation system (5) means finding the vector of activity levels $\vec{\lambda}$ that multiplied with the matrix A results in the vector $\overrightarrow{RF}$.

$$\begin{pmatrix} 1 & -7 & -4 & -2 \\ -0.05 & 1 & -1 & 0 \\ 0 & 0 & 1 & -1 \\ 0 & 0 & 0 & 10 \end{pmatrix} \begin{pmatrix} \lambda_1 \\ \lambda_2 \\ \lambda_3 \\ \lambda_4 \end{pmatrix} = \begin{pmatrix} 0 \\ 0 \\ 0 \\ 10 \end{pmatrix} \quad (7)$$

$$\begin{pmatrix} a_{11} & a_{12} & a_{13} & a_{14} \\ a_{21} & a_{22} & a_{23} & a_{24} \\ a_{31} & a_{32} & a_{33} & a_{34} \\ a_{41} & a_{42} & a_{43} & a_{44} \end{pmatrix} \begin{pmatrix} \lambda_1 \\ \lambda_2 \\ \lambda_3 \\ \lambda_4 \end{pmatrix} = \begin{pmatrix} q_1 \\ q_2 \\ q_3 \\ q_4 \end{pmatrix} \quad (8)$$

$$A \cdot \vec{\lambda} = \overrightarrow{RF} \quad (9)$$

Large linear equation systems can be solved with many methods. LCA software typically uses Gaussian elimination, which converts A into a triangular form or matrix inversion, and then calculates the inverse matrix A^{-1}, with $A^{-1} \cdot \overrightarrow{RF} = \vec{\lambda}$. Typically, matrix A is sparse (i.e., has many zero elements). Special solution methods exist for sparse matrices.

Once the vector of activity levels $\vec{\lambda}$ has been calculated, the LCI of elementary flows can be calculated by multiplying the product system's matrix of elementary flows, B, with $\vec{\lambda}$. Equa-

tion (10) shows this for the example from Figure 5.6, equation (11) for the example from Figure 5.6, and Equation (12) shows the matrix notation of a generic life cycle inventory model.

$$\begin{pmatrix} 1 & 0.8 & 0.5 & 0.2 \\ 0.2 & 0.1 & 0.1 & 0.05 \end{pmatrix} \begin{pmatrix} 20 \\ 2 \\ 1 \\ 1 \end{pmatrix} = \begin{pmatrix} 22.3 \\ 4.35 \end{pmatrix} \quad (10)$$

$$\begin{pmatrix} b_{11} & b_{12} & b_{13} & b_{14} \\ b_{21} & b_{22} & b_{23} & b_{24} \end{pmatrix} \begin{pmatrix} \lambda_1 \\ \lambda_2 \\ \lambda_3 \\ \lambda_4 \end{pmatrix} = \begin{pmatrix} LCI_1 \\ LCI_2 \end{pmatrix} \quad (11)$$

$$B \cdot \vec{\lambda} = LCI \quad (12)$$

In generic terms, once the reference flow vector $\overrightarrow{RF}$ has been determined, and the process inventories of all unit processes have been collected and organized in a matrix of intermediate flows A and a matrix of elementary flows B, the life cycle inventory of the modeled product system can be calculated as

$$B \cdot A^{-1} \cdot RF = B \cdot \vec{\lambda} = LCI. \quad (13)$$

5.5 SELECTING LCI DATA

When compiling the life cycle inventory, LCA practitioners frequently face two dilemmas: either specific data is unavailable, or more than one data set from different sources appears to be reasonable to represent the product or material flow in question. In both cases, the practitioner needs to select the most appropriate (proxy) data based on a set of data quality requirements as defined by ISO 14040/44, Section 4.2.3.6.2 (ISO 2006b, 2006c):

- time-related coverage
- geographical coverage
- technology coverage
- precision
- completeness
- representativeness
- consistency
- reproducibility
- sources of the data
- uncertainty of the information

The challenge when selecting data is that the ISO standards do not provide any guidance on how to prioritize data quality requirements, which means that the choice of data may depend on the order in which these requirements are taken into account. For example, the practitioner may have to make value choices as to whether technological, spatial, or temporal representativeness is the most important consideration, or whether precision is more relevant than consistency between data sets. These choices, just as any choices in an LCA, are to be made based on the goal and scope of the study. This in turn also means that the quality of a data set is not absolute. Data quality assessment in LCA is always performed with regard to the given context. This section will therefore only provide some general insights, rather than strict rules, on how to select data.

A variety of LCA databases are available today, both public and commercial.[1] The largest and most widely used ones are the ecoinvent database from the Ecoinvent Centre and the GaBi databases developed and maintained by PE INTERNATIONAL.[2] Relatively young databases include the National Renewable Energy Laboratory's (NREL) USLCI database and the USDA's LCA Digital Commons.[3]

Different databases can render somewhat different results (Simon et al. 2012). While these differences may sporadically be based on random errors, they can also be due to diverging methodological choices made by the providers of the data. Examples of such choices include the treatment of biogenic carbon uptake and re-release or allocation methods for multi-output processes. Due to these differences and their potential implications, consistency between data sets should be given a high priority when selecting LCI data. The goal should be to ensure that any differences in impacts between products, life cycle phases, or processes are based on actual differences in technology, rather than an artifact of methodological choices.

The next aspect to consider should be the time-related, geographical and technology coverage of the available data sets and their representativeness with regard to the product system under study. The acceptable age of the data depends on the maturity of the technology, its innovation cycles, or even on recent changes in relevant emission legislations. However, for a large variety of chemical products, the error introduced by choosing the appropriate country of origin, but a different manufacturing technology than the one under study, is significantly higher than the other way around (Koffler et al. 2012). In this sense, choosing the appropriate manufacturing technology, rather than the exact geography, may be considered as a first rule-of-thumb.

Completeness needs to be assessed both in terms of completeness of the value chain covered by the dataset as well as the completeness of the inventory flows. This is especially true of data sources that have been developed for a purpose other than LCA, and thus may not include a sufficient number

1. http://lca.jrc.ec.europa.eu/lcainfohub/databaseList.vm
2. www.ecoinvent.org/; http:// www.pe-international.com
3. www.nrel.gov/lci/; http://www.lcacommons.gov/

and type of inputs and outputs to allow for a comprehensive assessment of the impact categories chosen for the study. The Environmental Protection Agency's (EPA) eGRID database and the Argonne National Laboratory's GREET models are examples of data sources that report major greenhouse gases and criteria air pollutants, but may have to be combined with additional data sources (depending on the scope of the study).

In terms of precision of the LCI data, reliable uncertainty information on the level of inventory flows is scarce, as the data sources used to create LCI data often do not include uncertainty information either. Existing numerical uncertainty information on inventory flow level is therefore mostly an approximation based on best estimates (Ciroth et al. 2012). From a practitioner's perspective, the uncertainty introduced by either selecting unrepresentative inventory data (see above) or the primary data used to scale these individual inventories (e.g., electricity consumption, recycling rates, and transportation distances) introduces additional uncertainties on a high level of the hierarchic network of unit processes that represents the product system under study (the foreground system). These uncertainties can affect the overall results more immediately than individual flow uncertainties on lower levels of the network (the background system).

Reproducibility may also influence the choice of data, but is mostly about providing sufficient information about the data and sources in the study report. It encompasses a "qualitative assessment of the extent to which information about the methodology and data values would allow an independent practitioner to reproduce the results reported in the study" (ISO 14044, Section 4.2.3.6.2). If the data is taken from public or commercial databases, reporting the specific data used will be sufficient. On the other hand, data that is not available to third parties and cannot be reported due to confidentiality reasons will reduce the reproducibility. While confidential data can be excluded from being shared with third parties, it should be available in the review process as part of the study documentation (see ISO 14044, Section 5.2).

IN SUMMARY, THE FOLLOWING GUIDANCE CAN HELP YOU WHEN SELECTING LCI DATA:

- When selecting inventory data, strive for consistency first. Mixing data from a variety of data sources increases the chances of generating artificial differences in results.
- Prefer technological over geographical representativeness; factor in the age of the data based on the technology's maturity and innovation cycles, and update to recent legislation if necessary or feasible.

• Quantitative uncertainty information on unit process flow level is ideal, but mostly based on estimates. Primary data variance and assumptions, as well as the use of proxy data in the foreground system, will introduce additional uncertainties that have a more immediate effect on the final results. Focus your efforts on properly reducing or assessing these uncertainties first.

5.6 STREAMLINED LCA TECHNIQUES

Data collection and modeling the life cycle inventory has been recognized as the most elaborate, and therefore expensive, step of LCA studies (Rebitzer 2005). A variety of streamlining approaches have been developed to reduce the related workload, while still giving sufficiently robust results. The term "streamlining" was first formalized for LCA in 1999 by a working group organized by the Society of Environmental Toxicology and Chemistry (Todd and Curran 1999). Accordingly, streamlining approaches can be grouped as follows:

• Limit the scope of the study (removing upstream or downstream components of the product system)
• Apply cut-off criteria
• Use qualitative, less accurate or surrogate data
• Use specific entries to represent impacts or LCI

Based on the above criteria, streamlining is an intrinsic part of every LCA. Exclusion of less relevant components of the product system either when setting the system boundary or when defining cut-off criteria are common practice to reduce the overall data collection efforts. The same goes for the frequent use of surrogate (proxy) data when perfectly matching data is unavailable (see Section 5.7). Likewise, using single indicators like global-warming potential (GWP) or Cumulative Energy Demand (CED) as a screening indicator for overall environmental performance is sometimes applied (Huijbregts et al. 2006).

All of the above approaches to streamlining will result in a less complete inventory; their application, therefore, needs to be carefully evaluated against their ability to support the goal of the study. More recently, (Koffler 2008) proposed to extend the definition of streamlining to encompass approaches that aim to reduce the effort of the inventory phase by automating the creation of the LCI model using semi-automated interfaces between internal data systems and LCA software, without deliberately excluding any available information in terms of relevance thresholds (Meinrenken et al. 2012). These LCI

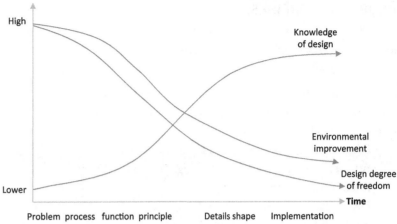

Figure 5.9. The Paradox of the Environmentally-Conscious Design Process (Bhander et al. 2003)

models, which are generated based on bill of material (BOM) or similar information, facilitate the consistent and efficient assessment of complex products and/or large product portfolios.

A third group of approaches to reduce the effort of data collection and analysis is characterized by the use of probabilistic metrics to represent inventories. This group of streamlining approaches tries to address what is known as the Paradox of the Environmentally-Conscious Design Process (Bhander et al. 2003). As can be seen from Figure 5.9, the degree of freedom to make changes to a product design decreases over time as the design matures.

The potential for substantial environmental improvements is therefore the greatest at the beginning of the design process. Detailed LCAs at the end of the design process may serve to inform the next product generation; significant changes to the current design are not possible anymore without incurring significant cost and potentially delaying the design process. One way to guide the design process therefore is to perform the analyses using high-level design data and probabilistic estimates of the resulting environmental impacts (Sun et al. 2003).

PROBLEM EXERCISES:

1. Based on the beer example, calculate the reference flows for the primary packaging materials (bottles and caps) for each beer.

2. Calculate the electricity necessary to cool the beer reference flows from room temperature down to drinking temperature. Use the following assumptions:

- Drinking temperatures: Amstel Light: 45° F, Samuel Adams Boston Lager: 46° F, Bud Light Platinum: 43° F
- Room temperature: 77° F
- Heat capacity of beer: 4.2 J/(g*K)
- Refrigerator coefficient of performance (COP): 3.5

3. Recalculate all reference flows for the three beers mentioned above for a functional unit that is based on a calorie content of your choice instead.

DATA QUALITY IN LIFE CYCLE ASSESSMENTS

ANDREAS CIROTH AND CHRISTOPH MEINRENKEN

6.1 MOTIVATION FOR DATA QUALITY IN LIFE CYCLE ASSESSMENT

Data quality in LCA has been critical since the emergence of life cycle approaches. As early as 1992, a Society of Environmental Toxicology (SETAC) working group presented a conceptual framework for LCA data quality as a result of a workshop (SETAC 1992). Today, data quality remains a key LCA topic.

The European reference Life Cycle Database (ELCD) database claims that "[for the database] focus is laid on data quality" [...]' (JRC 2012). The ecoinvent methodology report for version 2 mentions data quality nine times, while the report for version 3 mentions data quality 28 times (ecoinvent 2007, ecoinvent 2012). The subtitle of the latter report was 'Data quality guideline for the ecoinvent database version 3.' Similarly, the US-based National Renewable Energy Laboratory (NREL) database states that 'the goals of the U.S. LCI Database project are: to maintain data quality and transparency' (NREL 2012).

These examples show that data quality remains critically important for LCA. This chapter summarizes approaches for data-quality management in LCA today, starting with a definition of data quality in the context of LCA, and then presenting principles, approaches, and tools for quality management of LCA data.

6.2 DEFINITIONS OF DATA QUALITY IN LCA

For LCA, data quality is defined in the ISO standard

14040: "Data quality: characteristics of data that relate to their ability to satisfy stated requirements" (14040, 3.19).

According to this definition, data quality is not an absolute, inherent property of an LCA data set or an LCA study, but instead can only be evaluated in the context of given requirements. Therefore, whether the quality of the data at hand is "good" or "bad" depends on the stated requirements and on their fulfillment. One data set can have quite different data quality, depending on the requirements that are stated. This can be summarized as "beauty lies in the eyes of the beholder" (Ciroth 2010).

Quality requirements can cover a wide range of goals and applications. Still, when carrying out LCA in practice, two specific requirements—representativeness and (low) uncertainty—occur in a large number of LCA studies. Therefore, this chapter defines these two requirements in more detail. Readers will encounter them in the principles, approaches, tools, and specific examples for data quality management throughout Chapter 8: Advanced System Modeling.

Representativeness means that the acquired LCA inventory data shall be representative of (reflect and describe well) the specific product or process that is the subject of the LCA study. As shown later in this chapter in the context of pedigree matrices, the concept of representativeness can be further divided into temporal versus geographical versus technological representativeness. For example, suppose we need a data set for beer brewing in Germany for the year 2005. Two similar data sets are available from Germany: one from 2012, and the other from 2004. If no drastic changes have happened in the beer brewing process between 2004 and 2005, then the data set from 2004 has a higher data quality than the data set from 2012, both in general and under the given requirements.

Uncertainty, in the context of this chapter, is defined as the error margin associated with a given LCA data item. ISO standards refer to this as the dispersion of LCA data around their ideal and true (albeit usually unknown) values. For example, a study of US beer production concluded that 15 liters of water are used during cleaning. Could the true value be 14.9 liters? Probably. Could it be 10 liters? Maybe, but only if this particular datum carried a relatively large uncertainty or error margin; in this example, a five liter discrepancy exists from the true value of ten liters, implying a 33% error margin of the stated fifteen liters. The fifteen liters may have turned out to be different from the true value for many reasons. In measurement science and statistics, these are often categorized into types of error (e.g., random or bias/systematic or by the source and underlying reason of the error (e.g., parameter uncertainty, faulty measurement, wrong allocation method, or wrong underlying physical model). Furthermore, the concept of uncertainty itself can be further differentiated into limited accuracy versus limited precision. Chapter 16: Parallel Life Cycle Methods and Chapter 17: Project Management explore the concepts of accuracy and precision (including error types, standard deviation, and distributions) in more detail.

Uncertainty is also influenced by less tangible and quantifiable error sources, such as methodological choices and boundary settings. These, however, also contribute to deviations of calculated quantitative LCA figures. Therefore, all these cases can be comprised under the term uncertainty. We use this broad definition of uncertainty for this chapter.

The following example illustrates the requirements of representativeness and (low) uncertainty, in the context of the LCA of refrigerators, and how these two common requirements correspond to LCA's objectives of environmental impact reduction and comparison (see Chapter 1: Framework of Life Cycle Assessment). The requirement of representative data on one hand and (low) uncertainty on the other do not point to the same dataset as preferable. More often, more representative data has lower uncertainty uncertain. This is one of the reasons why LCA practitioners prefer site-specific (primary) data over secondary data. Still, the atypical example is chosen here to highlight the definitions that underlie the two requirements.

EXAMPLE: A household appliance store that operates in Ann Arbor, Michigan is concerned about the CO_2 emissions

created by transporting refrigerators from the factory to the store. These emissions were calculated for a specific brand and model that the store manager is interested in, the "Fresh Keeper." The store must choose between two data sources for the trucks' fuel consumption:

1. Site-specific* data: The trucking company that ships the Fresh Keeper units informs the household appliance store that shipping one refrigerator to the store will consume 2-3 gallons of diesel. A less uncertain figure (smaller range) is not available. While collecting this data, the store also learns that the trucks were retro-fitted last year and could, in theory, drive on a less carbon-intensive alternative fuel, which is available at gas stations on the trucks' routes.

2. Secondary data: Another trucking company, which ships freezers to stores in Denver, Colorado, offers the store access to a detailed LCA study from 2008. The freezers in this study are of the same size and weight as the Fresh Keeper refrigerators and would be transported on similar trucks. The fuel consumption, when extrapolated to the distance relevant for the store in Ann Arbor would be 2.2-2.3 gallons of diesel per freezer.

In the meantime, the marketing team of another refrigerator model, "Eat Healthy," has informed the store manager that trucking their model from the factory to the Ann Arbor store consumes 2.8-2.9 gallons of diesel.

Now, what tradeoffs with regards to data quality does the store manager face, and how are these tradeoffs affected by the underlying requirement of data quality? If the main focus of this study is to compare the trucking emissions of the Fresh Keeper model to those of the Eat Healthy model, the secondary data for the Fresh Keeper would be preferable. This information would enable the store manager to decide which of the two brands to carry in the store. However, using the secondary data would prevent the store manager from learning how to reduce the CO_2 emissions associated with transporting the Fresh Keepers. The store manager may not have learned that trucks were retrofitted because he was not focused on representative, site-specific data, and as a result he could not have lobbied the trucking company that ships the "Fresh Keeper to switch to the low-carbon fuel. The insight into one's own process, and thus the increased ability to improve environmental performance, is a key motivator to obtain site-specific, rather than secondary, data. Site-specific data is thus of higher quality with respect to this specific requirement, even if it is not always of lower uncertainty.

Concepts of site-specific (often referred to as "primary") versus secondary data are covered in detail in Chapter 6: Data Quality in Life Cycle Assessments and Chapter 16: Parallel Life Cycle Methods.

DATA QUALITY MANAGEMENT

Data quality is a rather abstract measure that is put into practice by data quality management. Data quality management should foresee efficient ways to measure data quality for given data sets. Furthermore, it should provide approaches to improve the quality where necessary. Finally, data quality management should establish an efficient and effective procedure for maintaining data quality, namely in light of changing requirements, additional data sets for a data base, and any existing data that becomes obsolete. This section explores principles and approaches for data quality management, and provides practical examples.

PRINCIPLES

According to its definition, data quality is based on the requirements that are implied by the goal and scope of the study. However, the relationship between these requirements, and the resulting data quality measures and management approaches, has rarely been explicitly described. Instead, current approaches often implicitly assume that LCA data should be of low uncertainty and, in particular, representative of the correct time and technology. Also, data should be complete, covering all materials and processes inside the chosen boundaries (Chapter 5: Life Cycle Inventory).

These assumptions capture obvious requirements that may be seen as common sense. However, when considering specific details of LCA modeling (e.g., system boundaries or allocation), data sources can differ significantly. Explicitly and transparently linking data management to requirements, including an approach to dealing with different requirements, would help users better understand the quality of different data under different requirements.

The following examples establish a link between requirements and data quality (and are described later in this chapter): a "crosswalks" approach by Cooper and McCarthy (2012); a "perspectives" approach by Ciroth (2012); and a "pragmatic and intuitive approach" (the pedigree matrix) developed by Weidema et al. (1996).

Data quality can be assessed for each input and output

flow of a process, for each unit process data set in a database and study, for product systems as interlinked process data sets, for LCA studies that often compare different product systems, and for complete LCA databases.

APPROACHES AND TOOLS

Many approaches for data management and data quality assurance exist in the LCA context. Usually, a set of data quality indicators is defined, along with rules to measure these indicators and minimum thresholds for each indicator to be considered satisfactory.

Common data quality indicators are related to (Ciroth 2008):

- data source and technique of data measurement
- completeness
- consistency of data
- reproducibility of data
- quantitative, time, technological, and geographical representativeness

These indicators can be applied on the level of input and output flows, process data sets, aggregated flows, full life cycles, and also complete databases. Approaches and tools can be differentiated into basic and more sophisticated ones.

BASIC APPROACHES

Several indicators can be assessed right away without any technical help. For example, the difference of the location or region of a data set to the desired region can simply be described. Likewise, the difference in time can also be described.

LCA models are models of mass and energy without any storage terms; a conversion of mass to energy is not relevant for products considered in LCA. Therefore, a common approach is to calculate mass balances per unit process or per aggregated system process. The calculated balance is a measure for completeness and consistency; "mass in" should equal "mass out" for a complete model, and any modeling decisions should preserve mass balances.

EXAMPLE: The focus is on the amount of natural gas delivered to a beer production facility (per functional unit 0.5 liter of beer consumed). How can the rule of mass balance be exploited in order to ensure the quality of the resulting GHG emission calculations?

1. CO_2: The amount of natural gas delivered (whether measured in cubic feet or BTU) contains a known number of methane molecules (CH_4). Therefore, one can infer the number of C (carbon) atoms that, once methane is burned in the furnace, will turn into CO_2. If the LCA inventory dataset lists more CO_2 (usually by weight) than the C atoms coming in

as natural gas could possibly provide (accounting for the added weight of oxygen, of course), then this raises a red flag that either one of the two data points is incorrect.

2. Methane and soot: What if the weight of CO_2 listed is lower? This may well happen, and it simply accounts for the fact that some of the delivered natural gas never burned (but was otherwise wasted) and that some of the methane molecules turned into soot rather than CO_2. In LCA in general, such waste or alternative flows (e.g.,scrap on the floor of a factory, leaks of water or gas to the atmosphere, or part of the mass going to a landfill) all have to be carefully tracked and included in the total flows. But the laws of physics prevail: If the "total in versus total out" in a LCA inventory do not match up, this means that the data quality needs improvement.

Simple logical checks are common and useful. Because every input and output flow represents mass or energy, negative flows are contradicting physical reality and therefore potentially violating the consistency criterion, with the exception negative flows that are sometimes used deliberately to account for the role of byproducts. Similarly, an easy check can determine whether units fit to flows; for example, electricity demand should not be provided in kilograms. Similarly, simple completeness checks can also be useful (e.g., determining if a data set contain information on its source, or if every flow contains an amount and unit). Finally, for a complete database, one should check whether units are duplicated with different spellings (e.g., liters written both an upper case "L" and a lower-case "l"). These basic approaches can often be applied in conjunction with software support, or are indeed directly integrated into LCA software (see Chapter 8: Advanced System Modeling).

MORE SOPHISTICATED APPROACHES

More sophisticated approaches apply data quality checks in a greater context or require more detailed analyses. Outliers are extreme values in relation to a group of otherwise comparable data sets. As defined by Ripley, they are "sample values that cause surprise in relation to the majority of the sample" (Ripley 2004). For example, outliers can be an indication of potentially incorrect data; they can only be detected in the context of other data sets through a joint analysis, or because they were detected an expert who is familiar with similar, comparable data sets and their values.

For these checks, it is required to first understand the group to which a certain data set belongs. This group can then be used as a benchmark. Figure 6. 1 below shows an example from the database, taken from (Ciroth 2009); all agricultural production processes, have CO_2 emissions between 0.1 and 2.8 kg per kg product. Within this group, it can be assumed that all products somehow cause a given range of carbon dioxide emissions per amount of product. These can be considered outliers with high or low emissions, and should be checked. However, such a strict causal relationship with carbon dioxide emissions does not apply to all processes. Instead, some processes have low carbon dioxide emissions, while others

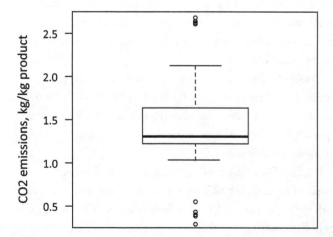

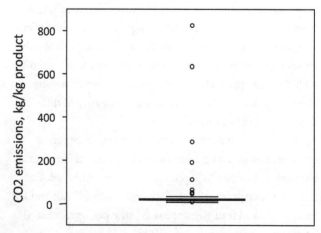

Figure 6.1. Boxplot of CO2 emissions of ecoinvent processes with a product in unit of kg. left: agricultural production processes; right: all other processes, from all data sets in ecoinvent v. 2.1. Note that the two plots have different scales.

have extremely high emissions of several hundred kg per kg product. Therefore, such outlier analysis can guide a user towards potentially suspect data items. But it cannot conclusively answer whether a data item is definitely wrong or not. For example, the analyzed sample group may not be homogenous and may contain a specific data set that does not fit to a group of generic data.

Another data quality issue, missing information, requires a more sophisticated approach that includes scrutinizing quantitative figures within their logical context, rather than just the numerical data itself. For example, every fossil fuel combustion process will have CO_2 emissions. Therefore, if a data set were to list a fossil combustion process without any CO_2, then this is most likely a mistake. Here as well, a benchmark group,

or a checklist, helps to identify missing information.

Similarly, some flows will not only indicate the existence of other flows, but actually allow us to infer their exact amounts. For example, with a certain amount of diesel consumed, a truck will have a predictable amount of CO_2 emissions, based on a basic underlying physical law (e.g., stoichiometry of the combustion). This way, it can be checked whether the relation between diesel and CO_2 matches, such as a data set describing truck transportation.

LCA experts can perform the quality checks described above. Still, more technical support—such as statistical analyses that identify comparable groups in data, perform tests, and plot graphs that help to find outliers and structures in data—can be helpful (see Chapter 17). The approaches described here

Indicator Score	1	2	3	4	5 (Default)
Reliability	Verified data based on measurements	Verified data partly based on assumptions **or** Non-verified data based on measurements	Non-verified data partly based on qualified estimate	Qualified estimate (e.g. by industrial expert)	Non-qualified estimate
Completeness	Representative data from all sites relevant for the market considered, over an adequate period to even out normal fluctuations	Representative data from >50% of the sites relevant for the market considered, over an adequate period to even out normal fluctuations	Representative data from only some sites (<<50%) relevant for the market **or** >50% of sites but from shorter periods	Representative data from only one site relevant for the market **or** some sites but from shorter periods	Representativeness unknown or data from a small number of sites **and** from shorter periods
Temporal Correlation	Less than 3 years difference to the time period of the dataset	Less than 6 years difference to the time period of the dataset	Less than 10 years difference to the time period of the dataset	Less than 15 years difference to the time period of the dataset	Age of data unknown **or** more than 15 years of difference to the time period of the dataset
Geographical Correlation	Data from area under study	Average data from larger area in which the area of the study is included	Data from area with similar production conditions	Data from area with slightly similar production conditions	Data from unknown or distinctly different area
Further technological Correlation	Data from enterprises, processes and materials under study	Data from processes and materials under study (i.e. identical technology, but from different enterprises	Data from processes and materials under study but from different technology	Data on related processes or materials	Data on related processes on laboratory scale or from different technology

Figure 6.2. Pedigree matrix in ecoinvent v3 (ecoinvent 2012)
3 Verification may take place in several ways, e.g. by on-site checking, by recalculation, through mass balances or cross-checks with other sources.
4 Includes calculated data (e.g. emissions calculated from inputs to an activity), when the basis for calculation is measurements (e.g. measured inputs). If the calculation is based partly on assumptions, the score would be 2 or 3.

can be used directly in a study, or used to check the quality of a study in a critical review.

INTEGRATED APPROACHES

Integrated approaches combine various individual quality indicators into an overall data quality result. One widely used example is the pedigree matrix, introduced into the LCA domain by Weidema and Wesnaes in 1996 and taken up by the ecoinvent database. This approach foresees five data quality indicators: reliability, completeness, temporal, geographical,

that is estimated using the uncertainty factors can be used as input to a Monte Carlo simulation or analytical uncertainty propagation, and can thereby be used to determine the composite uncertainty of the final quantitative results of an inventory calculation, or even a LCA comparative study.

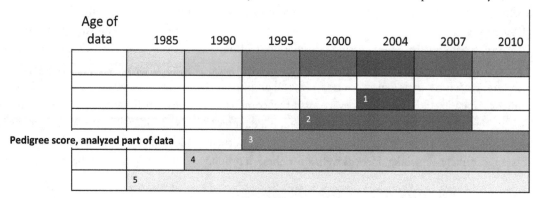

Figure 6.3. Pedigree scores for data sets with different reference time, for a desired, optimal time reference of 2004 (Ciroth et al. 2012)

and further technological correlation. These indicators are assessed according to rules that are given as threshold descriptions (Figure 6.2, page 65). The assessment is carried out separately for each input or output flow of a process data set.

For example, a flow whose source is not verified, and whose value is partly based on qualified estimates, obtains a score of 3 for the indicator reliability. The pedigree matrix is one of the few approaches where the assessment of data quality depends on the specific requirements that are defined for a data set. Therefore, the correlation indicators can only be assessed against a stated optimal time, technique, and location. This concept is illustrated in Figure 6.3.

This aligns with the definition of data quality in ISO 14040 that was introduced in the beginning of this chapter. For completeness and reliability, the matrix assumes that an optimal data set and study is reliable and complete; therefore, it is not necessary to state an optimal completeness level and an optimal reliability level.

For the ecoinvent database, the individual indicators are further processed, using so-called "uncertainty factors" into standard deviations for the analyzed input/output flow. These standard deviations represent levels of uncertainty or error margins (see Chapter 17: Project Management for a more detailed treatment of standard deviations). Assuming a log-normal distribution, the geometric standard deviation contributions are provided for each indicator score, and combined with geometric standard deviation contributions for each type of flow and type of process. Details can be found in the report (ecoinvent 2012). In principle, the uncertainty for the flows

Future versions of ecoinvent will make the approach available for distributions other than the lognormal distribution as well. So far, the factors used in ecoinvent are based on expert judgment; derivation of the factors is not documented. A first empirical foundation for the uncertainty factors is provided in (Ciroth et al. 2012). This foundation develops and proposes different factors than the ones currently used in ecoinvent (Ciroth 2013).

WORKING WITH LARGE DATASETS—PRIORITIZATION OF DATA QUALITY EFFORTS

Typical LCA studies require datasets with dozens, hundreds, or even thousands of individual data items (e.g., distances, masses, energy amounts, and volumes). It may be possible to meticulously review each data item and employ the approaches and tools stated above, thereby ensuring the quality of each data item vis-à-vis the stated requirements. In practice, however, the sheer amount of time and effort of such an exercise often becomes a real hurdle. The quality of LCA datasets is often inhomogeneous because the people conducting the study ran out of time, money, or energy (Meinrenken 2014).

This begs the question of whether a method exists for prioritizing data on a long list of data items in order to shorten the list. This would ensure higher data quality, while still serving the goals of the LCA as a whole. One such prioritization, using a concept first introduced by Heijungs in 1996, is often simply referred to as data screening (GHG Protocol 2011). Data screening works via the following steps:

- Assemble a first pass of the dataset for the LCA inventory with a focus on having at least estimates for all data items (e.g., distances, masses, energy amounts, and volumes), even if some of these items are still uncertain or less representative (or both) than would be ideal.
- Screen for contribution: Perform a screen of all data items, prioritizing them by descending relative contribution to the total environmental impact that is being studied. The first screen recognizes the fact that despite the often extremely long list of data items, a majority of environmental impacts are usually caused by just a few items. With the objective of impact reduction in mind, these data items represent the low hanging fruit that will be particularly useful.
- Screen for uncertainty: Lowering the uncertainty of the total environmental impact requires using a second screen in combination with the first. This second screen recognizes that data items that contribute greatly to the total environmental impact are not necessarily the ones that strongly affect the uncertainty of the total impact. Instead, data items with both high contribution and high relative uncertainty are the most likely to materially affect the uncertainty of the total impact. With the objective of impact comparison, and thus low uncertainty of the total, in mind, it is these data items whose lowered uncertainty would most benefit the uncertainty of the total. In contrast, data items with low relative contribution to the total usually do not materially affect the uncertainty of the total footprint, even if the uncertainty of the data item itself is low.

Carrying out above screening quantitatively requires dozens, hundreds, or thousands of sensitivity tests (one for each data item), which must then be prioritized accordingly (Meinrenken 2012). Sensitivity tests are explained in more detail in Chapter 15: Bias and Uncertainty in Life Cycle Assessment and Chapter 17: Project Management. In practice, however, simply organizing all inventory data into a simple 2x2 matrix according to the two screening criteria can save much time and effort, while still achieving high data quality for the overall LCA study.

EXAMPLE

Contribution: The largest relative contribution to the 65 liters of water use in our total life cycle of beer is due to agriculture (70% or 46 liters). The second largest is cleaning (23% or 15 liters). Various other data items have much lower individual contributions.

Uncertainty: Suppose the agriculture data has already been scrutinized and proven to be from a reliable source. We therefore assume a low relative uncertainty of this data item (i.e., small margin of error, 46±2 liters). Suppose we received conflicting data on two types of cleaning (10 and 20 liters) and a simple average is currently used as a placeholder. We therefore infer high relative uncertainty of this data item (large margin of error, 15±5 liters). In this example, screening would suggest the following prioritization: If the objective is impact reduction, first ensure that agriculture data is representative. If the objective is a comparison with another process for making beer (and hence a low error margin of the total impact), first lower the uncertainty of the cleaning data. The quality of all other data items is less critical, whether for impact reduction or comparison.

	Low — Uncertainty — High	
Contribution High	Agriculture *(Improve if focus on Reduction)*	Cleaning *(Improve if focus on comparison)*
Low	Various other *(least effort on quality improvement)*	Various other *(least effort on quality improvement)*

Low **Uncertainty** High

While the described prioritization approach helps to increase data quality for already identified LCA data, it cannot recognize the relevance of information that has not yet been identified. For example, a plutonium plant that is part of one's "real" system, but has not yet been added to an LCA model, will not turn up as relevant in the above 2x2 matrix. Therefore, the criterion of completeness remains a critical data quality indicator.

SOFTWARE SUPPORT FOR DATA QUALITY MANAGEMENT

Many tasks in data quality assessment are repetitive and deal with large amounts of data. Therefore, it makes sense to transfer some of these tasks to dedicated software. For example, in ecoinvent 3, an eco-editor has been created that can be used to submit data sets to the review process. The eco-editor contains some basic checks that make sure that data sets do not fail to meet simple validation checks, such as "text entered for the amount of an emission," "no unit provided," and so forth (ecoinvent 2012a).

Finally, more complex tools can be imagined, such as data quality analysis and assessment suites that identify structures and patterns in data sets. These, however require LCA data analysts to have training and expertise in statistical analyses (Ciroth 2009). Also, see Chapter 16: Statistics in LCA.

DATA QUALITY THROUGH CRITICAL REVIEW

None of the above approaches and tools are foolproof recipes for flawless data quality. To give one example, suppose a low-pedigree score is associated with the fact that an LCI dataset is less than three years old (see Figure 6.2). This may be cause for confidence when the underlying processes is making beer, a process in which underlying inventory data is unlikely to change much from one year to the next. However, if the underlying process were microprocessor manufacturing instead, even two-year old data may easily be a factor or two off from the process that is actually being studied (Meinrenken 2012). This has two implications: On one hand, above approaches and tools are constantly being improved and refined, such as, designing sector-specific pedigree matrices) (Ciroth 2012). On the other, vigilance and common sense should always prevail over a mere mechanical use of above approaches and tools.

A critical review typically addresses a complete LCA study, or, in some cases, only a single data set. For comparative LCA studies that are intended to be published, a critical review is mandatory (according to ISO 14040 and 14044). The critical review required in these cases follows the requirements defined in the ISO standards. In particular, the review needs to assess whether a study meets the aspects it has set out to perform and achieve, according to its own goal and scope.

Klöpffer describes the quality criteria

that are to be applied according to the ISO standards as follows: "There are no absolute quality criteria (except the consistency with the international standard), but the most important component 'Goal and Scope' is cited as the point of reference" (Klöpffer 2012). This directly aligns with the data quality definition provided at the beginning of this chapter.

The main consideration of a critical review is whether the study under review is performed according to ISO 14040 and 14044. Still, the review process, particularly the dialogue between reviewers and the practitioners performing the study, often leads to improvements in the study's design, data quality, and interpretation. More details about the critical review process according to ISO are provided in Chapter 2: The International Standard for LCA.

A critical review of a single data set as a single task is obviously less complex than the review of a full LCA study. Still, the effort for ensuring a consistent, comprehensive, peer-reviewed, and up-to date LCA data set can hardly be underestimated. For the ELCD database of the European Commission, it was estimated that for its 320 datasets, a transition period of at least three years would be required until the database meets the requirements defined in the ILCD handbooks (Wolf 2011).

The principle for a review of data sets is analogous to the review of full LCA studies. First, a quality guideline is required that defines how data sets for the database should look. This guideline includes nomenclature, units, and a set of default flows, as well as expected documentation, among other

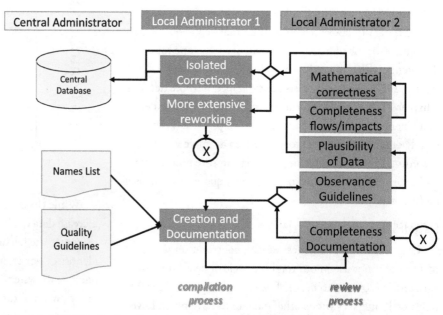

Figure 6.4. Dataset Review Procedure for the Ecoinvent Database 2 (Frischknecht and Jungbluth 2003, p. 54)

things. Also, modeling and measurement procedures, and possibly also testing routines, may be part of the guideline. Practical examples for LCA data set quality guidelines are the ecoinvent overview and methodology reports (ecoinvent 2007) and the ILCD handbooks, especially (JRC 2010). For the ecoinvent database 2, the following workflow is used for the dataset review (Figure 6.4).

A local administrator needs to assess the quality of a data set against published guidelines, as well as specific completeness and correctness criteria. This is similar to the principle of critical review for a study, and also to the pedigree matrix where stated requirements are used as a benchmark for evaluating the data quality.

6.3. OUTLOOK

Looking forward, two forces drive further improvements in the overall quality of LCA data. The first force is the continued application of previously mentioned tools and approaches, especially when the requirements of data quality are carefully calibrated to the goal and scope of a particular LCA study. The second force is related to the simple fact that a nearly infinite number of processes have not been studied in detail. As a result, LCA practitioners are forced to "settle" for non-representative data more often than they wish.

How can this lack of LCA studies be remedied? Continuously improving standards and guidelines for LCA will remove practical obstacles towards wider use and thus help grow the amount of available site-specific, representative data (Draucker 2011). Every student, such as yourself, who reads and learns about LCA can help increase both coverage and quality of LCA data.

PROBLEM EXERCISES

1. Describe, in your own words, the (i) concept and (ii) relevance of data quality in the context of LCA.

2. Give a concise definition of data quality in LCA.

3. List eight or more sources/types of possible errors in LCA inventory, and add a specific example for three of those types.

4. Explain the difference between systematic and random error, along with an LCA-relevant example for each of them.

5. "Data representativeness" in LCA: What is representativeness, what types of representativeness exist (give one example each), and how can this concept differ from the concept of error margins?

6. List three or more approaches to control data quality in LCA inventory datasets. Give examples for two of them.

7. Explain the motivation and approach of the "pedigree matrix" concept.

8. List two limitations of the pedigree matrix concept of which you should be aware.

9. Explain the concept of "data screening" and its use in building large LCI inventories.

10. What are the goal and basic steps of assessing data quality in a critical review process?

INPUT-OUTPUT MODELS FOR LIFE CYCLE ASSESSMENT

TROY HAWKINS AND CHRISTOPHER WEBER

7.1 INTRODUCTION TO EIO-LCA

Up to this point, this book has focused on building up life cycle inventory (LCI) data one unit process at a time or through the use of previously developed background inventory datasets. Two criticisms of carrying out LCA in this way are (1) compiling comprehensive data sets is time consuming and (2) when boundaries for inventory processes are drawn, they introduce the potential for cutoff error associated with omitted flows or processes. Decision makers rarely have a year or more to wait for a comprehensive LCA. On the other hand, presenting results from a streamlined LCA with a tightly defined boundary, which might exclude relevant upstream impacts, is undesirable.

One approach for dealing with these issues is to use an LCA model that incorporates publicly available economic input-output data. An economic input-output (EIO) data set provides a comprehensive description of transactions between sectors of the economy. While originally developed for national accounting purposes, an EIO dataset essentially provides the technosphere flows for each sector with a clear linkage to all other sectors. In an EIO-LCA model, the EIO data is coupled with environmental data containing the elementary flows describing resource use and environmental releases for each sector. Together, the EIO and environmental-extensions provide a complete set of unit processes for an entire economy.

The national level of the EIO data limits the analysis to within national value chains. Thus, if the value chain includes transfers between countries, there is currently no way to track those impacts. This is not necessarily important. For example, in the US about 90% of the economy is domestic. Work is underway to develop international EIO data that aims to address this issue.

A number of EIO-LCA models have been developed and made freely available. Together with the offer of quick yet comprehensive results, this makes EIO-LCA models popular for a number of applications. During the energy crisis of the 1970s, EIO models were used to track energy use and to look for opportunities to improve the energy efficiency of production and consumption activities (Bullard and Herendeen 1975). In the early days of the environmental movement, Wassily Leontief used EIO models to understand pollution and the costs of mitigation associated with goods (Leontief 1970). As computers became available for performing calculations, EIO-LCA models were used to understand the environmental implications of a wide variety of products, including batteries (Lankey and McMichael, 2000), automobiles (Lave et al. 2000) , and pavement options (Horvath and Hendrickson 1998), as well as service sectors (Norris, Croce, and Jolliet 2002), buildings (EPA 2013), waste management (Nakamura and Kondo 2002), and household consumption activities (Tukker et al. 2009), amongst others.

TERMINOLOGY

A number of terms and acronyms are often used in the context of input-output studies. First, while this chapter deals with an EIO-LCA, input-output analysis (IOA) is a broader field that seeks to understand the structure of the economy and relationships between sectors. Because most LCA input-output studies use economic data, the more general term input-output LCA (IO-LCA) is often used. Environmentally-extended input-output (EE-IO) analysis is another frequently used term.

EIO-LCA studies, particularly those based on national account data, are generally differentiated from process-based life cycle assessment (PLCA) studies, which rely on unit processes that use detailed estimates based on engineering data. A key distinction between EIO-LCA and PLCA models is the use of top-down data versus bottom-up data. Top-down data is provided as an aggregate total that is allocated to more specific processes. For example, national annual sector total releases to air and water provided in the National Emissions Inventory and Toxics Release Inventory are used to construct the US EIO-LCA models. Bottom-up data is built up from a combination of measurements, engineering estimates, and secondary data sources to construct detailed unit processes. For example, a unit process for driving a car could be constructed from measured emissions and reported fuel efficiency data.

In the context of IO-LCA, the term supply chain analysis is often used. This is the same as a cradle-to-gate LCA. Another term we may hear is hot spot. This refers to processes within the life cycle of a product or service that contribute significantly to a certain impact category. Finally, the term footprint analysis generally refers to an analysis that either tracks a specific type of impact, or an analysis that tries to pull together different impacts within a common metric. Carbon footprint and water footprint are examples of the first type. Ecological footprint is an example of the second type, which combines land use and global warming impacts under the unifying metric of human appropriation of biologically productive capacity.

The EIO-LCA approach is ideal in cases where a quick answer is needed. For example, it can be used in an initial screening/selection phase or for an initial evaluation of prospective technologies. While LCA is an iterative process, EIO-LCA could be used to inform where more specific data should be collected to improve the accuracy and precision of model results. Published EIO-LCA studies assessing the impact of product options, or comparing technologies, serve the purpose of structuring the discussion of alternatives around scientifically-based estimates of impacts that can be refined through further study. This idea has been incorporated by the US Environmental Protection Agency (EPA) in its efforts to promote the sustainable use of materials and a life-cycle perspective on decision-making (EPA 2009). In fact, EIO-LCA is specifically mentioned in the Greenhouse Gas Protocol Corporate Value Chain (Scope 3) Accounting and Reporting Standard as the recommended means of prioritizing activities for more detailed analysis.

The EIO-LCA approach is also ideal in situations where a wide variety of products and services are included in the functional unit, such that bottom-up data collection is impractical. For example, EIO-LCA is ideal for answering questions about the impacts of household consumption or any other collection of purchases. It has been applied for this purpose in projects sponsored by the European Union, the United Kingdom, the Netherlands, Japan, and the United States, amongst others. EIO-LCA models underlie many of the more sophisticated web-based carbon footprint calculators and are being incorporated in the calculation of national carbon, water, and ecological footprint accounts (Ewing et al. 2012). In response to Executive Order 13514, which requires federal agencies to advance sustainable acquisition, EIO-LCA approaches in the US are being applied by the General Services Administration and the EPA to understand the overall impact associated with federal procurement activities, and to identify hotspots within supply chains where improvement efforts could be focused.

Acronyms:

IOA Input-output analysis

IO-LCA Input-output life cycle assessment

EIO-LCA Economic input-output life cycle assessment

EE-IO Environmentally-extended input-output

MRIO Multi-regional input-output

PLCA Process-based life cycle assessment

Terms:

Bottom-up Combining measurements, estimates, and other sources to describe a process

Top-down Allocating aggregate total data to processes

Supply chain analysis Cradle-to-gate life cycle assessment

Hot spot High impact processes within a life cycle

Footprint A study dealing with a single impact, or one that pulls together impacts into one unifying metric

7.2 THE ECONOMIC INPUT OUTPUT MODEL

While IOA has recently seen growth in the field of environmental assessment (Tukker and Jansen 2006), its origins date back to the 1930s, with much of the groundwork for current approaches developed in the 1950s and 60s (Leontief 1970). This section provides a short overview of an environmentally-extended input-output model, its basic components, and its method of application. In-depth descriptions of input output models are provided at the end of this chapter.

SUPPLY AND USE TABLES

An EIO model has its basis in the United Nations System of National Accounts (UNSNA). The main two components used to compile the underlying model are called the supply (or make) table and the use table (UN 1999). Both tables represent the annual economic output of a nation, which are broken down into a number of producers and consumers. The productive sectors of the economy (agriculture, mining, manufacturing, and services) are broken into a series of indus-

tries (e.g., Industry1, Industry2, et al.), each of which make one or more commodities (e.g., Commodity1, Commodity2, et al.). Highly simplified schematics of these tables are shown in Figures 7.1 and 7.2. In short, the supply and use tables are used to track the economic transactions that make up a national economy. The supply table shows industries that produce certain types of commodities, while the use table shows industries (or final consumers) that use certain commodities.

Each entry in the Supply Table represents the amount of a commodity described on the row that is produced by the industry described on the column. For instance, X12 represents the amount of Commodity1 produced within the economy by Industry2. Imports to the economy are accounted for in a special column to the right. Each industry usually has one predominant commodity that corresponds to it. For example, the automobile manufacturing industry sector primarily produces the commodity automobiles. These industry and commodity pairs are usually listed in similar order on the rows and columns. Thus, the supply table is sparse, with the largest entries

INDUSTRY

		#1	#2 ...	# n	Imports	Total Supply	Total Industry Output
Commodities	Commodity1	X11	X12...	X1n	M1	S1=X11+X12+...+X1n+M1	Industry 1: I1=X11+X21+...+Xn1
	Commodity2	X21	X22...	X2n	M2	S2=X21+X22+...+X2n+M2	Industry 2: I2=X12+X22+...+Xn2 ...
	Commodity n	Xn1	Xn2...	Xnn	Mn	Sn=Xn1+Xn2+...+Xnn+Mn	Industry 3: In=X1n+X2n+...+Xnn

Figure 7.1. Supply Table

		Industries			Final Demand				
		Industry 1	Industry 2 ...	Industry n	Exports	Household	Capital	Gov	Total Supply
Commodities	Commodity1	U11	U12 ...	U1n	E1	HC1	K1	GC1	S1
	Commodity2	U21	U22 ...	U2n	E2	HC2	K2	GC2	S2
	Commodity n	Un1	Un2 ...	Unn	En	HCn	Kn	GCn	S3
Value Added		V1	V2 ...	Vn					
Total Industry Output		I1	I2 ...	In	E	HC	K	GC	

Figure 7.2. User Table

falling on the diagonal elements. However, some industries make more than one commodity. For instance, the automobile manufacturing sector also produces metal parts or machinery that other sectors may use. Thus, the make table contains a full description of all co-products produced by a sector. It can be used to allocate the sector's inputs and outputs to specific co-products. Summing along the columns of the supply table gives the total output of the industry (I1, I2, et al.), including all the commodities produced; summing along the rows yields the total supply (S1, S2, et al.), or, in other words, the total amount of each commodity available in the economy.

The use table is made up of three sections: inter-industry (U11-U33), final demand, and value added. Taken together, the inter-industry and final demand sections of the use table describe the use of each commodity by each industry sector or final consumer, including households, government, export markets, and capital formation. Take apples as an example: if we follow them through the table as Commodity1, the apples may be used by a food processor, Industry2, to make applesauce. Thus, U12 would represent the use of apples by the food processor. Apples may also be purchased directly by households for consumption (HC1), exported to other countries (E1), or purchased by the government for schools or the military (GC1)[20]. Value added is tracked below the inter-industry transactions section of the use table (I1 to In). Value added generally consists of wages paid to employees, profits, and other transactions with the government (e.g., taxes, subsidies, and resource royalties). Because the IO accounts are tracking payments made by the industry, taxes are accounted for as positive values, while subsidies are tracked as negatives. Value added makes up the difference between the value of an industry's produced output (e.g., I1-In) and the commodity inputs used to produce it (e.g., U11+U21+U31 for Industry1). Each row of the use table sums to the total supply or commodity output, and each column sums to total industry output, as shown by equations 1-2 for each commodity i and industry j. In this way, the row and column totals of the supply and use table are equal, and the tables are said to be balanced.

Equation 1: $\sum_{j=1}^{n} U_{ij} + E_i + HC_i + K_i + GC_i = S_i$

Equation 2: $\sum_{i=1}^{n} U_{ij} + V_j = I_j$

20 One additional type of final use, capital formulation, is possible but unlikely for apples. Capital formulation represents the consumption of a commodity by an industry when the commodity is used as capital equipment to produce other commodities. More detail is provided elsewhere (UN 1999).

All the entries in the supply and use tables represent monetary flows and are positive. This is different from the technology matrix developed in PLCA, in which entries represent different physical flows and have different units, and can be positive or negative following a certain sign of convention.

DIRECT AND TOTAL REQUIREMENTS

By themselves, the supply tables and use tables cannot be used for LCA. However, with some manipulation, they can be converted to the input-output model in either industry or commodity terms. Different models and different assumptions used to create the models are beyond the scope of this book [21]. This chapter will show the derivation of a generic input-output model using a symmetric input-output table (either commodity-by-commodity or industry-by-industry), after this advanced derivation has already occurred.

After accounting for co-products through the supply-use framework, a generic symmetrical input-output table is shown in Figure 7.3. Commodity or industry groups can also be called sectors, as the generic model shown here applies to either a commodity-based or an industry-based model. Both terms are in common use. In generic input-output modeling, the X always signifies the total output of the economy or sector, whereas Y represents final consumption or demand (which, in the use table, can represent final consumption by households, government, export markets, or capital purchases). Thus, the generic input-output model can be summarized in either sector or matrix form as:

Equation 3:
$$x_i = Z_{i1} + Z_{i2} + Z_{i3} + \cdots Z_{in} + y_i = \sum_{j=1}^{n} Z_{ij} + y_i$$

Equation 4: $x = Z + y$

where the total output of each sector is the sum of its inputs to other sectors (Z) and its sales to final consumers (Y). Note that this equation is essentially the same as Equation 1 above, with the exception that the symmetric input-output model assumes a one-to-one relationship between industries and commodities. As the result of the advanced derivation, production and use of commodities (as represented by Xij and Uij in the generic input-output model) are not directly used. Instead, Zij is used to represent inputs from Section i to Sector j in dollar value. Notably, no column for imports is provided. When developing the generic input-output model, it is commonly assumed that imported commodities have the same

21 For different derivations of input-output models from the supply/use framework see (Horowitz and Planting 2006).

production characteristics as compared to goods made domestically, although the imported goods may have different production characteristics.

	Input to Sectors			Final Demands	Total Output
Sector Output	1	2 ...	N		
1	Z11	Z12 ...	Z1n	Y1	X1
2	Z21	Z22 ...	Z2n	Y2	X2
N	Zn1	Zn2 ...	Znn	Yn	Xn
Value Added	V1	V2 ...	Vn	GDP	
Total Output	X1	X2 ...	Xn		

Figure 7.3. Generic Symmetric Input-Output Table

The input-output model was derived in a manner similar to the standard LCA model in that it uses production recipes. The major difference between the standard LCA model and the EIO model, however, is that the production recipe is in economic terms. Each sector produces an economic value (X_n) of product by utilizing inputs also in economic value $(Z_{1n}-Z_{nn})$. For instance, in the 2002 US input-output accounts, the breweries industry produced around $21.5 billion of output using the inputs (arranged from largest to smallest) shown in Table 7.1 below. As this table shows, the breweries sector uses much the same inputs as a process-based LCA would show, including metal, glass, and paperboard packaging, raw grains and pre-malted grains, and truck transportation and wholesale trade (warehousing and storage) for delivery. However, we also find sectors that would not typically show up in a process-based LCA, such as management of companies, which represents expenditures that brewing companies make for executive compensation and other business services. The total inputs to the sector ($12.3 billion) plus the value added of the sector ($9.1 billion) sum to the total output of the sector, which is $21.5 billion.

To utilize this input data for LCA requires normalization—that is to say, the value spent on each of the inputs must be normalized by the output of the final product, in a manner similar to the way technology matrix of process LCA often normalizes unit flow data. In this example, each of the input values to the breweries sector (i.e., $2.875 billion of metal containers) is normalized by the total output of the sector (i.e., $21.553 billion) to yield the normalized beer production recipe value (0.133). Table 7.1 also shows the resulting values for the A matrix column representing these normalized values. In matrix form, the entire economy can be normalized into such production recipes. The resulting matrix is typically called the direct requirements matrix, as signified by A in Equation 5. In Equation 5, the x with the "hat" above it indicates the matrix resulting from diagonalizing the total output vector (x). Each column of this matrix represents the inputs required from every other sector of the economy to produce a unit of output (e.g., $1 or $1 million). For example, to produce $1 million of beer, the sector requires $27,000 (0.027 x $1 million) of flour and malt manufacturing.

	Breweries' Use ($million)	A
Metal Containers	2875	0.133
Glass Containers	2042	0.095
Management of Companies	1247	0.058
Paperboard Containers	982	0.046
Grain Farming	619	0.029
Wholesale Trade	603	0.028
Flour and Malt	584	0.027
Truck Transport	417	0.019
Other Sectors	3016	0.140
Total Inputs	12385	0.575
Value Added	9168	-
Total Output	21553	-

Table 7.2. Inputs to Breweries Sector in 2002 US Benchmark Input-Output Model

Equation 5: $A = Z(\hat{x})^{-1}$ or Aij=Zij/xj

Here the ^ notation designates diagonalization: $\hat{x}$ is the diagonal matrix constructed using entries of vector x. Solving for Z and substituting Equation 5 into Equation 4 yields Equation 6, where the total output of an economy (x) can be expressed as the sum of intermediate consumption (Ax) and final consumption (y):

Equation 6: $x = Ax + y$

where A is the economy's direct requirements matrix, x is the vector of total output by sector, and y represents the sector's final demand. When solved for total output, this equation yields the following result (which is similar to the technology matrix in process LCA):

Equation 7: $x = (I - A)^{-1}y$

In equation 7, the term (I-A)$^{-1}$ is commonly known as the total requirements matrix, or Leontief inverse after the father of input-output economics, as it reflects not just the immediate inputs required to produce the sector's output (such as in the direct requirements matrix) but also the entire supply chain upstream from the sector. For instance, rather than just including purchases from a brewery's direct suppliers (aluminum cans, glass bottles, and grains), the total requirements matrix shows the total inputs required to produce these inputs as well (raw aluminum, all the way back to the mining of bauxite as in process LCA). Table 7.2 shows the largest contributing factors to the total requirements to produce a final demand of $1 million of beer. Many of the sectors are the same as a direct requirements matrix, but one can also see further upstream sectors (e.g., aluminum products representing finished aluminum before being converted into a can) and ubiquitous sectors such as power generation that are purchased all over the supply chain. The total requirements matrix includes the original $1 million final demand of beer; the value for the breweries sector is slightly larger, 1.005, implying the breweries sector purchasing from itself in its supply chain. The value 2.243, which represents the sum of all inputs required to make one unit of output, is sometimes referred to as the total multiplier, as it represents the total value created in the supply chain for a unit of demand.

For matrix (I-A) to have the capacity to be inverted, matrix A has to satisfy the Hawkins-Simon condition[22]. The mathematical proof of this condition is too complex to be discussed

22 Hawkins, D., and H. Simon. "Some Conditions of Macroeconomic Stability." Econometrica 17 (1949): 245-48.

	L = (I-A)$^{-1}$
Breweries	1.005
Metal Containers	0.152
Glass Containers	0.095
Management of Companies	0.091
Wholesale Trade	0.074
Paperboard Containers	0.056
Grain Farming	0.043
Truck Transport	0.033
Power Generation	0.028
Real Estate	0.027
Flour and Malt	0.027
Aluminum Products	0.026
Other Sectors	0.587
Total Inputs	2.243

Table 7.2. Total Requirements for Producing 1 unit of Output from the Breweries Sector (Using the US 2002 Benchmark Model)

here. For a typical economic system, this condition will be met.

The same basic result in Equation 7 can be achieved through a supply chain expansion that more explicitly delineates the supply chain required to produce a sector's output. Because the direct requirements matrix represents the inputs required by any single sector to produce one unit of output, A*y will represent the inputs required in the first tier of the supply chain. A*(A*y) will represent the inputs required for those inputs (the second tier of the supply chain), and so on, as seen in Equation 8:

Equation 8:

$$x = y + Ay + A(Ay) + A(A(Ay))) \ldots = y + Ay + A^2y + A^3 + A^ny$$

which is equal to (I-A)-1y as n approaches infinity. In this way, Equation 8 shows that the total requirements matrix is mathematically equivalent to analyzing a supply chain as far back as is possible (i.e., to infinity). As discussed below, this represents one of the largest advantages of using input-output modeling for LCA, as any errors associated with omitting processes are eliminated.

Environmental extensions

Of course, knowing only the economic supply chain to produce a sector's output is not helpful for LCA. They may help to clarify outcomes such as local economic impact and employment, but they do not model environmental and human health impacts. The model can, however, be extended to environmental LCA using energy consumption and environmental emissions data as described below. Generally such data represents the total energy use, resource consumption, or environmental emissions associated with the sector over the same year as the EIO table. If all emissions associated with productive sectors (e.g., omitting emissions directly attributable to households, such as passenger car emissions and emissions from fuels consumed at home) are attributed to each economic sector, a modeler can express the environmental damage associated with one unit of economic output by dividing total annual emissions by total economic output:

Equation 9: $F = E \cdot (\hat{x})^{-1}$

where E represents vectors of total environmental emissions or resource use (annual greenhouse gas emissions or annual water use), x is total output , and the ^ notation designates diagonalization. Environmental matrix (F) can then be used to convert economic output throughout the supply chain to environmental emissions so that we can evaluate the emissions impact of any amount of production (y):

Equation 10: $Total\ emissions\ (f) = F(I - A)^{-1}y$

ADVANCED MODELS: MULTIREGIONAL MODELS

Input-output models generally describe a national economy, as described in the data section below. However, situations can occur when a practitioner is more interested in different spatial boundaries, such as when analyzing local or global-scale environmental impacts. All the equations presented above describe a single-region input-output model, assuming that all supply chain inputs purchased by a sector are produced in the same economy as the good being analyzed or by the exact same production practices. In other words, single-region models assume that every time imported goods are used as an input for a sector's production, that they are produced by the same recipe. They also assume that they have the same environmental impacts as if the good were produced in the same economy. This may or may not be a good approximation, as environmentally-relevant parameters such as electricity mix, materials recycling rates, and energy efficiency vary considerably in different nations and regions (Wiedmann et al. 2007). This problem is similar to problems encountered with data

representativeness in process LCA.

An alternative, albeit more complicated, model can be derived using many regional or national inputoutput tables, where Equation 7 is generalized for an economy that trades with other economies (Peters and Hertwich 2007):

Equation 11:

$$x = (A_{11} + \sum_{j \neq 1} A_{j1})x + y_{11} + \sum_{j \neq 1} y_{1j} - \sum_{j \neq 1} y_{j1}$$

where A_{11} is the domestic portion of the production function (domestic inter-industry demand on domestic goods), $A_m = \sum_{j \neq 1} A_{j1}$ is the import matrix (domestic use of imports to make domestic output), and y_{11}, $y^m = \sum_{j \neq 1} y_{j1}$ and $y^{ex} = \sum_{j \neq 1} y_{1j}$ represent domestic final demand on domestic production, imports from all countries to final demand in country 1, and exports from country 1 to final demand in all other countries, respectively. This idea can be generalized further to the m-region multiregional case, where each of m countries imports from every other country, to both inter-industry demand as well as final demand:

Equation 12:

$$\begin{pmatrix} x_1 \\ x_2 \\ \vdots \\ x_m \end{pmatrix} \begin{pmatrix} A_{11} & A_{12} & \cdots & A_{1m} \\ A_{21} & A_{22} & \cdots & A_{2m} \\ \vdots & \vdots & \ddots & \vdots \\ A_{m1} & A_{m2} & \cdots & A_{mm} \end{pmatrix} \begin{pmatrix} x_1 \\ x_2 \\ \vdots \\ x_m \end{pmatrix} \begin{pmatrix} y_{11} & {}_{j\,1}y_{1j} \\ y_{21} & {}_{j\,2}y_{2j} \\ \vdots \\ y_{m1} & {}_{j\,m}y_{mj} \end{pmatrix}$$

This shows the relation between total production in each country (x_j,) and final demand in each country, both from domestic production (y_{mm}) and from imports ($\sum_{j \neq m} y_{mj}$). Each y_{j1} represents imports from country j to final demand in country 1, and y_{1j} represents country 1's exports to final demand in all other countries (Wiedmann 2009).

While multiregional models, in theory, solve the issues of assuming identical production practices in all regions, they are much more challenging to create than single region models and have uncertainties of their own, including dealing with multiple currencies and different sectoral structures (Wiedmann 2009). Nevertheless, as discussed in the following section, such models are increasingly being utilized for LCA and carbon footprinting applications.

7.3 DATA SOURCES

The key difference between economic input output and process-based LCA is the background data that are used and, more specifically, the way in which these data are produced. Process-based inventory data are developed through the assembly of information based on engineering considerations and process-specific measurements and/or approximations. The benefit of this bottom-up approach is that specific quantities could hypothetically be validated against real world measurements. EIO data are developed through the assignment of high-level, aggregated data to specific activities in order to represent regional averages. In contrast with process-based inventory data, EIO data are more difficult to validate through first principles. However, this top-down approach has the advantage of providing a clear connection between national- or regional-level totals and specific activities within the economy and society. In this way, an IO approach ensures that an LCA model representing all of the activity within a region will yield the full impacts associated with that region. Such a result is not guaranteed with a process-based inventory. This section describes some of the data sources used in EIO-LCA models and how they are combined to produce a model.

NATIONAL IO TABLES

The starting point of any EIO-LCA model is the economic data set. EIO data are produced regionally, most often at the national level. Because each country, region, and/or study has different objectives, economic structure, unique features, and capabilities, considerable variation exists in the conventions used in different IO datasets. Nonetheless, certain, fundamental aspects are common across IO tables. Section 7.2 discussed the aspects of IO accounts that are fairly common across tables for different regions. This section will indicate some of the ways in which they vary.

The first difference we will likely notice when comparing IO models for different regions is currency. EIO data sets are developed using the regional currency. This simple difference is connected to a number of underlying considerations that are important to take into account when using IO models for LCA. Prices can vary widely across regions and across sectors between regions. Because driving an IO model with inputs based on appropriate monetary values is important, a user should be aware of the local prices represented in an IO model. Converting a price from one region to another by using a currency conversion may not provide a reasonable estimate of the price in a local currency. Accounting techniques have been developed to introduce some consistency to the monetary values represented in different IO accounts. However, even these differ between studies and regions.

Another difference we will likely notice when comparing IO models for different regions is the number of sectors included and the classification scheme used to organize activities into categories. In the Benchmark Input-Output Accounts for the United States, sectors are currently defined by a classification scheme developed specifically for the IO accounts but which is closely related to the North American Industrial Classification System (NAICS). The Canadian IO tables are presented using the Canadian System of National Accounts for which the statistical office provides a correspondence to NAICS. While each European country may choose to define their own classification scheme for national use, Eurostat requires countries to submit IO data using the NACE[23]/CPA classification. Other classification schemes include the International Standard Industrial Classification (ISIC), region-specific systems, such as those described above, for Europe and North America and the associated Central Product Classification (CPC), which are promulgated by the United Nations Statistics Division and the Harmonized Commodity Description and Coding System, also known as the Harmonized System (HS) managed by the World Custom Organization.

Certain classification systems are paired in the above list. A key to sector classification is knowing how to distinguish between sectors. For purchasing and trade considerations, a scheme based on the characteristics of products and services is appropriate. However, for describing industries or other economic activities, the way in which products are produced or services are provided (the technology that is used) is the main concern. Thus, in the list above, NAICS, NACE, and ISIC describe activities or technologies, while CPA, CPC, and HS describe the characteristics of commodities and services. In practice, separating the two in classification systems can be challenging.

Together with classification system, another difference between different EIO tables is the number of sectors used. For LCA purposes, drawing upon tables with a higher degree of sector resolution is desirable. In the United States, the Benchmark Accounts provide roughly 429 industry sectors, plus additional sectors accounting for households, government, exports, and other activities that do not result in products or services sold to other sectors. The IO accounts published by Eurostat provide 65 industry sectors.

Other differences between regions are the release schedule and data quality used to produce IO tables. The US

23 Abbreviated from the French Nomenclature Statistique des activités économiques dans la Communauté européenne.

Benchmark IO Accounts rely on a large amount of underlying data. However, the large amount of work required to integrate these underlying data mean that only every five years is it feasible for the Bureau of Economic Analysis to produce these tables. Thus, by the time the detailed US tables are released, the data are already outdated by some standards. On the other hand, other IO accounts may be produced more frequently, utilizing less original data and relying more heavily on estimation techniques (and usually with fewer sectors). Keeping these differences in pricing schemes, sector resolution, data vintage, and underlying data quality in mind is essential when interpreting the results obtained from different IO models.

7.4 USING EIO MODELS

Section 7.2 showed the theory behind how a practitioner can utilize IO models for LCA. Much like in process LCA, in which the LCA practitioner uses unit processes and elementary flows to calculate the total environmental damage associated with a product in physical units, IOA utilizes economic data and economy-wide environmental emissions data to approximate the environmental damage associated with an amount of output y (an economic unit). Extending the brewery example, whereas a process LCA may estimate the emissions associated with one bottle or can of beer, input-output LCA estimates the emissions associated with $1 worth of output from the brewery sector. After reviewing the data sources and available models for input-output LCA in the previous section, this section provides further explanation on how such calculations are carried out in practice, including how to determine the input value for the model and how to estimate the contribution of different sectors to the total environmental impacts.

PRICES IN INPUT-OUTPUT MODELS

One of the main challenges in using a monetary IO table for LCA is determining the proper monetary input to the model. Ideally, one would easily determine the appropriate value by simply converting the modeled product to economic units (dollars) using an average price. However, prices do not obey the same laws of physics that mass and volume do—they can vary considerably for products or groups of products.

Three different types of valuations (prices) are typically utilized in input-output accounting (Horowitz and Planting 2006):

1. Basic prices: the amount received by a producer from the purchaser of that good or service minus any taxes, and plus any subsidies

2. Producer prices: the basic price received by a producer plus taxes, and minus subsidies on the product

3. Purchaser prices: the price paid by the purchaser of the good, including the cost of delivery (transportation costs) and wholesale/retail markups (margins).

Basic prices are the preferred scheme of the UN System of National Accounts, but some countries use producer prices instead (including the US). An LCA practitioner must know both which the pricing scheme that is used by the input-output table he/she is utilizing and how to convert from a purchaser price (which are the most commonly available prices) to this appropriate price system. Some input-output tables provide information that can be used to perform such conversions (average transportation costs and margins for different consumers of a sector's output), as shown in the example below. One can also use these values to estimate the impact of the gate to consumer portion of the life cycle of products (wholesaling, retailing, and final delivery to purchaser). Both of the US-based input-output LCA models (OpenIO and EIO-LCA) are available free of charge. Both provide models in producer and purchaser prices that allow the user to model life cycles to either the factory gate (producer) or the final household purchase.

To make matters more complicated, prices also vary with time and input-output tables are often produced by statistical agencies many years later; for instance, the 2002 US input-output model was released to the public in 2007 (Stewart et al. 2007). Thus, the practitioner often must inflate or deflate prices over several years. Two sources of data that most countries collect are helpful here: consumer price indices (CPIs) and producer price indices (PPIs). As their names imply, CPIs track changes in prices paid by households for individual commodities and baskets of goods, whereas PPIs track the prices of goods and services received by their producers. A detailed description of CPIs and PPIs is outside of the scope of this chapter, but an example is provided below.

One final issue with prices has to do with aggregation, which will be discussed more generally below. Aggregation is always an issue in input-output LCA, as often many types of products with different production practices are aggregated into a single producing sector. In terms of prices, an analyst must be aware of at least three types of aggregation can exist within a single sector:

1. Different products in the same producing sector: for instance, within the brewery sector dozens of different types of beer are produced

2. Quality differences: some products receive a higher price than others due to real or perceived quality differences

3. Price dependent on customer or quantity: some products may have differing retail, or even basic/producer, prices when sold to different customers or in different quantities; for instance, the same beer may have a different unit price at a corner store versus a large retailer, or when bought in a 6 pack versus a 30 pack

While no easy ways to deal with each of these aggregation issues exists, the analyst must be aware of them when estimating input price for the input-output model. Some countries supply a detailed breakout of the different types of products within a sector as part of their input-output accounts, and these tables can be highly useful for knowing what type of aggregation might be important for the product being modeled. For instance, the detailed item output table in the 2002 US input-output accounts shows that the grain farming sector includes a small amount of barley (2% of output), which is the predominant grain used in brewing, as compared to much larger quantities for corn (73%), wheat (22%), and other smaller contributions from minor grains.

To help manage the pricing problems listed here, we discuss some best practices for determining input values in an EEIO model.

- Utilize the most up-to-date input-output model available to minimize the need for price adjustments over long periods of time
- Use values, if available, from the use table to adjust between retail prices and basic or producer prices, as these adjustments already take into account the aggregation of different products and different retailers
- Estimate the price of the average product within a sector, rather than a single product of interest. For example, input-output can be used to estimate an LCI of the computer (the proper input price is the average computer of that vintage year), rather than a top-of-the-line or base model (Williams 2004). A good way to do this is to divide the sector's total output by the total production of goods or services in a sector in the model year
- Attempt to locate at least two sources of average price data, as any bias in prices will be linearly reflected in the model output

CONTRIBUTION ANALYSIS

Using the above equations, one can use different calculation techniques to determine the contribution of different sectors or different supply chain tiers to the total environmental impacts of a sector's output. For instance, the impacts of different supply chain tiers can be assessed by decomposing the total requirements matrix into the supply chain expansion, as shown in Equation 13:

Equation 13:

$$f = F(I - A)^{-1}y = F(y + Ay + A^2 y \dots) = Fy(I + A + A^2 \dots)$$

Similarly, by using matrix calculations such as diagonalization (a dot product yields the same answer for a nx1 F matrix), an analyst can assess the different impacts of any upstream supply chain portion. For example, to break down the total environmental effects into those attributable to each sector's production, an analyst simply diagonalizes the total output vector. Similarly, to determine the environmental effects embodied in each first tier supply chain purchase (the elements of Table 7.1), one would diagonalize the first tier supply chain purchases (Ay), as shown in Equation 13:

Equation 14:

$$f = F(\widehat{x}) = F((\widehat{I - A})^{-1}y)$$

Equation 15:

$$f = F(I - A)^{-1}\widehat{(Ay)}$$

EXAMPLE: BEER

This example will illustrate many of the concepts and data sources that have been discussed in the previous sections. The goal of the example will be to estimate the life cycle greenhouse gas (GHG) emissions (and total GWP/carbon footprint) associated with the production of beer in the United States, and to determine an appropriate price to compare the results with those from process LCA. This example utilizes two freely available models of the US economy in 2002: EIO-LCA and OpenIO.

The first step in using an input-output model is finding the appropriate sector and examining what makes up the sector's output. In the 2002 US input-output tables, beer is produced by a sector called "Breweries", code 312120. Examining the sector description shows that the sector only makes beer, malt liquor, and nonalcoholic beer, which means that aggregation is likely to be a smaller issue in this sector than others because only three highly similar types of commodities are produced by the industry.

The results obtained from EIO-LCA and OpenIO are shown in Table 7.3 after some analysis. Somewhat confusingly, these models show two different contribution analyses by default, with EIO-LCA showing the producer-based (Equation 12) and OpenIO showing the total consumption-based impacts of the first-tier purchases (Equation 13).

For instance, each row of EIO-LCA's default output represents the direct impact of the sector throughout the entire supply chain ("Power Generation" shows all electricity purchases throughout the supply chain), whereas OpenIO's output shows only the contribution of power generation purchased by the brewery itself. The values for EIO-LCA in Table 7.3 have been changed to match the OpenIO interpretation using Equations 12 and 13 (in Matlab, because this functionality is not available on the website). Further, the two models use different pricing assumptions by default: OpenIO takes input in producer prices and calculates the cradle to consumer total, whereas EIO-LCA's purchaser-based model takes input in purchaser prices and scales to the equivalent producer price (in this case, $1 purchaser prices = $0.502 producer prices, plus $0.26 wholesale, $0.22 retail, and $0.018 in delivery costs, all of which is taken from the 2002 Use Table for sales to households). Thus, to achieve the results shown here, one would input $1.99 million = $1 million/0.502 into the EIO-LCA purchaser price model.

As Table 7.3 shows, the main contributing sectors to carbon footprint (total GWP) in the breweries sector are cradle-to-consumer wholesale, as well as retailing and delivery, packaging (glass, metal cans, and paperboard), energy inputs in the sector itself and in purchased electricity, and grain and malt production. This is similar to process-based LCA results, although the effects of aggregation are clear. The average breweries production is a mix of bottled and canned beer.

To compare these results to process-based LCA results requires determining an appropriate average price (for model year 2002) for a function unit of beer (bottle, can, or case). We show three different methods for estimating this average price:

1. Using the sector's output: An internet search for "total beer production US 2002" leads one to the Brewer's Almanac 2010, which shows that in 2002 198 million barrels of beer were brewed in the US. Converting to L and dividing by the sector's total output in the input-output accounts ($21.5 billion in producer prices) yields an average price of $0.91/L in producer prices, or $0.91/L / 0.502 = $1.82/L in purchaser prices.

2. Using export statistics: Export and import statistics are often useful for estimating average prices, as they are assumed to be equivalent to producer prices (Horowitz and Planting 2006) and are often presented in both physical and economic units. In 2002, the US exported $171 million and 242 L of the commodity "beer made from malt," yielding an average price of $0.70/L in producer prices ($1.41/L purchaser prices).

3. Back of the envelope estimation: Sometimes estimation can yield consistent results and can act as a sanity check on other data sources. Assuming that the average beer produced in the US is sold in 12 packs, the Brewer's Almanac shows this is probably an adequate assumption; around 50% of cans were sold in 6 or 12 packs and 60% of bottles were in 6 or 12 packs, with 90% of all beer sold in cans or bottles. On average, it sold for $7 per 12 pack (the same data source shows average 6 pack price of $4.02 in 2002), which yields an assumption-based estimate of $1.64/L in purchaser prices or

mt/$M	OpenIO	EIO-LCA	EIO-LCA purch
Wholesale/Retail Trade	240		215
Glass container manufacturing	193	147	147
Metal can, box, and other container manufacturing	183	166	166
Breweries	109	76	76
Other Sectors	101	107	107
Power Generation	83	78	78
Delivery	65		50
Paperboard container manufacturing	62	47	47
Grain farming	51	128	128
Input Transport	50	38	38
Flour milling and malt manufacturing	39	64	64
Natural gas distribution	12	15	15
Total Cradle to Gate	884	866	866
Total Cradle to Consumer	1190		1131

Table 7.3. Results for $1 Million (Producer Prices) of "breweries" Production in OpenIO, EIO-LCA Default Model, and EIO-LCA Purchaser Model (g CO2e/$ = tons CO2e/$million)

$0.82/L in producer prices. Thus, three methods have led to prices $0.70-$0.90/L in producer prices. Using this range with the values shown in Table 7.3 (1130-1190 g CO2e/$) yields an overall range of physical unit impact from 800 to 1100 g CO2e/L.

7.5 HYBRID IO-LCA MODELING

IO-LCA models offer several advantages over traditional process LCA, including the speed with which analysis can be conducted once a model is completed, allowing us the ability to easily model baskets of consumed goods rather than single products or functional units. The relative strengths and weaknesses associated with process LCA and IO-LCA have been analyzed (Williams, Weber, and Hawkins 2009). Chapters XX and YY of this book also cover data quality and uncertainties. Generally speaking, these strengths and weaknesses are described below in Table 7.4 (adapted from Minx et al.).

As other chapters have shown, process LCA models can

	Strengths	Weaknesses
Process LCA	• Specificity to product and functional unit • Incorporation of primary data	• Cost • Labor intensity • System Boundary issue ("cut-off error") • Incomplete/unavailable data
IO-LCA	• Fast and low cost • Complete modeling of systems within a national economy • Ability to easily analyze capital goods and services	• High aggregation • Price uncertainty • Lack of use, end-of-life phases • Discrepancies due to international trade • Many materials and processes not specified

Table 7.4. Strengths and Weaknesses of Process LCA and IO-LCA

fully describe the production, use, and end-of-life of any product, if given the collection of primary data or use of secondary proxy data for relevant processes. The downside of this specificity, however, is the relatively high cost and effort involved with performing an LCA, particularly if multiple product lines or goods are involved. This cost and effort can be decreased if good proxy data exists, but this is not always the case. Further, process LCA suffers from the issue of cut-off error, whereby the analyst must cut off certain processes deemed to be unimportant by scoping assessment or product category rules due to time and resource constraints.

On the other hand, producing an IO-LCA is extremely fast once a model is constructed—as the example above shows, the primary time involved is in determining the correct input

price. Further, IO-LCA solves the system boundary problem by definition by including all interactions from cradle to gate within the economy, which is often called system completeness (see equations 7 and 8 above). Finally, a less commonly cited, but important, advantage of IO-LCA is the ability to model services or service inputs to products (e.g., corporate overhead and advertising), which are often ignored in process LCA due to assumed small contributions or difficulty in defining functional units. However, IO-LCA does not deal with use or end-of-life phases of products, as it is only cradle-to-gate or cradle-to-retail in purchaser price models. It uses cost as a functional unit, which is variable and uncertain, and has the critical issue of aggregation of several products into one sector.

One might notice that the strengths and weaknesses of process LCA and IO-LCA are rather complementary. For instance, process LCA has a high degree of specificity but requires significant time and effort investments, while IO-LCA suffers from low specificity (high aggregation) but is fast and inexpensive. IO-LCA can solve the significant system boundary issue of process LCA due to its inherent system completeness. Analysts have noted these compatibilities for some time, and have promoted various combinations of the two types of LCA known as hybrid LCA. The methods of IO-LCA and process LCA can be combined in several different ways, but generally have been classified into three different classes of hybrid LCA approaches (Minx et al.):

1. Tiered hybrid analysis: uses process LCA for critical upstream processes, use phase, and disposal phase, and IO-LCA for remaining upstream processes, including services, overheads, and capital inputs

2. IO-based hybrid analysis: starting with an input-output table, IO-based hybrid disaggregates important sectors using process LCA data to reduce aggregation issues in these sectors of interest

3. Integrated hybrid analysis: The most advanced version of hybrid analysis, integrated hybrid combines the input-output matrix in economic units with a technical

coefficients matrix in physical units to create a single, mixed unit matrix

From a practical standpoint, the tiered hybrid is likely the most commonly used method for hybrid analysis. It represents the simplest method for defining the boundary between inputs using process LCA and input-output LCA. In its most advanced form, tiered hybrid analysis has been called economic-balance hybrid LCA by some (Deng, Babbitt, and Williams 2004), as it utilizes the existing supply chain data embedded in input-output accounts, subtracting inputs covered by process-LCA to achieve a total balance between the portion of input costs covered by process LCA and the remaining value to be modeled using IO-LCA. As an example, Table 7.5 shows a hypothetical economic balance for the beer sector, where the highest impact inputs are subtracted out (and to be modeled using more specific process LCA), leaving in the third column the remaining value to be modeled by IO-LCA. In this case, the major inputs expected to contribute the most to the LCI (containers, grain and malt, and truck transport, which equal 61% of total input costs) are subtracted from total inputs and the remaining inputs (e.g., corporate overheads, wholesale trade, and all other sectors) are modeled by IO-LCA.

IO-based and integrated hybrid analysis represent much more difficult models due to the need to rebalance input-output tables after disaggregation (IO-based) or the construction of mixed-unit models (integrated hybrid). However, some LCA software packages include matrix-based process LCA and IO-LCA modules, allowing the easier construction of these types of models for practitioners.

FURTHER READING

Readers interested in exploring the development and use of EIO-LCA models further will find the following sources useful.

Hendrickson, C., L. Lave, and H. Matthews. Environmental Life Cycle Assessment of Goods and Services: An Input-Output Approach (1st ed.). RFF Press, 2005.

Horowitz, K., and M. Planting. Concepts and Methods of the U.S. Input-Output Accounts. Washington, D.C.: US Bureau of Economic Analysis, 2006.

Miller, R., and P Blair. Input-Output Analysis: Foundations and Extensions. Englewood Cliffs, NJ: Prentice-Hall, 1985.

ACKNOWLEDGEMENTS AND DISCLAIMER

This work has been made possible through the support of the US Environmental Protection Agency. The opinions expressed or statements made herein are solely those of the authors, and do not necessarily reflect the views of organizations mentioned above. This work has not been subject to review by the US Environmental Protection Agency and therefore does not necessarily reflect the views of the Agency. No official endorsement should be inferred.

	Breweries' Total Inputs per $million ($million input/ $million output)	Major Processes (process-LCA)	Remaining value (IO-LCA)
Metal Containers	0.133	0.133	0.000
Glass Containers	0.095	0.095	0.000
Management of Companies	0.058		0.058
Paperboard Containers	0.046	0.046	0.000
Grain Farming	0.029	0.029	0.000
Wholesale Trade	0.028		0.028
Flour and Malt	0.027	0.027	0.000
Truck Transport	0.019	0.019	0.000
Other Sectors	0.140		0.140
Total Inputs		0.349	0.226
Fraction modeled by each system		61%	39%

Table 7.6. Example of Economic-Balance Tiered Hybrid LCA

EXERCISE 1. DEALING WITH DATA VINTAGES

A hybrid LCA is being conducted for a liquid formulation product manufactured in the US. Process data are available to model the mass and energy inputs to the current operations using process LCA methods, but the project scope also includes capital goods and labor. Use the following information to calculate their contribution to the global warming potential for the 5 million kg production per year.

EXERCISE 2. COLLECTING EIO DATA FROM

Item	Cost, $	Year
Tanks and related equipment, installed	35,000,000	1995
Valves and pumps	300,000	1995
Analysis building	450,000	1995
Trucks	100,000	1995
Operating labor (yearly)	750,000	2010
Chemical analysis (yearly)	50,000	2010

Exercise 1

ONLINE

You want to know the impacts of purchasing $1000 of plastic and material resin manufacturing in 2002.

1. Go online to www.eiolca.net
2. Select US 2002 Producer
3. Select Plastics material and resin manufacturers
4. Type in $1000
5. Select TRACI (the site offers TRACI 2.0, not TRACI 2.2)
6. Run the model
What are the outputs in the 13 TRACI impact categories for the $1000 spent?
List by Impact category, value, and unit

EXERCISE 3. EIO LCA OF A PRODUCT

A new US-based company has proposed a route to a bio-based polyurethane and developed a cost estimate for the production of 10 million kg as shown below, which is estimated in 2002 dollars. What is the environmental footprint of the proposed product? Please use the TRACI method to characterize the footprint.

EXERCISE 4. CORPORATE FOOTPRINTS

Many corporations report their direct emissions of greenhouse gases and other materials for which they are directly responsible. Less often reported are values for upstream and downstream emissions related to their products. Using EIO-

LCA, one can estimate the upstream burdens for a company. Using the sales data below taken from the 2010 annual report of a chemical company, estimate the total greenhouse gas

Input	Cost per year, million US$ (2002)
Soy oil	8
Sodium hydroxide (caustic)	1
MDI (methylene diphenyl diisocyanate)	15
Flame retardant	0.5
Pentane	0.3
Electricity	0.2
Heat (from gas boiler)	1

Exercise 3

emissions for this company. The company also reported 29.2 million tons of CO2eq for direct ("scope 1") and 9 million tons of CO2eq for indirect ("scope 2" – for heat and power produced by utility companies and others, but used by the reported company), or a total of 38.2 million tons of CO2eq. What fraction of the total carbon footprint of the company came from its reported scope 1 and 2 emissions?

Market segment	Sales, 2010 (in millions of 2002 US Dollars)
Plastics	$9,587
Chemicals and Energy	$2,893
Hydrocarbons	$4,517
Corporate (ventures, insurance etc.)	$285
Electronics and Specialty	$4,183
Coatings and Infrastructure	$4,453
Health & Ag Sciences	$4,041
Performance Systems	$5,541
Performance Products	$9,049

Exercise 4

EXERCISE 5, COMBINING PROCESS LCI AND EIO LCI

You have been asked to conduct an LCA of a food product. Most of the inputs to the food product can be found in available LCI databases as process LCI. However, a few of the food ingredients are only available as EIO LCI. You estimate that collecting process LCI data on those ingredients by yourself would take too much time, so you decide to combine process LCI and EIO LCI in a tiered hybrid assessment. What information should you clearly document in the Inventory Analysis and Interpretation portion of your final LCA report?

EXERCISE 6. BEST TOOLS FOR THE JOB

You are the owner of LCASrUS, a consultancy focused on life cycle assessment and sustainability issues. You have six projects to assign to your staff, who are all proficient in both process LCA and IO-LCA. Which of those methods will you recommend for each of the projects and why?

1. Client 1 has given you bills of material for two products with the same function and would like your assessment of their environmental life cycle assessments by tomorrow morning.

2. Client 2 has given you bills of material for two products with the same function. The bills of material generally have the same amounts of the broad categories of inputs, but differ in the specific items within each category. They would like your assessment of their potential environmental life cycle impacts.

3. Client 3 needs an LCA to support an environmental product declaration.

4. Client 4 would like to compare two routes to make a new product: one that has very large capital cost and low operating cost with one that has low capital cost and large operating cost.

5. Client 5 wants to develop a better understanding of three products they make to decide if they want to make an environmental claim about any of them. They are a fully back-integrated producer of the products.

6. Client 6 makes a product that is sold globally, but whose end of life is much different in different countries and regions. They want to know the significance of the end of life scenarios on the life cycle attributes of the product

ADVANCED SYSTEM MODELING

DATU BUYUNG AGUSDINATA AND FU ZHAO

LCA has always been about system modeling. The systematic approach to managing data is evident in the interactions among energy, materials, and processes. These relationships are accounted for across all life cycle stages, from material extraction and manufacturing to product use and end of life. This chapter addresses several advanced system-modeling questions:

- Given the product system under consideration, where should we draw the system boundary and how do we determine which processes are required? (Section 8.1)
- Which system response principle should we use to evaluate flows and environmental impacts? Two systemic modeling approaches will be reviewed: attributional and consequential LCA. (Section 8.2)
- When multiple products (co-products or services) result from a single process, how can energy and material inputs, as well as environmental impacts such as emissions and wastes, be allocated to each product? This issue of

multi-functionality will be addressed through allocation and system expansion approaches. (Section 8.3)
- How can one model recycling and remanufacture processes at product end-of-life? (Section 8.4)

8.1 SYSTEM BOUNDARY DEFINITION

As has been explored in previous chapters, two of the most critical steps in the LCA framework are the definitions of the goal and scope of the study. These decisions will determine the system to be analyzed and require a proper setting of the system boundary.

A system boundary is the interface between a product system and the environment system or other product systems (ISO 14044). In establishing a system boundary, one decides which elements to include in the LCA model. Among the elements to be included are the following: life cycle stages, activity types, specific processes, and elementary flows. Setting the system boundary also means that one consciously decides to exclude other elements.

Setting a proper system boundary is essential because an LCA study needs to account for all relevant processes and potential environmental impacts. This is achieved by the inclusion of all relevant and significant processes involved in a product system. In order to have a clear perspective on how to properly establish a system boundary, a

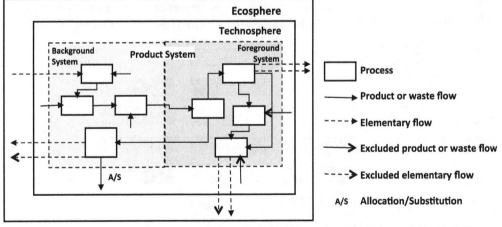

Figure 8.1. Overall Scope of System Definition (adapted from European Commission 2010)

distinction should be made among these four different system concepts: the technosphere, the ecosphere, the foreground, and the background system (European Commission 2010). Figure 8.1 illustrates the overall scope of system definition.

From the perspective of the role played by humans transforming elementary flows, a distinction is made between:

- The technosphere, which represents technological processes and systems developed by humans. It therefore includes both the product system under consideration and other relevant product systems.
- The ecosphere, which encompasses ecological processes and systems. It can be simply considered as "the environment." The ecosphere also physically encompasses the technosphere.

To the product system under consideration, a further distinction can be made about the specificity of data required to define it:

- A foreground system involves the processes of the system that require specific data from technologies and suppliers. For example, a product that is supplied by one or few suppliers using unique facilities belongs to a foreground system.
- A background system, by contrast, comprises processes that use generic industry average data to represent them. For example, in the end-of-life stage, where multiple recyclers exist, the data can be averaged to represent a homogenous market.

Figure 8.1 depicts a product system in which some processes are part of the foreground system and others that are part of the background system. These processes are connected through product and waste flows. The other interaction is between the product system and the other relevant products (i.e., the technosphere). Two kinds of flow occur here. The first flow comes from the technosphere into the product system in the form of excluded product and waste flows. In the other direction, product and waste flows move from the product system to the technosphere, resulting from allocation and substitution calculation (this subject is discussed in more detail in Section 8.3). Finally, interactions occur between the product system and the ecosphere. In addition to product or waste flows, elementary flows come in and out in both directions.

PART-SYSTEM AND SYSTEM-SYSTEM RELATIONSHIPS

Depending on the scope of the LCA as defined by ISO 14044, the system boundary should explicitly include the physical relationship between products of interest and the full system, including other parts/ components. This consideration is especially relevant when conducting product comparisons because it requires a consistent use of the functional unit.

We can distinguish two types of relationships: part-system and system-system. The part-system relationship refers to a subsystem that is a common part of another system and contributes to its function(s). For example, the use of water-saving shower heads results in reduced consumption of water and energy. This secondary effect will need to be accounted for in the use stage of the shower head when comparing its environmental impacts with those of a conventional shower head.

The system-system relationship refers to two independent systems that are related through co-function. For example, a product system such as refrigeration equipment in a supermarket affects the overall climate of the supermarket building. It does so through generated heat (i.e., the co-function). As a result, the heat becomes a co-product during a cold period and a waste in a warm period.

This consideration will be used to decide how to allocate environmental impacts to multiple products. This issue will be discussed in the later sub-chapter on attributional and consequential LCA.

CUT-OFF CRITERIA

In practice, the inclusion of all cradle-to-grave processes in an LCA study is almost impossible. A process to make a cup of coffee is a simple example (Figure 8.2). One will need items

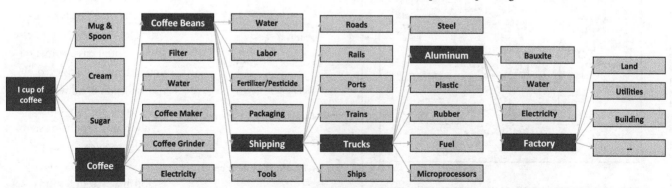

Figure 8.2. Chain of Processes Involved in Making of a Cup of Coffee (H. Scott Matthews of Carnegie Mellon University)

such as coffee, a mug, and a spoon, and possibly cream and sugar. The coffee itself is the product of processes involving coffee beans, a filter, and a coffee maker that runs on electricity. If one is to track and account for all the relevant processes, one could go a long distance up the value chain. One could end up having to determine the contribution of building a factory that smelts the aluminum material used for making the truck that ships the coffee beans. Such contribution to the environmental impacts could be so infinitesimally small that the efforts to collect the life cycle inventory data is not warranted.

To manage this problem, cut-off criteria are used to identify processes that are required within the system boundary. In some cases, the availability of data, time and budget make it necessary to limit the scope of considered flows at some points in the full system. A common approach is to use mass balance to determine the boundary. For example, for a product with hundreds of parts, one may develop a bill of materials and order all the parts by weight. The inventory developed may only consider parts that add up to 90-95% of total weight.

CONSEQUENTIAL AND ATTRIBUTIONAL LCA

In performing LCA, one can consider the environmental impacts of a certain product only within an isolated system. The environmental impacts as the outcomes of the processes are contained in a fixed system boundary. This is called attributional LCA. Alternatively, one can consider the impacts to include repercussions in a broader system beyond the fixed system boundary. This is called consequential LCA, which is based on the idea of expanding the system boundary.

The two LCA approaches differ in several respects. First, attributional LCA considers the immediate physical relationship/ flows normally accounted in terms of materials, energy, resources, and emissions flow. In attributional LCA, the product being analyzed is part of an existing system. A strict attributional LCA is essential for some purposes such as environmental product declarations and process improvement studies.

Attributional LCA performs an accounting of flows within the confinement of closed boundary system, whereas consequential LCA includes changes between alternative product systems. For accounting purposes, attributional LCA evaluates impacts under the control of the producer, while consequential LCA evaluates impacts that are not under the producers' control. By considering demand interaction between alternative products, consequential LCA uses economic data to expand the system boundary beyond immediate processes. Conse-

quential LCA is therefore useful for decisions large enough to perturb economic systems, such as government policy.

ATTRIBUTIONAL LCA

Attributional LCA determines the environmental impact of a product system from cradle to grave. In attributional LCA, the analyzed system generally covers a typical supply-chain of the product. The material flows begin from raw material extraction and end in end-of-life phase.

According to Sonnemann and Vigon, attributional modeling is defined as follows: "A system modeling approach in which inputs and outputs are attributed to the functional unit of a product system by linking and/or partitioning the unit processes of the system according to a normative rule" (Majeau-Bettez et al. 2011).

To illustrate, we compare an LCA of a traditional hybrid electric vehicle (HEV) and an internal combustion (IC) vehicle. An AHEV that combines lithium-ion batteries and an IC engine has a potential to reduce emissions by improving fuel economy. Ground transportation emissions, particularly those from combustion emissions from light duty trucks, are responsible for emitting 17 % of total greenhouse gas (GHG) emissions in the US (U.S. Environmental Protection Agency 2006). Therefore, this is an important issue.

HEVs are already on the market (e.g., Toyota Prius). In addition to the traditional hybrid vehicles, plug-In hybrid vehicles (PHEVs) are variants that can be recharged using the main electricity grid (e.g., Chevrolet Volt). Based on the source of power used, PHEVs sit between an IC-powered vehicle and an electric-run vehicle. They have, in effect, two power systems. In considering the life cycle environmental impacts of PHEVs, one should analyze both the life cycle of conventional automobiles as well as added elements. The life cycle stages of PHEVs involve these additional elements:
- production and end of life of the vehicle
- production, recycling, and disposal of storage batteries,
- production of liquid fuel to power the IC engine
- generation of electricity used to replenish or recharge the batteries

To fully appreciate PHEV environmental impacts, we should take a closer look at each individual product and component that makes up a PHEV (Figure 8.3). The total LCA impacts are the sum of the LCA impacts of the individual components.

The life cycle stages for each constituent product include logistics phases. Two types of logistics apply here: one for the raw materials and the other for the manufactured compo-

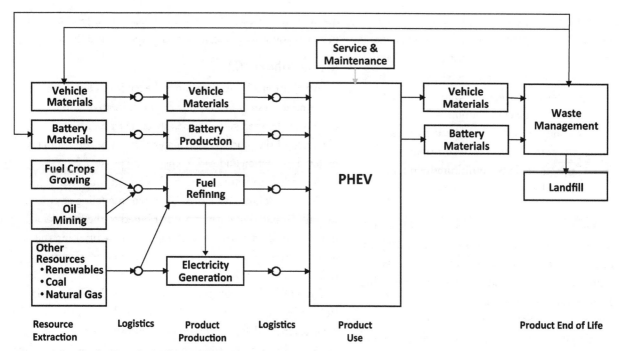

Figure 8.3. Attributional LCA for PHEV Product System

nents. In the production of biofuels, for instance, the logistical activities of the first kind include harvesting, storing, and transporting the energy crops from farming areas to the bio-refinery facilities. The second kind of logistical activities involve transporting and distributing refined biofuels to the points of use.

VEHICLE LIFE CYCLE

PHEVs are built surrounding a conventional IC engine vehicle. The materials used in the vehicle must also be part of the assessment. Measures have been taken to reduce fuel consumption by using lighter materials in the vehicle body design. These materials include polymers, aluminum, magnesium, and various composite materials that are used to replace heavier materials (e.g., iron and steel alloys). At the end of a vehicle service life, the materials will go through a variety of waste management activities to maximally reuse and recover some useful materials. Those materials that are not worth reusing or recovering will be discarded in a landfill.

BATTERY LIFE CYCLE

The LCA of a lithium-Ion battery involves accounting for the life cycle inventory of the battery. The battery is made of electrode substrates and paste. The electrode substrates of lithium ion are made of copper and aluminum foil. The production of copper is responsible for up to 50% of life cycle toxicity and ecotoxicity impacts. The production of electrode paste accounts for more than 97% of ozone depletion potential. The impact in terms of global warming potential, for instance,

can be attributed to the use of energy for battery manufacturing (Majeau-Bettez et al. 2011).

LIQUID FUEL LIFE CYCLE

Liquid fuels can be derived from two sources: fossil- and biological-based fuels. PHEV lifecycle impacts will partly depend on which fuel option is used. The environmental impacts of transportation fuels, both petroleum-based and renewables, have been widely studied (Mclean and Lave 2003). As conventional oil becomes more difficult to extract, alternative sources of liquid fuel are being sought. The quest for alternative fuels has sparked debate about their environmental impacts. Canadian oil sand, for example, has been studied extensively (Brandt 2012). Steam is injected into the sand formation to liquefy the petroleum product and permit pumping. In general, it takes more energy to produce a unit quantity of fuels from this unconventional source than from conventional petroleum oil deposits. More energy is required for resource extraction, as well as for processing and upgrading.

Biofuels have been considered "greener" fuel alternatives. Biofuel sources include oil-producing feedstock (camelina and algae) and lignocellulosic biomass (corn stover, switchgrass, and short rotation woody crops) (Agusdinata et al. 2011). At issue is whether these feedstocks create lower environmental burdens than conventional fuels.

ELECTRICITY LIFE CYCLE

The environmental impacts due to electricity use depend largely on the energy source that produces electricity. Sources such as coal produce much higher carbon footprints than cleaner ones like natural gas (Table 8.1). Electricity generated by renewables, such as hydropower and wind, is desired because it has the lowest carbon footprint.

An electricity generation mix depends on many factors: geography (e.g., coal, geothermal, and hydropower), political strategy (e.g., nuclear power), and market forces (e.g., petroleum imports). For example, in the US, coal is still the largest source of electricity generation (see Table 8.1). Its share, however, is declining, is being replaced, mostly by natural gas and, in smaller proportion, by renewables. Consequently, the overall environmental impacts are improving, though they are still far from ideal.

Electricity source	Share in net generation (percentage)		Carbon footprints (g CO₂ e/ kWh e)
	2005	2010	
Coal	49.6	44.8	863 – 1175
Natural Gas	18.8	23.9	577 – 751
Nuclear	19.3	19.6	60 – 65
Hydropower	6.5	6.3	15
Petroleum	3.0	0.9	893
Other renewables (incl. wind, biomass, geothermal, and solar)	2.1	4.1	Wind: 21 Photovoltaic: 106 Geothermal: 0 - 122
Other sources	0.7	0.4	

Table 8.1. US Electricity Generation Mix (Source: EPA, EIA)

USE PHASE

As in most energy using products, the use phase of a PHEV will dominate the environmental burdens. This is understandable because during the operations over the service life of the vehicle (say 10 years), it will consume a lot of energy, which comes from liquid fuels and electricity. The fact that the vehicle needs to be maintained (tires, spare parts) and serviced (oil change) will add up to the environmental impacts.

OVERALL IMPACTS

Overall, one study finds the following outcomes (Samaras and Meisterling 2008). Compared to conventional IC vehicles, the potential GHG emission reduction can be as great as 32%. However, compared to the HEV version, the potential reduction is small. In addition, the life cycle impacts depend on the distance scenario in which the vehicle is used in full battery mode. The longer the range, the bigger capacity battery is needed (and, hence, the greater the lifecycle impacts).

The nature of the future electricity infrastructure will influence how much greenhouse gas is emitted from PHEVs. As mentioned above, most environmental impacts will occur during the operations of the vehicle. Therefore, if cleaner sources of electricity are used, then the potential of GHG emissions reduction will be greater. The materials and production of storage batteries have a net impact of 2-5% over the total life cycle emissions of PHEVs.

The same conclusion can be also reached for the liquid fuel options. For biofuels, the net environmental impact in terms of CO₂ equivalents depends on factors such as land use change and fertilizer use.

CONSEQUENTIAL LCA

In contrast to attributional LCA, consequential LCA expands the system boundary to take into account selected economic relationships. These relationships can include (a) marginal demand, (b) substitution/rebound effects, (c) economies of scale, and (d) elasticity of supply and demand" (Marshall 1920). Referring to Figure 8.1, in both attributional and consequential LCA, the analyzed system generally covers the impact of changes in the foreground system to the background system. Specifically for consequential LCA, when the economic changes are applied to a product system, other product systems are also affected (see Figure 8.4).

According to Sonnemann and Vigon, consequential modeling is defined as a "system modeling approach in which activities in a product system are linked so that activities are included in the product system to the extent that they are expected to change as a consequence of a change in demand for the functional unit" (Majeau-Bettez et al. 2011).

Figure 8.4 shows conceptually how consequential LCA is made operational by expanding the system. During the production of product A, co-product B is also produced. Co-product B can either be a substitute or competing product of product C. In summary, consequential LCA not only considers the environmental impacts of product A, but also a larger system impact resulting from co-product B which affects the demand for product C and its associated environmental impacts.

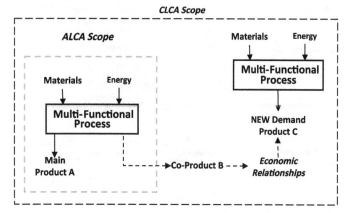

Figure 8.4. Conceptual Consequential LCA Scope through System Expansion

With few exceptions, the use of system expansion can avoid co-product allocation. When allocating emissions and wastes to co-products, one can use different allocation methods that often result in different environmental impacts (further in Section 8.3). Therefore, the system expansion approach can avoid this problematic disparity between allocation methods. By expanding the system boundary, the life cycle impact calculation is altered through adding credits. The credits, such as CO_2 emissions, can be positive or negative depending on how the expanded product system is affected. Each of the economic relationships is examined below.

ECONOMIC MARGINAL DEMAND CONCEPT/ELASTICITY

Marginal change is an important concept in understanding how the consequential LCA works. Marginal change is one of the several mechanisms that are responsible for expanding the system boundaries beyond the ones considered by attributional LCA.

Marginal change refers to a condition where a slight change (increase or decrease) occurs in the production and use of the analyzed product. This marginal change will create a repercussion in the other product systems that are substituted for the analyzed product. For example, in a soybean meal production, soybean oil is a co-product and a substitute for palm-oil and rapeseed oil (Dalgaard et al. 2008) because they provide the same function. When the supply of soybean oil increases slightly, the demand for other oil products decreases as a result. The calculation of the life cycle impact for soybean meal will be altered if production of palm-oil and rapeseed oil were avoided.

System expansion is made operational through a concept of marginal demand by addressing two questions: (1) what technologies or products will be affected by changes in the product system? and (2) how much will they be affected? To identify affected technologies and marginal demand relationship, several questions (Weidema et al. 1999) can be applied:

1. To what time horizon does the study apply?
2. Does the change only affect specific processes or a market?
3. What is the trend in the volume of the affected market?
4. Is there potential to provide an increase or reduction in production capacity?
5. Is the technology the most/least preferred?

As an illustration, we look at electricity production. Referring to the PHEV case in the previous section, electricity used for recharging the batteries can be produced using different technologies such as coal, hydropower, natural gas, and nuclear energy. As more electricity is generated using natural gas and renewables, how can we identify electricity-generating technologies that are affected?

The first question considers whether the effects are short term or long term. Long term effects include the consequences of investing in various technologies. For example, fewer new coal-powered plants will be built as a result of increased natural-gas power capacity, and some existing plants will need to be retired.

The second question looks at whether the electricity market is affected. Due to its intermittent nature, wind electricity, for example, will affect only the off-peak and not the baseload production capacity.

The third question deals with market trends as based on historical data generated through statistical time-series analysis. Is the overall demand increasing or decreasing? Consider a situation in which electricity demand increases at a faster rate than the replacement of old power plants. In this case, technology that is more likely to be phased out, such as coal, will still be used. By contrast, when electricity demand decreases at a slower rate than the replacement rate of existing production capacity, new and more expensive technologies, such solar and wind, will be affected. One can expect that under such circumstances, the share of renewables will increase.

The fourth question examines whether technologies are constrained in their ability to adjust to changing demand. For example, the availability of hydropower will be limited by the amount of water that is available in a particular region.

The fifth question explores the extent to which a technology will be adopted. The decision will typically be based on the unit production cost. The technology that will eventually be phased out is the one with highest short-term costs. For example, in a case when a carbon tax is implemented, coal-fueled power plants will be the costliest among technologies and therefore the most likely to be phased out.

ECONOMIES OF SCALE

Economies of scale occur when high production volumes result in lower product prices. This, in turn, stimulates more demand for the products. Generally, the more products are delivered, the more efficient the production process becomes. This improvement is achieved through, labor efficiency, better use of equipment, and standardization of processes and procedures. Such a relationship is captured in a learning curve (de Wit et al. 2010). A general form of learning curve is given as

$$Y = aX^b \quad (1)$$

where Y = the cumulative average cost per unit product, X = the cumulative number of units produced, a = cost required to produce the first unit, and b = log of the learning rate/log of 2. As the unit cost declines, it will be reflected in the product price. The additional demand resulting from the reduced price will be governed by the demand- price elasticity function, e:

$$e = \frac{\Delta D/D}{\Delta P/P} \quad (2)$$

where D = initial demand, P = initial price, ΔD = change in demand, and ΔP = change in price. Some products have very low demand elasticity, at least in the short term. For example, in cold climates during winter, the demand for natural gas is driven by heating-degree days. Even if the price is high, people will pay for the fuel needed to heat their homes. In the long term, this demand may change as people install better insulation or more efficient heating systems (e.g., heat pumps).

REBOUND EFFECT

The rebound effect refers to the condition in which certain measures, such as more energy efficient technologies, have unintended consequences that are opposite to the original intent. In such conditions gains in efficiency are offset by increased demand, resulting in an overall increase in undesirable impacts (e.g., carbon emissions).

Rebound effects can be direct or indirect. Each type follows a different chain of cause and effect. In general, the direct effect (otherwise known as the substitution effect or the pure price effect) is due to gains in energy efficiency that result in lower costs; in turn, this leads to higher demand and usage. The indirect effect (otherwise known as the rebound effect or the secondary effect) stems from higher efficiency that result

in lower costs, which, in turn, leads to savings. As a result, consumers have more money to spend on other products (Hofstetter and Norris 2003).

One study reveals the rebound effect of improvements in vehicle fuel efficiency to overall transportation demand—given in vehicle-miles traveled, or VMT, per year—and, hence, in energy consumption (Small and Van Dender 2007). Under normal circumstances, one would expect an increase in vehicle fuel efficiency to proportionately reduce overall transportation energy consumption. However, it has been observed that that is not what actually happens. Because travelling has become cheaper, travelling more frequently and/or across longer distances is incentivized. The net impact is not as large as normal driving patterns would suggest.

The extent to which the reduction deviates from the proportionality is considered a rebound effect. For instance, the study estimated that the short- and long-run rebound effects are 4.5% and 22.2%, respectively. This means that a 1% decrease in per-mile fuel cost would increase VMT by about 4.5% in the short term (Small and Van Dender 2007).

ROLE OF ECONOMIC MODELS IN CONSEQUENTIAL LCA

In modeling the indirect impacts in LCA, one can make use of economic models called partial equilibrium (PE) and computable general equilibrium (CGE) models. The details of these models (Earles and Halog 2011) are beyond the scope of this section. Originally, both of these models were used to model policy impacts by determining the equilibrium position of one or more markets. The difference is that PE models are based on an assumption of maximizing net social payoffs. Alternatively, CGE models use an assumption of maximizing agents and, unlike PE models, include all sectors within an economic system. CGE models, in particular, support consequential LCA in identifying and quantifying economic and/or contractual relations of affected sectors or markets.

CONSEQUENTIAL LCA EXAMPLES

An example of marginal demand can be found in an electricity production system. In a country like Norway, for example, around 99% of the electricity produced comes from hydropower. As the country intends to protect its environment by bringing back water streams to their natural states, its hydropower capacity is constrained. It cannot grow to keep up with rising demand. The alternative source of electricity will have to come from those that face no growth constraints, such as biomass or fossil fuels, as Norway is a big exporter of oil. A consequential LCA of an energy system must make sure that it excludes constrained sources in the electricity mix.

Another consequential LCA application example is the study on the environmental impacts of lead free solder (Earles and Halog 2011). Electronics manufacturers use solder pastes to provide electrical interconnections between electrical components and a printed circuit board (PWB). The use of solder paste containing lead was banned for use in products in the European Union (EU). Some technical aspects and considerations in assessing the environmental burdens of two kinds of solder are as follows:

- Two kinds of solder are compared: SnPb solder (62% tin, 36% lead, 2% silver) to a lead free solder (95.5% tin, 3.8% silver, 0.7% copper).
- The functional unit is the volume of solder paste needed to mount components on to a specific, common PWB.
- The choice of solder has little impact on the energy demand of the electronic product. Therefore, the use phase of electronic product is excluded from the analysis.

Now to analyze how the relevant markets are affected by the solder material switch. For the purpose of this chapter, we keep this analysis in qualitative terms on some of the relevant aspects.

At least two markets will be affected: (a) the alternative use of lead and (b) the market for virgin-lead metals. Alternative uses of lead have been identified. They include lead-acid batteries, leaded glass in television screens, lead sheet in the home construction industry for roofing and wall cladding, lead pipes in the chemical industry, and ammunition. Among all these, batteries consumed about 74% of lead worldwide in 1999.

What would happen to the demand of these products when the price of lead changes? If the lead price decreases, the price of lead-acid batteries would also decrease. As a result, the demand for these batteries would increase at the expense of other competing batteries. A lower lead price may also reduce the incentive to develop a better alternative and more environmentally friendly product such as the lithium-ion battery (refer back to the PHEVs example in the previous section).

The same effect of lead price changes will take place for other products such as lead sheet use in home construction. The impact of changes may vary. In the case of batteries, the price change impact can be great (high demand in price elasticity). In other products where the cost of lead is just a small fraction of the total costs, the impact might be small. In the same way, when lead has functional advantages compared to other materials, the demand for lead will be less sensitive to price changes.

Next is the lead metals market. The changes in the first market can be estimated by knowing how sensitive the supply

and demand of lead is to price fluctuations. Price changes can have a short- and long-term effect. In the short term, a price change has little impact on demand because of the difficulty of finding another material to substitute for lead (in cases of a price increase) or using lead as a substitute for other materials (in cases of a price decrease). In the long term, however, the demand will become more sensitive to price changes because alternative products that use lead (e.g., batteries, roofing and wall cladding) can replace competing products.

With regards to the supply of virgin lead metals, a different impact may result. Lead mining companies cannot react quickly to changing prices in the short term, so the elasticity of supply can be expected to be low. The long term elasticity is also expected to be low because lead is mostly a co-product of zinc and other metals. Overall, research established that "for each tonne of lead that is eliminated from the solders, the production of lead is reduced by 0.5 tonne and the use of lead in other products is increased by 0.5 tonne" (Earles and Halog 2011).

CONSEQUENTIAL LCA AND ATTRIBUTIONAL LCA COMPARATIVE CASE STUDIES

In this section, we contrast the two approaches by comparing the outcomes and conclusions that result from both attributional LCA and consequential LCA implementation using the lead free solder case study.

In a comparative study of the use of lead free solder, the closed supply chain of solder paste production and its use is modeled on both attributional LCA and consequential LCA (Ekvall and Andrae 2006). Attributional LCA shows that elimination of lead solder will result in a slight increase in GWP impact. In the case study, consequential LCA reveals that reduced use of lead in solder will cause an increased use of lead in batteries and other products. These results may, in effect, offset the overall global warming potential (GWP) impact because lead use in batteries produces less lead emissions. Care must be taken, however, in evaluating models of consequential analysis. In this example, one assumes that more lead batteries will be produced. However, lead batteries are primarily used in automobiles and replaced approximately every five years when internal chemical reactions "kill" the battery. The demand for lead batteries is therefore tied more to the number of automobiles on the road than to the cost of lead.

In summary, attributional LCA and consequential LCA differ in the kind of insights they provide and, hence, may reach different conclusions. Because attributional LCA models the average technology, whereas consequential LCA the marginal technology, environmental impacts obtained may be quite different. An attributional LCA approach often requires the use of an allocation procedure, which can greatly affect the results. In the consequential LCA approach, the allocation procedure is not needed. The approach can provide insights into indirect environmental consequences, which are important for decisions that have widespread economic consequences.

8.3 Dealing with Multi-Functionality

Many industrial processes produce more than one product or deliver more than one service. For example, the soybean-crushing process produces crude soybean oil and soybean meal (Pradhan et al. 2011). Depending on the technology used, there might also small amounts of soap produced (Pradhan et al. 2011). Another example is a modern crude oil refinery that may produce more than ten different products (Sheehan et al. 1998). These products include liquefied petroleum gas (LPG), gasoline, naphtha, kerosene and jet fuels, diesel fuel, fuel oils, lubricating oils, paraffin wax, asphalt and tar, petroleum coke, and sulfur. In addition to these products, oil refineries also produce intermediates for chemical industries. The multi-functionality can also show up on the input side. For example, a waste disposal process can handle two or more waste streams (e.g., waste paper and waste plastics). Although one can use a vector to represent a multifunctional process when using the unit-process approach, the multi-functionality issue has to be properly addressed before a complete life cycle inventory can be compiled and calculated. In the matrix-based approach, failing to do so will lead to a technology matrix that is not invertible. The sections below show several different approaches, including allocation, displacement, and disaggregation.

Allocation procedures

Allocation, also known as partitioning, splits a multifunctional process into a number of independent, monofunctional processes. The original unit process will be divided into several unit processes, with each delivering only one product or service. The splitting or allocation is usually done by multiplying the process inputs and outputs with a set of allocation coefficients. Mathematically, this can be done by:

where $\mathbf{p}_0$ is the original multi-functional process, $\mathbf{p}_1$, $\mathbf{p}_2$, ..., $\mathbf{p}_n$ are monofunctional processes after allocation, $\mathbf{a}$ represents material or energy input to a process, $\mathbf{e}$ represents emission or resource consumption, and $\mathbf{y}$ represents process function.

The allocation coefficients are usually chosen such that they lie between 0 and 1.

Because $\mathbf{p}_0 = \sum_{k=1}^{n} \mathbf{p}_k$ as required by mass and energy conservation, this results in:

$$\sum_{k=1}^{n} \lambda_{ki} = 1, \quad \forall i = 1, ..., m$$

$$\sum_{k=1}^{n} \eta_{kj} = 1, \quad \forall j = 1, ..., n$$

For simplicity, we have:

$$\lambda_{k1} = \lambda_{k2} = \cdots = \lambda_{km} = \eta_{k1} = \eta_{k2} = \cdots = \eta_{kl}, \quad \forall k = 1, ..., n$$

According to ISO 14044, allocation should be performed by following these steps:

Step 1. Allocation should be avoided if (i) a multifunctional process can be divided into two or more sub-processes and input/output data for these sub-processes can be collected or (ii) the product system can be expanded to include all the functions provided by the multi-functional process. However, the latter means that the system boundary has to be redefined and all the requirements on defining system boundary applied.

Step 2. If Step 1 is not possible to carry out, allocation should be done in a way that reflects the underlying physical

$$\mathbf{p}_0 = \begin{pmatrix} a_1 \\ a_2 \\ \vdots \\ a_m \\ y_1 \\ y_2 \\ \vdots \\ y_n \\ ---\\ e_1 \\ e_2 \\ \vdots \\ e_l \end{pmatrix} \quad \mathbf{p}_1 = \begin{pmatrix} \lambda_{11}a_1 \\ \lambda_{12}a_2 \\ \vdots \\ \lambda_{1m}a_m \\ y_1 \\ 0 \\ \vdots \\ 0 \\ ---\\ \eta_{11}e_1 \\ \eta_{12}e_2 \\ \vdots \\ \eta_{1l}e_l \end{pmatrix} \quad \mathbf{p}_2 = \begin{pmatrix} \lambda_{21}a_1 \\ \lambda_{22}a_2 \\ \vdots \\ \lambda_{2m}a_m \\ 0 \\ y_2 \\ \vdots \\ 0 \\ ---\\ \eta_{21}e_1 \\ \eta_{22}e_2 \\ \vdots \\ \eta_{2l}e_l \end{pmatrix} \quad \cdots \mathbf{p}_n = \begin{pmatrix} \lambda_{n1}a_1 \\ \lambda_{n2}a_2 \\ \vdots \\ \lambda_{nm}a_m \\ 0 \\ 0 \\ \vdots \\ y_n \\ ---\\ \eta_{n1}e_1 \\ \eta_{n2}e_2 \\ \vdots \\ \eta_{nl}e_l \end{pmatrix}$$

relationships. That is, if one changes the output of the product system considered (the combinations of products or functions), the changes should be reflected by changes on inputs/outputs of sub-processes.

Step 3. If Step 2 cannot be performed, allocation should be done based on other relationships (e.g., economic value).

MASS-BASED ALLOCATION

One potential way to determine the allocation coefficients is to use the mass or weight of the products coming out a multi-functional process. Taking soybean crushing process as an example (Ekvall and Andrae 2006), per 1 kg of soybean processed, an average of 0.17 kg of soybean oil and 0.76 kg of soybean meal are produced. The process consumes 0.07 kWh of electricity, 2MJ of natural gas, and 2 g hexane per kg soybean. In addition, per kg soybean the process emits 1.72 g hexane to air and generates 78 g wastewater with 0.85g soybean oil and 0.83 g triglycerides. The process information can be put into a table below (assuming natural gas has an emission factor of 0.093 $kgCO_2/MJ$):

	Substance	Quantity	Unit
Inputs	Soybean	1	kg
	Electricity	0.07	kW-hr
	Natural gas	2	MJ
	Hexane	2	g
Products	Soybean oil	0.17	kg
	Soybean meal	0.76	kg
Emissions	Hexane to air	1.72	g
	Wastewater	78	g
	Carbon dioxide	0.186 (=0.093 * 2)	Kg

With matrix-based approach, the following vector applies:

$$p_0 = \begin{pmatrix} -1 \\ -0.07 \\ -2 \\ -2 \\ 0.17 \\ 0.76 \\ --- \\ 1.72 \\ 78 \\ 0.186 \end{pmatrix} \text{ with linear space } \begin{pmatrix} \text{soybean, kg} \\ \text{electricity, kW hr} \\ \text{Naturalgas, MJ} \\ \text{Hexane, g} \\ \text{soybean oil, g} \\ \text{soybean meal, g} \\ --- \\ \text{hexane to air, g} \\ \text{wastewater, g} \\ \text{carbon dioxide, kg} \end{pmatrix}$$

Using mass-based allocation, the allocation coefficients are:

$$\lambda_{oil} = \frac{m_{oil}}{m_{oil} + m_{meal}} = \frac{0.17}{0.17 + 0.76} = 0.18$$

$$\lambda_{meal} = \frac{m_{meal}}{m_{oil} + m_{meal}} = \frac{0.76}{0.17 + 0.76} = 0.82$$

where m_{oil} and m_{meal} are weight of soybean oil and soybean meal produced out of 1 kg of soybean. The two mono-functional unit processes after allocation are represented in the tables below.

For soybean oil:

	Substance	Quantity	Unit
Inputs	Soybean	1*0.18=0.18	Kg
	Electricity	0.07*0.18=0.013	kW-hr
	Natural gas	2*0.18=0.36	MJ
	Hexane	2*0.18=0.36	G
Products	Soybean oil	0.17	Kg
Emissions	Hexane to air	1.72*0.18=0.31	G
	Wastewater	78*0.18=14	G
	Carbon dioxide	0.186*0.18=0.033	Kg

For soybean meal:

	Substance	Quantity	Unit
Inputs	Soybean	1*0.82=0.82	Kg
	Electricity	0.07*0.82=0.057	kW-hr
	Natural gas	2*0.82=1.64	MJ
	Hexane	2*0.82=1.64	G
Products	Soybean meal	0.76	Kg
Emissions	Hexane to air	1.72*0.82=1.41	G
	Wastewater	78*0.82=64	G
	Carbon dioxide	0.186*0.82=0.15	Kg

Or in matrix form, as:

$$p_{oil} = \begin{pmatrix} -0.18 \\ -0.013 \\ -0.36 \\ -0.36 \\ 0.17 \\ 0 \\ --- \\ 0.31 \\ 14 \\ 0.033 \end{pmatrix} \text{ and } p_{meal} = \begin{pmatrix} -0.82 \\ -0.057 \\ -1.64 \\ -1.64 \\ 0 \\ 0.76 \\ --- \\ 1.41 \\ 64 \\ 0.15 \end{pmatrix} \text{, with linear space } \begin{pmatrix} \text{soybean, kg} \\ \text{electricity, kW hr} \\ \text{Natural gas, MJ} \\ \text{Hexane, g} \\ \text{soybean oil, g} \\ \text{soybean meal, g} \\ ------ \\ \text{hexane to air, g} \\ \text{wastewater, g} \\ \text{carbon dioxide, kg} \end{pmatrix}$$

ENERGY-BASED ALLOCATION

Given the heating value of soybean oil (37 MJ/kg) and soybean meal (19 MJ/kg) [18], one can perform an energy-based allocation:

$$\lambda_{oil} = \frac{E_{oil}}{E_{oil} + E_{meal}} = \frac{0.17 * 37}{0.17 * 37 + 0.76 * 19} = 0.30$$

$$\lambda_{meal} = \frac{E_{meal}}{E_{oil} + E_{meal}} = \frac{0.76 * 19}{0.17 * 37 + 0.76 * 19} = 0.70$$

Energy-based allocation is more suitable for crude oil refining than mass-based allocation. However, energy-based allocation is not recommended for soybean meal and soybean oil because these are typically used as animal feed instead of fuel.

PRICE-BASED ALLOCATION

Using a physical unit (e.g., mass or energy) as a criterion for allocation may seem to be arbitrary under some scenarios, such as when the product and co-products serve different markets. In the case of soybean crushing, however, if the soybean oil is eventually used to produce biodiesel, one can argue both mass-based and energy-based allocation approaches have their own flaws. It could become even more challenging when the product and co-products consist of both energy and materials.

For example, in many countries fly ash from a coal-fired power plant can be sold as construction materials, with prices ranging from $10/ton to $40/ton. In this scenario, neither mass-based allocation nor energy-based allocation can be applied. To address the issues associated with physical unit based allocation, one can use price or revenue as the criterion. One may argue that economic allocation is more logical, as almost all processes exist because they generate revenue from selling products or services that the processes deliver.

Another benefit of using economic allocation is that the method can be used across the supply chain for any product, service, or process, and therefore meets ISO14040/4 requirements for consistency. This consistency may not achieved if physical unit-based allocation approaches are used, as mixed physical units are often used to deal with energy and material flows in and out of multifunctional processes.

Returning to the subject of soybean crushing, the current market prices of soybean oil and soybean meal are $1.1/kg and $0.49/kg, respectively. The allocation coefficients based on revenue can be determined as follows

$$\lambda_{oil} = \frac{R_{oil}}{R_{oil} + R_{meal}} = \frac{0.17 * 1.1}{0.17 * 1.1 + 0.76 * 0.49} = 0.33$$

$$\lambda_{meal} = \frac{R_{meal}}{R_{oil} + R_{meal}} = \frac{0.76 * 0.49}{0.17 * 1.1 + 0.76 * 0.49} = 0.67$$

A potential drawback of economic allocation is that prices of the product and the co-products change over time so the allocation coefficients are not constant. For example, in the past year soybean oil prices have stayed in the range of $1.1/kg to 1.23/kg, while soybean meal prices have changed radically. The lowest price was $0.32/kg in December 2011 and the highest was $0.59/kg in August 2012, meaning that the price almost doubled. The allocation coefficients vary accordingly, which increases LCA complexity. For some products the prices change from region to region, which further complicates the issue. For example, in UK the gasoline and diesel prices are close to one other while in the U.S. diesel sells at 20% higher. In practice, temporal and spatial average prices are commonly used as the basis for economic allocation.

DISPLACEMENT AND SUBSTITUTION

In some scenarios, the co-products from a multi-functional process can displace products from other stand-alone unit processes. For example, when dry milling technology is used to produce ethanol from corn grains, per bushel corn grain processes totaling 2.78 gallons of anhydrous ethanol are produced. In addition to ethanol, 14.7 lbs. of DDGS (moisture 11%) and 5.98 lbs. of WDGS (moisture 57%) are produced as co-products (Mueller 2010).

In the US, both DDGS and WDGS with different inclusion levels are fed to animals to displace corn, soybean meal, and urea in their diets. On average, 1 lb. of DGS can displace about 0.8 lb. of corn, 0.3 lb. of soybean meal, and 0.02 lb. of urea (Wang et al. 2011). Therefore, in order to determine the environmental credits from the production of co-products (or avoid the burden due to this displacement), one has to expand the system boundary to include soybean crushing (as reviewed earlier, this is also a multi-functional process) and urea synthesis. For this reason, the displacement/substitution method is also called system expansion. Mathematically, this can be expressed with the following equation (which designates vector **h** as the environmental impacts of a process):

$$h_{major\_product} = h_{multifunctional\_process} - \sum_{k=1}^{m} h_{kth\,coproduct}$$

In the case of dry mill corn ethanol production, this becomes

$$h_{ethanol} = h_{dry\_milling} - h_{corn} - h_{soybean\_meal} - h_{urea}$$

When allocation is used, the mass and energy balance must be maintained across all the derived mono-functional unit process. From this perspective, the displacement method is disadvantageous because some of the flows can become negative, which is a physical impossibility. Moreover, the outcome of system boundary expansion is that the system has mixed functions and functional units (e.g., kg ethanol for energy, plus kg feed for animals).

Although ISO14044 recommends system expansion over allocation, doing so is not always possible. For instance, in the case of crude oil refining, gasoline and diesel are the two major products (along with other products). Gasoline is the dominant fuel for passenger vehicles and light-duty trucks, while diesel is the dominant fuel for heavy-duty trucks and agricultural machinery. We cannot identify a product being displaced or substituted by gasoline or diesel on a large scale. In many cases, identifying avoided products are requires an analysis on material flows in the economic system. This is especially challenging when dealing with emerging technologies.

PROCESS DISAGGREGATION

According to ISO 14044, efforts should first be made for a multi-functional process to disaggregate the process into independent processes based on mechanisms that govern the process. Allocation and displacement should only be used when it is not possible or feasible to carry out process disaggregation. To disaggregate a multifunctional process, we need more knowledge about the process. An example is the automotive coating process, which is carried out in a paint shop. The car body goes through a series of cleaning, coating, and curing processes, and the coated car bodies can be of different colors.

With an allocation approach, one may partition the material and energy consumption, as well as pollutants generated based on the number of car bodies in different colors. If we take a deeper look into the paint shop, we may find that different colors are the result of different paints applied during the coating process. After collecting data about consumption of paints of different colors, we will be able to disaggregate the process into several processes. Of course, these disaggregated processes share many common steps (e.g., cleaning and curing).

Although we can avoid problems created by displacement and allocation, process disaggregation clearly requires more time, money, and labor. In addition, multi-functional processes exist in which production of the main product and co-products are inherently dependent on each other to the extent that disaggregation is not possible. One example is the Chloralkali process, in which electrolysis of sodium chloride solutions yields chlorine and sodium hydroxide as the products.

8.4 MATERIAL RECYCLING

According to the US EPA, recycling is the process that "minimizes waste generation by recovering and reprocessing usable products that might otherwise become waste (recycling of aluminum cans, paper and bottles)." In most scenarios, material recycling deals with post-consumer goods (the whole of a product, or a part of it, that has served its intended use). When a material is recycled, it could displace virgin material consumed by manufacturing processes. This displacement avoids environmental burden associated with producing the virgin materials as well as those caused by waste landfill or incineration) If the used products can be remanufactured to have the same technical performance as a new product, an even larger degree of environmental burden can be avoided.

Material recycling processes can be broadly classified into two different categories: open-loop and closed-loop. In closed-loop recycling, used products are processed to become materials that can be used in producing the same product. The material properties are not changed in comparison to the original primary material. In open-loop recycling, the recycled materials will be used to produce other products. For example, in the US, aluminum or steel cans are recycled to make new cans, representing a closed-loop recycling process.

Often, the inherent material properties are changed to such an extent that the recycled cannot be used in its original system. For instance, plastic products, when recycled, often do not generate plastics of the same quality as virgin materials and cannot be used to produce the same type of products, thus representing open-loop recycling. In reality, for a post-consumer product, closed-loop and open-loop recycling processes coexist. That is, a portion of the recycled materials are used to produce the same products, while the remaining is used for other purposes. One example is polyethylene terephthalate (PET) bottles. Recycled PET bottles are used to produce polyester fibers (which are then used to produce clothing, pillows, and carpets), polyester sheets, or strapping, in addition to making new PET bottles.

The recycling process (producing recycled material from post-consumer scraps) creates its own environmental impacts due to the materials and energy consumed as a result of steps such as collection, transport, sorting, grinding, washing, and melting. Therefore, we need to consider both the avoided burden due to displacement of virgin materials and avoided landfill, and the burden generated during recycling process when determining the environmental impacts of recycled materials. In addition, post-consumer scraps often have market value, which suggests that a portion of the environmental impacts associated with the product should be allocated to the recycled materials. All of these create some interesting issues when dealing with material recycling, especially in the case of open-loop recycling.

CLOSED-LOOP RECYCLING

For closed-loop recycling, the displacement/substitution method is often used. Because the recycled materials have the same properties as the virgin materials, one can argue that, due to the use of recycled materials, the need for virgin materials is reduced. Usually, the environmental impacts associated with closed-loop recycling operations are smaller than those associated with the production of virgin materials. In addition, via closed-loop recycling environmental impacts due to end-of-life treatment (e.g., incineration and landfill) can be avoided. This way, using recycled materials will bring environmental credits instead of incur environmental burden. It will also encourage recycling and use of recycled materials.

In a broad sense, we can consider remanufacturing as a special case of closed-loop recycling. Although additional efforts are required for remanufacturing, environmental impacts associated with product manufacturing, including material loss during machining, are avoided. Remanufacturing could be expected to bring higher environmental credits than closed-loop material recycling. However, if the remanufactured products suffer an efficiency loss and the products consume significant amount of energy during use phase, remanufacturing may be a poor choice from a life cycle perspective.

The displacement/substitution method is suitable when a product has a low recycling rate (most of the post-consumer products are treated as wastes). LCAs have demonstrated the environmental benefits of closed-loop recycling. If most of the post-consumer products instead go through closed-loop recycling as a common practice (as in the case of aluminum cans), a different approach is needed. In an ideal case, the original materials can be used infinitely so that the contribution to total environmental impacts from original material extraction and processing is minimal.

In reality, the recycling rate is always lower than 100% because of material losses during the recycling process. This requires the addition of virgin materials to the cycling loop. The recycled materials stay in the same production system so that no system expansion is needed (except unit processes involved in recycling steps). Mathematically, we can simply add

these unit processes to the original technology and intervention matrices. No allocation or displacement is needed and the resulting technology matrix is invertible.

OPEN-LOOP RECYCLING

The case of open-loop recycling is more complicated because the recycled materials are used to manufacture different products than the products that previously contained the recycled material. Recycled materials are sometimes of low quality due to mixing with, or contamination by, other materials. Among the challenges LCA analysts face is allocating the environmental burdens due to the original material extraction process and recycling process among the products produced at a series of life cycles. Current ISO standards do not explicitly address the issues in open-loop recycling. If the system expansion approach recommended by the ISO standard is adopted, one will have to consider all production systems affected by the flow of recycled secondary materials, in addition to the original production system. This means that the significant efforts might not be feasible due to resource constraints. As alternatives to system expansion, allocation methods have been proposed for open-loop recycling [foreground system, 25]. Below are the most commonly used approaches:

CUT-OFF METHOD

In this method, each product should be only assigned environmental impacts directly caused by that product. The product, which is manufactured using virgin materials, is responsible for all burdens from obtaining the virgin materials. The product using recycled materials is responsible for all the impact of recycling that material. Disposal can occur at any step of the process and at the end of life are assigned to the product, while recycling at any step of the process is assigned to the next product life cycle.

LOSS-OF-QUALITY METHOD

This method considers the loss of quality during the multiple life cycles of the virgin materials. After-use materials degrade and recycling processes are needed to make the materials usable again. Instead of assigning all burdens associated with virgin materials to the first product in the life cycles, one can argue that the first product is only responsible for part of the burdens corresponding to quality loss (plus burdens due to recycling). The last product in the life cycles will also be assigned the burden from waste management. In this approach, quality of material has to be determined. A common approach is to use material pricing data as a proxy.

A slightly different version of this method is to combine

environmental burdens due to virgin material production and final waste management, and then distribute them based on quality loss. This method is used in LCAs for regional or global average materials. It provides an average impact of a material irrespective of its source. This method is criticized because it does not reward either the reuse or recycling of the material.

CLOSED-LOOP METHOD

This method does not differentiate among products along the life cycles. Environmental impacts due to virgin materials production, recycling steps, and final waste management are evenly distributed to products produced along the material life cycle.

50/50 METHOD

This method attributes the impacts of producing virgin materials and waste management equally to the first and last products in the material life cycle. The same is done for recycling processes in the middle (burdens associated with recycling are assigned equally to the upstream and downstream product). The rationale is that supply and demand for recycled materials are both necessary to enable recycling. The 50/50 allocation is arbitrary.

PROBLEM EXERCISES

1. We discussed an attributional LCA for plug-in hybrid electric vehicles (PHEVs). Consider the life cycle stages for hydrogen cars versus PHEVs. Draw a similar life cycle stage diagram as illustrated in Figure 8.3 for a vehicle that is fueled by hydrogen produced from different energy sources.

2. Consider the possibility of configuring a PHEVs LCA with different options of internal combustion (IC) engine, battery types, and electricity source. Taking into account the unit emissions data provided in Table 8.1 and other sources of information:

   a Describe the best scenario of PHEV configuration with the least environmental burdens.

   b Describe the worst scenario of PHEV configuration with the most environmental burdens.

3. Use the five questions from Wiedema et al (1999) to determine affected technologies due to new electricity generation sources such as solar and wind energy.

4. Consider the consequential LCA from orange juice production with a by-product of orange peels that are typically used as animal feed. Determine affected markets and marginal suppliers.

5. A solvay process is used to produce soda ash, with calcium chloride as a co-product. With 1.5 kg sodium chloride and 1.2 kg limestone, 1 kg of soda ash and 1 kg of calcium chloride can be produced. 0.2 kg natural gas is burned for the heat that is required by the process. Ignoring other process inputs, develop an inventory for soda ash using mass based allocation. Repeat the process using price-based allocation if it is known that calcium chloride sells at \$160/ton, while soda ash sells at \$240/ton.

6. Combined heat and power (CHP) is a technology that simultaneously produces electricity and heat from a single fuel source. Here we consider a CHP system that consumes 72 GJ of coal per hour, while producing 6000 kW-hr electricity and 32.4 GJ heat. Assume that coal has a heating value of 25 MJ/kg. Using energy-based allocation, determine the carbon dioxide emission per kW-hr for this CHP system. How does it compare with a coal fired power plant that produces electricity only? For the coal fired power plant, it is assumed the thermal efficiency is 35%. Assume complete combustion and the coal used has a CO_2 emission factor 2.5 kg/kg.

7. In Problem 6, it is assumed that the heat generated by the CHP system replaces heat from a coal-fired boiler with 90% efficiency. If this is so, what is the carbon dioxide emission per kW-hr based on substitution?

8. In some EU countries, the collection rate of plastic packaging wastes has increased to above 90%. A significant portion of collected plastic wastes are sent for energy recovery. This approach can be treated as open-loop recycling. It is known that per kg plastic packaging manufactured, 90 MJ of energy (cumulative energy demand) is consumed. The packaging wastes have an average energy content of 20MJ/kg and are burned to generate electricity with a thermal efficiency of 30%. Energy consumed during transportation can be ignored. Determine the cumulative energy demand for plastic packaging and electricity generated from energy recovery using different approaches examined in this chapter.

NATURAL SCIENCE

RITA SCHENCK

9.1 NATURAL SCIENCE AND ITS RELATIONSHIP TO LCA

Natural science is the study of the physical world, and life cycle assessment, although a new science, is focused on the physical world. LCA, like other branches of environmental science, studies the impact of human economic activities on the biology, chemistry, physics, and geology of our world. This provides a basic understanding of the natural sciences from which life cycle assessment grew.

Sometimes LCA studies extend beyond natural science to aggregate environmental impacts into what are called damage categories, using social science methods. You will learn about these in Chapters 11 and 12.

Biology is clearly the central focus of environmental science. If no living thing is affected by human actions, then one cannot truly say that environmental impacts have occurred. Impacts on biological systems come from physical disturbance, from resource depletion, and from pollution release. Teasing out the anthropogenic sources of changes in nature can sometimes be a difficult process, but certainly that is not always the case. If water is removed from a river system to irrigate crops and then the habitat for spawning fish disappears, the connection is clear. On the other hand, tracing the linkages between emissions of small amounts of carcinogens to actual cancer cases is more difficult. The background level of cancer is relatively high, and often the affected population is too small to provide statistical correlation between the emissions and the illnesses. Nevertheless, we know that the biosphere surrounds us, ranging from thousands of meters below the earth's surface to more than a hundred kilometers above the earth's surface.

No matter where a toxic emission occurs, there are living organisms in the receiving environment.

A basic understanding of chemistry is essential to understand how emissions act in the environment. Chemical reactions underlie the nature of our world. Whether one considers how emissions transform in the environment, how minerals are formed and degraded, or the basics of how cells work, a basic understanding of chemistry is essential to understanding how the natural world is and how it changes.

The laws of physics are not changed when we evaluate life cycle systems. Water runs downhill, mass and energy are conserved, and force equals mass times acceleration. Nuclear physics rarely plays an important part in life cycle studies, but Newtonian physics always does.

Geology is part of life cycle assessment. Both hydrology and mineral geology play a part in understanding mineral resource depletion. The emission of pollutants into the environment affects what is called early diagenesis: the conversion of sedimented materials into sedimentary rocks. Diagenesis acts as a sink for pollutants, taking them away from the majority of the biosphere, as well as detoxifying heavy metals and many organic substances. Biogeochemistry combines a knowledge of all these sciences to better understand how materials cycle through the environment.

9.2 LAND, OCEAN AND ATMOSPHERE

One can think of the earth as divided into land, ocean, and atmosphere. The majority of the earth is a rocky mass, over 12,000 kilometers in diameter. But the surface of the earth is dominated by oceans and the atmosphere. Oceans cover about two-thirds of the planet's surface, and on average the world ocean is about 3700 meters deep, comprising at least 97% of the world's water. Most of the ocean is cold and dark, and moves slowly. The average temperature of the ocean is about 4 degrees centigrade. It has two major circulation systems: the upper system above the permanent thermocline (see below) and the lower system below the thermocline. The atmosphere can be thought of as an upside-down ocean. It has several layers, but only two major circulation systems divided by a temperature discontinuity. The atmosphere is more variable in density and has a much smaller mass, but a larger volume than the ocean. The atmosphere is more variable in temperature and moves much faster than the ocean.

The thermocline is the depth at which the ocean temperature changes sharply. The upper level of the ocean is warmer, and well-mixed by wind and wave action. During winter storms, mixing in the ocean tends to break up the thermocline, and this causes what is called the permanent thermocline, at about 500 meters below the surface. The seasonal thermocline is shallower, generally in the 100 to 200 meters depth. The warmer the atmosphere above the water, the stronger (steeper) the thermocline. Tropical seas have quite steep thermoclines. The thermocline provides a physical barrier to vertical exchange. Most motion in the ocean is lateral.

The upper circulation systems of the ocean include familiar ocean currents such as the Gulf Stream and the Kuroshio Current (also known as the "Japan Current"), as well as less familiar currents such as the equatorial counter currents and the monsoon currents of the Indian Ocean, and the Antarctic Circumpolar Current. These currents all move relatively briskly at a few kilometers per hour. In contrast, the ocean deep circulation is extremely slow, although it involves the majority of the ocean volume. The world ocean deep circulation starts where the Arctic Ocean flows into the Atlantic Ocean at Greenland. The colder, denser water sinks to the bottom, then it flows southward through the Atlantic Ocean, rounding Africa and South America before flowing northwards in the Pacific Ocean, rising and mixing slowly along the way. The travel time is about 2,000 years.

Chemicals enter the ocean from the atmosphere, from river and groundwater outflow from the land, and from hydrothermal activity in the deep ocean. The chemicals are removed from the ocean largely through sedimentation processes. These sediments can be sourced either from terrestrial sources or from local production. Sedimentation rates vary widely, with near shore sedimentation rates being on the order of centimeters per year, and deep ocean sedimentation rates being on

the order of millimeters per year (or even per decade). The residence time of elements in the ocean is highly variable, ranging from days to thousands of years.

Algal growth in the ocean occurs largely above the seasonal thermocline. Microalgae take up nutrients and grow to form blooms. These occur seasonally once or in some places twice per year, typically in the spring and in the fall. The algae are eaten by zooplankton, which are then eaten by larger animals (e.g., fish). Fecal material, as well as dead animal and algal materials, fall to the thermocline. If these materials are dense enough, they drop through the thermocline, providing food for animals living in deep ocean waters.

Eventually, the materials sediment and the organic matter in them degrades through microbial activity, consuming oxygen as it does so. What remains behind is largely a combination of clay minerals and the tests, or shells of microscopic plants and animals. Geologists study these sediments to understand what the world was like millions of years ago. Chemical, physical, and biological characteristics of sediments can yield substantial information and provide a picture of what Earth was like long ago.

The sedimentation rate is largely a function of the particle size, density, and shape of the particles. Many of the smaller particles are degraded or eaten before they reach the thermocline, and the nutrients they contain are recycled in the upper waters. Many lakes have a seasonal thermocline, which may be quite shallow and can be easily discovered by swimmers. Some lakes, particularly larger lakes in tropical areas, have a permanent thermocline. Sedimentation occurs in lakes and some rivers as well, but while sedimentation rates are high—sometimes as much as ten centimeters per year in highly eutrophied lakes—the total amount of freshwater sediments is a tiny fraction of the amount of marine sediments. The microbial degradation that occurs in sediments means that open ocean sediments are typically anaerobic at about one meter into the sediments. Where sedimentation rates are high (e.g., some lakes and nearshore environments), the sediments become anaerobic almost at the surface of the sediments. Microbial activity in anaerobic marine sediments is dominated by sulfur-reducing bacteria, which create hydrogen sulfide the source of the rotten-egg smell one can experience at low tide in the summertime.

The lowest level of the atmosphere is called the troposphere. It contains the highest level of biological activity. The birds and the bees, trees and people all live in the troposphere. The troposphere is warm. Its temperature decreases at a constant rate of about 6.7°C/km altitude. The top of the tropo-

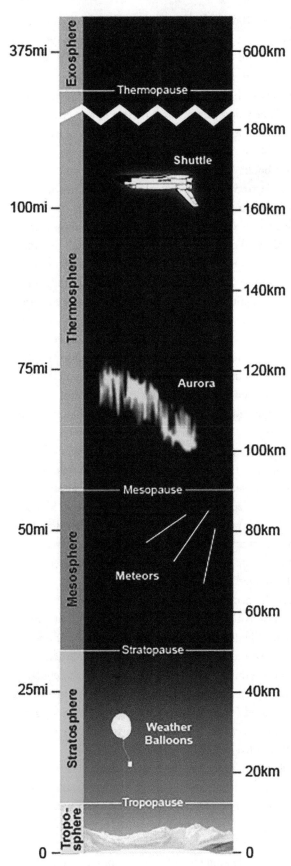

Figure 9.1: Layers of the Atmosphere
National Weather Service

sphere is called the tropopause, the level at which a temperature discontinuity occurs. Parallel to the thermocline in the ocean, the tropopause provides a physical barrier to transfer to higher levels of the atmosphere. The tropopause is the coldest part of the atmosphere. On hot summer days, one can often see the tropopause, demarcated by a brown layer of nitrogen oxides with clear air above it. The tropopause has its lowest elevation at the poles (about 9 km) and highest at the equator (about 17 km).

Weather and climate are driven by processes within the troposphere. It carries water from the ocean to the land—storm systems, droughts, and floods all depend on the movement of the air in the troposphere. Most clouds form in the troposphere when water vapor condenses around particles in the air. Like the ocean, movement in the atmosphere is driven by differences in temperature. But the atmosphere moves much faster than the ocean. Storm systems move thousands of kilometers per day, and hurricane winds can top 300 kilometers per hour. Tornadoes can generate even faster winds.

Above the tropopause is the stratosphere, so called because it is stratified. That is, the temperature of the stratosphere rises with altitude and little vertical mixing occurs in the stratosphere. The concentration of water in the stratosphere is almost zero, but occasionally clouds, known as nacreous clouds, form in the stratosphere. Nacreous clouds are implicated in forming the ozone holes at the poles.

Even higher than the stratosphere are the mesosphere and the thermosphere. At these levels, the atmosphere is extremely thin, and essentially no life exists there.

9.3 CHEMISTRY (THE QUEEN SCIENCE)

Chemistry is the science that studies the composition, structure, and properties of substances, as well as the transformations they undergo. It explains the material world and sits at the intersection of physics, biology, and geology. Without an understanding of chemistry, environmental studies become an exercise in weighing expert opinion, rather than a scientific endeavor.

Chemistry holds a few key concepts:
- **All matter is composed of atoms.** Atoms in turn are composed of protons (positively charged) and neutrons (no charge) in the nucleus, and an electron (negatively charged) cloud that surrounds and interpenetrates the nucleus. The shape and density of the electron cloud determines which chemical reactions can occur. Atoms with similarly shaped and charged electron clouds tend to act similarly.

- Thermodynamics:
  - Matter/energy is neither created nor destroyed, but different forms are interchangeable.
  - All transformations move towards greater disorder (increased entropy).
- Electrical charges must balance and this drives many reactions: opposite charges attract and same charges repel.
- Molecules are aggregations of atoms, bonded together several possible ways; strongest to weakest: ionic bonds (charges attracting), covalent bonds (equally shared electron clouds); hydrogen bonding (mediated by shared hydrogen atoms); and a weak kind of bonding mediated by Van der Wall's forces. Molecules have molecular electron clouds composed of the shared electrons of all the atoms in the molecule.

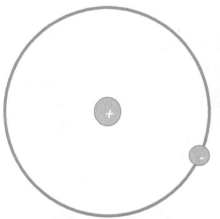

Figure 9.2 Hydrogen Atom

Hydrogen is the simplest atom, containing one proton and one electron. It can be thought of as a simple nucleus with an electron orbiting around it. The energy levels of electrons are called electron orbits, but actually the electron is not in a simple orbit, not even in a three-dimensional orbit. Rather the electron exists as a probabilistic, three- dimensional standing wave or electron cloud, where the highest probability is at the orbital distance. Electrons in orbitals tend to come in pairs, because this configuration is the most stable. The element helium is much more stable than hydrogen, because the electron orbit is full, with two electrons, charge balanced by two protons in the nucleus. The higher stability of paired electrons drives many chemical reactions. For example, hydrogen atoms rarely exist in nature: pure hydrogen gas is a molecule of two hydrogen atoms, where the electrons from the two atoms are paired and therefore more stable. $2H \rightarrow H_2$

More complicated atoms have more protons and electrons (always with a balanced numbers of each). The number of protons is called the atomic number. Each element has its own atomic number. As atomic number rises, lower energy orbitals become full and new orbitals are layered, with each representing the next higher energy level. The two lowest level orbitals are called the s-orbitals. The next highest energy orbitals are the p-orbitals, which form electron clouds in lobes shaped like dumbbells and oriented in three directions. The p-orbital contains six electrons. More complicated geometries exist for the d-orbital (which has ten electrons) and higher. Several excellent YouTube videos demonstrate these orbitals. The regular filling of electron orbitals causes periodic behavior of elements. A periodic table of the elements can be seen inside the back cover of this book.

Because electron orbitals are more stable when filled, chemical reactions occur that achieve paired electrons. Sometimes two bonded atoms share electrons equally, both filling their orbitals. This kind of chemical bond is called covalent bonding. The hydrogen molecule is an example of covalent bonding. Sometimes the atoms share more than one electron pair. For example, the nitrogen molecule (N_2) shares three electron pairs: this is called a triple bond. Covalent bonding also occurs in mixed molecules, for example, carbon-nitrogen bonds, carbon-oxygen bonds, and hydrogen-oxygen bonds are all covalent bonds. Carbon atoms can make up to four different covalent bonds, and it is this characteristic that allows formation of the large and complex organic molecules that make up living systems.

Ionic bonding occurs when one atom gives up an electron to another atom. This is most likely to occur when one atom has only a single electron in its highest orbital and another has all but one of its electrons paired in its highest orbital. Sodium chloride (NaCl) is a good example of an ionic bond. Here, the chloride atom takes the electron from the sodium atom and both atoms become ions, with the sodium ion (Na^+) being a positively charged cation, and the chloride ion (Cl^-) being a negatively charged anion. Some elements can gain or lose many electrons. Although ionic bonds are strong, most ionic substances readily dissolve in water and the ions physically separate from each other. This permits charge separation, the basis of everything from lightning to batteries to cellular metabolism. When an atom accepts one or more electrons, it is said to be reduced (it has more negative charge). When it gives up one or more electrons, it is said to be oxidized. The charge of an atom is called its oxidation state.

Some chemical bonds have both ionic and covalent character. In this case, the electron cloud is asymmetric and has a higher density at one end of a molecule than another. This provides a slight positive charge at one end of the molecule and a slight negative charge at the other end. Such molecules are said to have a dipole moment. The water molecule has a small dipole moment, and it is this characteristic that makes water a good solvent of almost everything.

When a molecule has one or more hydrogen atom, the electron cloud of the hydrogen atom can sometimes be shared with two atoms (e.g., nitrogen and oxygen), as well as with the hydrogen atom. This is called hydrogen bonding. Although it is a weak bond, it is essential in biological systems, where many hydrogen bonds together can make strong and stable molecules. DNA (deoxyribonucleic acid), the stuff of genes, is held together with hydrogen bonds. Many proteins include hydrogen bonds as well.

The weakest of bonds is the van der Waals bonding, which relies on the dipole moments of molecules. Molecules align so that their dipole moments match up, positive to negative. Van der Waals forces are even weaker than hydrogen bonds, but they are quite common, especially at surfaces. Contact cement relies on van der Waals forces, and so does the ability of geckos to walk on ceilings. Van der Waals forces are also the basis of partitioning of substances into sediments, an important biogeochemical phenomenon that drives the cycling of substances in the environment.

Chemical reactions create and destroy chemical bonds. These bonds, in turn, are driven by the laws of thermodynamics. The first law of thermodynamics states that matter/energy is neither created nor destroyed. Einstein's famous equation $e = mc^2$ describes the relationship between matter and energy, showing their interchangeability. However, as a practical matter this only happens on Earth during radioactive decay. Matter stays matter and energy stays energy. Elements remain the same element (except through radioactive decay), but energy can be converted to many forms; mechanical, kinetic, chemical, heat, radiation, and potential energy can all be interconverted, and all can be measured with the same units. Chemical equations describe chemical reactions that rely on the first law of thermodynamics; the equations balance with the same mass and energy in the reactants as in the products of the reaction. For example, the equation below describes the reaction of gaseous chlorine with solid sodium to create solid sodium chloride: $Cl_2(g) + 2Na(s) \rightarrow 2NaCl(s)$

The chlorine is shown in its elemental state, a diatomic gas molecule, and sodium in its elemental state, a solid metal. When a molecule has multiple atoms of the same elements, the number is denoted in a subscript to the right of the elemental symbol. The subscripts denoting the state of the substances (gas or solid) are informational, not essential to the equation. The number of chlorine atoms to the left of the arrow equals the number to the right, and this is also true of the sodium atoms. Insuring that this occurs is called balancing the equation.

The number of molecules of a substance required to balance the equation is denoted with a number before the molecular formula.

The reaction of chlorine and sodium is spontaneous when the two are brought into contact. Sodium chloride is common table salt, and its chemical bonds are more stable (lower energy) than the bonds of elemental sodium and chlorine. The reaction is exothermic: it gives off heat. The overall energy of the chemical bonds before the reaction equals the chemical bonds after the reaction, plus the heat energy released. Energy is conserved.

The reaction also follows the second law of thermodynamics, which states that for all reactions in closed systems, the end products are more disordered than the beginning products. The end products have higher entropy. The formation of sodium chloride releases heat, the most disordered form of energy. The second law of thermodynamics tells us that there is no free lunch. No reaction is 100% efficient. This concept is useful to keep in mind when considering recycling systems using LCA. No recycling system is 100% efficient.

The reaction illustrates an important concept in chemistry: changing chemical bonds and chemical composition creates substances that are different from the reactants. Chlorine gas is lethally poisonous. Elemental sodium is explosive in the presence of oxygen. Sodium chloride, on the other hand, is an essential nutrient for all life, and it represents about 2.5 percent of the mass of the ocean. Chemical reactions change things. Recognizing different chemicals in life cycle inventories is essential; otherwise, one is not able to properly characterize the impacts of the release of chemicals into the environment. Unfortunately, many chemicals have more than one name, and this leads to confusion in life cycle inventories. The American Chemical Society, through the Chemical Abstracts Service (CAS), provides unique identifiers for chemicals, called the CAS Registry Number (CASRN). Over 70 million chemicals have been registered. Wherever possible, the CASRN is preferred when identifying chemical compounds.

Many chemical reactions transfer electrons from one molecule or atom to another. These reactions are called redox reactions. These reactions are important in biology and in combustion. For example, burning coal (i.e., primarily carbon with some contaminants) can be expressed as: $C + O_2 \rightarrow CO_2$

Here, the oxidation state of the elements changes from neutral (the oxidation state of all elemental forms) to carbon at +4 and oxygen at -2. The most common redox reactions one may encounter in a life cycle study involve oxygen. Oxygen is reduced while the other substance is oxidized (the basis of the terminology). Note that the oxidation states on each side of the equation balance, from zero on the reactants to one +4 carbon atom and two -2 oxygen atoms in the products. Of course, burning coal releases a great deal of heat, which is then used for many purposes. Instead of burning coal one might burn wood, which is a polymer of glucose ($C_6H_{12}O_6$). This too is a redox reaction.

$$C_6H_{12}O_6 + 6O_2 \rightarrow 6\ CO_2 + 6\ H_2O$$

As before, the carbon changes from an oxidation state of zero to one of +4, and the oxygen changes from zero to -2. This reaction is also spontaneous and exothermic. This same reaction occurs in all living cells and is called respiration. The inverse of this reaction is photosynthesis.

$$6\ CO_2 + 6\ H_2O \overset{hv}{\rightarrow} C_6H_{12}O_6 + 6O_2$$

Plants use the energy of sunlight to drive this reaction. Energy from the sun (represented by hv) is supplied from outside the system to achieve this endpoint, which is not only higher energy but lower entropy because living organisms are highly ordered systems. The second law of thermodynamics is not broken because the system includes the sun, where entropy increases as radiation is produced.

Many times in LCA, a preliminary estimate of the emissions is derived from these simple chemical reactions. Each process in the life cycle inventory must balance for mass and, ideally, for energy as well.

Photosynthesis is not the only kind of reaction that is initiated by radiation. When radiation of the proper frequency strikes an atom, it can raise an electron to a higher energy orbital. Most of the time, this electron drops down to the lower energy state relatively quickly, releasing radiation whose energy is equivalent to the difference in the energy states of the orbitals. Sometimes, the electron stays in the higher energy state for a long time, and such atoms or molecules have relatively stable unpaired electrons. Often, the radiation serves to break apart a molecule into two parts, with one or both parts having unpaired electrons. These atoms or molecules are called free radicals. Free radicals are highly reactive, readily exchanging energy with other molecules to propagate more free radicals. Free radicals are common in the atmosphere and in biological systems.

Free radicals also occur inside of cells as a byproduct of metabolism. Free radicals cause damage to the cell's structure and are believed to be an important cause of ageing and cancer.

Free radicals are designated with a dot adjacent to the chemical formula. An important set of free radical reactions causes the destruction of ozone in the stratosphere, illustrated here using CFC-11 ($CFCl_3$).

$$CFCl_3 + hv \rightarrow Cl\bullet + \bullet CFCl_2$$

$$Cl\bullet + O_3 \rightarrow ClO\bullet + O_2$$

$$O_3 + hv \rightarrow O_2 + O$$

$$O + ClO\bullet \rightarrow Cl\bullet + O_2$$

Free radicals can also combine in what is called a termination reaction, such as:

$$Cl\bullet + \bullet CFCl_2 \rightarrow CFCl_3$$

Overall the reaction is:

$$2O_3 + hv \rightarrow 3O_2$$

9.4 THE WATER CYCLE

The Earth is a water planet: water covers almost three quarters of its surface. But only a small portion of the water (less than 3%) is fresh water. All of the life on land depends on

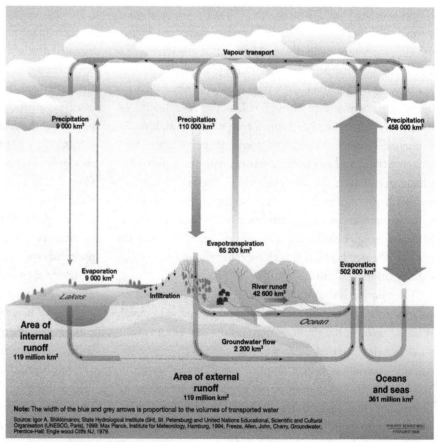

Figure 9.3 The Hydrological Cycle

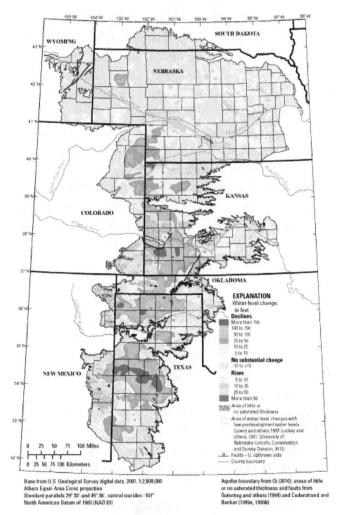

Figure 9.4 Losses in the High Plains Aquifer USGS

this small amount of water. Hydrologists study the cycling of water, which is shown in a much simplified version in figure 9.3. The figure shows both stocks of water (oceans atmosphere, glaciers and groundwater) and flows between them. The volume of lakes and rivers (not shown in the figure) is approximately 25,000 cubic miles (105,000 cubic kilometers) (United Nations Water). The major sources of water on land are lakes and rivers, precipitation and groundwater.

Groundwater is water in liquid form under the land's surface, like an underwater lake. Typically the surface of the groundwater is some distance below the surface of the ground, and its shape is a muted repetition of the ground surface without the abrupt elevation changes of the land surface. The ground between the land surface and the groundwater surface is called the unsaturated zone or the vadose zone. Water in a groundwater formation is called an aquifer.

Groundwater flow is variable and complex, depending on the surface and subsurface conditions. Groundwater can flow into streams (this is called a gaining stream) and it can flow from streams (this is called a losing stream). Some subsurface formations, such as clay layers, limit the vertical movement of the groundwater, thus creating separate aquifers. Multiple aquifers can exist in the same geographic location layered over each other. Some aquifers are quite small, while others are enormous, for example the Ogallala (also known as the High Plains Aquifer) extends from South Dakota to Texas as shown in the figure here. Melted snow from the Rocky Mountains provides its source.

After the Ogallala was used as a source of water for irrigation, its water level dropped as much as 60 meters in some places. This indicates that the flow of water out of the Ogallala is larger than the flow of water into the aquifer. Some aquifers no longer have any inputs. These are called fossil aquifers.

Removal of groundwater can cause earthquakes (Holzer et al. 1983). Likewise, pumping of water into aquifers (as is done during oil and gas exploration and extraction) also can cause earthquakes (Bolt 2006).

9.5 ECOLOGY

The goal of environmental science is to understand the impact of human activities on Earth's ecology, and the goal of LCA Is to link that impact to particular products and services. Ecosystems are defined as networks of organisms interacting with each other and with their physical environment. Ecosystems may be as small as the inside of a pitcher plant or as large as the Earth, but usually they are defined at a species assemblage level (i.e., at the scale of hundreds to thousands of square kilometers).

Within an ecosystem, many ecological niches exist, defined by how a particular species makes a living. Many dimensions define a niche: gradients of physical, chemical, temporal, and biological parameters can all be partitioned to provide many niches. For example, it might be a different niche to eat grass only in the morning versus at night. Some trees provide a canopy, while others are understory plants. Birds might use a tree for food, for shelter, or for both. In general, the larger the ecosystem, and the longer an ecosystem is relatively undisturbed, the more the available resources are partitioned among different species. The Galapagos Island finches partitioned their environment by evolving different beaks that could harvest different seeds. Eurasia, the world's largest land mass, has more species than North America, and so forth.

Organisms within an ecosystem interact either competitively or cooperatively. Plants compete for sunlight and nutrients. Their strategies vary, but often they relate to growing quickly whenever the resource becomes available, and slowing metabolism and growth in the absence of resources. Once introduced, invasive species take over a niche or niches rapidly. They tend to have a broad range of acceptable physical, chemical, and biological conditions.

Organisms often directly poison their competitors. For example, Western bracken fern is toxic to most trees, with the exception of Douglas fir. The fir tree and bracken fern are often found together. Many plants produce substances that are toxic to animals that might eat the plant. This kind of toxic competitive interaction is called allelopathy. One can consider the use of herbicides and pesticides by human beings as a kind of allelopathy, and human beings as an exceptionally effective invasive species present in all terrestrial ecosystems.

> **For more information on evolution, see the following:**
> http://evolution.berkeley.edu/evolibrary/news/
> 071201_adenovirus

Organisms also cooperate. Cleaner fish remove parasites from larger fish, often even cleaning inside their mouths. The cleaner fish are protected from other predators and make a meal of the parasites. Of course, the fish they clean enjoy a reduced parasite load. In many soils, especially in forests, mycorrhizae form a network of fungal hyphae underground integrated with plant roots. They bring nutrients and water to the plant roots, and in return the plant provides food in the form of the products of photosynthesis to the mycorrhyzae. They can represent as much as 90% of the biomass underground. Many plants rely on animal species to pollinate. Bees are common pollinators, as are other insects, birds, and even mammals such as bats. The pollinator collects nutrients provided by the plant, while the plant gains reproductive success. Similar partnerships exist for seed dispersal. The interaction of human beings with domesticated crops and animals is a cooperative one. Domesticated species are widely distributed by human actions.

The interactions among and between organisms constantly evolve. A prime example is the interactions between disease organisms and their hosts. Often, diseases are transferred from one species to another, and in the new host they are

virulent (i.e., deadly). Over time, the diseases become less virulent because the less virulent strains are more successful. Strains that kill the host tend to die, while those that merely make the host uncomfortable have the opportunity to be spread to other hosts. Eventually, disease organisms can become beneficial or become an integral part of the host. Many viruses integrate into host genomes and can be passed on to future generations. In fact, the virus-mediated movement of genes from one species to another is a driver of evolution.

The constant evolution of species forms the basis of biodiversity. An estimated 8.7 million eukaryote species exist on Earth (Mora et al. 2011), but the vast majority (nearly 90%) have not even been described. People value biodiversity for many reasons. All species have ethical value, but economic and cultural value can also derive from species diversity. The more diverse a plant assemblage is, the more productive the ecosystem (as measured by biomass production). Biodiversity exists at many levels: the number of species in an ecosystem, the number of species of a certain type (e.g., birds), and the genetic diversity within a species are all different ways of thinking about biodiversity.

A primary assumption of evolutionary theory is that all life on Earth is related and had an ultimate shared ancestor, near four billion years ago. The study of this interrelationship is called phylogeny, and that work is often depicted with phylogenetic trees.

The three major branches (or Kingdoms) of life are bacteria, eukaryotes (from which all plants and animals are descended) and archaea. Archaea were only discovered late in the 20th century and are identified by their unique biochemistry. Neither archaea nor bacteria have nuclear membranes or membrane-bound intracellular organelles, unlike all eukaryotes. The next lower level of taxonomy is the phylum, flowed

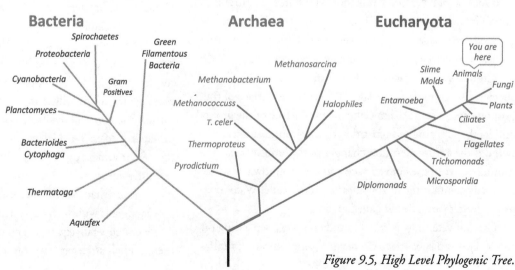

Figure 9.5, High Level Phylogenic Tree.

by the class, order, family, genus, and species. Normally, one refers to species with binomial nomenclature, or *Genus-species*. People are *Homo sapiens*. Members of a species are assumed to be able to interbreed, and all members of a genus are assumed to be closely related to each other. In the past, the taxonomy of species was determined by physical differences and similarities among organisms. With the advent of molecular genetics tools, phylogenetic trees have been much rearranged.

When an ecosystem is disturbed, e.g. through natural events such as a volcanic eruption or man-made events such as a clear cut of a forest, the ecology is severely disturbed. This creates a new opportunity for organisms to grow in the absence of competition from previously existing species. Organisms move opportunistically into the open space, and over time the habitat tends to grow back. The first organisms that move in are those that can survive in the disturbed environment. As they grow, they alter the environment in such a fashion that it becomes attractive for other species, thus making it less attractive for their own species. The assemblage of species changes over time until it returns to the species distribution similar to that found in the previous undisturbed space. This process is called succession, and the final state is called the climax state. The concept of a climax state is a useful one, but random processes, historical accident and other factors such as climate change mean that the climax state is not truly fixed. Random events drive the recovery of ecosystems from perturbations. For example, when a fish trawler scrapes the bottom of the ocean to gain its catch, it creates a new surface area on which organisms can settle and grow. However, only those organisms that happen to be reproducing at that time can take advantage of this opportunity. The available space will be taken up by those organisms and others may be excluded from the space.

The organisms that are first to appear after a disturbance tend to have relatively high growth rates and to be able to survive in harsh environments. They are sometimes called weed species, and in fact most garden weeds fall into this category. Weeds are said to be r-selected species because their presence is driven by their fast rate of population growth (r). They tend to be relatively short-lived, with few reproductive events, and have many offspring with low survivorship. R-selected plants make many seeds that are available for dispersal into disturbed environments. At the other extreme are K-selected (i.e., carrying capacity selected) species that live longer, have fewer offspring, and experience multiple reproductive events. Annual plants such as wheat are examples of r-selected species, while large mammals including people are K-selected species. This r/K selection is a great simplification of the reality of

reproductive success. The reality of the situation is that most species fall somewhere in between and variability exists within species as well.

The distribution of a given species is limited not only by the availability of niches for the species, but also by local events, such as extreme weather events, landslides, disease, and the arrival of more competitive species. For example, if a tree can only survive temperatures to -20 C, and one winter the temperature falls to -25 C, that tree will die, as will the others of the same species. This is a local extinction event. If the extreme weather event is widespread, and the species covers only a small area, it may lead to species extinction. Extinctions happen all the time. Extinctions are generally balanced by speciation, the creation of new species.

From time to time, global events occur that lead to mass extinction events. Nearly all life on Earth disappears during these extinction events, only to be replaced by newly evolved life forms. Because species extinction rates are high, and speciation rates are low during these extinction events, the species disappear from the fossil record. The kinds of organisms found in the fossil record are different before and after the extinction events, and geological periods are often identified as those between extinction events. Five major and several minor extinction events are recognized. Various hypotheses about the cause of the extinction events have been posited, but the most recent event (at the Cretaceous-Cenozoic border, 65 million years ago when the dinosaurs disappeared) was caused by a large meteoric strike in the Yucatan. This led to long-lasting global clouds that interrupted photosynthesis and caused global cooling for an extended period. Earlier events are not as well understood.

Some scientists believe that the Earth is currently undergoing a sixth extinction (Leaky and Lewin 1995) caused by human intervention. Large animals have experienced many species extinctions since the end of the last ice age (which occurred 10,000 to 12,000 years ago). Some of the evidence supports this conclusion, but there is also evidence supporting climate change as a primary cause of species extinctions. Clearly, human activities such as agriculture have displaced the habitat of many species and loss of habitat is a primary driver of local extinctions.

Natural habitats provide many essential services to human beings. These ecosystem services include a wide array of services that would otherwise have to be provided through economic activity. For example, in addition to providing wood products, a healthy forest cleans air and water, produces oxygen, sequesters carbon, and stabilizes soils. When forests are cut down to create cropland or roads, these ecosystem services are lost.

Perhaps the most important ecosystem services are the cycling of water and nutrients. While the elements are neither created nor destroyed, they are converted into forms useful for organisms, and then cycled into other forms that are not and back again. Nitrogen cycling is an important example. The air is over 70% nitrogen, but elemental nitrogen is not bioavailable to make the proteins and nucleic acids needed for all living things. Microorganisms called nitrogen fixers convert the elemental nitrogen into ammonia, NH_3, which is then used by plants and microorganisms to make amino acids and other biological compounds. Plants grow and are eaten by animals that excrete nitrogen compounds, and eventually they die and are degraded by microorganisms that release the nitrogen back into the atmosphere. Nitrogen fixation requires energy, while denitrification releases energy that microorganisms use to supply their energy needs.

Biological reactions are catalyzed by enzymes, which often have toxic metals at the site of the chemical reaction or are present as coenzymes. For example, nitrogen fixation requires molybdenum and iron. The conversion of sugar to bioenergy requires iron, zinc, and manganese. Organisms gather these micronutrients directly from the environment (in the case of microorganisms and plants) or through their food.

Another familiar example is the cycling of water. Evaporation from water bodies and transpiration by plants release water vapor into the atmosphere. When the atmosphere is saturated, the water is precipitated as dew, rain, or another form such as snow. That water is taken up by plants to be re-transpired, or else it finds its way into groundwater, rivers, lakes, and the ocean. In the Amazonian rainforest, the same water molecules may cycle six or seven times within the trees before joining a stream.

9.6 IMPORTANT ENVIRONMENTAL SCIENCE CONCEPTS

PARTITIONING

One can think of the environment as having different compartments: the atmosphere, oceans, freshwater, soil, and sediments. Substances in the environment tend to partition into these compartments based on physical chemical principles. Highly water-soluble substances tend to be found in the ocean. Chlorine, for example, makes up about 1.7 percent of the mass of the world ocean, even though the primary source of chlorine is in gaseous form in volcanic eruptions. Iron makes up most of the Earth's mass, but its high density has partitioned it into the Earth's core and relatively little is found in the crust. Iron in water reacts with any free (elemental) oxygen, forming iron oxyhydroxides, which precipitate, leaving little free iron in the water column, but relatively high iron in the sediments.

This behavior is illustrated in mid-ocean hydrothermal vents. Iron, magnesium, and other metals precipitate as sulfides (Zierenberg et al. 2000). Here one can see the active partitioning of materials from geothermal fluids to seawater. The microorganisms in and near the vents use sulfur, iron, and nitrogen redox chemistry to grow. This is called chemosynthesis, which requires no light.

Figure 9.6 Mid Ocean Ridge Black Smoker, courtesy of the National Atmospheric and Oceanic Administration http://www.pmel.noaa.gov/eoi/gallery/R852_DSC_092004_070004_03772.JPG

Sedimentation drives the partitioning of many substances. Sediments have a high surface area, as measured by surface area in a bulk volume of sediment, because the particles of the sediment are miniscule. The combination of the geometry of these small particles and the small surface charges they hold attracts almost all elements and compounds: organic and inorganic, charged and uncharged substances all tend to partition into the sediment due to van der Waals forces. As a result, substances that are found in parts per billion concentrations in the water column tend to be found in parts per million in the sediments.

Many soils derive from sediments. For example, the rich soils of the interior of the North American Continent are the remnants of the sediments of an ancient sea. Soils also derive from the weathering of rocks, driven primarily by carbonic acid in rain. To a great extent, soils act like sediments as a common place to partition substances. Both soils and sediments vary from place to place, based on the history of that location. The anthropogenic disturbance of soils is typically greater than the anthropogenic disturbance of sediments. Often, the concentration of a substance in the soil at one location may be an order of magnitude different from that found a few meters away.

Over time, sediments compress and can become a sedimentary rock. Sediments deposit in layers, with the most recent layers of sediment on top. Scientists use the layering of distinct strata to date rocks sediments and the fossil evidence of life found in them. This is called geostratigraphy, and until the development of isotopic dating, it was the primary method for estimating the age of fossils.

LCA practitioners use models based on the partitioning of substances into different compartments as a proxy for exposure estimates. However, the exposure of an organism to any substance is determined by a large set of factors, including bioavailability, site-specific dispersion of substances in the environment, and actions of organisms.

ACTIVITY COEFFICIENTS AND BIOAVAILABILITY

The concentration of an element does not tell one everything needed to understand its reaction in the environment. The form of the element, whether the element is bound to other substances and how strong that binding may be, as well as the chemical composition of the environment around the element all affect the reactivity (or chemical activity) of the element. Sometimes activity coefficients are developed for substances that reflect the relative availability of the substance to react. The activity coefficients are peculiar to the substance and to the environment in question. The chemical activity of the substance determines whether it is bioavailable. Only bioavailable fractions of a substance can participate in biological reactions.

9.7 ENVIRONMENTAL MECHANISMS

This chapter has described some of the underlying concepts that drive natural processes. In the field of LCA, we are interested in how human activities perturb these processes. Thus, when one speaks of environmental mechanisms, one describes how human interventions (emissions or resource use) affect ecosystems, based on and overlaying natural processes. Normally, several steps occur between the time when a pollutant is emitted and the time when some harmful endpoint occurs. Some environmental mechanisms are well understood while others remain to be clarified. The practice of LCA is only as good as the understanding we have of the environment, and our ability to model what happens in it.

CLIMATE CHANGE

Climate Change occurs as a result of the anthropogenic emissions of greenhouse gases. The gases act as a blanket in the atmosphere, reflecting infrared radiation back to the Earth's surface that would otherwise radiate into space. That radiation heats up the atmosphere and the ocean. It causes more extreme weather events and ultimately causes effects on organisms and ecosystems.

Climate change is primarily caused by emissions of carbon dioxide (CO_2), methane (CH_4), and nitrous oxide (N_2O), but it is also caused by emissions of industrial chemicals. CO_2 from burning fuels dominates greenhouse gas emissions, but CH_4 from leaks from natural gas and other fossil fuel production and from enteric (in-gut) fermentation by cows, sheep and other ruminants also contribute substantially.

STRATOSPHERIC OZONE DEPLETION

In the upper atmosphere, ozone acts to screen out ultraviolet radiation (UV). Ozone depletion is caused by the introduction of halogenated substances in the stratosphere. These substances convert the ozone (O_3) in the stratosphere into the more normal oxygen form (O_2). At high latitudes, the ozone layer is essentially gone, causing the seasonal "ozone holes." Depletion of ozone allows more UV to reach the Earth's surface, resulting in a higher incidence of sun burns, skin cancer, and cataracts. Higher UV also affects all organisms, causing mutations and other negative effects.

The greatest anthropogenic sources of ozone-depleting chemicals are halogenated solvents, whose production is regulated by an international treaty, the Montreal Protocol[1]. As a result of compliance to the treaty, the progress of ozone depletion has halted, although it will take some decades before background levels of stratospheric ozone are regained.

ACIDIFICATION

Acidification is the result of wet and dry deposition of acidifying substances from the atmosphere to the surface. The most common substances are oxides of sulfur and oxides of nitrogen. These substances derive primarily from combustion sources. In the atmosphere, they undergo redox reactions (which are sometimes facilitated by sunlight) and hydration to become nitric acid (HNO3) or sulfuric acid (H2SO4). Ammonia deposition in soils can also cause acidification through de-nitrification, which converts ammonia (NH3) to nitric acid. Industrial processes may also release other acid gases, such as hydrochloric and hydrofluoric acids.

The effect of acidification depends upon the neutralizing capacity of the receiving environment. Where considerable lime (calcium and magnesium carbonate) exists in the soils, the acid is neutralized. The neutralization releases CO_2 into the atmosphere.

$$CaCO_3 + 2HNO_3 \rightarrow CO_{2(g)} + H_2O + Ca^{2+} + 2NO_3^-$$

Where soils lack adequate neutralization capacity, the acid deposition acidifies the water in the soils (in the vadose zone) and the groundwater. This increased acidity mobilizes the naturally-occurring metals in soils, and these metals (chiefly aluminum) can become toxic in forests and in lakes. Acid deposition also causes degradation of the built environment, causing many millions of dollars of damage annually.

The high level of CO_2 in the atmosphere has led to ocean acidification. This has many potential biological effects, but the most notable is that it inhibits the deposition of calcium carbonate in sea shells. This affects the entire ocean ecosystem because many of the phytoplankton at the base of the food chain have cell walls made of calcium carbonate. This has already caused financial problems in the shellfish industry, as the development of early life stages of the shellfish is also inhibited.

EUTROPHICATION

Eutrophication is the overgrowth of algae as the result of the addition of the limiting nutrient. The limiting nutrient depends on the water chemistry and the needs of the individual alga. Different algae have different needs. For example, diatoms have shells made of silicon, while other algae do not. The chemistry of water bodies differ, too. Little silicon occurs in the middle of the ocean, and consequently few diatoms are found there.

In the 1930s, Alfred Redfield discovered that marine communities (as opposed to individual species) contain the major nutrients (carbon, nitrogen and phosphorus) at a fixed ratio of 106 Carbon:16 Nitrogen: 1 Phosphorus on an atom basis. This ratio has proven to hold in all aquatic environments and has become known as the Redfield Ratio. This ratio is an emergent community-level characteristic: individual species may use nutrients at different ratios, but entire communities use nutrients according to the Redfield Ratio.

Different water bodies have different limiting nutrients: typically, nearshore marine environments are nitrogen limited, freshwater bodies are phosphorus limited, and open ocean waters are iron limited. If one adds the limiting nutrient to natural waters, eventually, another nutrient becomes the limiting factor. Anthropogenic inputs have caused this outcome in many places. The Baltic Sea is sometimes nitrogen and sometimes phosphorus limited. The Everglades (a freshwater body) is nitrogen limited. The Gulf of Mexico and the Chesapeake Bay are phosphorus limited. The China Sea is phosphorus limited, at least part of the time (Bao, 2003).

The sources of nutrients that cause eutrophication are typically agriculture, wastewater treatment plants, and combustion sources. These sources are modeled in life cycle studies.

Once the algae bloom and die, their decomposition removes oxygen from the water column, eventually causing hypoxic or even anaerobic conditions. These conditions lead to fish kills and other ecological damage. The condition of hypoxia caused by eutrophication is common in nearshore environments globally. Some algal blooms are toxic, leading to paralytic shellfish poisoning and other illnesses in humans and other organisms. Toxic algal blooms (or red tides) are also common globally and are increasing in frequency and severity.

PHOTOCHEMICAL SMOG

Emissions of oxides of nitrogen in the presence of sunlight and volatile organic compounds (VOCs) leads to the creation of ground level ozone, the primary component of smog. The reaction proceeds through a free radical mechanism.

$$2NO + O_2 \rightarrow 2NO_2$$

$$NO_2 + hv \rightarrow \bullet O + NO$$

$$\bullet O + O_2 \rightarrow \bullet O_3$$

The ozone is naturally unstable and disappears in the absence of sunlight, but the presence of VOCs stabilizes free radicals in the following way:

$$VOC + \bullet O \rightarrow VOC\text{-}O$$

Where VOC-O represents an oxidized VOC such as an aldehyde or ketone:

$$VOC\text{-}O + O_2 \rightarrow \bullet VOC\text{-}O_3$$

Where $VOC - O_3$ is a peroxide radical.

Peroxide radicals stabilize both the production of ozone and the creation of NO_2, which itself creates O_3.

Photochemical smog has significant impacts on all organisms, causing crop yield reduction at low levels, and cancer, asthma, and many other negative environmental impacts at higher levels. Oxides of nitrogen are mostly derived from combustion sources, while VOCs are derived from both anthropogenic sources (e.g., petroleum) and from natural sources (e.g., plants).

RESPIRATORY EFFECTS

Many of the air emissions from combustion and industrial processes serve to form particulate matter in the air. The composition of the particulate matter is an extremely variable mixture of organic and inorganic compounds that, over time, become more homogeneous in the atmosphere. The tiniest particles in the air (those less than 2.5 micrometers in diameter) can be inhaled deep into the lungs, causing significant damage that can lead lung cancer, respiratory distress, and other symptoms. An estimated two to six million people die a year from the effects of particulate matter in the air, mostly in Asia. Respiratory effects can be seen in other organisms as well, but generally only the effects on humans are studied.

TOXICITY

"The dose makes the poison" is the primary tenet of toxicology. All substances are poisonous in the proper amounts: even drinking too much water can kill you. To speak of toxic and non-toxic compounds is therefore a gross simplification based on a desire to focus on the compounds that are most likely to cause significant harm to humans or other organisms.

To understand the toxicity of a substance, one must know:
- the chemical form in which it exists
- the mode of presentation (through the air, in water, or as a solid)
- the concentration
- the time course of the exposure
- the organism being exposed
- the life stage of the organism being exposed

Substances can affect any of the thousands of reactions that occur simultaneously in every cell, but first they have to get inside the cell. They have to be bioavailable and they have to be presented at a level that has an effect on that cell. In complex organisms, different cells have different responses to substances and different organisms have different sensitivities to any substance, either because they have different genetic makeups or because they live in different environments.

Substances get into cells in one of three ways:
1. They are non-polar substances that diffuse through the cell membrane.
2. They are ionic substances that passively respond to the electrochemical gradients that exist across all cell membranes (in living cells).

3. They are actively taken up by the cell, either deliberately because the cell needs the substance as a nutrient or as an intracellular signaling compound, or by accident as the substances resemble the desired substance.

Once inside the cell, the substance is likely to have many different modes of action.
- It can displace another substance in a biochemical reaction.
- It can disrupt electrochemical gradients in the cell by blocking the active transport sites in the cell.
- It can bind to a cellular component so that the component can no longer perform its normal function.
- It can induce actions within a cell that lead it to kill itself (this is called apoptosis),

Enough of any one of these actions will kill the cell and by extension, all of the cells of an organism, causing the organism's death. As noted above, different organisms and different life cycle stages can respond differently to a given substance. The similarity of response is highly correlated to the phylogenic relationship of the organisms. Mammals tend to respond like mammals, insects like their cousins the aquatic copepods and isopods, and so on. Living organisms exist in all locations, as the biosphere extends from kilometers underground to the stratosphere. No matter where a substance is placed on Earth, an organism probably exists that can be affected by it.

Toxicity is often measured using LC-50s or EC-50s. These measures identify when half of an exposed population responds. LC-50 represents the concentration at which half of an exposed population dies, while an EC-50 represents the concentration at which half of the maximal response occurs. For example, the concentration at which the growth rate of a population is halved. The range of these measures spans 18 orders of magnitude. To put that into context, that is the difference between the thickness of a coin and the distance to Pluto.

Because performing toxicity tests on humans is unethical, human toxicity is generally calculated as being the same as the toxicity in other mammals, notably mice and rats. Toxicity is not expressed as a lethal dose, but rather as the concentration at which the incidence of a response is noted (e.g., incidence of cancer).

LAND USE

Humans use land to provide food, fiber, and lumber. In doing so, they physically disturb soils and this leads to a release of stored carbon in the soils. Lands in agricultural use typically exhibit lowered primary productivity and much lower species diversity than wild habitats, as undesirable species are excluded through human intervention. Land use change, especially deforestation and conversion to agriculture or other use, is a major source of greenhouse gases that are responsible for the release of about 30% of the CO_2 in the atmosphere today. Relatively little land (two percent or less) is developed, while most of it is appropriated for agriculture and forestry.

The most significant impact of land use by humans is that it limits or eliminates its use by other species. About a quarter of the total primary production of the planet is used by humans. In addition, the land in use tends to be the most productive and, therefore, most capable to support high species diversity. In effect, human beings use almost all of the land that is not deserts, rocks, or ice.

9.8 Concluding Remarks

This chapter provided a simplified view of the natural science behind LCA. Every one of the topics is the subject of ongoing study by hundreds, if not thousands, of scientists. Textbooks exist on many of the subjects covered here. The student is encouraged to look further into topics of interest in order to better understand the relationship between humans and the rest of the natural world, including the ten million or more species it contains as measured in LCA.

Problem exercises

1. Calculating environmental mechanisms of sea level rise

People are discussing the possibility of sea levels rising and flooding coastal areas, even putting some island nations underwater. Emissions of greenhouse gases make the atmosphere warmer, and several steps must be accounted for going from that to sea level rise. The warmer atmosphere warms glaciers, causing melting. The warmer atmosphere also warms the sea, and this causes expansion of the volume of the sea. Melting the glaciers in Greenland and Antarctica means that water comes off these land masses into the sea, leading to more freshwater floating on and mixing into the ocean.

Which is the biggest source of sea level rise, currently about 3 millimeters per year?

To understand the answer to this question, one needs to have a handle on the size of the ocean and of glaciers, and how fast ice is melting and where it comes from. Many ice sheets, as well as the smaller ice floes, float on the ocean. When these sheets melt, the sea level is not affected because the ice floats and displaces seawater. Its melting will not make a difference. In contrast, ice melting off glaciers can raise sea level.

2. Write a balanced equation for the combustion of methane. How is this different from the combustion of carbon?

3. What are ecosystem services? Provide examples.

4. What are the four types of chemical bonds? List them from strongest to weakest.

5. What is the difference between the seasonal and the permanent thermocline?

6. If you have a contaminant in natural waters, what things can happen to it?

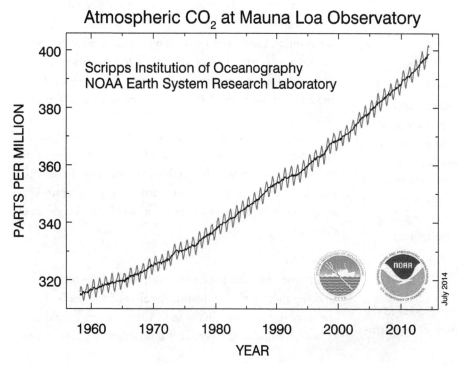

Atmospheric CO₂ at Mauna Loa Observatory

1 Nerem, R. S., D. Chambers, C. Choe, and G. T. Mitchum. "Estimating Mean Sea Level Change from the TOPEX and Jason Altimeter Missions." *Marine Geodesy* 33, no. 1 supp 1 (2010): 435

IMPACT ASSESSMENT AND MODELING

MATTHEW ECKELMAN AND ANDREW HENDERSON

This chapter builds on the environmental science described in the previous chapter on natural science to detail how to measure, model, and visualize how chemicals affect natural processes. This chapter also describes the ways in which these chemicals affect their environment across a range of environmental health, human health, and resource depletion categories. Impact assessment models are used to link emissions of a substance from the life cycle inventory to changes in the environment or public health. These models can be extremely complex. Learning how to use them is typically not part of the basic training for LCA practitioners. On the other hand, every life cycle impact assessment method relies on underlying physical models to provide characterization factors, so it behooves all LCA practitioners to know a bit about how these models operate.

We begin by presenting environmental sampling and measurement methods that gauge individual chemicals or general indicators of environmental quality. This is followed by an examination of environmental fate and transport models that try to explain how chemicals move through the environment. These models are crucial in understanding how an emission that happens over in one place (e.g., from a smokestack) migrates to other locations (e.g., downwind from the smokestack). Finally, the chapter describes, through exposure and effect modeling, how these substances affect people and organisms they come into contact with them, with a specific focus on toxicity to human health.

10.1 ENVIRONMENTAL SAMPLING AND MEASUREMENT METHODS

Impact assessment is primarily concerned with changes to the environment, as well as the resulting potential damages that occur from emissions and resource use over the product life cycle. Changes in the environment can be measured in terms of concentrations of substances in different media (e.g., air, water, and soil), but the measurement itself can be quite challenging. Some substances are of concern at extremely low levels (e.g., parts per trillion) or have properties that make them difficult to isolate. More importantly, one cannot measure everything, everywhere, all of the time, which has led to established protocols for environmental sampling, which is taking a limited series of individual measurements of a substance over time and/or space.

Sample measurements are then used to extrapolate information about the environment at large (e.g., the population, in a statistical sense). The focus of this section is to review the basics of environmental sampling, including measurement methods that are commonly employed to give baseline information to impact assessment models. This baseline information helps in two areas: first, to measure emissions directly; and secondly, to adjust and validate the output of environmental fate and transport models. Most LCA practitioners will not collect data from the natural environment, but understanding how these measurements are collected helps one understand the applicability and uncertainty associated with a given LCIA) method, and to communicate the meaning of an LCA to others.

SAMPLING POLLUTANT CONCENTRATIONS IN AIR, WATER, AND SOILS

Hundreds of methods and pieces of equipment can be used for environmental sampling, depending on the situation and the analyte, or substance under investigation. Luckily, standard methods exist for strictly defining many of these analytes in order to guide scientists and ensure comparability. These standard methods are typically developed by government agencies, such as the Environmental Protection Agency's Test Method Collections (www.epa.gov/fem/methcollectns.htm) or the U.S. Geological Survey's National Field Manual for the Collection of Water-Quality Data (http://water.usgs.gov/owq/FieldManual).

These methods define the number of samples and the frequency, or spacing, at which they should be taken, as well as define sample amounts, appropriate equipment and handling procedures, sample preparation (prior to analysis), and the analysis itself. The sampling procedure will also generally include quality control samples in order to quantify bias in the measurement method or equipment; if comparisons across time or among sites are intended, then control samples will also be taken to provide baseline measurements.

Sampling can either be active (Figure 10.1a), meaning that people are involved in the sampling activity, or passive (Figure 10.1b), meaning that samples are collected by the equipment itself and then later analyzed by people. In some cases, the actual analysis (determining analyte concentrations) can be carried out in the field using mobile equipment, but this is often impractical and field samples are sent to a laboratory for analysis. Samples can become contaminated or otherwise biased during this time (e.g., through reactions with the container walls), and so standard measurement methods specify how samples should be collected, stored, and transported.

Figure 10.1b. Passive Sampling of Air Quality on the Coast of Louisiana Following the Deepwater Horizon Oil Spill in 2010 (EPA photo by Eric Vance, www.flickr.com/photos/usepagov/)

DIRECT INDICATORS (pH, DO)

In many cases, measuring aggregate changes in the environment is more efficient than measuring the concentrations of each individual chemical included in a life cycle inventory (LCI). For example, one can measure changes in pH that are indicative of acidification in a lake, as opposed to trying to measure exactly how much deposition of various acidifying substances occurs and then modeling how this deposition will change the overall pH of the lake. The unit for concentration of hydrogen ions (pH) is on a log scale, similar to the units for acidification in some impact assessment methods. pH is an example of a direct indicator of environmental impact.

Similarly, dissolved oxygen (DO) is a direct indicator of eutrophication and is depleted after algal blooms that results from excess nutrients in water, leading to the death of aquatic organisms. DO is easily measur-

Figure 10.1a. Active Sampling of Suspended Solids in the Grand Canyon (photo by USGS, http://ga.water.usgs.gov/edu/2010/gallery/sediment-sampling.html)

able using hand-held probes with high accuracy. We could also measure the amount of nitrogen and phosphorus in different forms entering a water body, and then model how the resulting algae growth would affect DO. But this would be much more complex and introduces greater uncertainty than direct measurement. Such measurements would, however, give more information about the source of the pollutants, thus moving it up the cause-effect chain and providing points of control to reduce the pollution.

LCIA-RELATED MEASUREMENT METHODS AND UNITS

LCIA methods typically include impact categories that do not rely on physical measurements of environmental samples, including biological indicators such as biodiversity loss and human and ecosystem toxicity, or land-based metrics such as land occupation or transformation. The impact assessment models used to calculate characterization factors for these impact categories also rely on measurements that can be difficult to provide directly.

Assessing biodiversity loss, in theory, requires knowing how many species exist and how many become extinct over a certain period, but only ~2 million of the 5-30 million species on Earth have been formally identified (IUCN 2013). Instead, the metric of species richness is commonly used as a biodiversity metric, with a negative change in this metric being associated with a loss of biodiversity. Species richness is assessed by ecologists who count the number of species in a given area, while other metrics (e.g., species diversity) collect information about the quantity or value of each species. These methods are limited by the number and types of species counted, which tend to be conspicuous species (e.g., birds, mammals, or trees and other vascular plants).

Measurement of impacts can also be in the form of average statistics for the entire population, rather than for an individual site or sample. For example, modeling damages to human health (such as cancer caused by chemical exposures) must account for the average age of disease onset, the rate of mortality, any related morbidity, and the average longevity of the population. All of these statistics are then combined into an aggregate measure of disability-adjusted life year (DALY), codified by the World Health Organization (WHO) as a quantity of human life lost to disease or injury. This epidemiological data is difficult to correlate to the presence of particular pollutants.

We may assess environmental impacts by taking measurements remotely (using images from airplanes or satellites) under a discipline known as remote sensing. By analyzing the reflectivity of the surface of the Earth to different wavelengths of radiation, we can determine land cover, foliage type, the existence of water, and many other parameters. This is extremely useful in determining the total land area required for an industrial activity (e.g., surface mining). Taking a series of remote sensing measurement allows modelers to ascertain where and how land has been transformed, and again to associate this transformation to a particular unit process with a set of quantitative characterization factors.

10.2 HUMAN HEALTH AND DISEASES: TOXICOLOGY AND EPIDEMIOLOGY

In the context of classical LCIA, we are concerned about the transmission of pollutants via environmental media. Therefore, two main things need to be known: 1) what portion of an emitted substance will end up affecting humans or the environment (some receptor), and 2) the effect of that exposure. The former question deals with the fate and transport of a substance, and this is addressed in Section 10.5. This section investigates the latter question, which depends on toxicology and epidemiology.

Epidemiology and toxicology work hand-in-hand to provide the links between the emissions of an inventory and the impacts that are the output of the LCIA. Considering only the human side of impacts, this point is critical: 1.3 billion people breathe air that exceeds standards established by WHO. Overall estimates for the percentage of all human deaths that are related to environmental factors range from 25 to 40% (NIEHS 1999).

KEY TERMS (AFTER FRIIS 2007):

- **Agent:** energy (e.g., ionizing radiation), a substance (e.g., a pesticide) or an organism (e.g., influenza virus) whose presence is necessary to cause a certain disease

- **Dose:** the amount of an agent administered to (e.g., for substances or organisms) or received by (e.g., for radiation) a subject

- **Exposure:** coming into contact, or being in the vicinity of, a disease-causing agent to the extent that the deleterious effects of the agent may occur

- **Outcome:** the occurrence of an agent's effects; depending on the study, this may be a specific instance of a disease (morbidity) or a death (mortality)

TOXICOLOGY

To measure the effect of exposure, we could run some studies in which a test organism is dosed with a compound and measuring results. This works. If we ran such a study on tanks of water fleas (*Daphnia*), giving each tank a different dose of some substance, counting the number of fleas alive at the end of a week (this test would be a bioassay), the results might resemble those plotted on the curve in Figure 12.510.2:

Stressor-response curves (e.g. dose-% mortality

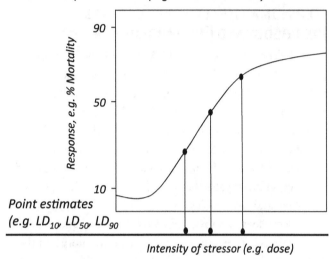

Figure 10.2. Example Dose-Response Curve

From such a curve, one could determine whether there was a threshold (a dose below which there was no measureable response). Such a point can also be called a no observable adverse effect level (NOAEL) or a no observable effect level (NOEL). Definitions of these terms can vary from study to study, so read data reports carefully is essential. For the purposes of this chapter, the dose-response curve would indicate at which concentration (i.e., dose) 50% of the population was affected. This would be the LD_{50}, which is the lethal dose that kills 50% of test subjects. One could run similar studies with different compounds or different exposure routes (e.g., inhalation versus ingestion).

If one were interested in humans, one would need to work out how to extrapolate from animal subjects to humans (see Olson et al. (2000) for a general discussion, and USEtox documentation for such factors (Rosenbaum et al. 2008)). But over time, one would start to develop a sizeable toxicology dataset. Toxicology literally means the study of poisons, but more accurately it investigates the detrimental effects of substances on living things. The EPA's Integrated Risk Information System (IRIS) compiles decades of toxicology research (US EPA 2013a).

Problems can occur once gathering real-world data becomes the objective. Are data representative of chronic (longer term and lower level) or acute (short term and higher dose) exposures. Was the study long enough to capture the desired effects? Some agents have long latency periods; for example, mesothelioma, a cancer caused by asbestos exposure, may take 40 years to develop (Friis et al. 2009). Did the study have enough subjects to capture the trend of interest? Finally, ethics should be considered because animals are used in these studies. Thankfully, studies on humans are rare and done with great care; historic violations of human rights gave rise to codes of conduct (US Congress 1991).

The cost of thorough toxicological studies is a serious consideration. To build enough statistical power, especially with low incidence rates, large subject numbers and long study times are needed. Indeed, it takes about five years and approximately $10 million to study one compound (Perkel 2012). Consider this in light of the fact that eight million compounds are commercially available, but some level of toxicity data is available for ~ 100,000 chemical compounds (Egeghy et al. 2012). That is a bit over 1%.

Furthermore, many environmental concentrations are low and exposure is long; many compounds must be considered, and there maybe synergistic, antagonistic, or other combinatory effects. The number of combinations is staggering. Recognizing this imbalance, the EU and the US EPA programs are working on computational toxicology and exposure approaches that will reduce cost and time (US EPA 2013b; EC/CO-LIPA 2013). In addition, advanced understanding of genetics (e.g., the "omic" revolution) is moving towards opening up the black box of toxicology (where a dose creates some outcome, without knowing exactly why). To have a picture of the many tens or hundreds of thousands of compounds in some LCAs, it is necessary to supplement the toxicology approach.

ENVIRONMENTAL EPIDEMIOLOGY

Epidemiology concerns the health of populations, specifically focusing on the way illness and death are distributed among people and on the factors that influence these patterns (Friis et al. 2009). The Greek roots give a hint here; this word must deal with the study of (-logy) something that is upon (epi-) the people (demos). Epidemiology is a wide-ranging field of study, covering disease outbreaks, bioterrorism, obesity patterns, and asthma. We can draw from the sub-discipline of environmental epidemiology, which is defined as the "study of diseases and health outcomes (occurring in the population) that are linked to environmental factors" (Pekkanen et al. 2001).

Environmental epidemiologists use ambient environmental levels of contaminants, working conditions, and accidents to provide insight into human dose-response (e.g., PM_{10} and $PM_{2.5}$ levels in urban air, agricultural workers' chronic exposure to pesticides, or mercury contamination in Minamata, Japan) (Friis and Sellers 2009). The list of acute exposure via accidental releases is likewise tragic, with Bhopal, the site of release of methyl isocyanate in 1984 that killed thousands, being the most notorious.

HUMAN TOXICITY IN LCIA

Impact assessment relies on characterization factors that translate emissions into impact; these factors depend on both the fate of a substance in the environment as well as its potency towards a receptor. Section 10.3 provided some grounding in the potency question. Section 10.5 addresses the fate question. In this section, the exposure and potency, as applied in LCIA, will be explored.

In the structure of USEtox (Rosenbaum et al. 2008) and several of its underlying models, potency is expressed as the effect factor (EF), which indicates an increase in a disease midpoint (e.g., cases) or endpoint (e.g., years of life affected or lost) over a lifetime. The EF is a function of the exposure pathway. In addition to the potency, we need to know about the exposure of humans to the compound in various environmental compartments. This is expressed as the exposure factor (XF). Put together, the characterization factor (CF) is a function of fate, exposure, and effect:

$$CF = FF*XF*EF \qquad (10.1)$$

EFFECT FACTORS

To connect doses to effects, we use dose-response curves, which could be constructed from toxicological or epidemiological data. Figure 12.4 shows a typical, non-linear curve. To use such a curve in LCIA, we would need to know, at a minimum, whether we were above threshold and, ideally, where a population already exists on the curve. Knowing this requires either a full knowledge of background concentrations or a complete model of all industrial activity in a region. Both are impractical. In keeping with the precautionary principle—LCIA aims to provide likely estimates, not the most conservative—LCIA pragmatically assumes a linear model with no threshold (Krewitt et al. 2002). Figure 10.3 shows the approach to linearize a dose-response curve. The working point on this curve is 0.5; this effect factor is based on the point at which 50% of the population suffers from an incidence of the disease.

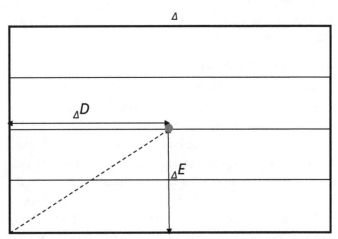

Figure 10.3. Example Extrapolation for Human-Toxicological Effect Factors, After USEtox (Rosenbaum et al. 2008)

The effect factor is then calculated as the risk / effect dose, in this case the ED_{50}:

$$EF\left[cases / kg_{intake,\,lifetime}\right] = \frac{0.5\left[cases/person\right]}{ED_{50}\left[kg_{intake,\,lifetime}/person\right]} \quad (10.2)$$

The next step is to link outcomes—the occurrence of a disease, expressed in cases—to impacts. Impacts can be defined by the study goals. For example, LCIA can operate at the midpoint level, where only the number of cases is reported (e.g., the USEtox model) or where the number of cases is normalized to some reference substance (e.g., benzene). Further down the cause-effect chain, human health endpoints considered are morbidity (i.e., disease) and mortality (i.e., death) due to cancer and non-cancerous diseases. Certainly, Overall, this framework provides a useful starting point in the complex world of human health.

Once we have data about the probability of disease, we need to link the type of disease to either morbidity or mortality in order to determine the severity of the disease. A widely-used metric is the disability-adjusted life year (DALY) concept (Murray et al. 1996). DALYs combine both morbidity and mortality into a single scale. The weighting of different health effects requires some subjectivity, but DALYs are widely used and supported by the WHO and the World Bank (Krewitt et al. 2002).

Endpoint methods may have their own factors to translate cases of disease to DALYs, though the distinction between cancer and non-cancer diseases is common. For ReCiPe, the factors are 11.5 DALYs / cancer incidence and 2.7 DALYs / non-cancer incidence (Huijbregts et al. 2005). IMPACT uses values of 13 and 1.3 DALYs / incidence for cancer and non-cancer (IMPACT World+ Team 2013); previous estimates were 6.7 and 0.67 (Crettaz et al. 2002). As a result, we can then link an exposure (dose) to an impact in terms of cases, a reference substance, or DALYs.

Exposure pathways and factors

Consider for a moment all of the different ways that environmental, or at least external, compounds can move from outside the body to inside. Typical pathways modeled in LCIA include inhalation of air and ingestion of water and food. The latter has one entry point but connects to a diverse of sources: drinking water, plant crops, animal products (e.g., dairy), and animals (e.g., fish and meat). Less common routes include dermal exposure to contaminated soil (McKone 1993). One thing that such classical LCIA methods have in common is that they consider transport and exposure of contaminants through the natural environment to human (or ecosystem) receptors. For many compounds, these exposure routes are not efficient, as ample opportunity for dilution and degradation resides in environmental media.

However, other routes can lead to direct exposure. For example, make-up, lotions, and other personal care products (PCPs), when applied directly to the skin, can result in direct exposure. Many times, workers have been exposed to high levels of substances in manufacturing facilities. Inside buildings, off-gassing from PCPs, furniture, or paint can create high air concentrations, leading to higher exposure than might be predicted from an environmental transport model (Ernstoff et al. in prep.).

Between the FF and EF of Equation (12.1) is the exposure factor (XF), which connects concentrations in environmental media to exposure. For example, if one lived downwind of a plant emitting benzene, there would likely be high concentrations of benzene in the air to be breathed in by everyone in the area. If one were to hold one's breath all day, it would cut exposure drastically. One might still have some exposure via dermal adsorption or ingestion, but one would probably be asphyxiated within twenty minutes. For each of the exposure pathways discussed above, exposure factors can be calculated as follows:

$$XF_{inh} \ (1/day) = Rate\_inhalation \ (vol/day\text{-}pers) \\ * \ population \ (pers) \ / \ volume\_air \ (vol) \quad (10.3)$$

In the case of inhalation, Equation (12.3), the XF shows the fraction of the air environmental compartment that the population inhales every day; the inverse is the number of days it takes the population to consume (i.e., cycle through) all of the compartment.

$$XF_{ing\_water} \ (1/day) = Rate\_drinking \ (vol/day\text{-}pers) \\ * \ population \ (pers) \ / \ volume\_water \ (vol) \quad (10.4)$$

Drinking water is similar:

Water and air are special cases, in which the bioaccumulation factor (BAF) is unity. The BAF describes the extent to which the concentration in a substrate is higher than the concentration in a medium. When water is the substrate humans consume, the concentration is the same in the medium (water). The concentration in fish is not necessarily identical to that in water. The units of the BAF are [kg water / kg fish] = [(kg chemical / kg fish) / (kg chemical / kg water)]. In the case of food items, the XF depends on the bioaccumulation of compounds in the food, as well as their production rate (production and consumption must be roughly equal over sufficiently long time scales).

Example BAF calculations were shown in Figure 10.1 and Figure 10.2.

$$XF_{ing\_fish} \ (1/day) = fraction\_dissolved \ (\text{-}) \\ * \ BAF \ (kg \ water \ / \ kg \ fish) \ * \ fish\_intake \ (kg \ fish/day\text{-}pers) \ * \ population \ (pers) \ / \ volume\_water \ (vol) \quad (10.5)$$

Putting it all together, we return to Equation (12.1):

$$CF = FF*XF*EF \quad (10.1)$$

High-characterization factors depend on three criteria being met: 1) a substance must be persistent in the environment, 2) it must be persistent in media to which humans are exposed, and 3) it must have a high EF. High impact requires some combination of high CF and high emissions.

Characterization factors depend on environmental fate, transport models, and toxicity models. There can be large uncertainty associated with each of these models. According to Humbert et al. (2009), an LCIA practitioner should assume a factor of 100 uncertainty on toxicity-related impacts. For other categories where the modeled uncertainty is smaller (e.g., global warming), the uncertainty can be as low as 10%.

10.4 Fate and Transport Models

The natural environment is an extremely complex system from a physics and chemistry point of view (to say nothing of biology!). Substances that are released into the atmosphere, for example, might be blown by winds to a different continent, broken down by sunlight, swept out of the air by rain, or react with other chemicals to form new substances. Almost always, multiple processes are at work.

Fate and transport models try to predict how a particle or a chemical will move around the environment (through air, water, and soils and sediments) while undergoing physical processes and chemical transformations. As with all modeling, the goal is to create a simplified representation of reality that offers insights, without requiring us to model the entire system at hand. Capturing the major physical processes and transformations in LCIA modeling is essential, but as Einstein advised, we strive to "make everything as simple as possible, but not simpler."

DESCRIPTION OF MAIN MODELED PROCESSES

The following primary environmental processes comprise the fate and transport models that underpin life cycle impact assessment methods.

Advection is movement of a substance due to movement of the overall media (bulk flow). Typical examples include movement of water pollutants downstream from natural river flow, or transport of air pollutants from wind currents.

Mass diffusion is movement of a substance from a region of high concentration to a region of low concentration, such as a concentrated plume of pollution spreading to contaminate an entire lake (but at much lower concentration).

Aggregation occurs when inter-molecular or inter-particle forces cause substances to attract each other, overcoming any repulsive forces that might be present and leading to the formation of larger clusters that are subject to physical process of flotation or settling.

Flotation or settling are familiar processes that apply to aggregated particles, typically suspended in water. Flotation occurs for less dense materials once the volume becomes large enough that buoyant forces (from displacement of liquid) pushes aggregates to the surface. Conversely, settling occurs for dense materials once the size of the aggregate becomes large enough that gravitational pull, or applied centrifugal forces, dominate and the aggregate moves to the edge or the bottom of the system (as with sediment at the bottom of a lake).

Chemical reactions will also occur in most environmental systems, as substances react with other constituents in air, water, or soils, or with the media itself. One of the most important classes of reactions is redox (reduction-oxidation) reactions in which the oxidation state of a chemical is changed, a reaction that can greatly affect its reactivity or toxicity.

Degradation is a type of chemical reaction in which a substance is irreversibly altered by breaking it into smaller compounds that may have different environmental characteristics than the original substance. Photodegradation occurs when radiation from the sun is absorbed by a molecule, causing bonds to break. A common example is polymers, which become brittle or change color when exposed to the sun as their bonds break down and the polymer chains become

shorter or less ordered.

Finally, chemicals that move from one media to another undergo phase transformation. This can occur if the substance in question naturally partitions into a different phase; for example, hydrophobic substances do not mix in water but can readily dissolve in a different solvent. A substance might also change its own phase (e.g., by volatilizing and turning into a gas from a liquid).

Several other physical and chemical processes take place in environmental systems, but those listed above are commonly included in LCIA methods. The processes rarely act alone and substances in the environment often undergo multiple processes simultaneously.

COMBINATIONS OF PROCESSES IN ATMOSPHERIC AND AQUATIC MODELS

This section traces substances in three environmental media to explain how different physical and chemical processes act in concert. First, consider a molecule of sulfur dioxide (SO_2) being emitted from the smokestack of a coal-fired power plant. Upon entering the atmosphere, the SO_2 will be carried downwind of the plant through advection, and the plume will widen and disperse through diffusion into the surrounding air. The SO_2 may chemically react with water and other atmospheric constituents to form sulfuric acid, or it may aggregate into small droplets called aerosols (which is a type of particulate matter).

Consider an emission of a copper particle from automobile brake wear released into a neighboring stream. This copper particle undergoes advection (transport through a fluid) as it moves with the water current. It undergoes chemical transformation as some copper atoms on the surface of the particle dissociate into copper ions, and the particle itself may aggregate with larger organic molecules (e.g., from the decomposition of leaves). Where advection slows (e.g., at a calm point in the stream), the aggregates may settle down to the bottom of the stream bed and become part of the sediment.

EQUILIBRIUM MODELING AND MULTI-MEDIA FUGACITY MODELS

In environmental modeling, many types of physical and chemical processes can be treated in a unified framework that employs equilibrium and multi-media fugacity modeling. The concept of equilibrium is central to fate and transport models, and to environmental chemistry in general. Equilibrium is, in essence, an equality or balance in a physical or chemical system, and nature will always push systems toward equilibria. For example, If a hot room exists next to a cold room, and someone were to open a door between the two, heat will naturally move into the cold room until the temperatures are equal—or, in other words, until the system has reached equilibrium. (This heat diffusion is analogous to the mass diffusion described in Section 10.5) Similarly, in a chemical system, a balance exists between the concentrations of the reagents and the concentration of the products; the ratio of these concentrations is called the equilibrium constant. If we introduce disequilibrium, either by adding reagents or by removing products from the system, we can induce a chemical reaction as the systems attempt to find equilibrium again.

In the case of phase transformations, the ratios at which a substance exists in its different phases are called partitioning coefficients, which depend on the concentrations of the substance in each phase. The tendency of a substance to move from one medium to another (e.g., from water to air) can also be expressed using the notion of fugacity, or an effective pressure that the concentration of a chemical in each phase exerts on the other. When fugacities are equal, then the system is in equilibrium; when they are not, substances will move from one phase to another, potentially leading to increased exposure.

The most common way to model environmental fate and transport is through the use of multi-media fugacity models, which can predict how a chemical will partition among several phases. These models can also incorporate other processes (e.g., chemical reactions or degradation) that affect the concentration of a chemical in each environmental compartment. Common fugacity models include the LEV3EPI model that is part of the USEPA's EPI Suite tool, and QWASI for evaluating fate and transport of chemicals in lake systems.

FINITE ELEMENT MODELING OF CONTAMINANT TRANSPORT

Another modeling technique that can be used to predict fate and transport behavior of chemicals in the environment is the finite element method, particularly with application to systems undergoing fluid flow, heat transfer, and chemical reactions. In this framework, a 2-D or 3-D region (e.g., an estuary) is converted into a "mesh" of individual geometric elements (triangles or boxes). The physical and chemical equations that govern fate and transport of fluids are much easier to solve for regions with a simple shape. The benefit of the finite element method is that it can convert complex geometry like an estuary into a mesh of simple shapes (whereas solving these equations for the entire region would be impossible). These equations can then be solved for each individual element and then "stitched" back together to give a detailed picture of behavior over the entire region.

Examples of finite element analysis relevant to life cycle impact assessment include modeling the movement of pollutants in a river system, or monitoring how a substance moves through a particular type of soil (as in the case of a chemical spill).

10.5 RESOURCE DEPLETION MODELS AND NON-RENEWABLE RESOURCE ECONOMICS

Several impact assessment methods include impact categories for the use of non-renewable resources (abiotic resource depletion). Like the impact categories of cumulative energy demand or water use, this impact category is concerned with the use of resources, rather than with emissions during a product's life cycle. A non-renewable resource is one that has a finite supply that is not replenished (e.g., fossil fuels). Using that resource today means that it will not be available for use in the future. In the context of natural resource economics, this means that the future supply will be decreased, thus increasing the future price of that resource (even if environmental implications of that resource use exist as well).

Impact assessment models treat resource depletion in one of two ways. In the first model, the amount of each non-renewable resource in the LCI is compared to its overall reserve R (the total supply of that resource in the environment). Each resource (e.g., each element in the periodic table) has a different reserve level, which is taken into account. The ratio between current use and reserves is then added up for a total of all physical inputs into the product system. This provides a midpoint indicator that represents the overall use of non-renewable resources. Impact assessment makes use of reference substances to provide a common unit. In the case of abiotic resource depletion, early work used antimony as a reference; therefore, many impact assessment methods now report results in kg antimony (Sb) equivalent.

The second model examines the future consequences of using a unit of non-renewable resource today. Resources are usually extracted from areas where it is easiest or cheapest to do so, while extracting these accessible resources may make it more difficult or expensive to extract the next unit of that resource. This method then calculates the increase in energy or money that will be required to extract resources in the future because of non-renewable resource used today.

While the basic method is fairly simple, several complexities occur in creating robust models of non-renewable resource depletion. For one, we do not in many cases know exactly how much of a given resource resource exists, or how much of it is practically available to us given our current extraction technologies. We question how long these resources should be available to society, a concern that has led some researchers to pursue distance-to-target methods that compare current resource use to the level that would maintain availability for 1,000 years. In addition, not all resources are created equal—they have different technological functions, different values in our lives and in the economy, and different potential substitutes. Because of these and other complexities, non-renewable resource use is sometimes omitted from the suite of indicators chosen for an LCA, although development in this area of impact assessment modeling continues.

PROBLEM EXERCISES

1. What is the difference between active and passive sampling? What factors might we consider when deciding between active and passive sampling?

2. What are the pros and cons of field versus lab analyses?

3. How are environmental measurements used in the development of LCIA methods?

4. For each of the following, decide on the scale of possible environmental and human health impacts (e.g., global, regional, or local), and explain why.

Life Cycle Impact Assessment

Tom Gloria

11.1 Introduction

Chapter 10 introduced the basic approaches to environmental impact assessment methods and models development. In this chapter, midpoint life cycle impact assessment (LCIA) impact categories, and the methods and models available to practitioners, are introduced and explored.

11.2 Life Cycle Impact Assessment Methodologies and Models

The methodologies and underlying models of life cycle impact assessment (LCIA) contain the scientific knowledge of how to assess the environmental impact of a system. The state of the practice has evolved to the point that a collection of impact categories (e.g., climate change, acidification, or ozone depletion), which are referred to as an LCIA methodology, are typically selected when conducting a study. One such example is the US Environmental Protection Agency's (EPA) Tool for the Reduction and Assessment of Chemical and other environmental Impacts (TRACI) methodology. The classification and characterization of chemicals represented by the characterization models are contained within the LCIA methodologies.

A practitioner follows three mandatory LCIA steps (ISO 14044) when conducting a study. The first step is the determination of relevant impact categories and characterization

models to be included in a study. The other two steps, classification and characterization, are part of LCIA models created by academic institutions, government agencies, and non-governmental organizations. The remaining steps are optional, and deal with reducing the many impact category results into a single score (or a small group of scores) and a final data quality check. In many cases, the current methodologies do not contain guidance on estimating the uncertainty of the results obtained.

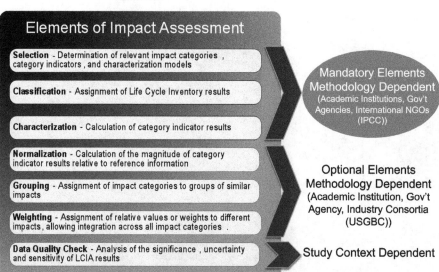

Figure 11.1. Elements of Life Cycle Impact Assessment

Portfolios of LCIA methodologies have emerged for many regions of the world, such as the US EPA's TRACI methodology (Bare 2011), the LIME method in Japan, CML (Leiden University) and ReCiPe methods from the Netherlands, and the collection of methodologies chosen for the European

Commission's (EC) Product Environmental Footprint (PEF) guidelines (EC 2013). Portfolios of LCIA methodologies are a collection of impact assessment methodologies developed by a variety of institutions (e.g., academic, governmental and non-governmental) that cover a broad spectrum of impact categories. This chapter will familiarize the reader with the state of LCIA practice, as well as the underlying models and their appropriate use.

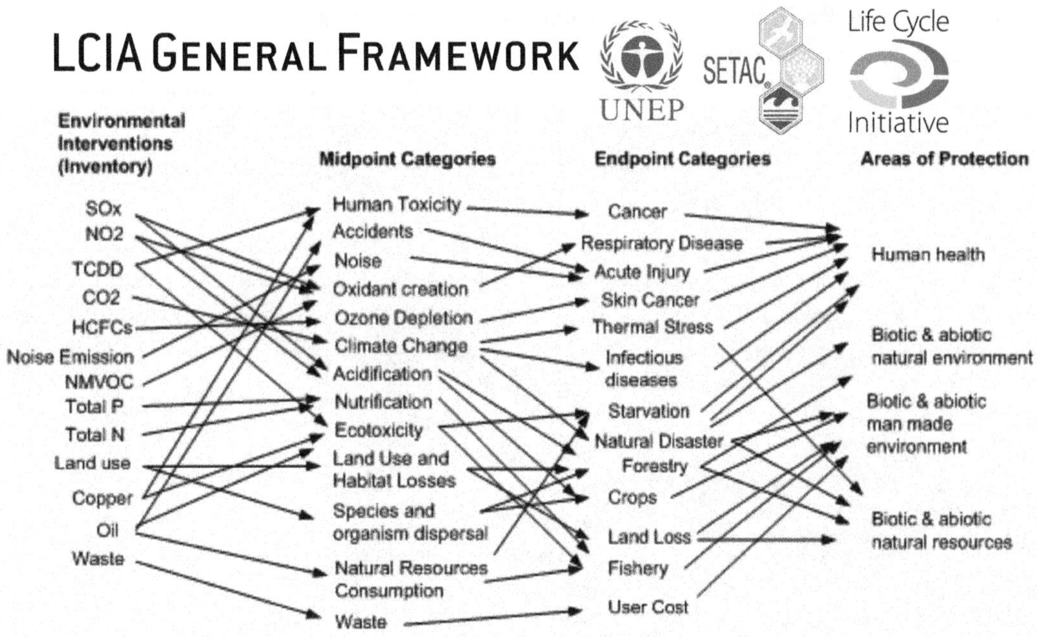

Figure 11.2. LCIA General Framework. Adapted from Jolliet et al. (2004) The LCIA Midpoint-damage Framework of the UNEP/SETAC Life Cycle Initiative, IJLCA 9 (5) 394-404.

LCIA Application: A Practitioner's Perspective

Impact assessment is the step in LCA where the life cycle inventory (LCI) data, also known as environmental interventions, are classified into categories of impact and characterized to estimate relative potential harm to both humans and ecosystems. As shown in Figure 11.2, the general framework in LCIA is to convert inventory items into impact measures using characterization factors. The inventory includes emissions to air, water, and soil, and the consumption or transformation of resources, including minerals, fossil fuels, land, and water. Note that three out of four of these endpoints are related to human wellbeing, rather than to the wellbeing of the other 10 million species on the planet.

The interventions are the beginning of the environmental mechanism that ultimately are related to endpoints, which is when damage occurs to areas we humans wish to protect, such as the natural environment (e.g., plants and animals), and the non-living, or abiotic, environment (e.g., water and minerals). Damages to the man-made environment (e.g., buildings, roadways, and crops), and to natural resources (e.g., metals, minerals, forests, domesticated animals, and crops) can also be evaluated. Normally, the environmental mechanism is better understood near the intervention than it is near the endpoint, so most common LCIA methods are evaluated at the midpoint (somewhere between the intervention and the endpoint). These LCIA models are called midpoint models. A few impact

models calculate impacts at the endpoint, such as the USE-tox models calculate impacts due to toxicity at the endpoint. However, there are also models that calculate an endpoint or damage category by aggregating impacts to a few areas of protection or a single score (e.g. Ecoindicator 99 and LIME). These models depend on gross value judgments.

To estimate specific environmental and human health loads within an impact category, the categorized LCI flows are characterized using one of many possible LCIA methodologies into common equivalence units, which are then added together to provide an overall impact category total (Figure 11.3). This equivalency conversion is based on characterization factors of the selected LCIA methodology. In their simplest form, characterization factors are relative rankings of impact of various chemicals by mass and are typically converted to a reference chemical. For example, when assessing climate change impacts at the 100 year time horizon, methane has a potency 25 times that of carbon dioxide; therefore, the characterization factor for methane is 25 CO_2 kg equivalents per 1 kg of CH_4. At the practitioner level, the application of impact assessment involves multiplication and addition. Applying the chemical rankings of an impact assessment method involves multiplying the amount of mass of a chemical emitted by its relative rank and then adding the figures. Similar rankings against reference chemicals are made throughout all impact categories.

The characterization factors are derived from the primary and applied research performed by scientists working in environmental or natural science fields. Where natural science has not reduced the relationship between emissions and measured impacts into a readily adaptable form, or where it cannot find consensus on that relationship, the given impact category cannot be evaluated by LCA. For this reason, LCA does not represent the cutting edge of environmental science. In practice, the practitioner uses a software package and the only decision to be made is the choice of impact models to be used, typically a portfolio model.

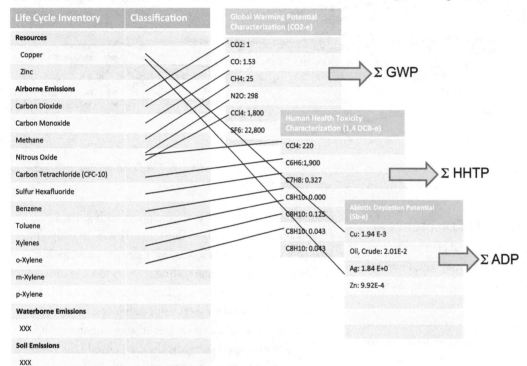

Figure 11.3. Classification and Characterization of LCI Inputs and Outputs

public to understand, and they do not attempt to describe the final impact, which is always local in nature.

A midpoint impact measure is not precisely at a point equidistant from when a chemical is emitted to the target of impact. It resides somewhere between the two. For reasons of comprehensiveness, the general approach is to measure impact at a point as far from the source of emission as can be backed by proven scientific study. The diverse impacts related to ozone depletion (e.g., skin cancer, crop damage, and cataracts) are due to ultraviolet light at the Earth's surface; reducing ozone allows UV through the atmosphere. To evaluate the effect of ozone depletion, one measures a chemical's potential to deplete the ozone in the stratosphere. These mechanisms are well-known and laboratory tested. LCIA borrows the CFs developed through this practice of environmental science. Most often, one characterizes the impact of an inventory item by using reference compounds, as in Figure 11.2, where the reference compound for Global Warming Potential is CO_2, while for Human Health it is 1,4-dichlorobenzene and for Abiotic Resource depletion it is antimony (Sb).

One typically thinks of the impacts that can occur at the endpoints. (e.g., people, plants, and animals). Although assessing impacts at the endpoint can provide us with tangible indicators, such as morbidity (sickness or injury) and mortality (death) to humans, one perspective in LCA is to look at impacts at the midpoint, which is where impact occurs along the chain of environmental mechanisms that is scientifically quantified prior to the endpoint (Figure 11.3). For example, a chemical's ability to induce radiative forcing (absorb and re-release infrared radiation in the atmosphere) is called its global warming potential (GWP). Midpoint indicators have the advantage that their relationships with the elementary flows are well-known and reproducible,. They are typically valid regardless of where those flows occur although some midpoint indicators are only applicable in some locations. Their disadvantage is that they are difficult for the general

It is also possible to assess the impacts at the level of the inventory. This is done when good models do not exist for evaluating the environmental mechanism. It is particularly common when impacts are local in nature, because life cycle inventory data is rarely available at the local level. Finally, some impact methods are end-point methods. Most endpoint methods require gross value judgments. For example, we could aggregate the impacts of many different categories to yield an ecosystem indicator, or use disability-adjusted life years (DALYs) to measure human health indicators. The World Health Organization (WHO) has developed equivalencies between morbidity effects (e.g., asthma) and mortality, which are the source of DALYs. Such equivalencies can only be developed using value judgments. In contrast, the USEtox method does not depend on such value judgments, since it uses the probability of disease incidence as its primary metric.

The LCIA methods that are used today are generally risk-based. They typically attempt to include not only the hazard potency of a chemical, but also its likelihood that it will travel to the target (its fate and transport) that will ultimately create the impact. For example, to assess the human health impact of a pesticide, one must understand where the chemical partitions into the environment, how long it remains there, and how people might be exposed. This is in contrast to hazard-based methods that do not include fate and transport or exposure mechanisms.

Some substances can participate in more than one environmental mechanism. For example, some ozone depleters are also greenhouse gases (GHGs). They are characterized in both impact categories.

The current state of the science does not cover many (or most) types of chemicals for many of the impact assessment methodologies. More than 70,000 chemicals are in use by industry, but fewer than 5,000 chemicals have been classified and characterized for their contribution to human health and ecological toxicity.

Unlike traditional risk assessment (RA- see Chapter 1: Framework of LCA), whereby potential impact is communicated by the probability of occurrence, LCIA can communicate potential impact in relative terms. For example, in RA the goal might be to reduce risk to a probability of one in a million. In LCIA, impact may be communicated relative to the results of another system or life cycle stage. For example, LCAs of two products could compare the energy or water use; within one of those product systems, LCA results could compare the contribution of climate change from material production versus from product use.

LCIA methods generally assume an infinite or long time horizon, such as 100 years for GWP. This means that emissions will be assessed for potential impact until either the chemical changes form (e.g., speciates) or it no longer creates harm due to fate and transport mechanisms (e.g., destruction of methane in the atmosphere). Of course, matter is neither created nor destroyed, so the elements in a pollutant will never actually disappear (barring radioactive decay). This creates challenges in evaluating the risks of some chemicals (e.g. heavy metals) whose toxicity is dependent on their elemental characteristics. Heavy metals can be sequestered due to biogeochemical processes, but these processes are poorly characterized by current LCIA methods.

The following sections are organized as short summaries of the most widely-used LCIA methodologies. They provide a brief summary, external publication references, and a table outlining the essential aspects of each methodology. The scope of the material includes the LCIA methodology developed by the two most active government agencies: the US EPA's TRACI[1] and the recently published EC PEFs[2]. The EC's PEF methodology is a collection of LCIA models, selected from several LCIA methodologies (mostly specific to Europe), with a few emergent characterization models that, until the EC's PEF publication, were not part of any LCIA portfolio.

TRACI PORTFOLIO OF METHODOLOGIES

The TRACI portfolio of LCIA methodologies was originally developed by the US EPA, specifically its Office of Research and Development (ORD) and National Risk Management Research Laboratory (NRMRL) divisions, as a midpoint method that represents the environmental conditions in the US as a whole or per state. Version 2.1 is the most current version of TRACI. Many of the impact methods are also used by the US Department of Commerce's Building for Environmental and Economic Sustainability (BEES)[3], which is widely applied in the US building sector. However, BEES utilizes version 1.0 of TRACI with some additional impact categories (e.g., indoor air quality, habitat alteration, and water intake).

Initially, consistency with previous modeling assumptions (especially those of the US EPA) was necessary for the development of every impact category. The initial human health cancer and non-cancer categories were fundamentally based on the assumptions made for the US EPA's Risk Assessment Guidance for Superfund and the US EPA's Exposure Factors Handbook that uses the CalTOX methodology developed at UC Berkeley[4]. The toxicity categories in TRACI Version 2.1 have since been replaced by the USEtox method. For categories such as acidification, particulate matter and smog formation, TRACI Version 1.0 utilized detailed US empirical models, such as those developed by the US National Acid Precipitation Assessment Program (NAPAP) and the California Air Resources Board (CARB), that allow the inclusion of sophisticated, location-specific approaches and location-specific characterization factors. This regionally-specific approach was abandoned with the Version 2.1 release. Instead the characterization factors are intended to cover the entire region

1 www.epa.gov/nrmrl/std/traci/traci.html
2 http://ec.europa.eu/environment/eussd/product_footprint.htm
3 www.nist.gov/el/economics/BEESSoftware.cfm
4 www.dtsc.ca.gov/AssessingRisk/caltox.cfm

of the continental US. However, research is moving forward with highly specific characterization of land use and water consumption in a future version of TRACI. The original version of TRACI was released in August 2002 (Bare et al. 2003) followed by a release of TRACI Version 2.0 in 2011 (Bare 2011) and an updated TRACI Version 2.1 in 2012 (Bare 2012) with a complete set of the Intergovernmental Panel on Climate Change (IPCC) GHG emission factors.

EUROPEAN COMMISSION'S PRODUCT ENVIRONMENTAL FOOTPRINT PORTFOLIO OF METHODOLOGIES

The EC's PEF LCIA portfolio of methodologies are a collection of methodologies from several European and broader international institutions. They include:

- the CML 2002 methodology, developed by Leiden University, Institute of Environmental Science (CML)
- the Impact Assessment of Chemical Toxics (IMPACT) 2002+ methodologies created through the collaboration

of the Centre interuniversitaire de recherché sur le cycle de vie des produits, procédés et services (CIRAIG), Polytechnique Montreal, University of Michigan, Quantis International, and Ecole Polytechnique de Lausanne (EPFL)

- the ReCiPe[5] methodology, created by Dutch Government National Institute for Public Health and the Environment (RIVM), Radboud University, CML, and PRé;
- the Environmental Design of Industrial Products (EDIP) methodology developed at the Technical University of Denmark (DTU).

In addition to LCIA models within the respective LCIA methodologies, the EC's PEF has identified several additional stand-alone LCIA models as preferred approaches. The stand-alone models cover impact areas related to water consumption, ionizing radiation, land use, acidification potential, and eutrophication potential.

Impact Category	Methodology	Midpoint Units
Climate Change	Intergovernmental Panel on Climate Change 2007 (revised 2011) (Solomon et al. 2011)	kg CO2 eq
Ozone Depletion	World Meteorological Organization (WMO 2003) and the US EPA (2008a, 2008b, 2008c)	kg CFC-11 eq
Ecotoxicity	USEtox model (Rosenbaum et al. 2011) and USEPA (Bare 2012)	Comparative Toxic Unit – Ecotoxicity (CTUe)
Human Health Toxicity (cancer and non-cancer)	USEtox model (Rosenbaum et al. 2011) and USEPA (Bare 2012)	Comparative Toxic Unit – Human (CTUe)
Human Health Toxicity (cancer)	USEtox model (Rosenbaum et al. 2011) and USEPA (Bare 2012)	Comparative Toxic Unit – Ecotoxicity (CTUe)
Particulate Matter Respiratory Effects	Humbert (2009) adjusted for North America	kg PM2.5 eq
Photochemical Ozone (Smog) Formation	Maximum Incremental Reactivity (MIR) method, Carter (2012)	kg O3 eq
Acidification	US EPA (Bare et al. 2003)	kg SO2 eq
Eutrophication	US EPA (Bare et al. 2003)	kg N eq
Fossil Fuel Depletion	US EPA (Bare et al. 2003) and Eco-Indicator 99 (Goedkoop and Spriensma 2001)	MJ Surplus

Table 11.1. TRACI Portfolio of Impact Assessment Methodologies

5 The acronym also represents the initials of the institutes that were the main contributors to this project and the major collaborators in its design: Dutch Government National Institute for Public Health and the Environment (RIVM), Radboud University, CML, and PRé.

Impact Category	Methodology	Midpoint Units
Climate Change	Intergovernmental Panel on Climate Change 2007 (revised 2011)	kg CO2 eq
Ozone Depletion	Environmental Design of Industrial Products (EDIP) (based on World Meteorological Organization (WMO 2003))	kg CFC-11 eq
Ecotoxicity (Freshwater)	USEtox model (Rosenbaum et al. 2011)	Comparative Toxic Unit – Ecotoxicity (CTUe)
Human Health Toxicity (Cancer and Non-Cancer)	USEtox model (Rosenbaum et al. 2011)	Comparative Toxic Unit – Human (CTUe)
Human Health Toxicity	USEtox model (Rosenbaum et al. 2011)	Comparative Toxic Unit – Ecotoxicity (CTUe)
Particulate Matter	RiskPoll model in IMPACT 2002+/Ecoindicator 99	kg PM2.5 eq
Ionizing Radiation (Human Health)	Human health effects model (Frischknecht et al. 2000) and (Dreicer 1995)	kg U235 eq
Photochemical Ozone Formation	Van Zelm et al., 2008 as applied in ReCiPe	Kg NMVOC eq
Acidification	Accumulated Exceedance (Seppälä et al. 2006) (Posch et al. 2008)	mol H+ eq
Eutrophication (Terrestrial)	Accumulated Exceedance (Seppälä et al. 2006) (Posch et al. 2008)	mol N eq
Eutrophication (Aquatic)	Struijs et al., 2009 as implemented in ReCiPe	fresh water: kg P equivalent marine: kg N equivalent
Resource Depletion (Water)	Swiss Ecoscarcity (Frischknecht et al. 2008)	m3 water use related to local scarcity of water
Resource Depletion (Mineral and Fossil)	Leiden University (CML 2002) (Guinée et al. 2002)	kg antimony (Sb) equivalent
Land Transformation	Soil Organic Matter (SOM) model (Milà i Canals et al. 2007)	kg (deficit)

Table 11.2. European Commission Product Environmental Footprint Portfolio of Impact Assessment Methodologies

11.3 DESCRIPTIONS OF THE METHODOLOGIES USED IN TRACI AND THE EC'S PEF

CLIMATE CHANGE

Both the US EPA and the EC's PEF portfolios use the same basic method to assess climate change. The method is based on a measure of anthropogenic (man-made) influences based on GWPs of emitted GHG relative to carbon dioxide. The six main GHGs as identified by the Kyoto Protocol include: carbon dioxide (CO_2), methane (CH_4), nitrous oxide (N_2O), hydrofluorocarbons (HFCs), perfluorocarbons (PFCs), and sulfur hexafluoride (SF_6). (UNFCCC 2008). Sources for CO_2 are mainly related to the combustion of fossil fuels for electricity generation and transportation, and can also be caused by certain chemical reactions and agricultural activities

The GWP method employed in LCIA is the globally recognized model (i.e., the Bern model), developed by the IPCC to calculate the radiative forcing of all GHGs and brand them as GWPs. The factors used by the IPCC are published in (Solomon et al. 2011).

As shown in Figure 11.4, the midpoint GWP measure is based on global atmospheric mixing (which takes about two years) of GHGs, the inherent ability of each gas to absorb and re-radiate infrared radiation, and the residence time of each gas in the atmosphere. The point of emission is irrelevant to this measure. All environmental impacts are local in nature and climate change induces many impacts at the endpoint, including reducing net primary production and decreasing biodiversity, water stress, wild fires, and human health damages, including heat stress, increase in infectious diseases, higher frequency of flooding events, and malnutrition. Some endpoint models have attempted to capture these endpoints but they have not generally been useful, in part because of the need to aggregate impacts that occur in many different locations.

OZONE DEPLETION

The impact category of ozone depletion considers the reduction in the total volume of ozone in the Earth's stratospheric stratosphere (the ozone layer). The ozone layer absorbs 97 to 99% of the sun's medium-frequency ultraviolet light, which can damage both humans and other forms of life on Earth (NASA 2013). The CFs for ozone depletion accounts for the destruction of the stratospheric ozone layer by anthropogenic emissions of ozone depleting substances (ODSs). The major ODSs are chlorofluorocarbons (CFCs), halons (halogen containing hydrocarbons), methane (CH4) and nitrous oxide (N2O). Chlorofluorocarbons are used in refrigeration, foam blowing, solvents, and specialized aerosol propellants. Halons, which are similar to chlorofluorocarbons, are used in fire extinguishers. Methane is a product of agricultural, industrial, and mining activities, while nitrous oxide is from combustion and fertilizer use.

The basic measure of the potency of a chemical to deplete the ozone layers is its ability to form equivalent effective stratospheric chlorine (EECS) relative to CFC-11. As shown in Figure 11.5, the ozone destruction cycle involves a catalytic reaction of ozone (O3) with a chlorine radical (Cl·). The chlo-

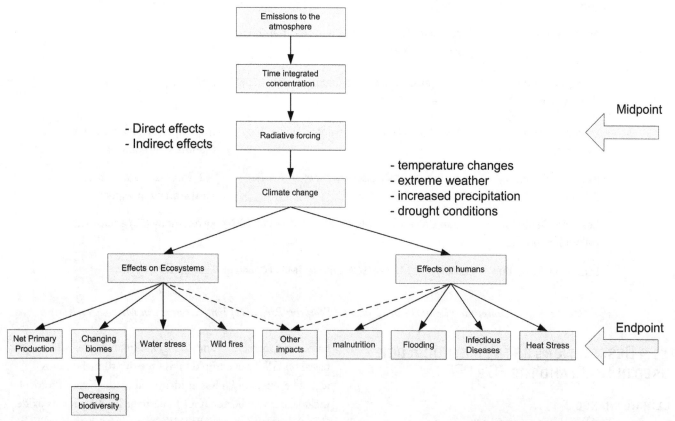

Figure 11.4. Midpoint and Endpoint Model of Global Warming (adapted from ILCD Handbook: Framework and requirements for LCIA models and indicators (JRC 2010))

rine radical breaks apart the ozone, forming chlorine monoxide and diatomic oxygen. The chlorine radical is broken free from the oxygen molecule by a second reaction and is available to start the cycle again.

The scientific community has been able to identify key

chemical reactions and also the transport properties that bring ODSs to the stratosphere. Hence, the ozone depletion potential (ODPs) equivalency factors incorporate atmospheric residence time, the formation of reactive free radicals, and the resulting stratospheric ozone depletion.

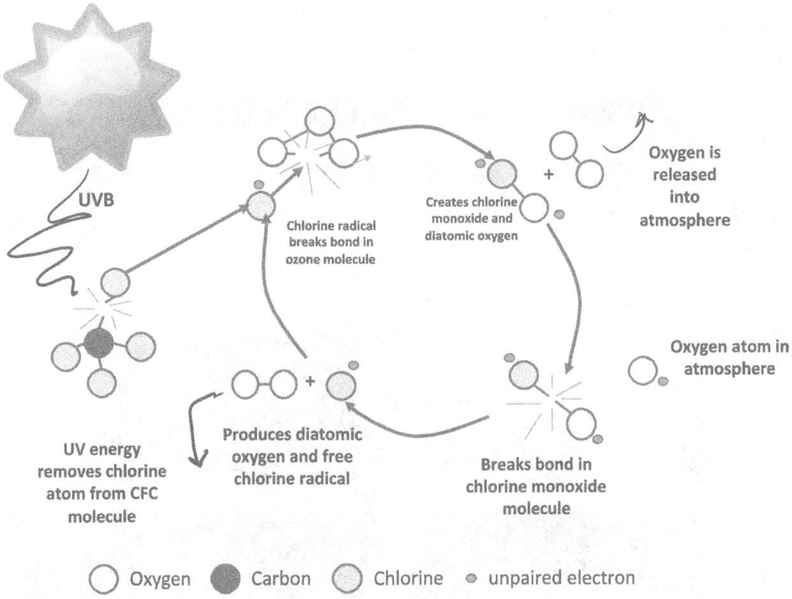

As shown in Figure 11.6, the midpoint measure of the ODPs is based on global atmospheric mixing, the inherent potency of the chemicals in question, and their residence time in the atmosphere, all of which are fairly close to the point of emission along the environmental mechanism chain. Similar to GWP, the ODP impact indicator incorporates a wide range of impacts that affect both ecosystems and humans. For ecosystems, this can include effects to primary production, forests, crops, aquatic life (e.g., amphibian and fish populations), and the general destruction of any polymer- or biopolymer-based materials. The ODP factors are published by the World Meteorological Organization (WMO) and the US EPA (WMO 2003, USEPA: 2008a, 2008b, and 2008c). All portfolio standards that include ozone depletion use these ODPs to characterize ozone depletion.

Figure 11.5 Ozone Destruction Cycle

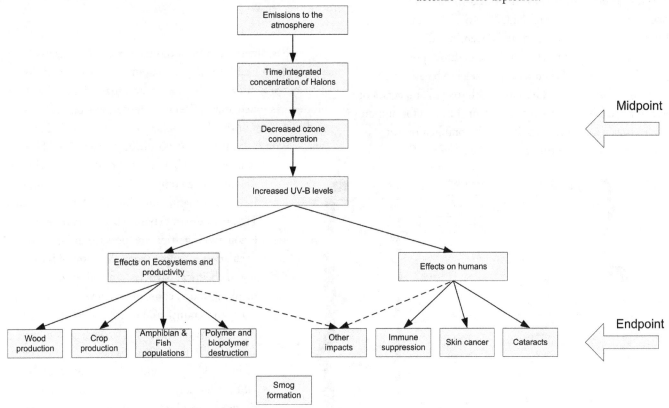

Figure 11.6. Midpoint and Endpoint Model of Ozone Depletion (adopted from ILCD Handbook: Framework and requirements for LCIA models and indicators (JRC 2010))

Toxicity – human health and ecological

The state of the LCIA practice, which assesses potential chemical toxicity impacts to humans and ecosystems, is a risk-based approach. In contrast to a hazard-based approach, which only considers the intrinsic potential for a chemical to cause harm, the risk-based approach in LCIA takes into account the probability that a chemical will actually cause harm. This means that we consider not only the toxic potency of a chemical but also its fate. How far and fast will the chemical travel in air, water and soil once it is released? What potential route will the exposure take? Will the chemical come in contact with the skin, be ingested, or be inhaled)?

Unlike many of the other impact categories used in LCIA, toxicity is particularly challenging due to the need to cover the more than 70,000 chemicals currently used by industry, as well as the broad range of degree and severity of effects that are to be captured. As such, the most widely used toxicity method in LCIA is the consensus model USEtox developed by a team of researchers from the Task Force on Toxic Impacts under the UNEP-SETAC Life Cycle Initiative. Their work grew out of the many different toxicity methods already in use. USEtox's risk-based approach is designed to describe the fate, exposure, and effects of chemicals. Because the USEtox model was developed under the auspices of the UNEP-SETAC, it will be further evaluated, developed, and disseminated.

As shown in Figure 11.7, the basic USEtox approach provides a combined chemical fate model to be used for both assessing ecotoxicity and human health toxicity impacts. For ecotoxicity, this becomes the fate factor (FF) and for human health toxicity chemical fate is used in combination with exposure route to determine the intake fraction (iF).

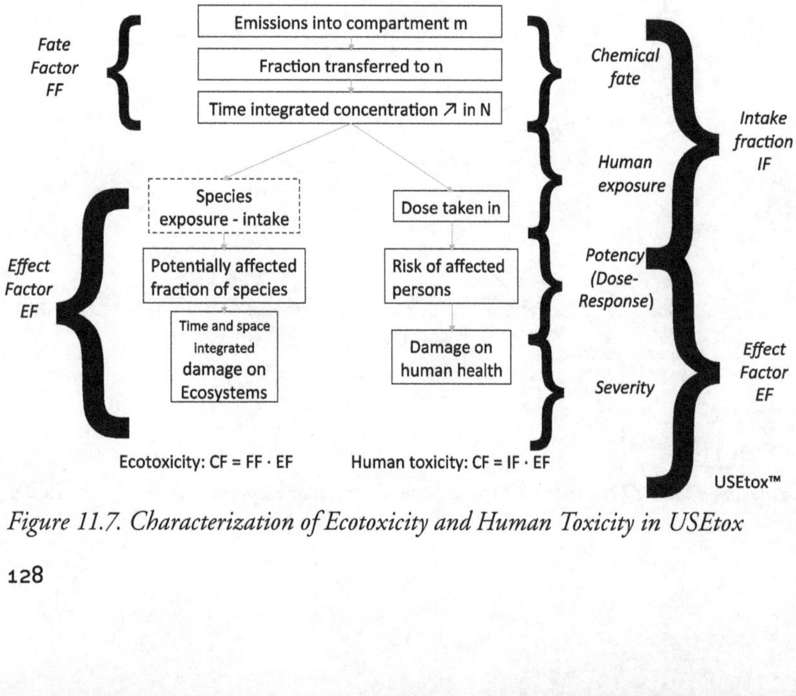

To estimate chemical fate in USEtox, two geographical scales are specified: continental and global scale. The continental scale has the compartments of urban air, rural air, freshwater, sea, natural soil, and agricultural soil; the global scale has air, freshwater, ocean, natural soil, and agricultural soil. The continental scale is nested in the global scale as shown in Figure 11.8. A nested chemical can be transported from one scale to the other.

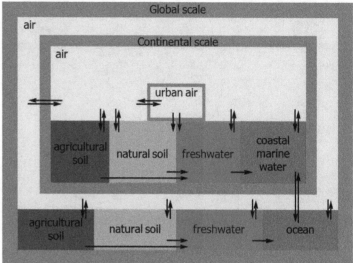

Figure 11.8. Nested Fate Structure of the USEtox. (Huijbregts et al. 2010)

The basic method in most human toxicity models used to assess human health toxicity is to examine incidences of cases of cancer or non-cancer disease per unit mass of a chemical released. As stated above, the characterization factor is based on the quotient of the iF multiplied by the FF. The iF is a combination of the fate and exposure of the chemical. Human exposure is a measure of the rate at which a pollutant in a compartment is able to transport into the human population. As shown in Figure 11.9, the human exposure pathways of inhalation and ingestion are modeled in USEtox. Inhalation is simply the mechanism of breathing in chemical particles. Ingestion includes drinking water, consuming above-ground (e.g., fruits and grains), and below-ground produce (e.g., root vegetables), meats, dairy products, and fish. Dermal contact is not currently included in the USEtox model. The elements of the USEtox human health exposure model are very similar to standard human health-risk assessment approaches.

Figure 11.7. Characterization of Ecotoxicity and Human Toxicity in USEtox

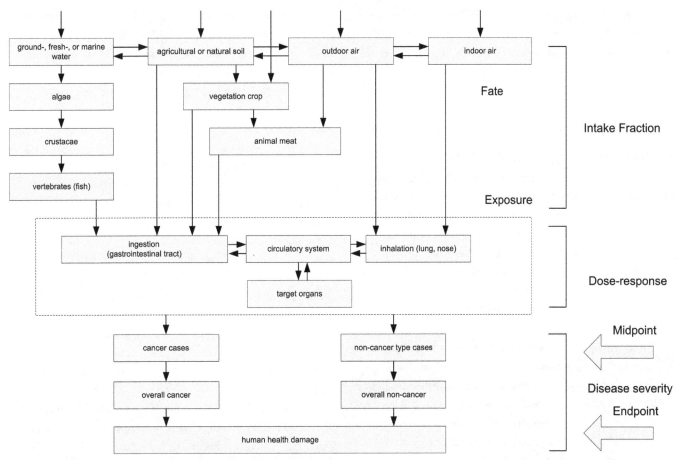

Figure 11.9. Midpoint and Endpoint Model of Toxicity (adopted from ILCD Handbook: Framework and requirements for LCIA models and indicators (JRC 2010))

The human effect factor (EF) is based on the change in the lifetime disease probability due to change in lifetime intake of a pollutant (cases/kg$_{intake}$). For USEtox, the response curve for the effective dose at 50% response level (ED50) for a given chemical is used. The dose-response curve assumes linearity in concentration-response up to the point at which the lifetime disease probability is 0.5. Factors of ED50s for inhalation and ingestion were developed for both cancer and non-cancer based on chronic ED50s for humans.

The approach taken to estimate the ecological toxicity of chemicals is based on the measure of potentially disappeared fraction (PAF) of species per mass of a chemical emitted. The characterization factor is based on the Fate Factor (FF) X Effect Factor (EF). More precisely, the Effect Factor is based on a measure of the hazardous concentration (kg/m3) of a chemical at which the species in a freshwater aquatic ecosystem (HC50) are exposed to a concentration above the concentration at which 50% of a population dies in a laboratory test (EC50).

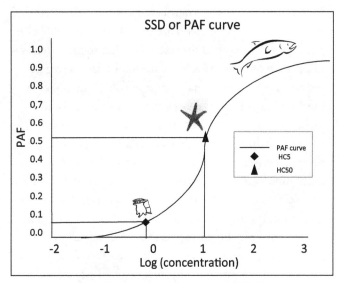

Figure 11.10. Species Sensitivity Distribution (SSD) Curve for USEtox (Rosenbaum et al. 2008) The geometric mean of the chronic HC50s from 3 species from 3 trophic levels are used.

At least three different EC50 values (species) from at least three different trophic levels (algae, crustacea, and invertebrates). Chronic (long-term) EC50s are preferred with population-relevant endpoints (e.g. reproduction, growth, and mortality). If a sufficient number of $EC50_{chronic}$ factors are available (≥ 3), an $HC50_{chronic}$ is calculated and used directly in the EF calculation. If the number of $EC50_{chronic}$ values are insufficient (< 3), but sufficient $EC50_{acute}$ values (≥ 3) are available, an $HC50_{acute}$ is calculated. A $HC50_{chronic}$ is calculated by use of an assessment factor of 2: $HC50_{chronic} = HC50_{acute}/2$. In the majority of cases, only sufficient acute data are available.

Respiratory effects (particulate matter)

Particulate matter, also known as aerosols, is a leading global cause of human health problems. Small particles less than 10 micrometers in diameter pose the greatest problems because they can get deep into lungs and, in extreme cases, even enter one's bloodstream and potentially harm heart function. Types of particulate matter include inhalable coarse particles (e.g., from roadways and dusty industries), which are between 2.5 and 10 micrometers in diameter, and fine parti-

A word of caution when using human health and ecological toxicity models is clearly stated in the USEtox User's Manual: As described by Rosenbaum et al. (2008), the characterization factors must be used in a way that reflects the large variation, often orders of magnitude, between chemical characterization factors as well as the three orders of magnitude uncertainty on the individual factors. This means that contributions of 1%, 5%, or 90% to the total human toxicity score are essentially equal but significantly larger than those of a chemical contributing to less than one per thousand of the total score. Disregarding this fact has been a major cause of complaints about the variability of these factors across impact assessment methods, whereas the most critical chemicals were often the same within a factor 1,000 across methods. In practice, these toxicity factors are useful to identify the 10 or 20 most important toxics pertinent for their applications. The life cycle toxicity scores thus enable the identification of all chemicals contributing more than, e.g. one thousandth to the total score. In most applications this will allow the practitioner to identify 10 to 30 chemicals to look at in priority and, perhaps more importantly, to disregard 400 other substances whose impacts are not significant for the considered application. Once these most critical substances have been identified further analysis can be carried out on the life cycle phase, application components responsible for these emissions, and the respective importance of fate, exposure and effect in determining the impacts of this chemical (Huijbregts et al. 2010).

cles (e.g., found in smoke and haze), which are 2.5 micrometers in diameter and smaller (US EPA 2013).

Many scientific studies have linked particle pollution exposure to a variety of problems, including:

* premature death in people with heart or lung disease
* nonfatal heart attacks
* irregular heartbeat
* aggravated asthma
* decreased lung function
* increased respiratory symptoms (e.g., irritation of the airways, coughing or difficulty breathing).

The state of the practice of LCIA is to assess a measure of human health effects per emission of particulate matter by mass, which includes the following: ammonia (NH3), carbon monoxide (CO), sulfur dioxide (SO2), and nitrogen oxides (NOx). All of these particulates are precursors to particulate matter, including the following: particulate matter less than 2.5 µm (PM2.5); particulate matter less than 10µm (PM10); and total suspended particulate (TSP). The measures of impact are all normalized to PM2.5 µm emissions as the reference substance.

As shown in Figure 11.11, similar to human health toxicity developed in USEtox, the Characterization Factor = Intake Fraction (iF) X Effect Factor (EF). The methodology has been developed for both European and North American analysis for TRACI 2.1. Therefore, the (i) is based on spatially resolved, multimedia, multipathway, fate, exposure, and effect models for North America and Europe that have been developed to evaluate intake fractions (Humbert et al. 2009).

The EFs are based on DALYs from the Eco-Indicator99 Method, widely used by the World Health Organization (WHO). The DALYs are constructed by adding Years of Life Lost (YLL) + Years Life Disabled (YLD). The DALY effect factors used in the respiratory effects model are based on European conditions as developed by Hofstetter (1998). The significance of the development of characterization factors based on DALYs occurs on two levels. First the method is not a pure mid-point method, as it examines endpoint effects on humans using DALYs. Second, since the method uses DALYs, it inherent weights inherently weighs years of life lost (YLL) with years of life disabled (YLD), which is not permitted when disclosing a comparative assertion to the public; it also is not permitted, nor in the selection of impact categories to be used for environmental product declarations (EPDs).

Respiratory effects (ionizing radiation)

The EU added a respiratory effects model that includes

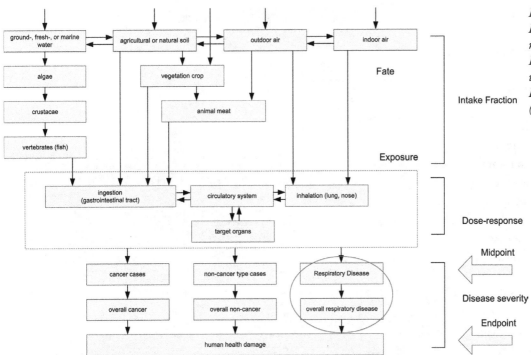

Figure 11.11 Midpoint and Endpoint Model for Respiratory Effects (adopted from ILCD Handbook: Framework and requirements for LCIA models and indicators (JRC 2010))

ionizing radiation, which goes back to Frischknecht et al. (2000). This method is also used in Eco-indicator 99 (Goedkoop and Spriensma 2001), CML 2002 (Guinée et al. 2002), Impact 2002+ (Jolliet 2003), ReCiPe (Goedkoop et al. 2008) and Swiss Ecoscarcity 2006 (Frischknecht 2009).

Similar to the human health toxicity model developed

for USEtox and the respiratory effects model developed by Humbert et al. (2011), the ionizing radiation model provides a method for comparative evaluation of impacts of radioactive substances on human health due to respiration (see Figure 11.12).

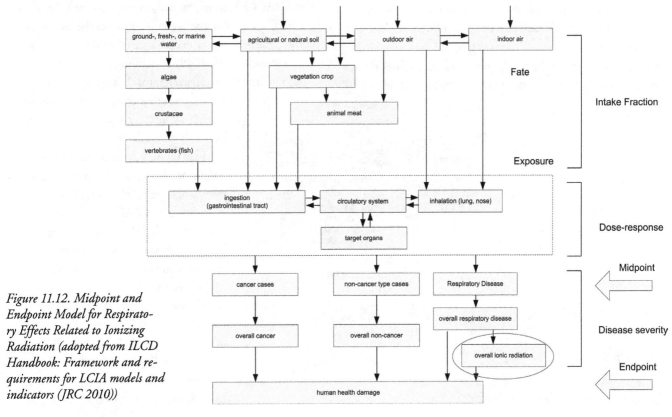

Figure 11.12. Midpoint and Endpoint Model for Respiratory Effects Related to Ionizing Radiation (adopted from ILCD Handbook: Framework and requirements for LCIA models and indicators (JRC 2010))

The method is applicable on a global scale and is compatible with the human toxicity category. Similar to the respiratory method by Humbert et al. (2009), it provides comparative results utilizing DALYs at endpoint. The method is fairly robust and is based on site-specific fate and exposure models for French nuclear facilities generalized for a site-independent assessment. No spatially differentiated factors are presently available. No independent midpoint exists with this category, but intermediary data on fate and exposure, and on the number of cases, could be used to develop such an approach. Any DALY-based method is inherently values-based, because the CFs are developed by panels of experts within WHO. Nevertheless, the methodology is an accepted endpoint method.

ACIDIFICATION POTENTIAL

Acidification potential is the basic measure of increased acids in the environment from air emissions. Sources, particularly from the combustion of fossil fuels that contain sulfur and nitrogen compounds, are emitted and travel as gaseous and particulate pollutants in the atmosphere. The gases and particulates then come in contact, or precipitate out of the atmosphere, in wet

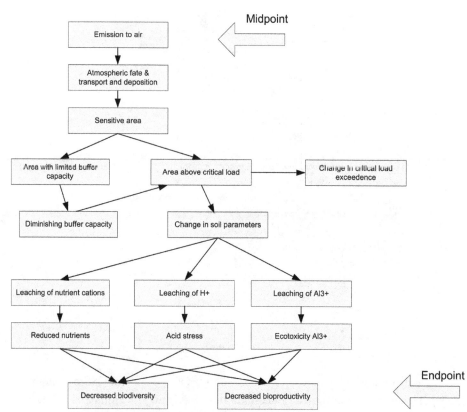

Figure 11.14. Midpoint and Endpoint Model for Acidification Potential TRACI Version 2.1 (adopted from ILCD Handbook: Framework and requirements for LCIA models and indicators (JRC 2010))

or dry form to receiving bodies, both land and water.

For TRACI 2.1, receiving body conditions are ignored. In the EC's PEF portfolio, the method considers the location of deposition, identifying the exceedance of local neutralizing capacity (Seppälä, et al. 2006). It is valid only in Western Europe. The more generalized approach is currently favored for regions other than Western Europe as global continental impact factors are currently unavailable.

The TRACI 2.1 method is based simply on the increasing concentration of hydrogen ions (H+) within a local environment. TRACI 2.1 uses an acidification model which incorporates the increasing hydrogen ion potential within the environment without incorporation of site-specific characteristics such as the ability for certain environments to provide buffering capability (Wenzel *et al.* 1997, Wenzel & Hauschild 1997). As shown in Figure 11.14, the midpoint measure is at the point of emission to air (Acidification Potential does not include

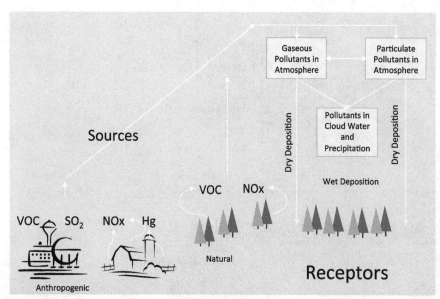

Figure 11.13. Fate and Transport for Acidification

emissions to the other compartments of water and soil). As such, the fate and transport, and receiving bodies of both water and soil are excluded in this impact method. These two approaches illustrate the trade-offs between models with universal application but lower prediction of potential harm versus models with better local prediction but low global applicability.

Similar to the method selected for TRACI Version 2.1, the accumulated exceedance (AE) approach is a measure of increased acids in the environment from air emissions that also includes the consideration of exceeding a critical threshold load (Seppälä et al. 2006). Critical load indicates the amount of a given substance (per defined unit of area and time) that can be introduced into an ecosystem without bringing about long-term environmental damage. The AE approach provides European country-dependent Characterization Factors for both aquatic and terrestrial environments. The atmospheric transport and deposition model to land area and major lakes and rivers is determined using the European Monitoring and Evaluation Programme (EMEP) model combined with a European critical-load database.

A shown in Figure 11.15, the AE method mid-

point is further down along the environmental mechanism chain in comparison to the TRACI 2.1 method. This allows for greater accuracy, but comes at the expense of model sophistication and limited applicability.

A recent publication (Posch et al. 2008) updated the factors of the AE method using the latest 2006 version of the EMEP Eulerian atmospheric dispersion model (Tarrason et al. 2006). This provides depositions onto different land cover categories, as well as the newest critical load database (Hettelingh et al. 2007) consisting of about 1.2 million different ecosystem such as forests, surface waters, and semi-natural vegetation.

The AE approach has high stakeholder acceptance as in Europe, as AE-type calculations are used for policy purposes by the EC and by the United Nations Economic Commission for Europe's (UNECE) Convention on Long-range Transboundary Air Pollution (LRTAP). This method includes atmospheric and soil-fate factors sensitive to emission scenarios, and distinguishes between load to non-sensitive and sensitive areas. AE is the most readily adaptable method that can be used in further research to generate global default Characterization Factors. It could also be used to generate a set of consistent CFs for each continent, if complemented by a set of re-

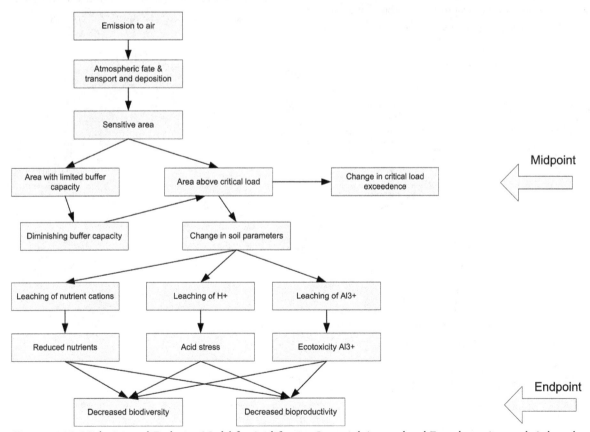

Figure 11.15. Midpoint and Endpoint Model for Acidification Potential Accumulated Exceedance Approach (adopted from ILCD Handbook: Framework and requirements for LCIA models and indicators (JRC 2010))

gional/continental models that are consistent with each other and can eventually be integrated in one global model and expert estimate on soil-sensitive areas.

PHOTOCHEMICAL OZONE (SMOG) FORMATION POTENTIAL

Photochemical ozone, also known as smog or ground-level ozone, forms in the troposphere (the region from the Earth's surface to 12 to 20 km). Ground-level ozone represents a serious air-quality risk to human health (e.g., emphysema, bronchitis and asthma), especially to the elderly, children, and people with heart and lung conditions. The majority of smog formation occurs when nitrogen oxides (NOx), carbon monoxide (CO), and volatile organic compounds (VOCs) react in the atmosphere in the presence of sunlight. Chemical such as NOx, CO, and VOCs are called ozone precursors. The combustion of motor-vehicles fuels, industrial emissions, and chemical solvents are the major sources of smog.

Two general approaches can estimate the potential impacts associated with smog. The TRACI 2.1 method involves a chemical's potential to react with NOx to form tropospheric ozone based on the maximum incremental reactivity (MIR) scale by Carter (1994). The EC's PEF preferred method is to measure of a chemical's potential to degrade organic compounds and the production of tropospheric ozone based on the Master Chemical Mechanism (MCM) method (Derwent

et al.) and the European LOTOS-EUROS fate and transport modeling.

As shown in Figure 11.16, both midpoint methods are focused at the point in the environmental mechanism chain related to increased tropospheric ozone concentration. As such, both methods include impacts related to increased exposure of ozone to humans and to vegetation.

The Maximum Incremental Reactivity (MIR) approach incorporated into the TRACI Version 2.1 portfolio of methods involves considering a specified scenario of meteorological conditions, emissions, and initial concentrations. An incremental reactivity (IR) of an organic compound is the change in the peak ozone concentration, divided by an incremental change in the initial concentration and emissions of the VOC. The maximum IRs are calculated for reference scenarios covering the United States, consisting of a specified meteorological situation, initial pollutant concentrations, and emission rates of NOx and VOC. The TRACI 2.1 method uses the most up-to-date MIR factors by Carter (2012). It is, again, an example of a method with more global application.

The MCM approach selected by the EC's PEF is a characterization factor for ozone formation of substance x expressed as non-methane volatile organic compounds (NMVOC) equivalents. The method is representative for both potential ecosys-

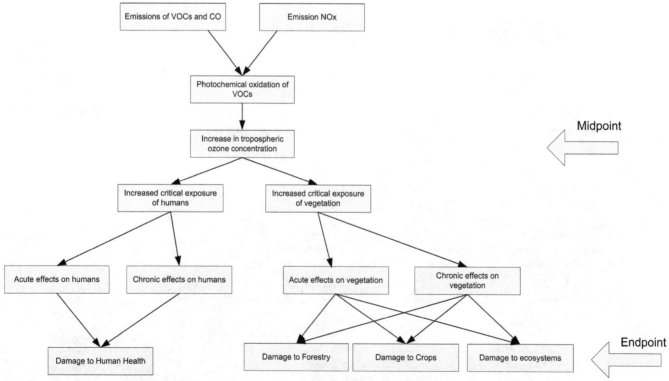

Figure 11.16. Midpoint and Endpoint Model for Photochemical Ozone Formation (adopted from ILCD Handbook: Framework and requirements for LCIA models and indicators (JRC 2010))

tem and human health effects, and is defined as the marginal change in the 24h-average European concentration of ozones (Struijs et al. 2009). The method has been adopted by ReCiPe (Goedkoop et al. 2008) and Impact 2002+ (Jolliet 2003).

EUTROPHICATION POTENTIAL

Similar to acidification potential, the current state of the practice of LCIA takes two approaches to assessing the eutrophication potential: the TRACI 2.1 Redfield ratio approach, and the AE approach. In general, eutrophication is the process by which a body of water acquires a high concentration of nutrients, especially phosphates and nitrates. These typically promote an excessive growth of algae. As a result, the algae dies and decomposes, and high levels of organic matter and decomposing organisms deplete the water of available oxygen, causing the death of other organisms (e.g., fish and live fish eggs).

The basic method in LCIA is to measure of increased macronutrients in bioavailable forms on aquatic and terrestrial ecosystems. For TRACI 2.1, the characterization factor is based on the increasing concentration of limiting nutrients of nitrogen (N) and phosphorus (P), biological oxygen demand 5-day period (BOD 5), and Chemical Oxygen Demand (COD). The N and P equivalents are then aggregated based on the Redfield ratio (Norris 2003). Site-specific characterization factors based on fate and transport modeling are not available, as factors are averaged for all of North America. As show in Figure 11.17, the TRACI 2.1 midpoint method is at the point of emission of chemicals to both aquatic and terrestrial environments, allowing for a broad use of the characterization factors.

The eutrophication AE approach follows the same AE approach to assessing impacts related to the emission of acid causing chemicals, whereby receiving body critical loads are considered. As such, the eutrophication AE method involves two separate models: one for terrestrial eutrophication and the other for aquatic eutrophication. The units of measure are different and cannot be combined.

For terrestrial eutrophication, the AE method is a measure of increased macro-nutrients in bio-available forms on terrestrial ecosystems that exceed a critical load (Seppälä et al. 2006). The characterization factors are based on the increasing concentration of nitrogen (N) nutrients with consideration of local environment AE over critical load. The terrestrial AE provides European country-dependent Characterization Factors based on the atmospheric transport and deposition model to land

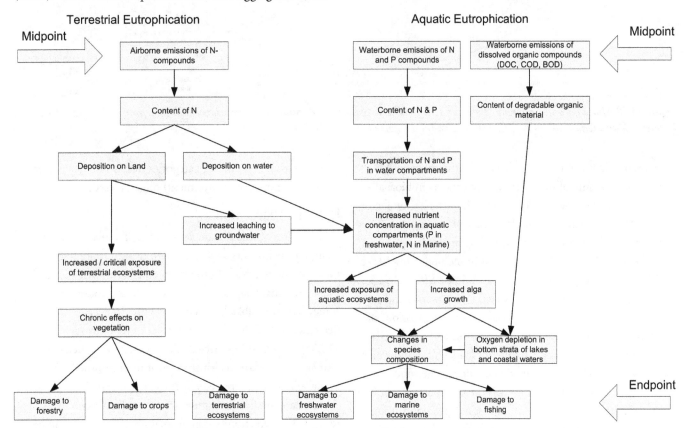

Figure 11.17. Midpoint and Endpoint Model for TRACI 2.1 Eutrophication (adapted from ILCD Handbook: Framework and requirements for LCIA models and indicators (JRC 2010))

areas and major lakes and rivers, while also using the EMEP (www.emep.int) model combined with the European critical load database. As shown in Figure 11.18, the terrestrial eutrophication AE method includes impacts only to chronic effects on vegetation (e.g., forests, crops and other terrestrial ecosystems). One must use the aquatic eutrophication AE method to assess freshwater and marine environments impacts.

32 countries where emission takes place) is subdivided into 101 river catchments and 41 coastal seas. As shown in Figure 11.19, the aquatic eutrophication AE method distinguishes freshwater systems (only P emissions considered) and marine systems (only N emissions considered). The N and P equivalents are not aggregated by the Redfield ratio. Therefore, in order to apply the AE method to assess total eutrophication

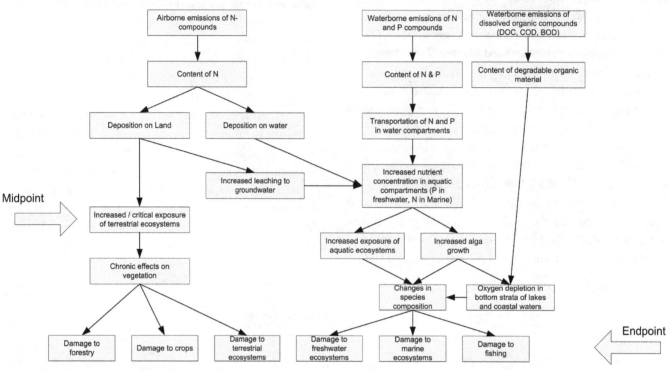

Figure 11.18 Midpoint and Endpoint Model for Terrestrial Eutrophication Accumulated Exceedance (adopted from ILCD Handbook: Framework and requirements for LCIA models and indicators (JRC 2010))

The aquatic (marine and freshwater) eutrophication AE method is a measure of increased macronutrients in bioavailable forms on aquatic ecosystems. The characterization factors are based on the increasing concentration of limiting nutrients of nitrogen (N) and phosphorus (P). The methodology is from the ReCiPe (Goedkoop et al. 2008) method using EUTREND for atmospheric emissions and Cause Effect Relation Model (CARMEN) for ground water, inland waters and coastal areas. EUTREND is applied to convert emissions of NH3 and NOx into air into deposition on sea and soil. CARMEN calculates the change in nutrient loads in ground water, inland waters, and coastal seas from changes in net nutrient emissions or gross supplies. It models the transport of nutrients from agricultural supply and atmospheric deposition through groundwater drainage and surface runoff spatially resolved over 124320 grid elements. The European continent (i.e., the

potential, the three impact categories (terrestrial, freshwater aquatic, and marine environment) must be used.

RESOURCE DEPLETION

Measures of resource depletion in LCIA involves understanding the impacts associated with using a finite resource found in the Earth's lithosphere (metals, minerals, and fossil fuels). Resource depletion can also include resources that are somewhat renewable, but that may, on a time scale, result in decreased availability (e.g., freshwater). As shown in Figure 11.20, the focus of resource depletion is primarily related to damages to human wealth that result in diminished future availability or an increased effort to satisfy a need. To some extent, damages to ecosystems are covered, particularly as they relate to the depletion or contamination of freshwater bodies.

The TRACI Version 2.1 portfolio of methods contains

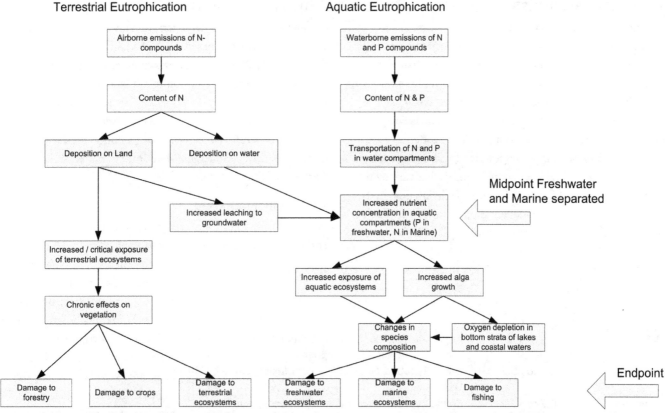

Figure 11.19. Midpoint and Endpoint Model for Aquatic Eutrophication Accumulated Exceedance (adopted from ILCD Handbook: Framework and requirements for LCIA models and indicators (JRC 2010))

only one resource depletion impact category, fossil fuel depletion. The method is based on the concept that the continued extraction and production of fossil fuels tends to consume the most economically recoverable reserves first, so continued extraction will become more energy intensive in the future (assuming that technology is fixed). The characterization factors by fuel type (coal, oil, and natural gas) estimate the amount of energy required for extraction for the scenarios at a point in the future when total cumulative consumption is five times the present cumulative consumption. The increase in unit energy requirements per unit of consumption for each fuel provides an estimate of the incremental energy input cost per unit of consumption (MJ/MJ). The method was originally developed and introduced by Müller-Wenk (1998) and integrated into the Eco-Indicator 99 portfolio of impact assessment methods (Goedkoop and Spriensma 2001).

The EC's PEF portfolio includes two resource depletion categories: abiotic depletion of minerals and fossil fuels, and water scarcity.

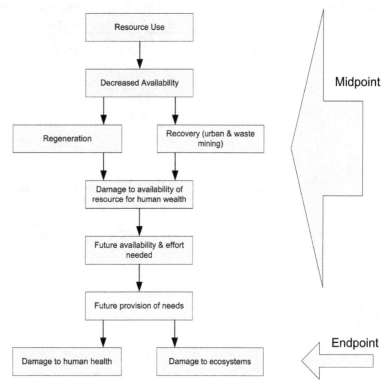

Figure 11.20. Midpoint and Endpoint Model for Resource Depletion (adopted from ILCD Handbook: Framework and requirements for LCIA models and indicators (JRC 2010))

ABIOTIC RESOURCE DEPLETION

The basic method to measure of the depletion of minerals and fossil fuels is based on ultimate stock reserves and extraction rates. Ultimate stock reserves are estimated by multiplying the average natural concentration of the resources in the Earth's crust by the mass of the crust. Fossil energy carriers are assumed to be full substitutes (both as energy carriers and as materials). As a result, the ADPs for energy are all equivalent. The heating values for each fossil source are then used to determine ADPs, resulting in ADPs that vary by mass. The characterization factors are abiotic depletion potentials (ADPs) and expressed in kg of antimony equivalent, which is the adopted reference element (Guinée et al. 2002).

WATER RESOURCE DEPLETION

The water resource depletion method contained with the EC's PEF portfolio is based on the Swiss Ecoscarcity approach (Frischknecht 2009). Characterization Factors are based on the ratio of water consumption to available water resources, with consideration of recovery rates. The resulting ratios are calculated at the country level and delineated into six categories based on OECD water-scarcity groupings (as shown in Table 11.3): Low, Moderate, Medium, High, Very High, and Extreme (OECD 2004).

LAND TRANSFORMATION

The estimation of land use impacts is highly complex and still in nascent development within the field of LCIA. At present, the EC's PEF portfolio has selected a land-use impact method, one that partially addresses land transformation. The TRACI 2.1 portfolio does not include any estimation of impact associated with land use. From Figure 11.21, land use impacts are separated into land transformation and land occupation. Land transformation can involve physical changes to the soil surface and subsurface, physical

Classification of several countries in the water scarcity categories

	Water scarcity ratio $\left(\frac{\text{Water Consumption}}{\text{Available Resource}}\right)$	Typical countries
Low	<0.1	Argentina, Austria, Estonia, Iceland, Ireland, Madagascar, Russia, Switzerland
Moderate	0.1 to <0.2	Czech Republic, Greece, France, Mexico, Turkey, USA
Medium	0.2 to <0.4	China, Cyprus, Germany, Italy, Japan, Spain, Thailand
High	0.4 to <0.6	Algeria, Bulgaria, Morocco, Sudan, Tunisia
Very High	0.6 to <1.0	Pakistan, Syria, Tadzhikistan, Turkmenistan
Extreme	≥1.0	Israel, Kuwait, Oman, Qatar, Saudi Arabia, Yemen

Table 11.3. Water Scarcity Classification - Swiss Ecoscarcity Method

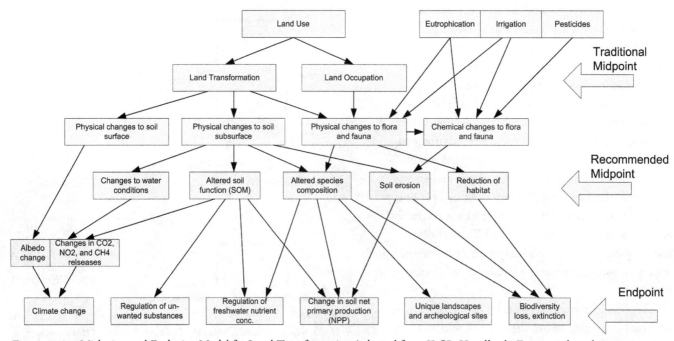

Figure 11.21. Midpoint and Endpoint Model for Land Transformation (adapted from ILCD Handbook: Framework and requirements for LCIA models and indicators (JRC 2010))

changes to flora and fauna, and chemical changes to flora and fauna (from the application of pesticides, increased nutrients, and irrigation).

The land transformation method selected the EC's PEF is focused on alteration of soil function through the depletion of soil organic matter (SOM). The method was developed under the United Nations Environmental Programme (UNEP) and the Society of Environmental Toxicology and Chemistry (SE-TAC) LCI (Milà i Canals et al. 2007a, 2007b).

In this method, SOM is considered a soil quality indicator, particularly for assessing the impacts on fertile land use (agriculture and forestry systems). The SOM provides buffer capacity, soil structure and fertility. However, weaknesses exist in the model, whereby in highly acidified or waterlogged soils the SOM may not correlate directly with soil quality.

In order to apply this methodology, the LCA practitioner is expected to know the location, the timeframe, and the SOM values before and after the land occupation, as well as the SOM value of the reference land system, the relaxation rate, and associated SOM values. Based on this, the LCA practitioner is expected to calculate the characterization factors for the foreground system. Characterization factors for certain land-use flows in the background system are provided in Milà i Canals et al. (2007c). Practitioners are cautioned, that, when applied to LCIA, the methodology should be combined with biodiversity indicators.

PROBLEM EXERCISES

1. What does MIR mean, and what impact category does it refer to?

2. USEtox models human health exposure using what parameters?

3. Explain the difference between a midpoint and an endpoint LCIA method.

4. Some impact methods are regional and others are global. Explain the pros and cons of these different approaches

Decision Support Calculations

Philip White and Bengt Steen

12.1 Introduction

Life cycle assessment (LCA) strives to objectively model and calculate substance and energy flows, and their potential environmental impacts, within a system. The impact characterization process provides this modeling within a uniformly structured and scientifically based life cycle framework. This chapter describes approaches and calculations that can assist in analyzing and better understanding life cycle inventory (LCI) data and characterized impact data. Four primary decision support methods are reviewed:

a. grouping

b. mid-point calculations, including normalization and weighting

c. damage assessment calculations, which integrate weighting

d. multi-criteria aggregation procedures

LCA practitioners need to keep in mind that all of these decision-support approaches involve subjective decisions to varying degrees. We need to clearly communicate to the commissioners and recipients of each assessment about the type of subjective decisions that we have made in the decision-support process.

12.2 Grouping

Grouping sorts impact categories in an LCA study into category groups. It assigns impact categories into one or more sets as predefined in the goal and scope definition, and it may or may not involve ranking of the category groups. Like post-characterization calculations, grouping is an optional step in an LCA (ISO 14044.4.4.3.1). Grouping may be applied in addition to any of the other decision-support calculations described in this chapter. According to ISO (14044.4.4.3.3), grouping can be accomplished in two ways. The first approach sorts the impact categories on a non-ranked, nominal basis. In the second approach, the impact categories are sorted on a ranked, ordinal basis.

Nominal grouping

An LCA practitioner may want to group characterized impact results according to the qualities of the impact categories. For instance, impact categories can be placed in groups according to global versus regional impact scales, or according to a general impact category type (Table 12.1). Nominal grouping requires an understanding of the qualities by which impact categories can be grouped.

Grouping	Impact category
Geographic scale grouping:	
Global impacts	climate change, fossil fuel depletion
Regional impacts	acidification, photochemical smog
Impact typology grouping:	
Ecological health	climate change, acidification
Human health	photochemical smog
Resource use	fossil fuel depletion

Table 12.1 Examples of Nominal Grouping of Impact Categories

Nominal grouping can also be applied to LCI data, differentiating whether inventory data refer to resource consumption, or to emissions into the air, water, or soil. As with nominal grouping applied to impact results, nominal grouping applied to LCI helps organize and contextualize the information.

Nominal grouping can also be used in complex products to combine the impact results of multiple parts into subassemblies, or to combine constituents in a material formulation. This can consolidate what would otherwise be large number of smaller elements into easier-to-comprehend groups.

Indicator-result grouping can alter the perceived significance of normalized or weighted results (or can help enhance understanding of the significance of the results). This may cause confusion if you are focusing on "hot spots" identified by the largest stacked bars in a group, as the following graphs demonstrate. Figure 12.1 shows the category indicator results for an entire tube of toothpaste grouped in life cycle stages. In this example, the water used for tooth cleaning and for the subsequent wastewater treatment of the used water is excluded.

Figure 12.2 shows the same results grouped into categories as defined by the manufacturer who commissioned the study. In this case, the second largest impacting category after toothpaste constituents is transportation. This example demonstrates the importance of providing the individual values within any grouped result, so that the reader understands which parts create the largest and smallest impacts.

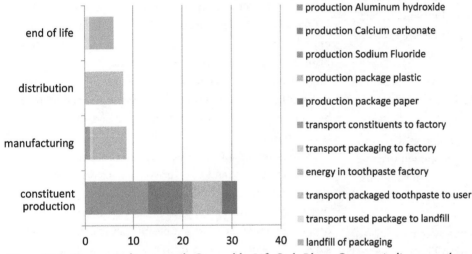

Figure 12.1. Category Indicator result Grouped by Life Cycle Phase. Category indicator results have been grouped in four life cycle phases.

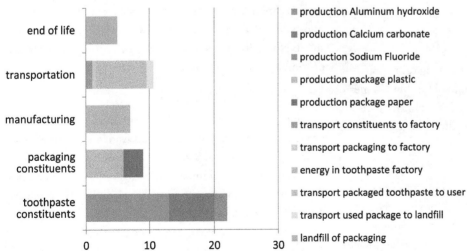

Figure 12.2. Category Indicator Result in Other Grouping. The same category indicator results are grouped into other sets defined by the commissioner of the study.

ORDINAL GROUPING

An LCA practitioner may wish to organize impact categories according to ranked preferences. Ordinal grouping sorts the impact categories in a preferred hierarchy. Creation of ordinal rankings is a social and political process, based on value-choices. Like weighting, which is discussed subsequently, ordinal grouping is sometimes rejected by proponents of a strictly scientific use of LCA. Ordinal ranking may be desired by a study commissioner who wishes to organize the results based on value choices.

People who create the ranking define the process by which ordinal rankings are established. Table 12.2 shows importance rankings for impact categories by two different commissioners or stakeholder groups, indicating their preferences. Different individuals, organizations, and societies have different preferences. Therefore, different parties can reach divergent ordinally ranked results based on the same indicator results or normalized indicator results.

Various methods can

Importance Ranking	Category rank by commissioner A	Category rank by commissioner B
highest	human toxicity	climate change
high	ozone layer depletion	marine ecotoxicity
low	climate change	human toxicity
lowest	marine ecotoxicity	ozone layer depletion

Table 12.2. Examples of Ordinal Grouping

be used to create ordinal group rankings, depending on the expressed Goal and Scope, and intended communication strategy for the study. The number of people participating in the rank definition process and the degree to which they understand impact categories are factors that influence this process. Participants can create the ratings by voting, by applying an averaging process, and by repeating the process, if necessary, for categories in tied positions. Likewise, if the participants are not familiar with environmental science, it may be instructive to first provide a tutorial on the basic science of the pertinent impact categories.

CONDITIONS FOR APPLYING GROUPING

The LCA practitioner holds considerable leeway in deciding whether to apply nominal or ordinal groupings. Grouping may be a useful approach for the study if grouping impact categories according to relevant characteristics can help the study's stakeholders better understand and interpret the indicator results.

Nominal grouping may be an efficient way to communicate some common impact characteristics. This may be helpful in evaluating ways for improvement. For example, local impacts may be regulated by permits, while regional and global impacts require international agreements.

Because the process of defining ordinal rankings inherently applies value choices that can be influenced by personal, societal, or political factors, and are not scientific in nature, ordinal grouping should usually be avoided unless a logical reason for it exists. If ordinal grouping is used, the LCA practitioner should conform to the above-mentioned ISO standards for communicating the study results, especially if the results of the study are to be published.

COMMUNICATING GROUPED RESULTS

Few mandatory guidelines exist for nominal grouping in an LCA study; however, ISO (14044.3.2.2) provides requirements for communicating the results of ordinal grouping. 14040 and 14044 do not support the underlying value choices used to ordinally group impact categories. Resulting ordinal groupings are the sole responsibilities of the commissioners of

the study. Grouping methods need to be described in, and be consistent with, the goal and scope of the LCA. The grouping methods used in the LCA should be documented to supply transparency. Further, if ordinal grouping is included in the LCA, the study report should clearly describe:

a. how grouped results were used in the study

b. which recommendations derived from ordinal grouping are based on value-choices

c. a justification of the criteria (i.e., personal, organizational, or national choices) used for ordinal grouping

12.3 MID-POINT CALCULATIONS

Mid-point decisions support calculations that can occur after mid-point impacts have been characterized. Normalization or weighting after characterization are not mandatory; ISO 14044.4.4.3.1 defines these operations as "optional elements of LCIA." Characterized mid-point indicators are reported in distinct units for each characterized impact category. To most decision makers (the recipients of an LCA report), characterized results in different impact categories with non-comparable category units are difficult to interpret and compare in a balanced way.

Normalization and/or weighting can contextualize the relative significance and relevance of characterized impact results. Such mid-point calculations can deliver LCA results that allow more direct comparison between impact categories or can determine the most significant impact categories in the study. Consequently, many decisions makers apply normalization and/or weighting to support their assessment related decisions. Particularly for stakeholders without an extensive background in environmental science, or who lack sufficient time to carefully study detailed characterized impact results, mid-point calculations can make LCA results easier to understand. Normalization and weighting can be singularly applied or combined with each other.

12.4 NORMALIZATION

Normalization compares the size of impacts from the product system to the size of impacts from a selected reference system (ISO 14044.4.4.3.2.1). It calculates, for example, the percentage of the total impact that the product system creates. For each impact category, normalization divides the product system's impacts by a reference system's impacts. Like grouping, normalization is not a required part of an LCA. However, normalized impact results exhibit unique qualities within LCA methodology. Normalization renders results with the same (often dimensionless) unit, enabling comparison or further aggregation among impact categories. A frequent motivation to normalize characterized impact results is to enable comparison of impacts in the divergent impact categories, which makes the overall LCA results easier to comprehend and evaluate.

Characterized impacts can be normalized with either of two techniques (Reap et al. 2008). The first method, internal normalization, references data internal to the system studied. The second method, external normalization, references data external to the system studied. It uses either the total estimated annual impacts for a geographical area, or the total estimated annual per capita impacts for a geographical area. Both internal normalization and external normalization can be useful in different situations.

INTERNAL NORMALIZATION

Internal normalization divides the impacts of the product system with the impacts of an analogous product system or one of the systems in the study (i.e., the alternative, which often had the highest impact in a specific impact category in the

Impact category	Impact Unit	shoe A	shoe B	shoe C
Abiotic depletion	kg Sb eq	0.053	0.044	0.065
Acidification	kg SO$_2$ eq	0.054	0.062	0.041
Eutrophication	kg PO$_4$ eq	0.010	0.011	0.019
Global warming	kg CO$_2$ eq	6.701	5.820	5.459
Human toxicity	kg 1,4-DB eq	1.898	2.372	1.168

Table 12.3. Characterized Impacts of Three Types of Running Shoes

assessment). To demonstrate the process of internal normalization, we employ an LCA of three pairs of running shoes. We use one of the shoes as a reference value for internal normalization. Some of the key system components and life cycle stages of the shoes include molded elastomers, woven cotton fabric, cardboard box, oceanic container shipping, and land-filling at the shoes' end of life. The functional unit is 1,000 hours of use

of the running shoes. Three shoes have been characterized by applying the CML 2000 model, as demonstrated in Table 12.3 Only five impact categories were selected to simplify the example.

Impact category	normalized impact unit	shoe a	shoe b	shoe c
Abiotic depletion		1.00	0.83	1.22
Acidification		1.00	1.15	0.76
Eutrophication		1.00	1.12	1.94
Global warming		1.00	0.87	0.81
Human toxicity		1.00	1.25	0.62

Table 12.4 Internally Normalized Impacts of Three Types of Running Shoes

We can internally normalize these shoes by dividing each shoe impact value by the impact values of shoe A (Table 12.4 and Figure 12.3). Because the characterized impact unit (for instance, kg Sb equivalent for abiotic depletion potential, or (ADP) is divided by the identical unit of shoe A, the resulting impact unit is dimensionless.

The internally normalized value for shoe A is 1.0 in all categories, while the other shoe impact values are often greater or less than 1.0. If we had normalized with shoe B or shoe C, the relative scale of the values in the table would remain proportional to all the values in Table 12.4.

EXTERNAL NORMALIZATION

External normalization uses data from outside the immediate product system to contextualize the characterized results. External normalization values are typically scientifically based estimates of the total amount of annual

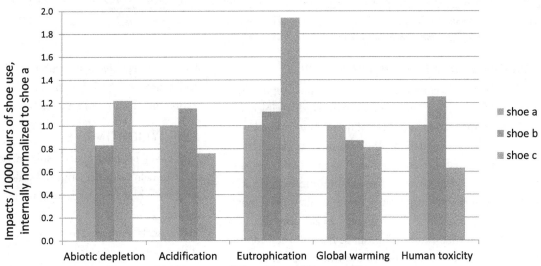

Figure 12.3. Internally Normalized Impacts of Three Types of Running Shoes

impact, per impact category, created in a geographical region. ISO specifies that the selection of the reference system should "consider the consistency of the spatial and temporal scales of the environmental mechanism for each impact category and the reference value." The environmental mechanism is defined as the "system of physical, chemical and biological processes for a given impact category linking the life cycle inventory analysis results to category indicators" (14044.3.38).

These recommendations imply that impact categories with global scales (such as global warming) should ideally be normalized with global values, while local or regional impact categories (such as acidification) should ideally be normalized with local or regional values. This practice is often limited by the deficiency of local and regional normalization data; thus, national or global normalization is often the most commonly applied approach.

Returning to our previous example, the running shoes can be normalized using external normalization values. Table 12.5 lists the annual normalization values per person globally in the year 1995 for the five impact categories, as well as the resulting normalized impact units and values.

LCA nomenclature. Internal normalization, however, cannot provide information about the relative importance of the different scales of the impact categories prior to normalization.

External normalization, like internal normalization, communicates the relative scales of each of the impact categories and provides direct comparison of the percentage contribution in different impact categories. External normalization data provides contextual perspective because it indicates the scale or relevance of contributions to impacts occurring in the natural environment (Norris 2001). Conversely, external normalization data is only as useful as its degree of accuracy and can contribute to bias in normalized category indicator results (White 2010). Methods to reduce normalization bias are discussed in Chapter 15, Bias and Uncertainty.

COMMUNICATING NORMALIZED RESULTS

Normalized impact results should be described in, and be consistent with, the Goal and Scope of the LCA. Further, all methods and calculations used should be documented to provide transparency (ISO 14044.5.5.1). Selection of the reference system normalization data included in an LCA should be described and justified in the assessment text. The normalization method should be clearly documented in the text of the study, and in all tables and charts that visualize the normalized impact results. Further, data and indicator results, or normalized indicator results reached prior to weighting, should be provided with the weighting results to ensure that trade-offs remain available to decision makers. Characterized results should be reported along with normalized results in public studies, especially with LCA studies used as a basis for public comparative assertions about competing products or services (ISO 14044.5.2).

impact category	World 1995 annual per capita normalization values & units*		normalized impact unit	normalized impacts shoe A	normalized impacts shoe B	normalized impacts shoe C
Abiotic depletion	1.6E+11	kg Sb eq./person-year	1 /person-year	3.4E-13	2.8E-13	4.2E-13
Acidification	3.2E+11	kg SO2 eq./ person-year	1 /person-year	1.7E-13	1.9E-13	1.3E-13
Eutrophication	1.3E+11	kg PO4 eq./ person-year	1 /person-year	7.4E-14	8.3E-14	1.4E-13
Global warming	4.1E+13	kg CO2 eq./ person-year	1 /person-year	1.6E-13	1.4E-13	1.3E-13
Human toxicity	5.7E+13	kg 1,4-DB eq./ person-year	1 /person-year	3.3E-14	4.2E-14	2.0E-14

Table 12.5. Externally Normalized Impacts of Three Types of Running Shoes. Frischnecht, et al, 2007, CML 2001, Centre of Environmental Science of Leiden University

CONDITIONS FOR APPLYING NORMALIZATION

The decision to employ normalization should support the Goal and Scope and meet the needs of the study recipients. Internal normalization and external normalization can be applied in different situations. If two or more product systems employ the same functional unit and system boundaries, internal normalization can effectively communicate the relative scales of each of the impact categories, and allows for a more direct comparison of the different impact categories than without the normalization. This can be helpful when communicating to consumers or other groups who are unfamiliar with

12.5 WEIGHTING

A characterized impact result provides a quantified value that can help us understand the scale of the impact in a given category, while a normalized impact result provides a contextualized value that can also help one understand the impact's relative scale compared to other impact categories. But neither of these results necessarily expresses the degree to which a particular impact category is a problem that needs addressing.

Weighting is an optional post-characterization step that incorporates stakeholder perceptions about the relative importance of each of the impact categories in the LCA.

Weighing has two meanings among LCA practitioners. The first meaning is a simple multiplication factor that signifies the perceived importance of each impact category. We call this simple weighting. The other meaning of weighting is a process by which the characterized impacts are multiplied by values with category units specific to that impact category. This cancels the category specific units in the weighted result. We call this aggregative weighting, which is often used in damage-assessment methods that are explained later in this chapter.

Simple weighting converts indicator results by multiplying each result with a weighting factor that is based on value choices. The term weighting was originally called valuation, but was renamed weighting by ISO in the 1990s. Individuals, organizations, and societies hold different preferences; therefore, different parties will reach different weighting results based on the same indicator results or normalized indicator results. Weighting is mainly performed for two purposes: 1) to support interpretation of the indicator results and 2) to quantify consequences when defining cut-off criteria in the goal and scope phase (ISO 14044.3.4 2006). Weighting methods are often revised, as new knowledge of impacts and values develop.

Once we apply weighting factors and obtained weighting results for all impact categories – be it for emissions and resources at the inventory level, at the mid-point level, or at the damage level – one can compare and aggregate different impacts. Comparisons of overall impacts can also be made between different unit processes or groups of unit processes. The

choice between two options can be made even if trade-offs are made between different impact category indicator results.

WEIGHTING PRINCIPLES

Weighting involves trade-offs between different category indicator results. The process of making trade-offs employs two principles: 1) define a number of criteria to be met or 2) use a common measure. Sometimes both principles may be used.

The first principle is often used in environmental policy, where goals and standards for emissions and ambient concentrations are applied. The first weighting method in LCA that was published used ambient air and water standards as criteria, and calculated how many m^3 were required to dilute an emission to the standard level (Ahbe 1990). Then all impacts could be expressed in m^3. In LCA literature, this weighting principle is referred to as a "distance to target method."

The second principle is used in economics. Everything has an economic value and can be compared according to things with other economic values. A product system with lower environmental cost is better than one with a higher cost.

At the mid-point level, economic value is sometimes used together with the distance to target method. A target level for an impact category (e.g., acidification) is set and the control cost to reach that level is used as a weighting factor (Vogtländer et al. 2000). A criticism of economic weighting is that, by equating the value of environmental impacts to money, it facilitates substitution between ecological capital and financial capital. Such a substitution represents a form of weak sustainability (see insert).

Weighting factors are often determined by a representatively selected group of people who are consulted to gauge their perceptions about the relative importance of environmental health, human health, and resource use impact categories. Many panel methods are available, based on traditions from product development, marketing, and decision-making. An example of panel weighting at the mid-point level is BEES (Lippiatt 2007). More about different weighting methods can be found in Alroth et al. (2011).

EXAMPLE OF CHARACTERIZED RESULT WEIGHTING

Weighting values for a given set of impact categories are often apportioned so that they sum to a value of 1.0 (or 100%). Because the weighting values are lower than 1.0, and the characterized impacts are multiplied by their respective weight values, the resulting weighted impact results are smaller than the un-weighted results. This disparity is of minor concern because the results are evaluated in relationship to each other.

Weak sustainability allows for the substitution of economic capital and ecological capital, and thereby allows natural capital (habitats, species, and resources) to be systematically degraded. Strong sustainability prohibits the substitution of ecological and financial capital (Hediger 2000) and requires preservation of natural capital. The concepts of weak and strong sustainability have evolved from attempts to define sustainable development in more detail than what was provided by the initial Brundtland Commission (Brundtland, et al. 1987). A vivid body of literature on the matter now describes the sliding scale from weak to strong. Some authors even argue that weak is not so weak, because a strong economy leads to more sustainable development.

impact category	weighting values	weighted unit	weighted impacts shoe A	weighted impacts shoe B	weighted impacts shoe C
Abiotic depletion	0.10	kg Sb eq.	0.005	0.004	0.007
Acidification	0.12	kg SO$_2$ eq.	0.006	0.007	0.005
Eutrophication	0.18	kg PO$_4$ eq.	0.002	0.002	0.003
Global warming	0.39	kg CO$_2$ eq.	2.613	2.270	2.129
Human toxicity	0.21	kg 1,4-DB eq.	0.399	0.498	0.245
total of weights	1.00				

Table 12.6 Weighted Impacts of Three Types of Running Shoes

To demonstrate this process, Table 12.6 shows the weighting values used in this example, indicating how they sum to a value of 1.0. The resulting weighted impact values still include the same units from the characterization process. These values relate directly to the original shoe values in Table 12.3. Because of the disparate units, the weighted values cannot be summed or aggregated. The weighted values reflect the scale of their individual category weight values.

Aggregation: Creating a Single-Figure Score

One approach to interpreting results is to sum the impact categories that have the same units to make a single-figure

based on a value choice, just as deciding to use different weight values in aggregation is a value judgment.

Normalized Result Weighting and Aggregation

Weighting can be readily applied to normalized impact results, whether the normalization is internal or external. The process is identical to that of applying weighting to characterized impacts, except that the normalized results are multiplied by the respective weighting values for each impact category.

Table 12.7 and Figure 12.4 show an example of normalized impacts of running shoes, similar to Table 12.5, but with equal weighting values applied to each of the five impact categories. The units in the weighted result are identical (1 / person-year), as they were in the pre-weighted normalized impact results. Because the units are identical, the impacts can be summed, and in this case the weighted result shows that shoe B has the lowest aggregated impact. However, the difference

impact category	weighting values	normalized and weighted unit	weighted and normalized impact shoe A	weighted and normalized impact shoe B	weighted and normalized impact shoe C
Abiotic depletion	0.2	1 /person-year	6.8E-14	5.6E-14	8.4E-14
Acidification	0.2	1 /person-year	3.4E-14	3.8E-14	2.6E-14
Eutrophication	0.2	1 /person-year	1.5E-14	1.7E-14	2.8E-14
Global warming	0.2	1 /person-year	3.2E-14	2.8E-14	2.6E-14
Human toxicity	0.2	1 /person-year	6.6E-15	8.4E-15	4.0E-15
total	1.0		1.6E-13	1.5E-13	1.7E-13

Table 12.7. Externally Normalized and Equally Weighted Impacts of Three Types of Running Shoes

score. We can also refer to this process as aggregation. ISO 14044.4.1 notes that a scientific basis for aggregating LCIA results from different indicators does not exist. This reinforces the understanding that all single figure scores involve value judgments that are not entirely scientifically justified. A single-figure score generated by aggregation is not suitable to use in public comparisons, marketing, and eco-labeling because they lack the necessary transparency (Goedkoop and Springsma 1999). The decision to apply equal weighting values is

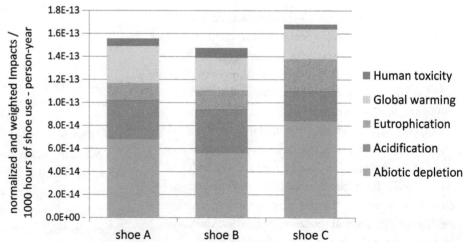

Figure 12.4. Externally Normalized and Equally Weighted Impacts of Three Types of Running Shoes

is small, and it is fair to say that no pair of shoes is definitely better than another pair.

Table 12.8 uses unequal weight values. These values are just for illustration, but they could represent the result of a poll to a group of stakeholders. Table 12.8 and Figure 12.5 demonstrate how different weighting values are applied to the externally normalized impacts of the running shoes. Shoe B still has the lowest aggregated impact, as was the case in the externally normalized results with equal weighting. Different weighting values could significantly change the resulting impact values. This could identify a different shoe as having the lowest normalized and weighted impacts.

impact category	weighting values	normalized and weighted unit	weighted and normalized impact shoe A	weighted and normalized impact shoe B	weighted and normalized impact shoe C
Abiotic depletion	0.10	1 /person-year	3.41E-14	2.81E-14	4.15E-14
Acidification	0.12	1 /person-year	2.00E-14	2.31E-14	1.53E-14
Eutrophication	0.18	1 /person-year	1.33E-14	1.50E-14	2.59E-14
Global warming	0.39	1 /person-year	6.30E-14	5.46E-14	5.13E-14
Human toxicity	0.21	1 /person-year	6.97E-15	8.72E-15	4.29E-15
			1.37E-13	1.30E-13	1.38E-13

Table 12.8. Externally Normalized and Unequally Weighted Impacts of Three Types of Running Shoes

SENSITIVITY AND UNCERTAINTY ANALYSIS WITH WEIGHTED VALUES

Once we determine an aggregated weighting value expressing the overall environmental impact of a product system, we obtain a new capacity in making sensitivity analyses. The contribution of any number used in the calculation to the total impact value can be determined. When comparing two alternatives, the consequences of uncertain input data for the ranking of the alternatives can be determined. This may help in identifying which input data need to be improved (Steen 1997).

CONDITIONS FOR APPLYING WEIGHTING

Deliberate consideration of the Goal and Scope and the LCA recipients' needs should inform the decision to apply weighting factors. The commissioners of the study may want to apply their or other stakeholders' opinions about the relative importance of the impact categories in the weighting values. The decision is best made in close communication with the recipients of the study. According to ISO (14044.5), weighting should not be used in LCA studies intended to support comparative assertions that are disclosed to the public. Further, if damage assessment is used, weighting factors will usually be multiplied after the damage assessment calculation, thereby doubly weighting the assessment

COMMUNICATING WEIGHTED IMPACT RESULTS

If a third-party report is prepared, the reasons for selecting the weighting factors should be justified (ISO 14044:2006 5.2). The study should clearly indicate whether additional weighting values were used and, if so, whether the weighting system was created especially for the study or if it was created by a separate organization (such as a scientific panel) for a particular purpose. Further, if we created new weighting values that reflect the commissioner's opinions, the study results should not be published in order to avoid accusations of partiality or conflict of interest.

If no weighting has been applied, but the study results have been normalized and aggregated, a clear statement that equal weighing was

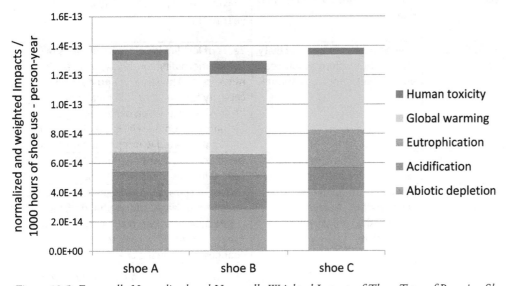

Figure 12.5. Externally Normalized and Unequally Weighted Impacts of Three Types of Running Shoes

applied to each impact category can be provided. Although not required by ISO, such a disclosure will avoid confusion about weighting.

Data and indicator results, or normalized indicator results reached prior to weighting, should be presented together with the weighting results. Providing the pre-weighted results ensures that trade-offs remain available to decision-makers and to other readers of the study in order to help them appreciate the ramifications of the results. At a minimum, these results should be presented in the form of tables in the appendices of a study. Depending on the intended audience, this could also include graphs of the characterized impact results, both before and after weighting, in the body of the text. If significant differences arise between un-weighted and weighted results, the practitioner should highlight these differences to the commissioner.

All tables and graphs of weighted impact results should be clearly labeled that they display weighted results. ISO (14044.3.4.3) requires the reporting of characterized results along with weighted results in public studies, especially with LCA studies used as a basis for public comparative assertions about competing products or services.

12.6 DAMAGE ASSESSMENT

Damage assessment is characterization at the end-point level. According to ISO 14044.5, "the category indicator may be chosen anywhere along the environmental mechanism between the LCA result and the category endpoints." Damage assessments are assessments of negative changes of environmental, economic, and social qualities. In Chapter 10, environmental impacts are described in terms of contributions to environmental threats, like global warming and acidification. In damage assessment, one can describe environmental impacts in terms of observable degradation of values, such as decreased life expectancy, crop harvest, or mineral reserves.

Damage assessment can be based on scientific understanding, however it often requires value judgments that are not based purely on scientific facts. The USEtox models are exam-

ples of a damage assessment that is science based. In contrast, the IMPACT World model uses disability –adjusted life years (DALYs) that are based on the value judgments of panels within the World Health Organization (WHO). This means that mid-point characterized impacts are typically more scientifically grounded and have lower uncertainty than end-point characterized impacts, which are more subjectively determined and have greater uncertainty.

The challenge is that we often cannot definitely link observed degradation to observed inventory data. The source of human health impacts is notoriously difficult to prove. For example, is life expectancy reduced due to poor diet, chemical exposure, or inadequate health care? Linking emissions and resource use to impact categories on the mid-point level is relatively accurate, while modeling damage at an end-point level is more uncertain. On the other hand, it is easier for people to assign value to generic damage categories such as human health, environmental health, and resource use than more specific threats like acidification, eutrophication, and human respiratory health.

The differences between mid-point and end-point level impact categories can be illustrated by CFC emissions leading to ozone layer depletion (mid-point) and then following to end-points of skin cancer, crop damage, and damage to materials like plastics (Bare et al. 2000). Damages are described by impact categories, category indicators, and environmental threats at the end-point level. If an end-point level truly exists, where can we stop the analysis?

Method name	Safeguard subjects	End-point category indicators
Eco Indicator 99	Mineral and fossil resources, eco-system quality, and human health	MJ surplus energy, % vascular plant species*km2 *year, disability adjusted life year (DALY)
EPS 2000d	Human health, bio-productivity, biodiversity, abiotic resources, and aesthetic and recreational values	Years of lost life (YOLL), person-years of severe morbidity, morbidity, severe nuisance and nuisance, kg production capacity for crop, fish and meat, and wood, normalized extinction of species, kg of elements in commercial minerals depleted, and kg decreased water production capacity
Recipe 2008	Human health, ecosystem health, and resources	Disability-adjusted life-year (DALY), species year, $ increased resources cost
LIME	Human life, human health, social welfare, ecosystem, biodiversity, and primary productivity	Cancer, respiratory disease, thermal stress, infectious diseases, starvation, disaster, cataract, skin cancer, plant, crop, land loss, fishery, energy loss, cost, terrestrial changes, and aquatic changes

Table 12.9. Safeguard Subjects, Impact Categories, and Category Indicators for Selected Damage Assessment Methods

Second-order effects from the damages may occur, such as wars over decreased resources. Environmental impacts may not be discrete changes in environmental assets, but rather pathway changes with multiple and interdependent impacts. Such pathway changes may occur in species extinction or introduction of invasive species. However, in LCA one needs to define system boundaries not only for the technical system modeled in the LCI, but also for the environmental system modeled by environmental impact methods.

Table 12.9 gives examples on some damage assessment methods, and addresses areas of environmental qualities (safeguard subjects) and impact category indicators applied at the end-point level. End-point impact category indicators describe impacts on the safeguard subjects from emissions and resource use.

VALUES AND WEIGHTING IN DAMAGE ASSESSMENT

Damage assessment can be made in several ways. The outcome depends primarily on the damage types and in terms of temporal, spatial coverage, and degree of resolution that are included in the model. The creation of damage-assessment models requires subjective decisions about the comparative value of different impact categories. Chapter 11: LCIA Methods describes characterization methods that have been developed for use at the mid-point level and the end-point level. Some damage-assessment weighting methods are shown in Table 12.10. In choices between alternative products designs or similar choices, damage assessment is difficult to use without aggregative weighting. Most damage assessment methods are developed for use with aggregative weighting sets that are particular to the damage assessment method.

Method name	Weighting principle (see 14.4)	Way of measuring	Reference
Ecoindicator99	Not defined	Panel	Goedkoop 1999
EPS 2000d	External cost	Willingness to pay	Steen 1999
LIME	External cost	Conjoint analysis	Itsubu 2003
ReCiPe	Not defined	Panel	Goedkoop, et al 2012

Table 12.10 Damage Assessment Weighting Principles

Many weighting methods use panels of experts or stakeholders to define the aggregative weighting values. In those cases, the weighting principles depend on the opinions of the panel. Panel methods measure attitudes that can evolve over time with emerging knowledge of environmental and social problems. If a panel is organized to create a new set of weighting values, the participants on the panel should be carefully chosen. A range of methods can be used to elicit weight values from the panel. (Gloria et al 2007)

DAMAGE ASSESSMENT MODELS

Damage models describe environmental mechanisms linking emissions to end-point category indicators. The description is both qualitative, explaining what is causing what, and quantitative, counting how much is changed. The result from a damage model is a characterization factor. Quantitative models may be both of a mechanistic and a statistic character. For instance, dispersion models may be used to describe exposure to people for a certain substance in the surroundings of an emission source, and a dose-response function may be used to estimate the damage. Or, for a certain area, the sum of all emissions of a substance may be determined, and from the correlation between the concentration of the substance and, for instance, excess sick-days, the number of sick-days per mass unit of emissions may be calculated. Once the characterization factor has been determined for a certain substance and a certain environmental mechanism (e.g. SO_2 and acidification), equivalency factors may be used for modeling damage via acidification from other substances such as hydrochloric acid (HCl).

Damage models are simplified representations of reality. They always have some degree of uncertainty. Not knowing where the source is located and not knowing the source strength are factors that add to uncertainty. It may seem pointless to try to rank two LCA results in terms of damage with such large uncertainty. LCI values seldom differ more than 10 to 50%, while the uncertainty in damage assessment may be 10% - 300%. A damage model with high uncertainty, however, is meaningful because product systems often are dispersed over large areas and, when using damage assessment together with weighting, local effects tend to weigh less than global effects. This may partly be explained by the history of environmental science, where local impacts have been addressed more effectively than global impacts. An example of a damage model for sulfur dioxide is shown in Figure 12.6 and Table 12.11.

EXAMPLE OF APPLIED DAMAGE ASSESSMENT

Suppose that we are choosing between two pairs of shoes and would like to know which one has the lowest overall environmental impact. In this example, we employ three different damage assessment methods: Ecoindicator99 with hierarchist perspective (Goedkoop 1999), EPS2000d (Steen 1999) and the ReCiPe method (Goedkoop et al. 2012). Table 12.12

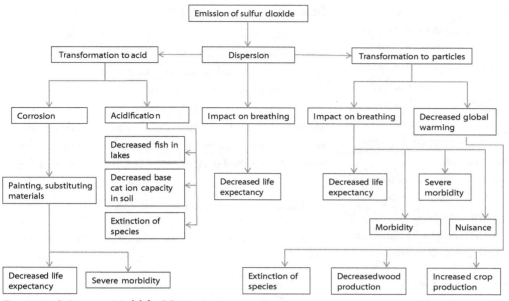

Figure 12.6. Damage Model for SO2

contains inventory data for the two pairs of shoes. In reality, a more extensive data set would be used, but these emissions are sufficient to illustrate the damage assessment procedure.

The inventory data are multiplied with the combined characterization and weighting factors for each elementary flow. We then find that two of the methods indicate that shoe 1 has the lowest overall impact, while one method indicates that shoe 2 has the lowest impact (see Figure 12.7). At a first sight,

this seems to depend on the fact that different methods give different weights to different emissions and resources. But behind this lie different weighting principles and system boundaries for the environment. EI99, HA, and Recipe are both based on stated preferences from a panel to which damage impacts are presented. EPS2000d is based on monetary values, partly revealed, and has a long time perspective. The different results represent differences in how impacts are perceived, not as shortcomings in the weighting procedures. When using damage assessment with a weighting method, practitioners should understand the weighting principles and the context in which the methods were developed.

CONDITIONS TO APPLY DAMAGE ASSESSMENT

Damage assessment enables an understanding of the significance and character of environmental impacts at the

Substance flow group	Indicator	Pathway	Model type	Pathway specific charcterisation factor	Characterisation factor for all pathways	Uncertainty factor
SO2	YOLL	direct exposure	statistical	0,000000191		10
SO2	YOLL	secondary particles	equivalency to PM10	3,73522E-05		3
SO2	YOLL	corrosion *)	equivalency to induced emissions	2,80815E-08		4
SO2	YOLL	all			3,75712E-05	
SO2	severe morbidity	secondary particles	equivalency to PM10	-0,00000659		4,2
SO2	severe morbidity	corrosion*)	equivalency to induced emissions	1,27E-08		4
SO2	severe morbidity	all			-6,5773E-06	
SO2	morbidity	secondary particles	equivalency to PM10	0,0000102	0,0000102	4,2
SO2	nuisance	secondary particles	equivalency to PM11	0,00645	0,00645	2,4
SO2	crop	secondary particles-global warming	equivalency to CO2	-0,0183	-0,0183	2,6
SO2	fish&meat	acidification	statistical	0,00118	0,00118	3
SO2	wood	secondary particles-global warming	equivalency to CO2	0,0281	0,0281	2,4
SO2	base cat-ion capacity	acidification	statistical	1,56	1,56	3
SO2	NEX **)	acidification	statistical	1,18E-14		3
SO2	NEX	secondary particles-global warming	equivalency	-3,06E-13		4,2
SO2	NEX	all			-2,942E-13	

Table 12.11. Quantitative Damage Model for Sulfur Dioxide (Steen 1999)
Characterization factors are expressed in indicator units per kg of SO2. Indicator units are in person-years and kg except for normalized extinction of species (NEX).
*) Secondary impacts from painting and substituting constructions
**) Normalized extinction of species

LCI data			Weighting factors			Weighted results			Weighted and normalized results		
			EI99, HA	EPS 2000d	Recipe	EI99, HA	EPS 2000d	Recipe	EI99, HA	EPS 2000d	Recipe
	emission,resource	kg	Ecopoints/kg	ELU*/kg	units/kg	Ecopoints	ELU	units	% of total	% of total	% of total
shoe 1	CO_2	5.69	0.0297	0.108	0.1968	0.169	0.615	1.120	49.7	39.4	81.0
	NOx	0.0186	1.39	2.13	1.703	0.0259	0.0396	0.0317	7.6	2.5	2.3
	SOx	0.0333	0.66	3.27	1.551	0.0220	0.109	0.0516	6.5	7.0	3.7
	Oil from ground	0.868	0.14	0.5	0.206	0.122	0.434	0.179	35.8	27.8	12.9
	Cr from ground	0.000399	0.0218	84.9	0	0.00001	0.0339	0	0.0	2.2	0.0
	Cu from ground	0.00158	0.873	208	0.027	0.00138	0.329	0.00004	0.4	21.1	0.0
	Sum					0.340	1.559	1.382	100	100	100
shoe 2	CO_2	4	0.0297	0.108	0.1968	0.119	0.432	0.787	28.9	23.8	64.2
	NOx	0.015	1.39	2.13	1.703	0.0208	0.0319	0.0255	5.0	1.7	2.1
	SOx	0.028	0.66	3.27	1.551	0.0185	0.0916	0.0434	4.5	5.0	3.5
	Oil from ground	1.8	0.14	0.5	0.206	0.252	0.9	0.371	61.2	49.5	30.2
	Cr from ground	0.000399	0.0218	84.9	0	0.00001	0.0339	0	0.0	1.8	0.0
	Cu from ground	0.00158	0.873	208	0.027	0.00138	0.329	0.00004	0.3	18.1	0.0
	Sum					0.412	1.818	1.227	100	100	100

Table 12.12. Comparing Two Shoe Types with Different Damage Assessment Methods
* ELU stands for environmental load units. 1 ELU is equal to 1 Euro in damage cost

end-point. The underpinning science for damage assessment is less robust than the science for mid-point characterized impacts. Consequently, end-point characterized impacts have higher uncertainty than impacts computed by mid-point impact characterization methods (Bare & Gloria 2006). One can choose to use damage assessment when the scientific basis for LCA is less important than the ability to communicate the results in terms that the intended audience can easily understand. For instance, if the LCA recipients are the public or an audience who is not well educated about environmental impact categories, end-point indicators may be easier to comprehend than mid-point indicators. This decision should be made in consultation with the commissioners and recipients of the study.

12.7. MULTI-CRITERIA AGGREGATION PROCEDURES

Multiple criteria aggregation procedures (MCAP) are complex and powerful decision support methods, comprising a branch of multi-criteria decision analysis (MCDA). MCAP can be applied at the mid-point or the end-point. The common types of MCAP applied in LCA are Type 1, which uses synthesizing criteria, and Type 2, which uses outranking normalization (see Table 12.13).

Type 1 MCAP applies synthesizing criteria to evaluate alternatives independently from each other. Originating from "the American school", Type 1 enables positioning of an alternative on a pre-defined scale. It allows offsetting the increase in one criterion by a sufficient decrease in another criterion.

Type 2 MCAP originated from "the European school." It applies outranking normalization to evaluate one alternative in relation to another. Type 2 is a form of internal normalization by means of pair-wise comparisons, followed by a process of ranking the alternatives. This process separates less desirable alternatives from more desirable alternatives.

Most LCA practitioners are currently unfamiliar with

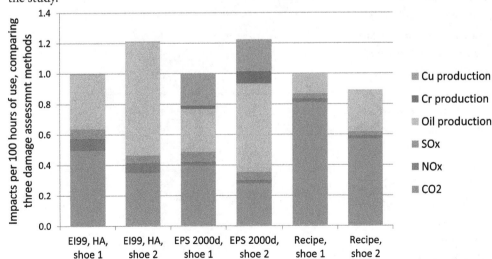

Figure 12.7. Comparison of Two Shoes with Three Types of Damage Assessment Data from table L, internally normalized to the total weighted impact of shoe 1

Multiple Criteria Aggregation Procedures (MCAP)	
Type 1: Using Synthesizing Criteria	**Type 2: Using Outranking Normalization**
Multi-Attribute Utility Theory (MAUT)	ELECTRE
SMART	MELCHIOR
TOPSIS	Trichotomic Segmentation
MACBETH	PROMETHEE
Analytical Hierarchy/network Process (AHP)	SMAA

Table 12.13. Outline of Multiple Criteria Aggregation Procedures
Explicit MCAP explanations, which are excessively complex for this text, are found in the literature
(Figueira et al. 2005; Munda 2005; Rowley et al 2012)

MCAP methods. Some practitioners consider MCAP to be a time consuming and overly precise methods for LCA—the equivalent to cutting butter with an industrial-scale laser—and note the subjective decisions made in some MCAP-process steps. Proponents of MCAP methods, on the other hand, note their ability to evaluate impact categories independent of each other.

Knowing how to identify the subjective, unscientific characteristics of these methods is essential for sound LCA practice. Some of the subjective decisions in these methods are easy to identify. For instance, the criteria for grouping are usually quite apparent. Others, such as the subjective judgments embedded within most end-point characterization methods, are more difficult to identify. Because of the subjective nature of some steps in decision support calculations, practitioners must be aware of these subjective qualities and clearly communicate them to the assessment recipients.

12.8 USING DECISION SUPPORT METHODS

This chapter describes the application of decision-support methods on environmental and human health impacts. The basic rules for grouping, normalization, and weighting also apply to social and economic life-cycle studies. If possible, use of the same type of decision-support calculations across the economic and social assessments will make the results of the parallel methods more comparable. Further, the qualitative nature of many social phenomena may make some of them unable to fit in a decision-support framework that was developed for environmental and human-health assessment. For further information about decision-support calculations and social and economic assessment, see Chapter 16: Parallel Methods.

Reiterating points made earlier, grouping, normalization, weighting, and/or damage assessment are not required to part of an LCA. Many practitioners employ the methods because they can be helpful in allowing us to better understand the characterized indicator results from an LCA.

As a beginning LCA practitioner, one may feel somewhat overwhelmed by the many different decision support methods that can be applied to an LCA and confused about which methods, if any, should be used for a given project. This problem is completely understandable. Even experienced LCA experts often disagree about which methods are most appropriate for a particular study. With practice in applying different decision support methods in different projects, one will becomes more comfortable and confident about the judicious use of these methods.

PROBLEM EXERCISES

1. GROUPING

Life cycle environmental impacts of storing 50 MJ of electrical energy in NiMH (Nickel Metal Hybrid), NCM (Li-ion battery based on Ni, Co and Mn), and LFP (Li Fe-Phosphate) traction batteries and delivering it to a plug-in hybrid electric vehicle (PHEV) or a battery electric vehicle (BEV) powertrain was determined by Majeau-Bettez et al. (2011). They obtained the following results:

		NiMeH	NMC	LFP
Global warming potential, 100 years	kg CO_2-eq	3.5	1.9	1.4
Fossil resource depletion potential	kg oil-eq	0.99	0.45	0.37
Freshwater ecotoxicity potential, infinity	kg 1,4-DCB-eq	0.13	5.1×10^{-2}	3.4×10^{-2}
Freshwater eutrophication potential	kg P-eq	4.5×10^{-3}	2.7×10^{-3}	2.0×10^{-3}
Human toxicity potential, infinity	kg 1,4-DCB-eq	5.6	4.1	2.7
Marine ecotoxicity potential, infinity	kg 1,4-DCB-eq	0.13	5.6×10^{-3}	3.7×10^{-2}
Marine eutrophication potential	kg N-eq	4.0×10^{-3}	2.5×10^{-3}	1.9×10^{-3}
Metal depletion potential	kg Fe-eq	1.1	0.85	0.30
Ozone depletion potential, infinity	kg CFC-11-eq	1.0×10^{-5}	1.1×10^{-5}	7.5×10^{-6}
Particulate matter formation potential	kg PM10-eq	2.3×10^{-2}	3.6×10^{-3}	2.1×10^{-3}
Photooxidant formation potential	kg NMVOC	1.7×10^{-2}	4.5×10^{-3}	3.0×10^{-3}
Terrestial acidification potential, 100 years	kg SO_2-eq	9.8×10^{-2}	1.2×10^{-2}	6.5×10^{-3}
Terrestrial ecotoxicity	kg 1,4-DCB-eq	1.2×10^{-3}	3.1×10^{-4}	1.7×10^{-4}

Make a nominal grouping based on impact scale

2. INTERNAL NORMALIZATION OF LARGE-SCALE PROCESSES

Internal normalization can be applied on many scales—on a product, service, factory, or company level, as well as on larger regional, national, or global scales. Here, you apply internal normalization at a factory level and a national level. Two factories are in production. The steel foundry (S) that makes sheet steel, the refrigerator manufacturing plant (R) that cuts and bends the sheet steel, and combined with the other components to constructs the refrigerator; on the national level, it is the distribution process for the total number of refrigerators sold in Mexico (M) in the year 2010.

Steel Foundry (S)

inputs		outputs	
Scrap iron	11 million lb.	Carbon dioxide (CO_2)	169 million lb.
Pig iron (from ore)	83 million lb.	Nitrous oxides (NOx)	0.41 million lb.
Electricity	900,000 kW-hr	Particulates	1.17 million lb.
Natural gas	1.5 million MJ	Sulfur dioxide (SO_2)	0.44 million lb.
		Scrap iron	4 million lb.
		Primary product	90 million lb. sheet steel

Refrigerator Manufacturing Plant (R)

inputs		outputs	
Sheet metal	5.5 million lb.	Carbon dioxide (CO_2)	2.49 million lb.
Other components	6.2 million lb.	Nitrous oxides (NOx)	5000 lb.
Electricity	3.42 million kW-hr	Sulfur dioxide (SO_2)	13700 lb.
Fuel oil	21,000 lb.	Scrap (iron) steel	392,000 lb.
		Primary product	56,000 refrigerators

Refrigerators sold in Mexico (M)

inputs		outputs	
Refrigerators	465million lb.	Carbon dioxide (CO_2)	14.7 million lb.
Transport, 6 ton truck	4600 million miles	Nitrous oxides (NOx)	69,000 lb
		Sulfur dioxide (SO_2)	16,100 lb
		Particulates	9,200 lb
		Primary product	2.3 million refrigerators sold

a. On each of the three levels, normalize each input and output according to the primary product of that level. Include the new normalized unit for each normalized input and output (unit/primary product unit).

b. What percent of the steel foundry's sheet steel does the refrigerator plant consume?

c. What percent of the annual sales of refrigerators sold in Mexico does the refrigerator plant represent?

3. INTERNAL NORMALIZATION OF PRODUCTS

You apply internal normalization to the characterized impacts of four refrigerators. Their life cycle impacts, including refrigerator production, use, and end-of-life treatment, have been calculated. The impacts have been apportioned to the functional unit of 0.7 m3 of refrigerated space for one year. The table below shows the impact categories and the quantified impacts of each of the four refrigerators.

Table K, Impacts of four refrigerators, per 0.7 m3 –year of refrigerated volume.

TRACI 2.1 Impact category	Impact unit	Refrig. 1	Refrig. 2	Refrig 3	Refrig. 4
Ozone depletion	kg CFC-11 eq.	4.5E-05	1.9E-05	6.4E-05	3.0E-05
Global warming	kg CO2 eq.	1000	390	1400	720
Smog	kg O3 eq.	47	21	65	36
Acidification	kg SO2 eq.	6.7	2.9	9.1	4.9
Eutrophication	kg N eq.	0.67	0.31	0.92	0.51
Carcinogens	CTUh	1.8E-05	9.4E-06	2.8E-05	3.9E-05
Non carcinogens	CTUh	7.3E-05	3.5E-05	0.0001	5.2E-05
Respiratory effects	kg PM2.5 eq.	0.43	0.19	0.62	0.31
Ecotoxicity	CTUe	810	190	600	290
Fossil fuel depletion	MJ surplus	14	5.8	19	9.8

a. Normalize each of the four refrigerators by the impact values of refrigerator 1.

b. If you had normalized by a different refrigerator than refrigerator 1, do you think that the proportional relationship among the normalized results of the four refrigerators would be different? (In other words, for the normalized impacts in each impact category, would the numerical proportion of refrigerator 2 impacts compared to refrigerator 1 impacts change if a different refrigerator was used for the internal normalization?)

4. EXTERNAL NORMALIZATION

You now externally normalize the characterized impacts of the same refrigerators in Exercise 3. The table below shows the normalization values per person per year for the US in the year 2008 for each impact category.

a. Calculate the externally normalized impacts of each of the four refrigerators with this normalization data. Make sure you also include the new normalized impact units in your normalized impacts table.

b. Explain in one sentence what the externally normalized value represents.

TRACI 2.1 Impact category	2008 US Normalization Values*	Normalization unit
Ozone depletion	0.16	kg CFC-11 eq./ yr-capita
Global warming	24000	kg CO2 eq./yr-capita
Smog	1400	kg O3 eq./yr-capita
Acidification	91	kg SO2 eq./yr-capita
Eutrophication	22	kg N eq./yr-capita
Carcinogens	0.000051	CTUh/yr-capita
Non carcinogens	0.0011	CTUh/yr-capita
Respiratory effects	24	kg PM2.5 eq./year capita
Ecotoxicity	11000	CTUe/yr-capita
Fossil fuel depletion	17000	MJ surplus/yr-capita

5. WEIGHTING 1

a. Apply equal weighting to each impact category in the internally normalized impact results from Question 3. That means that each of the ten impact categories receives a weighting value of 10%. When that step is complete, sum all of the normalized and weighted results in a new row at the bottom of the table. Which refrigerator had the largest (i.e. worst) impacts and which had the smallest (i.e., best) impacts?

b. Apply the same equal weighting to the results of the externally normalized results from Question 4. Which refrigerator had the largest (i.e., worst) impacts and which had the smallest (i.e., best) impacts?

c. Try to perform this exercise with a group of other students. Individually adjust each of the ten impact category weights, either increasing the percentage above 10% or reducing it below 10% (according to your perceptions of how important each of the ten impact categories). The total of the ten weighted values must sum to 1.0. Once each person has created her or his ideal weighting scheme, share them all with each other to compare the differences. First, calculate the mean and deviation of each weight value for the entire group, and place in a table. Which impact categories had the largest differences of opinion (as indicated by the largest deviation)?

d. Next, discuss the impact categories in question with the group. Why do you disagree about individual impact category weight values? Are the arguments scientific or purely perceptual? After hearing arguments for and against the new weighted values for each of the categories, the group decides upon the group weight values. Apply these and show them in a table. Which refrigerator had the largest (i.e., worst) impacts and which had the smallest (i.e., best) impacts with the new weighted values?

6. WEIGHTING 2

The management of a power plant wants to improve the company's environmental performance. The management is considering whether install a new desulphurization plant at the oil fired boiler. The LCI data, (which is mainly from the energy loss) for the cleaner is as follows:

Calculate the weighted amount of the elementary flows using weighting factors in the table above.

a. Discuss the sources of the differences in the weighted results.

Elementary flow	Unit	Amount
CO_2 to air	kg/100000m3 flue gas cleaned	1500
NOx to air	kg/100000m3 flue gas cleaned	6
SOx to air	kg/100000m3 flue gas cleaned	-200
Oil from ground	kg/100000m3 flue gas cleaned	480
Cr from ground	kg/100000m3 flue gas cleaned	0.125
Cu from ground	kg/100000m3 flue gas cleaned	0.0025

b. Make a recommendation to the management whether to install the flue-gas cleaner or not.

c. Selecting an LCIA method for a particular client

A manufacturer that produces internationally and sells internationally asked you to conduct an LCA of one of their products. You are considering characteristics of different LCIA methods in the following table to select the most appropriate method. The company indicates that normalized results will help them better understand the relative scale of impacts among different impact categories, and they also need an internationally relevant result. Considering these parameters, which of the following combinations of characterization and normalization methods would best meet the needs of your client?

LCIA method	Characterization level		Normalization	
	Mid-point	End-point	Region	Year
TRACI	yes	no	North America	2001
CML IA	yes	no	World	1995
ReCiPe	yes	yes	World	2008
Impact2002+	yes	yes	Europe	2012

INTERPRETATION OF RESULTS

SHAWN HUNTER AND RICH HELLING

13.1 INTRODUCTION TO INTERPRETATION

In the interpretation phase of LCA, the LCA analyst brings together all of the work done in the previous stages of the LCA and applies LCA expertise to draw conclusions based on the data. The interpretation phase answers the questions posed in the goal and scope, and transforms the life cycle inventory (LCI) data and life cycle impact assessment (LCIA) results into knowledge and insight. Interpretation is where the holistic life-cycle picture becomes clear for the system under investigation.

In some cases, the conclusions reached may be quantitative in nature. For example: Product A exhibits 20% fewer greenhouse gas (GHG) emissions over its life cycle when compared to Product B. Due to the quantitative nature of LCA, Some people may superficially view LCAs as merely producing a series of bar graphs as the final output. However, the bar graphs are the means, not the end. LCA results are about much more than just a number or series of numbers.

LCAs should instead be viewed as producing knowledge about the relative environmental impacts of products and services. The insights provided by LCA are far more valuable than any numerical value derived in an LCA study. According to ISO (2006), the numbers are relative and typically have little meaning as absolute values. Rather, the insight gained from doing the study is most valuable and provides the information required to support decision-making. For example, not only can LCA help quantify a difference in life cycle GHG emissions among options, it can also help us understand the following:

- Why does the difference in GHG emissions exist? Are

there inherent technical differences in the life-cycle efficiencies of competing technologies?
- What are the significant issues on which the difference depends? Are there one or two unit processes that account for a majority of the GHG emissions for a given product?
- How robust is the conclusion? If we had chosen a different dataset, or made different assumptions, would we have reached the same conclusion?

Understanding issues like these is critical to the interpretation phase of any LCA. In this sense, interpretation will allow us to understand what the life-cycle data states and, therefore, what the LCA analyst can state with science-based credibility, as based on the study's results.

CORE CONCEPTS OF INTERPRETATION

The idea behind interpretation is fairly simple. In this stage, the LCA analyst uses the results of the LCI and LCIA stages to draw conclusions about the environmental or sustainability performance of the system being studied. These conclusions must be supported by the data and results of the study, as in any science-based study. In order to gain sufficient knowledge to support the conclusions, the LCA analyst typically has to ask (and answer) several questions about the system, which can include, but are not limited to, the following:

- Where are the environmental "hotspots", or stages/processes in which the most significant life cycle impacts occur, for each indicator examined?
- How do the results change if different choices are made about system boundaries, allocation, or data?
- Have all of the relevant indicators been considered in

order to meet the goal and scope?

- What are the most important issues that influence the results?
- Where are the largest opportunities for improvement?
- Is more or less uncertain data needed to fully address the goal and scope?
- Are the recommendations and claims that the LCA practitioner intends to make fully supported by the data and analysis?
- Is further iteration required?

To address these types of questions, the LCA analyst uses several approaches, such as those outlined in ISO 14044: contribution analysis, completeness check, sensitivity analysis, and consistency check. In this chapter, each of these approaches will be reviewed in detail. As LCA is an iterative process, information gained by addressing any of these questions could point the analyst to a need for further work on the project before the final conclusions can be drawn.

This chapter presents an introduction to the interpretation phase of LCA. It begins with a brief discussion of LCIA method expertise as a foundation for interpretation. The discussion is then organized according to the major areas of interpretation as outlined by the ISO standards: significant issues, evaluation, and conclusions/limitations/recommendations. The goal of this chapter is to provide sufficient information to enable the LCA analyst to perform a rigorous, ethical interpretation and to develop credible, science-based messages that can be used to support decision-making.

13.2 Understanding the LCIA Method is Essential

Before starting to write the interpretation process, we should prepare for the task. A solid understanding of the environmental mechanisms involved in the categories select-

ed, and of the details and limitations of the models used to derive the LCIA method, are key to performing interpretation and knowing which conclusions are supported by the results. These prerequisites help make LCA an expert field, with LCAs requiring a level of expertise to perform adequately. For example, after consider the LCIA results presented in Figure 13.1: what conclusions can be drawn?

Considering the figure, one can ask critical questions:

- Does Option B have a lower impact in category X than Option A? Or is there so much uncertainty in the indicator that an orders-of-magnitude difference is required in order to claim that one has lower impact than the other?
- Does the indicator used reflect theoretical potential impacts? Does the indicator reflect average values for a particular geographic region?
- Does the indicator reflect an appropriate time frame for addressing the goal and scope?
- Should other indicators for the same environmental impact category be considered before drawing a conclusion?

Knowing the answers to these types of questions provides the LCA practitioner with a deeper meaning behind the indicator results, and allows the LCA practitioner to draw strong conclusions backed by the science of LCA. Without this understanding, we could draw a conclusion that is simply not supported by the data.

We must be familiar with the LCIA methods and models to successfully interpret LCA results. Chapters 9 (Natural Science), 10 (Impact Assessment and Modeling), and 11 (LCIA Methods) of this book provide an excellent overview of the scientific expertise needed to perform interpretation in LCA, and provide the foundation for interpreting LCIA results correctly. The LCA analyst should ensure that he/she has a solid understanding of these chapters before proceeding to interpret LCA results.

What does an LCIA result mean?

When dealing with LCA and LCIA results, scientific curiosity can give rise to a stimulating question: what do LCIA results actually mean? Are they just numerical results, or is there some greater meaning?

This question has likely confronted many LCA analysts. According to the ISO standards, "LCIA results do not predict impacts on category endpoints, exceeding thresholds, safety margins, or risks" (ISO 2006). This statement, and the meaning of an LCIA result, can be understood through the use of averaged LCI data in calculating the LCIA results, as well as the nature of environmental mechanism complexity as

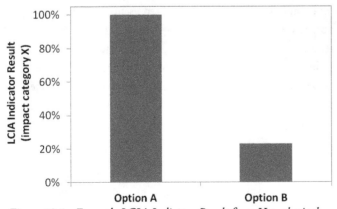

Figure 13.1. Example LCIA Indicator Result for a Hypothetical Comparison of Two Options.

it relates to LCIA model development.

One aspect concerns the representation of LCIA results on a per-functional-unit basis. For example, when we express a global-warming potential (GWP) as a certain number of kg CO_2-eq per kg, per t-km, or per MJ, remembering that no causal relationship exists between the GWP and the functional unit is essential. If an LCA determines that a certain material exhibits a cradle-to-gate GWP of 3.4 kg CO_2-eq/kg, it does not mean that exactly 3.4 kg of CO_2-eq emissions will occur as the result of producing one kg of material, or that exactly 3.4 MT of CO_2-eqs will be emitted as result of providing 1000 kg of the material. Due to geographical, technological, and temporal variations, the production of a single kg of the material may have a GWP that is different than 3.4 kg CO_2-eq.

The goal (in this example) is not to determine exactly how much CO_2 is emitted for the production of 1 kg, or even how that quantity might vary from kg to kg. Rather, because the LCA is intended for application across all kg produced, understanding, on average, how much CO_2 is emitted for the production of the material is critical. This information can then be used reliably to inform decision-making so that if a certain decision is taken, the system will perform on average as indicated in the LCA.

A second aspect relevant to the meaning of LCIA results is the fact that LCIA results are not absolute, intrinsic quantities that can be rigorously measured. For example, while the mass density of a certain polymer can be measured to be 0.7 kg/m3, its cradle-to-gate GWP might be estimated (not measured) at 1.3 kg CO_2-eq/kg polymer. Because they cannot be directly measured, LCIA results are instead calculated based on environmental models. These LCIA models seek to relate an elementary flow to a unit of environmental damage, according to complex environmental mechanisms as described in Chapter 12. Accordingly, LCIA results are reported on the basis of potentials. The potential exists that a certain amount of environmental damage will occur as the result of the system under analysis, but this is not guaranteed.

If LCIA results are not real, physical quantities that are tied directly to the products and systems that one evaluates in LCA, then what are they? The best answer to this question is simply that LCIA results are indicators that are intended for use in decision-making. Thinking about the literal meaning of an LCIA indicator result, we could summarize the meaning as such: according to this data used, these assumptions made, and this environmental model, the relative potential impact of this system is X. More practically, we could say this: according to this LCA, we can expect this product/option/life cycle stage to have a higher/lower environmental impact than another product/option/life cycle stage.

For a given LCIA result, a smaller value is generally expected to correspond to a smaller environmental impact than a larger value. This has been referred to as the Principle of Correspondence (Heijungs et al. 2003). The fact that LCIA results are related to environmental impacts allows them to be used to support decision making. When a decision maker takes an action that has an LCIA result which is clearly lower in a given impact category, he/she can feel confident that the result of his/her action will be to lower the environmental impact of the function in question for that category.

For example, while the material mentioned earlier may not emit exactly 3.4 kg CO_2-eq from cradle-to-gate for every kg of material produced, or have exactly the equivalent environmental damage caused by the emission of 3.4 kg of CO_2, it can be expected to have a lower impact on climate change than one that exhibits 5.5 kg CO_2-eq/kg and result in a much lower impact than one that exhibits 9.1 kg CO_2-eq/kg, assuming 1 kg of each material is functionally equivalent. In the context of supporting decision making, choosing the product that exhibits 3.4 kg CO_2-eq/functional unit over the other two options is expected to be the best choice related to climate change.

As the chief goal of LCA is to support decision-making, we see that LCIA results are useful calculated quantities that, when appropriately interpreted in LCA, allow us one to support decisions from an environmental sustainability perspective.

SIGNIFICANT ISSUES

The first element of interpretation is the identification of significant issues, which is a major objective of the interpretation. A significant issue, as defined contextually in ISO 14044, simply refers to an aspect of the life cycle that contributes significantly to the results of the study. Significant issues can be life cycle stages, processes, elementary flows, and/or environmental impact categories that represent a relatively large potential environmental impact. These significant issues can represent areas of focus for improvement, and their identification can help the practitioner to understand why a system performs as it does from a life cycle perspective. Techniques discussed here for identifying significant issues are contribution analysis, normalization, and weighting.

CONTRIBUTION ANALYSIS

Contribution analysis is the quantitative determination of how much any particular input contributes to a specific impact assessment metric for a product or service. There can be numerous inputs to the system under investigation in the

LCA. Typically, a much smaller number of these inputs create the majority of a potential impact. Understanding the few that are important helps focus future work and decisions on those aspects that have a large impact, and helps us to avoid spending time and effort on issues that have little or no impact. Contribution analysis is the way to gain this understanding.

Calculation of contributions requires keeping track of material and energy flows for each process block in the model of a product life cycle, in addition to using the sum of all these flows to calculate a specific potential impact. This can be done for simple processes by hand or in a spreadsheet, but using software designed for this purpose aids in most studies significantly. Imagine a product made from two materials, A and B, for which a LCA report sponsor wants to know the GWP. Figure 13.2 shows a simple, hypothetical life cycle flow diagram for this product. The GHG emissions are given in kg CO_2-eq for each step, as is the main product flow, but no other inputs or outputs are shown:

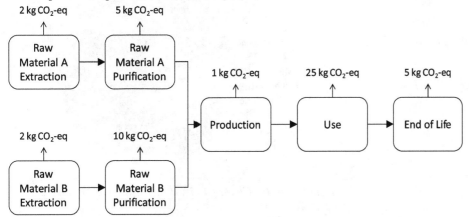

Figure 13.2. Simple Life Cycle Flow Diagram for Hypothetical Product

In this example, the total GWP per functional unit is 50 kg CO_2-eq. One way to look at the contributions is by life cycle stage, which in this example is:

- 4 kg or 8% of GWP from raw material extraction
- 15 kg or 30% of GWP from raw material purification
- 1 kg or 2% of GWP from product production
- 25 kg or 50% of GWP from product use
- 5 kg or 10% of GWP from product end of life

The contributions can also be described for specific portions of the upstream processes:

- 7 kg or 14% of GWP from raw material A (sum of extraction and purification)

1 The term "waste product" refers to a product that has little or no economic value.

- 12 kg or 24% of GWP from raw material B (sum of extraction and purification)

Although none of the inputs to this hypothetical example were negligibly small, the greatest contributions to the total GWP clearly come from the use phase and from raw-material purification. If a goal is to reduce the GWP of the process, these would be logical stages to look at first for improvement opportunities.

One can calculate and display the results of a contribution analysis in ways other than the simple text used above. Stacked bar charts, sorted bar charts, and pie charts can all be used to convey the basic contribution analysis results. Consider a process used to produce polyurethane shoe soles using waste biomaterials[1] as a process input, as shown in Figure 13.3:

The potential cradle-to-gate impacts for this process were compared to those for a conventional process using only fossil-derived feedstocks (Helling et al. 2012). The contributions to the GWP of the different inputs for the two routes to functionally equivalent shoe soles are shown in Figures 13.4, 13.5 and 13.6. The information in Figures 13.4 and 13.5 is identical. Both give the numeric values of the GWP per input and a visual indication of the relative proportion. In Figure 13.4 (stacked bars), seeing the numeric value of the total GWP is easier. In Figure 13.5 (sorted bars), seeing the numeric value for each input is

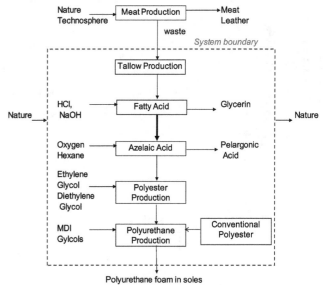

Figure 13.3. Simple Block Flow Diagram for Polyurethane Shoe Sole Production

easier. Figure 13.6 (pie chart) gives the clearest depiction of the relative proportion of the different inputs, but no indication of the total GWP. All these graphs clearly show that the top three contributions to the GWP of the shoe soles are two of the raw materials, adipic acid and methylene diphenyl diisocyanate (MDI), as well as the natural gas burned for heat in the production processes.

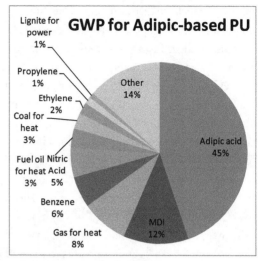

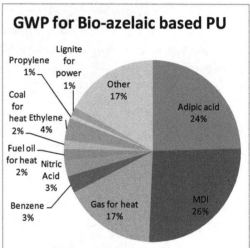

Figure 13.6. Pie Charts for GWP for Polyurethane Production Options

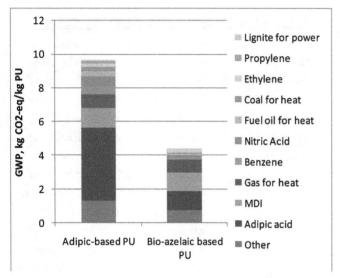

Figure 13.4. Stacked Bar Chart of GWP for Polyurethane Production Options

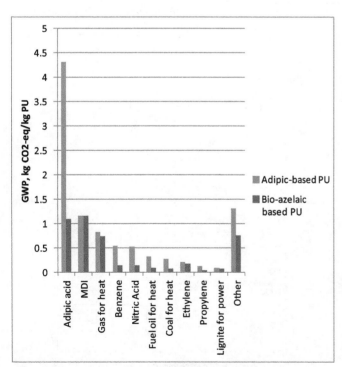

Figure 13.5. Sorted Bar Chart of GWP for Polyurethane Production Options

A common and powerful graphing technique to display and understand a contribution analysis is a Sankey diagram (Schmidt 2006), in which lines are used to show the dependence of one block in a process on another block, with the relative thickness of the line representing its relative contribution to the indicator. An example of this for the GWP per kg of polyurethane shoe soles made from fossil-based adipic acid is shown in Figure 13.7. The graphs have been truncated to show only those operations that contribute 6.5% (based on cumulated results) or more to the total, but in principle can be expanded to any level of detail. The software used for this diagram also represents the life cycle contributions by a "thermometer" at the side of each process operation block.

The results shown in Figure 13.7a represent cumulated results, where the value shown for each process accounts for the cradle-to-gate GWP generated by that process and by its up-

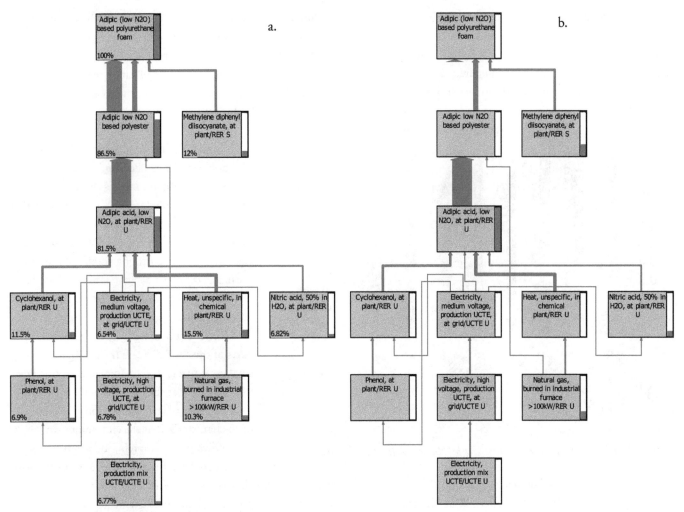

Figure 13.7 Sankey Diagram for GWP of Fossil-Based Polyurethane.
a. Cumulated indicators are shown., b. Non-cumulated indicators are shown.

stream processes. The results shown in Figure 13.7b represent non-cumulated results, where the values shown for each process represent the relative contribution of GWP emissions directly and exclusively from that process. The results shown in Figure 13.7b are identical to those shown in Figures 13.4 to 13.6.

Figure 13.7a clearly shows how various inputs combine to form an increasing percentage of the GWP in 1 kg of polyurethane foam. Using the width of the arrows to gain insight, one can deduce, for example, that the adipic acid production block inputs with largest cradle-to-gate GHG emissions are cyclohexanol production and heat production, as these two arrows are the largest of the four that enter the adipic acid production block. Looking at Figure 13.7b, one can quickly see that the most significant process from a GHG emissions perspective is the production of adipic acid, low N_2O, at plant. This block has the largest thermometer value in the figure, more than three times the size of the next largest contribution of methylene diphenyl diisocyanate production.

By doing the contribution analysis, the emissions from adipic acid production are clearly critical to the overall results from a technical perspective (to guarantee confidence in the model and results), and because it potentially represents an opportunity for process improvements if the goal is to reduce the GWP of the product. The evolution of LCA software has enabled rapid and detailed contribution analysis, so that the analysis is now readily done at a unit-process level in addition to the higher, life cycle stage level, which enables more thorough analysis of options and of detailed decisions to be made.

NORMALIZATION AND WEIGHTING

Although some stakeholders may be interested in only one or two impact categories in an LCA, one of the great powers of LCA is the ability to look at a wide range of potential impacts. Inclusion of multiple impact categories is an LCA requirement under the ISO standards because it enables a more holistic view of the options. Internally normalized results are shown for the polyurethane shoe sole example (Helling et al. 2012) in Figure 13.8, which includes six additional impact categories besides GWP. For this example, the bio-azelaic material has similar or lower impact potential compared to the adipic acid material in all the categories except for water use. A contribution analysis using Sankey diagrams was performed in order to show that the higher water use was due to water inputs to the

al. 2001), which was the LCIA method used in the polyurethane shoe sole example, includes normalization factors based on global data for 1990. This normalization process yields the results shown in Figure 13.9 for all the categories except water and cumulative energy demand (these categories are not included in the CML 2001 impact assessment method, Figure 13.9).

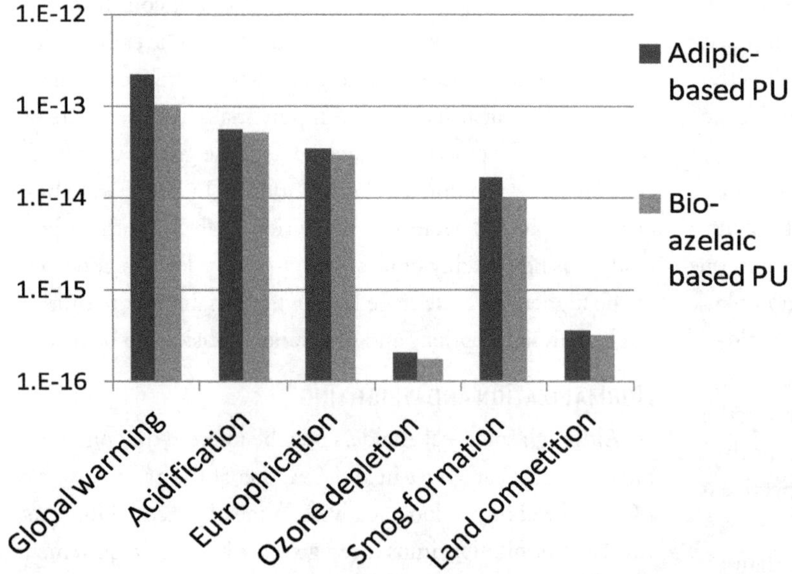

A logarithmic scale is used on the vertical axis in Figure 13.9. The implication from this approach to normalization is that, from the perspective of global impacts in 1990, GWP is clearly the most significant impact category for this project, with the possible exception of acidification and eutrophication. All the other normalized impact categories are <1% of that for GWP and are not considered significant according to this method of analysis.

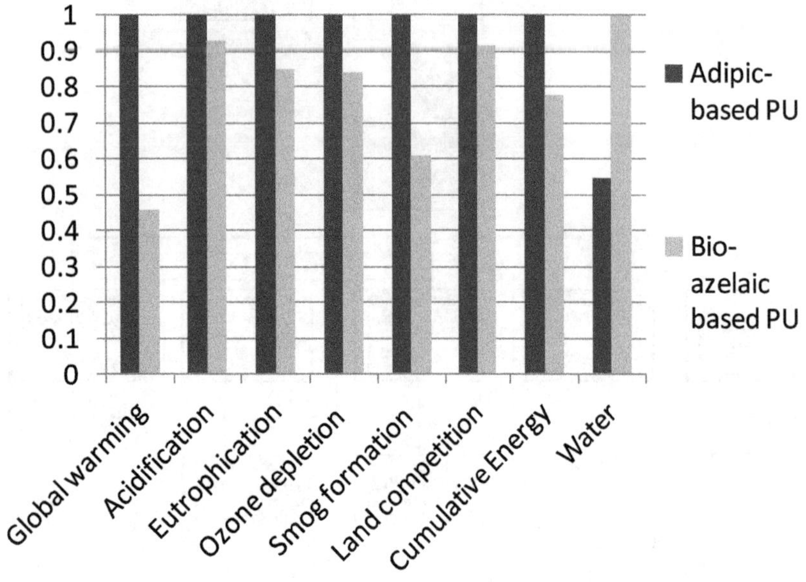

Figure 13.8. *Impact Categories Normalized by Highest Value Per Category*

conversion process for azelaic acid.

As the Chapter 12: Decision Support Calculations explained, characterized results can also be externally normalized. The CML 2001 impact assessment method (Guinée et

Figure 13.9 *Impact Categories Normalized Using 1990 Global Factors (Logarithmic Axis)*

Contribution analysis identifies which parts of the system create the largest and smallest impacts in each impact category. For example, look at the contribution analysis on the two types of acid production that have been characterized at the end-point using Eco-indicator 99. Applying Eco-indicator 99 to the polyurethane shoe sole example leads to 556 mPt for the fossil-based material and 379 mPt for the bio-based option, based on a choice of impact categories similar to that made in Figures 13.8 and 13.9. This result indicates that the bio-based option is the preferred option, according to the weighting assumptions of the Eco-indicator 99 method.

One can also perform a contribution analysis with respect to mPt, as shown in Figure 13.10. This graph shows that the use of fossil resources and climate change contribute most to the ecopoints score, and by definition provide the largest contributions for this material towards the burden created by an average European. The contributions from the other impact categories are insignificant. Weighting is therefore another way to identify impact categories where the most significant impacts are occurring, so that improvements can be focused on those areas. In this example, the bio-based material uses significantly less fossil fuel resources, but does not eliminate their use (which is quite difficult to do.).

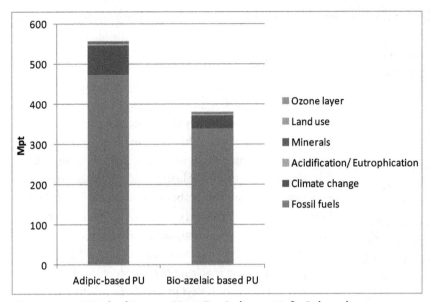

Figure 13.10. Weighted Impacts Using Eco-Indicator 99 for Polyurethanes

13.4 EVALUATION

Evaluation is the second element of interpretation, and is performed to establish confidence in the conclusions drawn. Evaluation tests the robustness of the results and conclusions. Three elements of evaluation are outlined by ISO: the completeness check, the sensitivity check, and the consistency check (ISO 2006).

COMPLETENESS CHECK

The completeness check is performed to ensure that the amount of information collected during the course of the LCA is sufficient for answering the questions posed in the goal and scope. This stage reminds the practitioner to take a step back during the course of the study (which may be short on time) and consider whether any additional information needs to be obtained in order to support the conclusions drawn. In some cases, the LCA analyst may find that the information needed to draw some conclusions is missing and may be difficult or impossible to obtain. In these cases, the goal and scope of the study may need to be revised, or the conclusion of the study may simply be that the existing information is insufficient to reach a conclusion.

QUALITATIVE CHECK

To perform the completeness check, one needs to systematically consider each stage of each product, and consider whether the collected information is complete or whether additional information may be needed. ISO suggests the use of a table

Unit process / Life cycle stage	Mass inputs		Energy inputs		GHG emissions		Additional LCIA category flows	
	Complete?	Action required	Complete?	Action required	Complete?	Action required	Complete?	Action required
Resource extraction	Yes	none	No	Need energy required for resource Y extraction	No	none, direct GHG emissions insignificant for this stage based on reference Z	...	...
Resource refining	Yes	none	Yes	none	Yes	none	...	...
Product manufacturing	No	Need upstream model for input X	Yes	none	Yes	none	...	...
Transportation	Yes	none	Yes	none	Yes	none	...	...
Additional stages	...	...	...	...	...	...	...	...

Table 13.1. Suggested Format for Conducting a Completeness Check (With a Hypothetical Example)

for conducting a completeness check (ISO 2006), which is expanded upon in Table 13.1.

Table 13.1 illustrates the consideration of the data completeness of each life cycle stage/unit process for all relevant elementary flows. The mass and energy inputs are considered, as are the inputs or outputs that are relevant to each LCIA category included in the study. By completing a table similar to Table 13.1, one is able to address whether sufficient information has been gathered to meet the goal and scope of the study. If all entries are complete, then clearly enough information is available to meet the goal and scope. If some information is missing, there still may be enough information collected to meet the goal and scope. Strong evidence can be provided to suggest that the missing information is immaterial to the conclusion reached. For example, if multiple options are being compared, it might be possible to demonstrate that the missing information is the same or similar for each option. Therefore, the conclusions will not be affected. However, if it is not clear whether the missing data is significant, then that data should either be collected or explored further in order to enable a sound conclusion.

Quantitative check: cut-off criteria

As part of the completeness check, the LCA analyst also determines whether the cut-off criteria stated in the goal and scope have been met. To do this, one must confront a "cut-off criteria paradox,, in which one attempts to understand whether X% of the mass/energy/impact has been accounted for without knowing what the full 100% value actually is.

To overcome this paradox, an estimate must be made for the missing information, and expert judgment applied to the expected importance of the missing information. Estimates may be obtained from the following sources:

- engineering calculations
- similar or average unit-process datasets
- input-output databases

Expert judgment can be obtained only through building LCA knowledge as a result of conducting and reviewing several LCAs. For the new LCA analyst, the formal or informal critical review of

one's work by a senior LCA colleague can provide the expertise needed to determine the significance of missing information.

Sensitivity analysis

Sensitivity analysis is a critical part of every LCA when the practitioner looks at all of the decisions and choices made in the project, and quantitatively determines the impact of the most significant choices on the results of the study. It differs from uncertainty analysis, in which one looks at the impact of data quality and may be able to perform a precise error calculation using knowledge about the statistical distribution of input variables. In sensitivity analysis, the focus is on the impact of choices, such as analysis boundaries, allocation methods, data sources, treatment of data gaps, and process assumptions. Because analyzing the impact of every choice made in a project is not always practical, the practitioner uses results from preliminary calculations, and has the experience and judgment to determine which specific analyses to conduct. A high degree of ethics is required to do this. One must take a critical view of the project and be prepared to challenge aspects of the results that appear less certain or that could readily be challenged by others. No set number of sensitivity analyses are required for a given project. Rather, the approach is to focus on an appropriate number of sensitivity analyses, allowing the practitioner to explore the aspects of the system with the largest potential impacts.

The most common way to conduct a sensitivity analysis is to perform a scenario analysis: one calculates the life-cycle results for additional scenarios in which alternative choices are made (hat are different from those made in the base case of

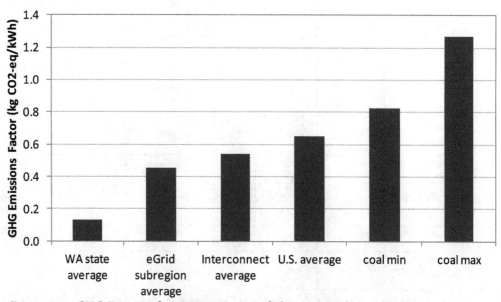

Figure 13.11. GHG Emissions from Various Sources of Electricity (Weber et al. 2010)

the study). For example, one could examine the sensitivity of the results to the choice of allocation method for a particular unit process by performing the calculations with both a mass-based and an economic value-based approach to the allocation.

As another example, one could look at the impact of the choice of model for electricity in a unit process by doing the calculations, using in one scenario the average grid electricity for a geographic area and in another scenario using the production technology at the power plant across the fence from the location of a manufacturing step in the product-supply chain. The potentially large impact of the choice of electricity model is shown in Figure 13.11: the GWP for various options for electric power generation span an eight-fold range between the average electricity production in Washington state and a worst case for electricity from coal (Weber et al. 2010). If a significant portion of the supply chain for a product included operations in Centralia, Washington (the location of the state's one-and-only coal-fired power plant), then one may need to do a sensitivity analysis on the choice of electricity model for that part of the life cycle.

In many cases, appropriate scenarios for sensitivity analysis can be suggested based on the significant issues identified during contribution analysis. In the polyurethane shoe sole example, the contribution analysis seen in the Sankey diagram in Figure 13.7 showed that the key unit process for the GWP was adipic acid, low N_2O, at plant. N_2O is a byproduct from the adipic acid production that was initially vented to the atmosphere and is now typically removed using thermal or catalytic oxidation for vent treatment. The model in ecoinvent for adipic acid, at plant contained the assumption of 80% destruction of N_2O by using vent-treatment technology. State-of-the-art plants can achieve >95% destruction, so the study authors assumed a 90% destruction efficiency to represent a reasonable average of current production, which would include plants of different ages and vent-treatment technologies.

Because the modeling choice of a 90% destruction efficiency could have such a large impact on the GWP calculation, a sensitivity analysis was done by considering an alternative scenario that assumed 100% destruction efficiency, or no direct emissions of N_2O from the adipic acid unit process. The

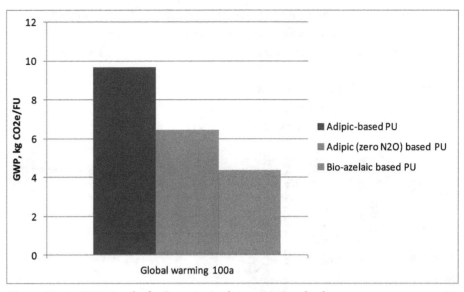

Figure 13.12. GWP Results for Scenario Analysis on N2O technology

results of this scenario analysis, as shown in Figure 13.12, are that the use of an ideal N_2O destruction technology would reduce the calculated GWP by 33%. However, even under this scenario, the GWP of the adipic acid-based PU would still be 46% higher than the GWP of the partially bio-based material. In this case, the partially bio-based material has lower GWP than the adipic acid-based PU, and this lower GWP is not affected by the alternative scenario to adipic acid production.

Experience is also an excellent guide in selecting topics to explore with sensitivity analysis. LCAs of bio-based materials have often shown strong sensitivity to the particular geographic source of the bio-feedstock and the feedstock's production methods. For example, sugarcane-based ethanol from Brazil can have much different life cycle impacts as compared with ethanol from US corn grain, due to the much different climate, fertilizer application, yield, process energy, and process technology for these two routes to bio-based ethanol (Hoefnagels et al. 2010). Use of waste streams as inputs to a process also brings up a critical allocation issue. Should the waste bear all, some, or none of the burdens of the process used to create it? No universally correct, science-based answer to this question exists, but waste allocation is usually important to explore with sensitivity analysis.

The feedstock for the bio-based material in the polyurethane shoe sole example was a waste product (tallow) from meat production. The study authors assumed this waste was "burden free" in the LCA, because the demand for meat drives the production of meat. Without the demand for meat, no tallow production would occur. Although a reasonable assumption, it is clearly open to question and a good choice for sensitivity analysis. Two alternative scenarios were considered:

- What if animal fat is not a waste, but has value and upstream burdens modeled using economic input-output (EIO) data?
- What if the feedstock was vegetable oil (modeled as a European blend for fatty acids) and not animal fat?

The sensitivity results from these scenarios are shown in

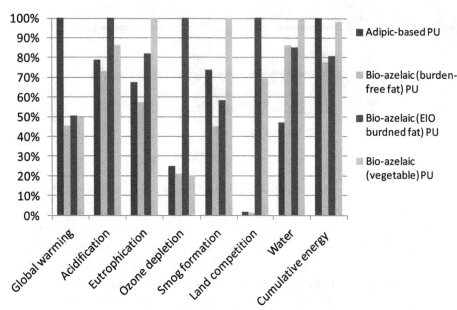

Figure 13.13 Sensitivity Results for Source of Bio-Feedstock

Figure 13.13, using the same impact categories as in Figure 13.8, internally normalized by the largest value in each impact category. The sensitivity analysis shows that the advantage in GWP for the partially bio-based polyurethane is clearly robust to changes in bio-source (all bio-options have much lower GWP than the fossil-based adipic acid PU). The advantages for the bio-based material in the other metrics are less robust and show potential trade-offs, depending on the choice of bio-feedstock.

The base case (burden-free animal fat) azelaic acid polyurethane is the most advantaged in all metrics amongst all scenarios, except for water use. The adipic polyurethanes are the best and the use of non-waste vegetable oil is the worst. The fossil adipic acid polyurethane has the maximum category value for GWP and CED. For all other metrics, the maximum comes from one of two alternatives for modeling the renewable feedstock. Using vegetable oil based azelaic acid gives the maximum of eutrophication potential, photochemical ozone (smog) creation potential and water consumption. Using animal fat with upstream burdens (from EIO) gives the maximum acidification potential, ozone depletion potential, and

land competition. Choice of this material or other bio-feedstocks would produce tradeoffs among the potential impacts.

This example shows that alternate choices can sometimes lead to alternate results and illustrates the importance of exploring alternate choices in a sensitivity analysis. In these cases, the conclusions drawn must reflect the fact that some of the results may depend on the choices made.

Sensitivity analysis is part of the iterative process used in LCA. Significant issues identified by contribution analysis should lead to specific assumptions to be explored by sensitivity analysis, or to improvements to be made in the process models or data quality. Results from a first round of sensitivity analysis may lead to a second round that explores some issues in more detail. When does one stop? Ideally, one stops when one has met the goal and scope of the project, and can provide the decision maker with the information he or she needs to make a well-informed decision. In practical terms, one may also be forced to stop

Assumptions	allocation rules
	system boundaries
	cut-off criteria
Methods	attributional vs. consequential
	process LCI vs. input-output LCI vs. hybrid
	choice of LCIA methods
	choice of software tool
Data	data sources
	data accuracy
	technology coverage
	time-related coverage
	data age
	geographical coverage

Table 13.2 Summary of Assumptions, Methods, and Data Choices Considered in a Consistency Check

when time or money for the project runs out. If that happens before the analysis is as complete as desired, then the LCA practitioner needs to clearly define what, if any, conclusions and comparisons can be made, what the key unresolved issues are, and how they might be addressed.

CONSISTENCY CHECK

The consistency check is performed to ensure that the information collected during the study is consistent in the choices made, and therefore suitable for drawing conclusions and making recommendations. From the selection of secondary data to support the primary data collected, to the choice of impact assessment methods applied to the LCI results, several choices are made by the practitioner during the course of an LCA. This step allows the practitioner to consider specifically whether the choices made have been applied consistently. If or when they differ, the analyst must consider whether the differences are permissible in accordance with the goal and scope. If the differences violate the goal and scope, then further work is necessary before drawing conclusions from the information collected.

The consistency of choices made in three areas is considered in the consistency check: assumptions, methods, and data. Table 13.2 summarizes topics to be considered during this exercise.

Similar to the completeness check, a table can be used to organize one's findings about the consistency check. The following example illustrates the use of a table. Consider an LCA in which a product produced from corn-based sugar is compared with the same product produced from sugarcane-based sugar. The LCI data for corn-based sugar and sugarcane-based sugar are likely to be significant for influencing the results of the study, and warrant thorough consideration during the con-

Assumption, Method, and Data considerations	Datasets under evaluation		Consistent with goal and scope?	Action required?
	Corn sugar dataset	Sugarcane sugar dataset		
Allocation	Mass	Economic	No	Recast data to use same approach. Use both methods in sensitivity analysis
Biogenic carbon accounting	Included as negative CO_2 emission	Not included	No	Modify data used to consistently account for biogenic carbon
Direct land use change	Included	Not included	No	Remove direct land use change from corn sugar dataset to be consistent with goal and scope
Indirect land use change	Not included	Not included	Yes	None
Type of dataset	Process-based	Process-based	Yes	None
Data source	LCI database X	LCI database Y	Maybe	Review summaries of databases X and Y to ensure no major methodological differences exist between the databases
Data accuracy	Addressed quantitatively during examination of the data quality indicators (Chapter 8) and during uncertainty analysis (Chapter 17)			
Technology coverage	US Midwest average	Single specific sugar cane plantation	Maybe	Consider whether any Brazil average sugarcane sugar datasets exist, else use this one with proper interpretation
Time-related coverage	Average of three years production	Single year of production	Maybe	Consider whether single year sugarcane data is representative of average production
Data age	10 years old	3 years old	Yes	Consider whether any significant advances in corn production have occurred over the last ten years
Geographical coverage	US	Brazil	Yes	None

Table 13.3. Application of a Consistency Check Table to a Specific Example

sistency check. For this example, Table 13.3 is prepared based on the data selected for the study.

Table 13.3 illustrates that several areas were considered for consistency, including some that are specific to the agricultural processes under investigation (e.g., biogenic carbon and land use change). The consistency check revealed that some choices need to be refined (allocation choice) and some questions need to be answered (is a dataset available that represents average sugarcane sugar production in Brazil, rather than the single-site data currently used?) before the data can be used to draw conclusions for this study.

In theory, a table similar to that in Table 13.3 would be prepared for all assumptions, methods, and data used in the study. In practice, however, this would be impossible to do for all datasets used in the study. Hundreds, or even thousands, of unit process datasets are typically applied in an LCA, when considering the full scope of the secondary data used. When using data from a single large database, we should check whether its authors have already taken some measures to help ensure consistency throughout the database. In some cases (Frischknecht et al. 2007), this has been done to some extent by the database authors. However, examining the choices made by the authors (and confirming that they were, in fact, consistent) is good LCA practice. To make the consistency check practical, this exercise can be combined with the contribution analysis so that only the most influential datasets can be examined for consistency. This approach can help the LCA analyst understand best where to spend time applying the consistency check.

Data accuracy is included during the consistency check, and its goal is to make sure that all data used is of sufficient quality to address the goal and scope of the study. While the quality of the data used is a focus of the consistency check, the quality of the data is examined quantitatively according to the data quality indicators as described in Chapter 6. The data accuracy can also be described according to uncertainty analysis, which is described in Chapter 15. These chapters should be considered in conjunction with the questions asked during the completeness check in order to fully understand the accuracy of the data, and to ascertain whether the data is sufficiently accurate to draw conclusions.

13.5 Conclusions, Limitations, and Recommendations

The final element of interpretation is to draw conclusions, identify limitations, and make recommendations based on the identification of significant issues and evaluation of the

results. In this stage, the work done in the significant issues and evaluation stages is considered concurrently, and the questions posed in the goal and scope are addressed. Life-cycle insight is gained, and the life-cycle story being told by the data is revealed.

Conclusions must be firmly supported by the data and results of the study. The limitations of the study must be clearly identified, and presented alongside the conclusions, in order to maintain an ethical presentation of the results. Recommendations can be made based on the conclusions, which are often suggestions for actions that can be taken to improve the life-cycle performance of the system under investigation.

The significant issues and evaluation stages of Interpretation can involve a lot of information and many results, all of which can be difficult to keep track of. To make the task of interpretation easier, an interpretation summary table can be used. An interpretation summary table can contain:

* concise statement of the goal and scope
* learning from significant issues identification
* learning from evaluation

Organizing this information as such can facilitate the forming of conclusions, limitations, and recommendations. To illustrate this process, we consider the bio-azaleic acid example that has been discussed in this chapter. Table 13.4 shows a hypothetical summary of the work that may have been performed during the interpretation stage for this study.

From the information collected in Table 13.4, we can make the following interpretation from the study about GWP:

* Based on the information collected in the study, the bio-based option has lower GWP compared to the fossil based option.

The technical reason for this difference is related to the process technology: both options have adipic acid, MDI, and process heat as major contributions to the GWP. Looking at the information in Figure 13.4, one can understand the difference based on these major contributions without even seeing the LCI. The two have similar process heat and MDI inputs, but the bio-based option has a lower adipic acid input, corresponding to lower adipic acid burden.

* The figure shows a lower total of the minor contributions for the bio-based option compared to the fossil based option, which helps to accentuate the difference between the two even further.
* This conclusion is not impacted by alternate allocation/modeling choices. The base case approach was to assume the bio-based raw material to be a waste material carrying no upstream burdens. Treating the bio-based

Goal and Scope		Goal: to understand the life cycle advantages and tradeoffs of PU shoe soles using waste biomaterials, in comparison with conventional fossil-based materials	
Significant Issues	**Contribution Analysis**	• Top 3 contributors to GWP for bio-based option: adipic acid production, MDI production, and process heat • Top 3 contributors to GWP for fossil based option: adipic acid production, MDI production, and process heat • MDI and process heat contributions are fairly similar for both options • Adipic acid production is the most significant GWP process for both options, and has a much lower GWP contribution for the bio-based option	
	Normalization	• The bio-based option has lower impact than fossil based option for each impact category, except for water • GWP is the largest impact when normalized against country totals	
	Weighting	• The bio-based option has lower weighted impact than fossil based option, using the Eco-indicator 99 method	
Evaluation	**Completeness Check**	• Not discussed in the chapter text for this problem. Assume that completeness check was conducted and all needed data has been collected to address the goal and scope	
	Sensitivity Check	• N$_2$O destruction efficiency: for the scenario in which the N$_2$O destruction efficiency is 100% in the adipic acid production process, the bio-based option retains a lower GWP than the fossil based option • Allocation: when the bio-based material is modeled as a valued material using EIO data, or is modeled as vegetable oil, the bio-based option retains a lower GWP than the fossil based option. However, tradeoffs occur in the other impact categories, depending on the feedstock source/model.	
	Consistency Check	• Not discussed in this chapter text for this problem. Assume that consistency check was performed and assumptions, methods, and data were judged to be consistent	

Table 13.4. Sample Interpretation Summary Table

raw material instead as a valued product, while using two different modeling approaches, led to GWPs for the bio-based option that were still lower than the fossil based option and similar to the "burden-free" GWP.

• This conclusion is also not affected by improved N$_2$O destruction efficiency in the adipic acid production process. Even at 100% destruction efficiency, the bio-based option maintains a lower GWP

Based on this summary, the LCA analyst can conclude robustly that the bio-based option has lower GWP. We can make the following interpretation about the other impact categories:

• The bio-based option has a higher water demand compared to the fossil-based option. This conclusion was not impacted by the allocation/modeling choices examined. However, drawing on our LCA expertise and knowledge of the indicator used, we observe that this particular indicator is not an impact potential as such. Rather, it simply represents a sum of the total water inputs throughout the life cycle. The actual impact related to the difference in water usage may not be adequately captured by this indicator. The impact is instead only a preliminary indication of potential water resource impact. Nevertheless, the information acquired in the study suggests that water use may be a tradeoff for the bio-based material.

• The bio-based option has lower impacts in all other categories examined when the bio-based raw material is treated as a waste material with zero burden. However, when the bio-based raw material was modeled differently, tradeoffs in these impact categories emerged. Tradeoffs may be present for the bio-based option, depending on the modeling choice made for the bio-based raw material.

• The other impact categories were not explored in as much detail (based on the information presented in this chapter) as the GWP. Contribution analysis was not done to understand which processes contribute the most to each category. Without contribution analysis, a technical understanding and explanation for any potential difference between the two options cannot be provided. Thus, with the information available in this chapter, the results should be treated as preliminary results. A deeper dive into the data and life cycle processes would help to clarify why the tradeoffs exist and would allow for a more robust conclusion.

Combining all of this information, the conclusions and

limitations that can be made for this example study are as follows:

- The bio-based option has lower GWP than the fossil-based option, based on the scenarios considered in the study. The GWP for the bio-based option is roughly half of the fossil-based option.
- Water usage is a tradeoff for the bio-based option. This metric does not represent well the impact of the water usage, but it is a first-level indicator that water usage could be a concern for the bio-based material. Further analysis of the potential water impacts, including the use of more sophisticated methods, is required for a detailed discussion of the water performance of these options.
- Impacts in other environmental categories may be a tradeoff for the bio-based option, depending on the feedstock choice and modeling approach. Further analysis is required to completely understand the nature of the tradeoffs.

Recommendations for this study are as follows:

- To maximize the environmental benefits of the bio-based option, a waste material should be used as the raw material. The magnitude of the benefit will depend on any burdens allocated to the waste product.
- The life cycle GHG emissions can be addressed significantly by implementing improved N2O destruction technology in the adipic acid process.
- The potential impact of water use should be evaluated in detail to understand better the relative water impact of each option.
- Further analysis should be performed to understand the significance of the life cycle tradeoffs in categories other than GWP and water use.
- Any bio-based material options outside of those considered in this study should be further evaluated for life cycle performance. For example, a bio-based raw material that is associated with a large amount of energy-intensive farming or processing may not carry a lower GWP when compared to the fossil-based option.

In drawing these conclusions, we made statements that are supported by the data and analysis available, and cautiously described the limitations of the analysis where appropriate. The recommendations highlighted some of the life-cycle improvement opportunities, as well as suggested opportunities for additional LCA work on this system. They also transparently warned against using this study to support decisions that involve feedstocks not considered in the study, which is

a potential pitfall that some enthusiastic stakeholders could make when considering bio-based alternatives.

INTERPRETATION AND COMPARING INDEPENDENTLY-CONDUCTED STUDIES

LCA practitioners should recognize that the results from two independently-conducted LCAs are unlikely to be directly comparable. One cannot, for example, simply take the GWP results for various biofuels from five different LCAs reported by different authors, compare the numbers, and quickly draw conclusions without looking critically at the details and assumptions behind each study. The fact that independently conducted studies may not be directly comparable is acceptable, and even expected, because each LCA is conducted according to its own goal and scope. The goal and scope drives all decisions made in the study. Different decisions may therefore be made in different studies that have different goals and scopes. A brief list of some of the aspects that may make independently conducted LCAs unsuitable for direct comparison is provided in the box below.

In order to use the results from different studies to reach new conclusions beyond those presented by the original

Independently-conducted LCAs may be unsuitable for direct comparison if they have:
- different goals and scope
- different system boundaries
- different assumptions (justified according to the goal and scope)
- different impact assessment methods
- different environmental impact categories
- normalization and/or weighting

authors, the underlying assumptions of the original work must be examined carefully and evaluated for comparability. In effect, one needs to define a new goal and scope, and assess whether the previous information can be used to answer the new questions posed. This point emphasizes again the need to apply LCA expertise when interpreting studies and drawing conclusions.

One approach that seeks to make results from independently conducted studies comparable is the use of Product Category Rules (PCRs) to guide the development of LCA-based Type III environmental product declarations (EPDs). PCRs provide specific guidance for conducting LCAs of a particular type of product, and seek to eliminate the potential causes of incomparability as listed in the box. Chapter 19 provides further discussion on the use of PCRs to generate EPDs, and on the comparability of independently conducted LCAs published in EPDs.

13.6 CONCLUDING THOUGHTS ON INTERPRETATION

Chapters 1 and 2 describe how the interpretation phase cuts across the other three phases of an LCA: goal and scope, inventory assessment, and impact assessment. he LCA practitioner must be familiar with the tools used for interpretation, and apply them often in a project, in order to help the study fulfill its goals and to enable decisions to be made. Successful interpretation requires knowledge of strengths and weaknesses of all the aspects of LCA covered in the other chapters, such as data sources, data quality, impact assessment methods, and especially the process or product under study. Interpretation weaves together quantitative calculations, qualitative observations, LCA experience, and the goals of the project.

Interpretation is an iterative process. In the quest to identify the most important inputs or aspects of a product or process, one can develop a deeper understanding of the system, or better data for it, and then recalculate the results and refine the interpretation. Perfect interpretation would require infinite time and resources to perform infinite iterations, so LCAs always stop short of perfection. However, LCAs do not need to be perfect in order to answer the questions and support the decisions that the studies are intended to address. Ideally, the decision to stop the iterative process can be taken when the results, including consideration of uncertainty, are sufficiently clear to allow the project goal to be met. Less ideally, a project deadline or budget may require an end to the iterations before the results are completely clear. In this case, any presentation of the results must clearly describe the limitations and the known data gaps, or logical next steps.

The interpretation phase is also a time when ethical considerations and behavior are paramount. LCAs are often commissioned to understand whether perceived environmental advantages really exist when viewed from a life cycle perspective. The actual results may end up being all or partially contrary to the assumed results, or the conclusions may be critically dependent on a highly uncertain input or a key assumption. Maintaining a commitment to strong ethical performance as discussed in Chapter 3, the LCA practitioner must be able to deliver the "bad news" to the project commissioner when necessary, and also must make sure that any communication of the results reflects the limitations, sensitivities, and uncertainties.

LCA is an expert field, and some level of expertise is required to properly interpret LCA results. This may seem daunting to the new LCA analyst, yet one should remember that awareness of the need for expertise is already an accomplishment and is the first step toward developing the expertise. With this awareness, one can be cautious in developing conclusions and recommendations, taking care to ask the right questions during the interpretation, and seeking out help from others in the field when necessary.

A high-quality interpretation is necessary for any successful LCA. The LCA practitioner must understand how to properly interpret LCA results. Just as an experimental research scientist must understand how to distill the results of a complex experimental investigation into key results and conclusions, the LCA practitioner must transform the LCI

EXAMPLE: PACIFIC NORTHWEST CRAFT BEER

"IT'S THE WATER....AND A LOT MORE"

Imagine you look out over the water on a warm summer evening with a cool glass of your favorite craft beer, as you and some friends discuss the health benefits and risks of beer consumption (Keiji 2004). As someone familiar with LCA, you also wonder about the environmental life-cycle burdens of the products. Fortunately, several craft breweries in the US Pacific Northwest region collected data so that they could create EPDs and label information for their products. The data was combined to create a single dataset that we can use to explore the life cycle impacts of craft beer.

To start the analysis, you select a functional unit, analysis boundaries, and impact assessment methods: 0.5 liters of consumed beverage, cradle-to-grave (including wastewater treatment). Because this study occurs in the US, you select the TRACI 2 characterization, plus water because you know that the water is essential to beer production. These fit exactly the data available, so you quickly use your LCA software to generate results, starting with water.

Much to your surprise, the LCA found that 65 liters of water were used over the life cycle of the 0.5 liter glass of draft beer, so you create a Sankey diagram (Figure 13.14) for water use, showing only those operations that contribute 5% or more to the burden. This simplifies the diagram, and allows one to focus on the most important things. (see pages 174-175)

and LCIA results into key results and conclusions that address the goal and scope of the study and can be used to support decision-making. When viewing LCA interpretation in this way, one understands that the results of an LCA reach far beyond the typical series of bar graphs or charts that are generated during the LCIA phase, and instead start to form an insightful story. This insight comes from understanding the benefits, tradeoffs, and opportunities for improvement of the system under analysis. With proper interpretation, the understanding gained in the study will allow the practitioner to provide science-based insight about the system under evaluation, to avoid greenwashing in developing environmentally related messages, and to deliver information that can help advance sustainable development.

About 70% of the water burden comes from agriculture (irrigation of barley and hops) and 23% from the use phase

(cleaning the glass and disposal of the "used beer"), with the balance being production steps and other small inputs. The critical water use in beer production is not so much where the beer is brewed, but where the crops are grown and the product is used.

Because climate change is an important topic, you decide to look next at the GWP and find a result of 0.89 kg CO_2-eq/0.5 liter of consumed beer. To find if the agriculture and use are as important to GWP as they are to water, you construct a Sankey diagram for GWP, using again a 5% display cut-off and lopping off much of the aluminum process chain:

Although agriculture is still an essential input for GWP (barley production contributes 13% of the total), also import-

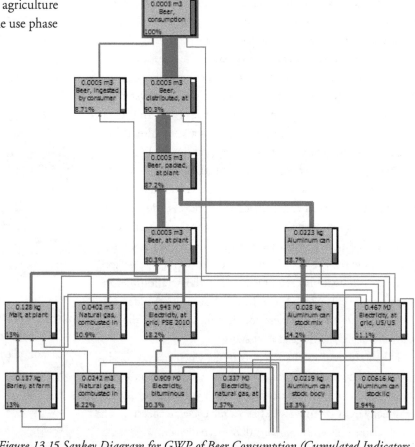

Figure 13.15 Sankey Diagram for GWP of Beer Consumption (Cumulated Indicators are Shown)

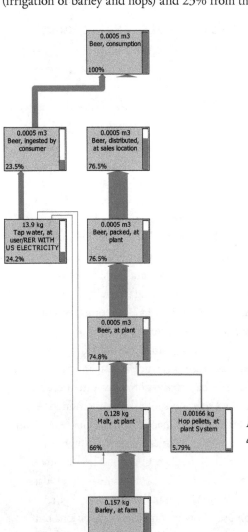

Figure 13.14 Sankey Diagram of Water Use for 0.5 Liters of Beer (Cumulated Indicators are Shown)

ant are the natural gas (heat) and electricity inputs to beer production (inputs to "Beer, at plant," roughly 25% combined, as some of the natural gas shown as 10.9% in the diagram is used to produce the malt), and the aluminum keg production (28.7% - labeled as "can" in the diagram). The use phase ("ingestion by consumer") contributes 9%, larger for electricity for refrigeration for the (brief) storage time. With a broadened awareness of the breadth of activities than contribute to beer and its burdens, you use your LCA software to generate the LCIA for all the metrics in TRACI 2:

Without a benchmark, an alternative product for comparison, or great experience, it is difficult to determine which of these impact categories is most important. TRACI 2.1 included normalization factors based on 2008 data for the US, and by applying these to the data above suggestions we find that the three most critical impact categories are acidification, eutrophication, and carcinogens (which is a human health impact category).

Impact category	Unit	Total
Ozone depletion	kg CFC-11 eq	1.08E-07
Global warming	kg CO2 eq	0.889
Smog	kg O3 eq	0.064
Acidification	mol H+ eq	0.363
Eutrophication	kg N eq	7.62E-03
Carcinogenics	CTUh	1.68E-08
Non carcinogenics	CTUh	7.50E-08
Respiratory effects	kg PM10 eq	8.66E-04
Ecotoxicity	CTUe	3.000

Table 13.5. TRACI Impact for Beer Consumption

You then created Sankey diagrams for each of these impact areas and studied them in great detail, looking at both the higher level descriptions of the largest contributors and the specific emissions that played the largest roles in the calculated impacts. By looking at both what was emitted and from where, you could create a simple table that explains 80% or more of the three impact areas in terms of what and where:

The last topic you decide to investigate before enjoying your beverage is a "what-if" or sensitivity analysis. Because GWP and acidification potential are strongly influenced by electricity production, you wonder what improvement would be possible if the beer manufacturer changed all the electricity inputs from various Northwest utility grids to 100% wind power. Figure 13.16 shows the comparison of these two scenarios. A potential benefit in GWP, acidification potential, and smog formation potential, with minimal other trade-offs would be possible, if it were indeed possible to use 100% wind power.

You report your findings to your friends, who happen to be starting a batch of homebrew beer the following week. With your LCA insight, they decide to select barley and hops only from local farms that practice good water management, and brew at a house that is completely powered by renewable energy. Thrilled that they are taking steps to reduce the environmental burden of their beer, your friends offer you a few free bottles of their upcoming batch. After thanking them for their generosity, you remind them that the average dataset used in the analysis is for Pacific Northwest craft beer, which is produced at a much different scale than their home brewing process, so the results of the analysis may not be completely relevant for their batches. There may be additional steps in their own activities that could have an even greater impact on the environmental impact of their beer. However, for an entire keg you may be willing to conduct another LCA.

As you pause to reflect on the insights you have gained, you realize that you have developed the skills necessary to perform a credible LCA interpretation. In exploring the life cycle of beer consumption, you have defined the scope, boundaries, and impact assessment methods, performed a contribution analysis including external normalization, examined the sensitivity of the results to a key input, and

Impact category	Unit	Total	Normalization factor	Normalized impact
Acidification	mol H+ eq	0.363	0.011	3.99E-03
Eutrophication	kg N eq	7.62E-03	0.0463	3.53E-04
Carcinogenics	CTUh	1.68E-08	19706	3.30E-04
Ecotoxicity	CTUe	3.000	0.0000905	2.71E-04
Non carcinogenics	CTUh	7.50E-08	952	7.14E-05
Smog	kg O3 eq	0.064	0.000718	4.56E-05
Global warming	kg CO2 eq	0.889	0.0000413	3.67E-05
Respiratory effects	kg PM10 eq	8.66E-04	0.0412	3.57E-05
Ozone depletion	kg CFC-11 eq	1.08E-07	6.2	6.67E-07

Table 13.6. Normalized Impacts for Beer Consumption

Process Area	Contribution of process area to:		
	Acidification	Eutrophication	Carcinogenics
Farm equipment			36%
Stainless steel used in wood pallets			16%
Aluminum can production	27%	80%	6%
US Electricity use	35%		12%
Crude oil production			9%
Barley production	19%	10%	
Notes	SO2 and NOx emisions from fossil fuel combustion for electricity	Significant phosporus emissions from Al production, and from fertilizer use	Essentially all due to chromium in disposed slags, tailing, and dusts

Table 13.7. Contribution Analysis Summary for Beer Consumption

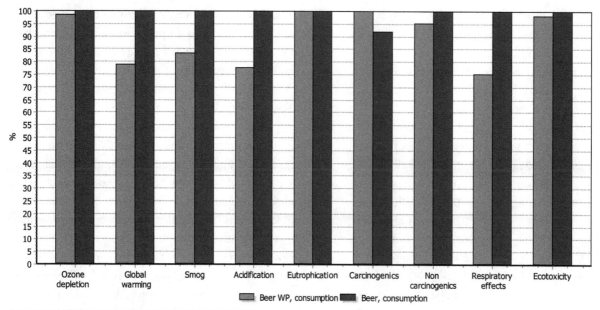

Figure 13.16. TRACI Impacts with Conventional and 100% Wind-Powered Beer Production

articulated both the conclusions and the limitations to your friends. With a quick look at the beer data, you learned that just a few steps in a long process can contribute the greatest burdens and that those important steps may be different steps depending on the particular impact category. The work brought not only insights into beer production, but also LCA methodology and some common features of many product supply chains. And, of course, free beer.

Cheers!

ACKNOWLEDGMENTS

The authors thank David Russell at The Dow Chemical Company for mentorship and leadership in LCA at Dow and in Europe.

PROBLEM EXERCISES

1. LCA INTERPRETATION KEY IDEAS
TRUE OR FALSE?

1.1 LCIA results such as GWP and CED are intrinsic quantities of a material that can be rigorously measured.

1.2 LCIA results are indicators that are intended for use in the decision-making process.

1.3 Contribution analysis can be used to identify the most significant processes in the life cycle.

1.4 The most important outputs obtained from an LCA are the LCIA bar graphs and numerical results.

1.5 When a decision-maker chooses an option that has a lower LCIA result in a given impact category, he or she can have good confidence that the result of his or her decision will be to lower the environmental impact of the function in question for that category.

1.6 The term "significant issues" refers to a group of common problems encountered during LCA interpretation when using LCA software to analyze LCIA results.

1.7 A Sankey diagram depicts a flow of materials or environmental burdens throughout the life cycle, with relative contributions of the material or environmental burden being depicted by the relative thickness of the lines.

1.8 An advantage of LCA is that results from independently conducted studies can be easily compared.

1.9 In sensitivity analysis, the focus is on the impact of choices made during the study, and not on the uncertainty related to the data used in the study.

1.10 Successful interpretation of LCIA results is possible with only a mild introduction to LCA methods and with little environmental science background.

1.11 Conclusions must be firmly supported by the data and results of the LCA.

1.12 During the Interpretation phase of an LCA, the LCA analyst transforms the LCI data and LCIA results into knowledge and insight, and uses the results of the LCA to answer the questions posed in the goal and scope.

2. CONTRIBUTION ANALYSIS

Reverse osmosis (RO) is a technology used to obtain purified water from brackish water, salt water, and even wastewater. Use of RO requires installation-specific equipment ("modules"), electricity for operation, and periodic use of chemicals such as ethylenediaminetetraacetic acid (EDTA), sodium phosphate, sodium perchlorate, and an assortment of other chemicals for cleaning the modules. Production of modules requires inputs of plastics, electricity from the grid, and other materials. An RO module manufacturer has conducted an LCA to examine the life-cycle impacts of an RO-based water recovery unit installed in a wastewater treatment plant in Singapore. The study included the cradle-to-grave production of the RO unit, the treatment of wastewater over the three-year life of a module, and disposal of the module at the end of life. The reference flow used in the study was 1 m³ of purified water. The pie chart below shows GWP results from the analysis.

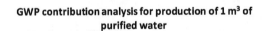

GWP contribution analysis for production of 1 m³ of purified water

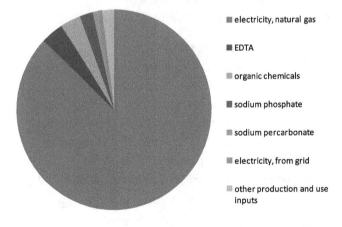

- electricity, natural gas
- EDTA
- organic chemicals
- sodium phosphate
- sodium percarbonate
- electricity, from grid
- other production and use inputs

2.1 Can you explain why the results are dominated by a single contribution?

2.2 What source of electricity was assumed for the operation of the wastewater treatment plant in Singapore? Do you agree with this selection?

2.3 The wastewater treatment plant in Singapore is undertaking a new sustainability initiative and seeks to reduce the impacts from their operation of the RO unit. What actions would you advise them to pursue?

2.4 What actions could the RO manufacturing company take to improve the life-cycle GHG performance of their RO units?

2.5 The purchasing manager for the wastewater treatment plant in Singapore has found an EDTA supplier who claims to offer EDTA that has 20% fewer cradle-to-gate GHG emissions when compared to market-average EDTA. The EDTA is offered for a 10% premium over the standard EDTA. The plant managers ask for your opinion on whether they should pay the premium for this EDTA source to help them meet their GHG reduction goals. How would you advise them?

3. LCIA INDICATOR EVALUATION

You have run across a recent press release from an insulation foam manufacturing company that claims to have invented a new type of insulation that exhibits lower human and environmental toxicity based on an LCA. The company's press release states that this new foam is "better for the planet" based on the LCA.

The company made a comparative assertion and is obligated by ISO to provide a copy of the LCA upon request, so you take the opportunity to request a copy of the study and evaluate their claim. You find that the study boundaries included production of the foam, use of the foam in a refrigerator for twenty years, and the end of life of the foam, but it does not include production of the other parts of the refrigerator as they were assumed to be the same for two foaming technologies. USEtox was the LCIA method used to evaluate toxicity

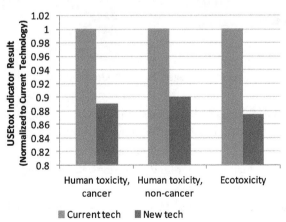

(Rosenbaum et al. 2008). The USEtox results shown in the report are presented below:

3.1 Do you agree with the conclusion? Why or why not?
3.2 Is there a better way to present this data?
3.3 Is the new foam truly "better for the planet"?

4. SENSITIVITY ANALYSIS

An LCA colleague was asked to compare the sustainability performance of widgets produced by PacNorWest widgets, located in Seattle, Washington, with average widgets produced on the market. After some quick analysis, they used the following data, made the following assumptions, and generated the following results:

Assumptions:

* cradle-to-gate analysis conducted on functionally equivalent production of 1 widget
* data obtained from one-month of operation of the Seattle production plant
* US average electricity grid used
* 95% of raw material inputs accounted for
* average widget data is US industry-average data, collected 5 years ago

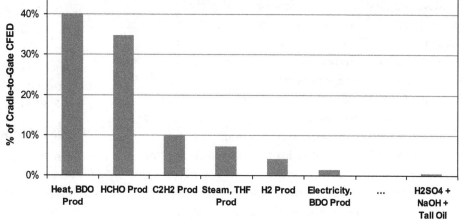

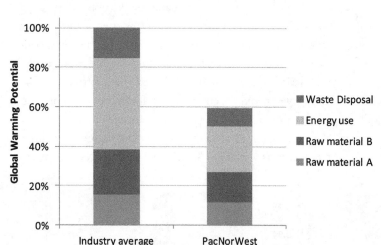

* IPCC 2007 GWP 100a used as indicator

The LCA analyst reached the following conclusion:

The PacNorWest option is clearly advantaged over the industry average option for GHG emissions, having roughly 40% fewer GHG emissions.

Do you agree with this conclusion? What sensitivity analysis work would you want to do in order to examine this claim? Would you draw any different conclusions?

5. CONTRIBUTION ANALYSIS

You have just come up with an idea for a new process to make tetrahydrofuran (THF), which is a chemical used to make elastomeric polymers such as Spandex. The new process could eliminate the uses of acid and base common in the current process, but with increases in the steam and electricity required. Before you begin working on development of this project, you do a quick LCA that looks at the cumulative fossil energy demand (CFED) for THF produced via the new potential process. The bar chart below shows the contributions to the CFED. The abbreviations BDO, HCHO, C2H2, and H2 in the graph represent the main raw materials in the process, which will not change if your new route concept is successful. What inputs may be more beneficial to focus on than acid and base use? Might your analysis change by considering other impact categories?

6. ALLOCATION EXAMPLE

A steam cracker is typically used to transform mixtures of "light" hydrocarbons, such as liquids from natural gas or light fractions from crude oil processing, into mixtures of lower molecular weight compounds, especially olefins such as ethylene and propylene. These lower molecular weight compounds are used as building blocks for several

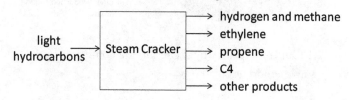

other industrially crucial chemicals and plastics. For example, polyethylene and polypropylene are two common plastics made from the ethylene and propylene that is produced by steam cracking.

Because stream crackers produce many more products than just ethylene and propylene, allocating the inputs and environmental burdens to these products is necessary. We have many ways to allocate these burdens to the cracker products and, subsequently, to everything made from them. Potential allocation methods include:

- equally by mass to all the cracker products (every kg of material leaving the cracker has the same environmental burden)
- allocation by mass but only to the higher valued fraction (e.g., the olefins)
- allocation by economic value to all of the cracker products

A comparison of the life-cycle GHG emissions for a given application using plastic or glass is shown below, using these three methods of allocation for the plastic product (assuming that the plastic is produced from the steam cracker product ethylene). The graphs are normalized to the glass GWP in each case.

What conclusions can you draw from these figures?

7. COMMUNICATING DIFFERENCES

You work as LCA analyst for an innovative bottled water company focused on delivering low-cost drinking water to Sub-Saharan Africa. Recently, your Chief Executive Officer has come across an environmental product declaration for bottled water in Italy[2] that claims to have a carbon footprint of 95.8 g CO_2 per liter of water delivered in 2L quantities. The CEO compared that published number with a cradle-to-gate GWP that you had estimated for your company's 2L bottle, which you found to be 81 g CO_2 per liter of water. He concludes that your water is GHG-advantaged, and asks you to write up a quick press release touting a 15.4% GHG savings associated with your water for use at a conference the next day. How do you respond?

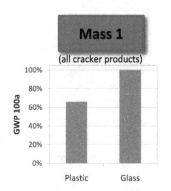

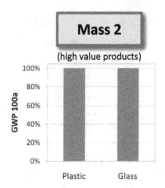

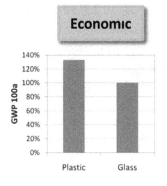

2 http://gryphon.environdec.com/data/files/6/8867/epd212_San%20Benedetto_aqua_minerale_2012.pdf

STATISTICS IN LCA

MATT PIETRZYKOWSKI AND RONALD WROCZYNSKI

14.1 WHAT IS STATISTICS?

Merriam-Webster defines statistics as "a branch of mathematics dealing with the collection, analysis, interpretation, and presentation of masses of numerical data." All statistical analysis depends on having multiple measures of the same item. This chapter discusses how to understand, and best utilize, the input data and the outputs from LCA studies. It also explores some of the fundamental concepts, and application of statistical tools, in order to help us understand and interpret the data. The focus will be on primary data and the collection of data from the actual processes.

Primary data collection is the basis of all unit process LCI data and, in turn, the basis of all LCAs regardless of the models used, so one needs to properly collect and evaluate primary data. Specifically, one wants to know if the life cycle impact of one product or process is different (better or worse) than another. Unfortunately, most of the LCI data available to the practitioner through commercial and public databases present results as a single measurement. The information needed to evaluate the uncertainty of the data is not available to answer this fundamental question. This can change over time as new practitioners develop and publish new unit process data sets that include the statistical information needed. As new primary information is collected, this statistical information should be collected as a standard practice.

This chapter provides information on methods to evaluate both primary data and the secondary data, provides a core set of quantitative analysis tools that assess the implications of modeling choices and data quality, that tease trends from highly uncertain data, and that clearly support conclusions and assertions.

14.2 FUNDAMENTAL CONCEPTS

PRIMARY AND SECONDARY DATA

Primary data is the term used to describe site-specific data collected directly for the materials or processes that are being modeled in the life cycle assessment. The following standards from the International Standards Organization (ISO) should be reviewed for the explicit definition of primary and secondary data (ISO14040 2006; ISO14044 2006). Primary data is generally the most accurate data because it represents the actual system under study, and often is well documented and traceable.

For example, in the production of beer, the brewmaster most likely has accurate records of the raw materials (e.g., malt, hops, yeast, and water) used for each batch of beer and the corresponding yields of beer for each batch. This data would be considered primary data because it traces directly the beer process under study.

However, in many cases the collection of data for the actual materials or processes is not possible, either because it is too costly in time or labor, or because the material or process are not accessible to the LCA practitioner. In this case, the data may be obtained indirectly from other LCA studies, or from suppliers that have studied, or produced, the materials or processes of interest. These data are referred to as secondary data. A special case of secondary data is proxy data. Often, no direct primary or available secondary data is available for a material or process, but data exist for a process that can be considered similar. This often occurs for many chemical substances. In this case, the data for the similar process can be used and evaluated to see if it has a significant effect on the outcome of the results. If the proxy substance/process does have a significant effect, then a greater effort may need to be expended to obtain the primary data for the actual process/substance.

For example, secondary data could be used for the production of hops and malt used in the beer production process. Data for agricultural products are often obtained from United States Department of Agriculture (USDA) databases. These extensive data are organized by state/county and constructed from surveys of farmers cataloging their yields per area farmed, as well as the amounts of fertilizers, fuels, and water used to grow the crops. However, data used in the beer production are secondary data because they are not obtained for the process being studied by the practitioner. Another example of secondary data relates to transportation. Although the distances traveled and the quantity of produced beer transported to customers is directly available, the exact emissions from gasoline or diesel fuel combustion probably are not directly available. Secondary data from a validated database, for example, will be used to describe the typical emissions for a particular size truck that operates with a particular fuel.

The quality of the data, as will be discussed in some of the following sections, is usually expressed in mathematical terms that are the same for both primary and secondary data. However, the values of the mathematical terms that describe the data will vary based on the quality of the data, and whether it is of a primary or secondary nature. Secondary data often, but not always, is considered to be of lower quality, or to have less traceability, than primary data. These factors depend on the representativeness of the data as well as the documentation of the data collection process.

OTHER TYPES OF DATA

Statistics classifies data into a number of different types. Quantitative data is numeric data and typically results from some form of measurement (e.g., weights, temperature, and speed). Categorical (or qualitative) data are data in which things are grouped by type according to a common property or set of properties. For example, gender, color, size and blood type are all categorical data. Categorical data is typically described by the number of occurrences for each categorical description (e.g., the number of females and the number of males). A number value does not necessarily mean that the data are quantitative. For example, shoe sizes are often described with numbers from 5 to 20, but these are categorical. Likewise, 4, 6, 8, 10, and 12 cylinder engines and telephone area codes are categorical, and not quantitative, in nature.

Discrete data have a finite number of possible values.

They can be counted (e.g., the number of correct answers on a test, or the number of bales of hay collected from a field). Continuous data can potentially have an infinite number of steps in values over a range. The range may be finite or infinite (e.g., the height of women, or the time it takes to complete a task). Sometimes, discrete data can be converted into continuous data and it may be useful in data analysis to do so. For example, rather than counting the bales of hay obtained from a harvest, each bale could be weighed and the weights from a harvest could be used. Similarly, rather than counting the correct answers on a test, each answer can be given a point value and partial credit used, and then the test scores can become continuous data.

DATA COLLECTION

Data collection is an important element of LCA because it plays a part in determining the data quality. Good data quality is essential to insure the best possible results. Poor quality data can lead to misleading results and misinterpretation. The data used in LCA can be single point (based on only one measured value) or multi-point (based on a collection of a number of values). In general, multipoint data is preferred because it allows for a more realistic estimation of the variation of the measured value and, ultimately, a more realistic estimation of the error in the LCA model's results. The term error in statistics does not imply that the individual sample measurement is incorrect. It merely measures the spread of the data around the average of the population.

Occasionally, a single sample may be the only quantified piece of information one has. In this case, one must estimate the distribution from which the data point was sampled from expert knowledge, literature review, or other methods. Another method that is often used in LCA is to use a pedigree matrix (Weidema and Wesnæs 1996) to evaluate aspects of uncertainty when using data. The result of using the pedigree matrix is an estimated variance for a lognormal distribution. No matter whether single point or multipoint data is collected, structuring the data collection so that the most representative data is obtained is essential. For example, if data are obtained from a survey, the survey should be carefully structured so that the participants clearly understand the questions, the data that is desired, and the operational conditions that are relevant to the data.

DOCUMENTATION

Keeping a close track of the source(s) of the data and the conditions under which they are collected is essential. Examples of some of the types of information that should be documented include the following: the time and date; an account of whether the data was directly measured, calculated from other measurements, or estimated; the units of measurement; any process conditions (temperature, pressure, mixing speeds, and mass flow rates); and the conditions under which the measurements occurred, a factor that can influence the measurements. For agricultural data, for example, we should note the geographic location of the crop (continent, country, state, or region), the year of harvest, and the area over which the measurements were made. In general, the data collected will have to be normalized to the basis of the functional unit. Clearly documenting which data has been normalized, as well as the actual normalization method used, is crucial.

If the data being used is derived from multipoint data, then we should clearly note how the multipoint data was treated for entry into the LCA model and whether the entered data is the numerical average (mean) or median, or whether it was derived by another method (e.g., interpolation or extrapolation). In addition, the variation of the multipoint data should be documented by an appropriate method.

INITIAL ERROR ASSESSMENT

We should consider potential sources of error from the beginning of the data-collection process. These errors could come in the form of systematic errors or random errors. Systematic errors (bias) can come from the wrong settings or calibration on an instrument, poorly worded questions in a survey (which can lead to a consistent under- or over-reporting of a value), or the wrong units used for a specified value (e.g., pounds instead of kilograms, degrees Fahrenheit instead of degrees Centigrade, or miles instead of kilometers). We should carefully inspect the data and data collection methods to avoid systematic errors. Random errors typically result from everyday variation in a variety of uncontrolled (or unmeasured) factors. Random errors are unpredictable. For example, variations in estimating measured values that fall between the marked gradations, electronic noise in controls, and fluctuations in temperature and humidity can all induce random errors.

Systematic errors can have a significant effect on the accuracy of a set of values without affecting their precision. Random errors affect the precision of a value without necessarily affecting the accuracy of a set of values. We will discuss the concepts of precision and accuracy in greater detail in Section 14.3:

STATISTICAL MEASURES.

By carefully considering where errors may arise, it may be possible to take early action to reduce the number or types of errors. One of the easiest ways to assess the magnitude of random errors on a dataset comprised of multiple values is to construct simple frequency plots such as dot plots or histograms. The dot plot is easily constructed manually with pen and paper, with each dot representing an occurrence of a particular value (or value range) within the dataset. A histogram is a bar chart where the height of the bar represents the total number of occurrences of a particular value (or value range) within the dataset. Each of these types of plots shows the range of values for the dataset and the frequency with which these values occur within the dataset.

As an example, consider the data in Table 14.1. This data represents the daily production of kegs of beer produced in a craft brewery over a three-week period. The LCA analyst plotted the data in both dot plot and histogram formats, as shown in Figure 14.1.

Kegs of craft beer produced per day			
Day 1	200	Day 12	300
Day 2	700	Day 13	700
Day 3	500	Day 14	500
Day 4	500	Day 15	400
Day 5	400	Day 16	700
Day 6	600	Day 17	400
Day 7	600	Day 18	700
Day 8	600	Day 19	400
Day 9	500	Day 20	500
Day 10	600	Day 21	1500
Day 11	500		

Table 14.1. Table of Beer Production

Both charts clearly show the production is clustered around 500 kegs/day and that the production for Day 21 is much larger than the rest. These charts cannot determine if the Day 21 production is clearly an outlier—more sophisticated tests would be needed—but they do indicate that this data point should be looked at closely. Perhaps this error occurred during data entry.

A number of statistical packages (JMP Statistical Discovery Software, Minitab 16 2013, and Statgraphics Centurion XVI.I) incorporate tools to construct both dot plots and histograms. The Data Analysis add-in that comes with Microsoft Excel can be used to construct histograms. Dot plots can also be simply constructed in MS Excel.[1]

1 The simple procedure is to sort the data in ascending order, and then in an adjacent column place numbers (1, 2, 3, et al.) for each occurrence of a particular value in the corresponding row of the adjacent column. Then construct a scatter plot with the values as the" x's" and the adjacent column as the" y's."

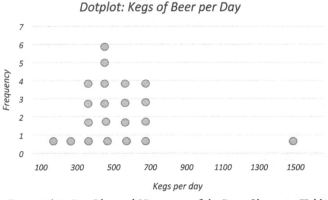

Dotplot: Kegs of Beer per Day

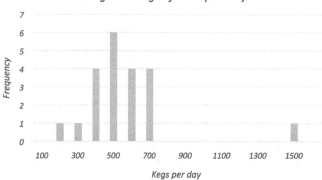

Histogram: Kegs of Beer per Day

Figure 14.2. Dot Plot and Histogram of the Data Shown in Table 14.1

A somewhat more sophisticated plotting technique, box plots (Kiemele, Schmidt, and Berdine 1999; Levine, Smidt, and Ramsey 2001), can also be used to derive more information from the data and to more easily compare multiple datasets. Most general statistics software packages will construct box plots from a set of data. The same data in Table 14.1 is plotted in a box plot given in Figure 14.1. The first quartile (Q1) marks the value for which 25% of the values in the dataset are less than it and 75% are greater than the value. The third quartile (Q3) marks the value for which 25% of the values are greater and 75% of the values are less than the value. The median represents the value for which 50% of the values in the dataset are above and 50% of the values are below it. The lines ("whiskers") extending to the left and right of the box represent the lowest and highest value within the lower and upper limit, respectively. These limits are calculated as 1.5 times the interquartile range (IQR = Q3 − Q1), subtracted from the first quartile and added to the third quartile. Potential outliers are estimated as those data points that fall outside the calculated lower and upper limits.

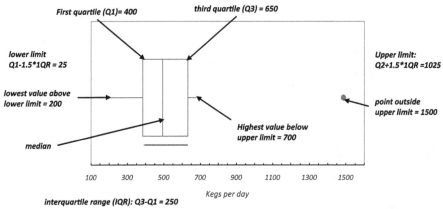

Whiskerplot: Kegs of Beer per Day

Figure 14.2: Anatomy of a Boxplot

14.3 STATISTICAL MEASURES

PRECISION AND ACCURACY

For a dataset, precision refers to how closely the individual values in the data set agree with each other. In contrast, accuracy refers to how closely these individual values match their "true" value. The true, or "expected," value is often approximated by the mean value of the data set. The concepts are often depicted by using a target used by a marksman. An accurate marksman will have shots on the target that are randomly distributed around center of the bullseye. A less accurate but more precise marksman will have a tight cluster of shots on the target, but they will be off center. An accurate and precise marksman will have a tight cluster of shots, all of which are on or near the bullseye.

In a similar manner, an estimation of the average can be obtained from a data set with individual values evenly spread around the true average (they are on target). However, if the number of available data points is small, or the precision is low, the averaging of a small subset of these values will have a larger spread and the average will be less certain. Both of these concepts are essential for understanding the concept of uncertainty.

CENTRAL TENDENCY

Central tendency is a term used in statistics to describe the observation that most values within a data set tend to cluster around a single, central value or location. Thus, we can usually select a single value to represent the entire data set. Many measures can be used for this central tendency and four primary measures that are most pertinent for LCA studies are described here.

The most common and well known measure of central tendency is the mean (or, more exactly, the arithmetic mean). The mean is the mathematical average of all the values in the data set. For a data set of n observed values, it can be written as:

$$mean = \bar{x} = \frac{x_1 + x_2 + \cdots + x_n}{n} = \frac{\sum_{i=1}^n x_i}{n} \quad \text{Eq (14.1)}$$

For a data set where n is large, the mean is represented by symbol for the population mean, μ.

The median represents the data value for an ordered version of this data set in which half of the data set values are less than and half the data set values are greater than this value. We call this the 50th percentile. For example, if the values in the data set are listed in increasing order as x_1, x_2, x_3, ..., x_n then the median is the middle value of that set.

Although the arithmetic mean is easily calculated, extremely high or low values can affect the average significantly. In contrast, the median is less sensitive to extreme values, but it is not a purely mathematical operation in that it requires sorting and counting operations. When the mean and median are equivalent, the values in the data set are symmetrically distributed around the center point. However, if the mean is significantly less than the median, then the data is skewed to the left. Conversely, if the mean is significantly greater than the median, then the data is skewed to the right. As an example, the data in Table 14.2 is used to illustrate the comparison of

mode that corresponds with the most frequent value in the data set. In the case of the data in Table 14.2, the mode for "Data set 1" corresponds to a value of 5.4 and the mode for "Data set 2" corresponds to a value of 4.5.

The geometric mean is a measure of central tendency often used for data that has a lognormal distribution. This distribution will be discussed in more detail in the following section. For a data set consisting of n values, the geometric mean is defined as the n-th root of the product of all the values in the data set.

$$geometric\ mean = G = \sqrt[n]{x_1 * x_2 * \dots * x_n} = (\prod_i^n x_i)^{\frac{1}{n}} \quad \text{Eq (14.2)}$$

$$\ln G = \frac{1}{n} \sum_{i=1}^n \ln x_i \quad \text{Eq (14.3)}$$

VARIABILITY

The spread or dispersion of the data set around the central value is an indication of the variability of the data. The range is the difference between the lowest and highest values in a data set and is a simple indicator of the spread of the data. However, it depends on only two (possibly extreme) values in the data set and ignores the rest of the data set members.

On the other hand, the variance and standard deviation are common measures of dispersion that take into account all of the values of the data set. Both of these measures are based upon the distance of each point from a central value, usually the mean. The variance is defined as the average squared distance of each value in a data set from the mean. In a population, or a data set that is representative of all possible values, the variance is represented by Equation 14.4 in which N is the number of members in the population and m is the population mean[2]. For smaller data sets, the variance is better represented by Equation 14.5, in which s2 is the sample variance and is the sample mean.

$$Population\ variance = \sigma^2 = \frac{\sum_{i=1}^N (x_i - \mu)^2}{N} \quad \text{Eq (14.4)}$$

$$Sample\ variance = s^2 = \frac{\sum_{i=1}^n (x_i - x)^2}{n-1} \quad \text{Eq (14.5)}$$

The standard deviation is derived by taking the square root of the quantities in equations 14.4 and 14.5. The standard

[2] The summation quantity in the numerator of equations 14.4 and 14.5 is called the sum of squares. The sum of squares is a less useful comparative estimate of dispersion because its value depends on the number of values being considered. A better comparative measure of dispersion is obtained by dividing the sum of squares by the number of values.

Data set 1						Dataset 2					
2.7	3.8	8.4	5.5	6.0	5.4	4.5	8.7	6.3	4.5	2.9	5.1
4.7	4.7	7.7	5.4	5.8	5.4	16.2	3.6	4.8	4.8	2.4	6.6
6.6	4.9	6.1	9.5	6.1	5.8	4.1	4.0	6.7	3.8	4.7	7.2
4.3	4.7	4.3	4.4	0.8	5.9	1.9	4.9	8.3	4.5	3.9	4.0
6.6	5.5	6.8	6.4	6.0	3.6	4.2	5.0	2.4	6.1	2.7	3.6
6.6	6.6	7.0	5.4	2.9	4.7	5.5	3.8	9.6	3.5	6.3	3.2
4.6	7.3	7.2	4.7	4.1	5.0	3.4	2.6	5.2	7.1	4.1	8.6
5.4	0.4	4.0	5.3	7.1	6.0	3.6	6.6	3.2	2.5	3.4	6.8
6.7	6.1					3.8	4.2				

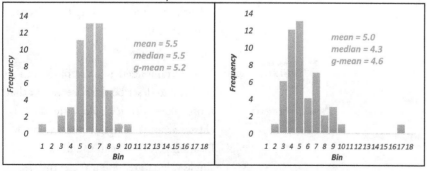

Table 14.2. Illustration of Difference Measures of Central Tendency

differences in means and medians in some example data sets.

Another method of estimating the central tendency is the

deviation is commonly used because the units of the resulting quantities are the same as those in the original data set, allowing an easier comparison of the dispersion to the mean.

Population standard deviation $= \sigma = \sqrt{\sigma^2} = \sqrt{\frac{\sum_{i=1}^{N}(x_i - \mu)^2}{N}}$

Eq (14.6)

Sample standard deviation $= s = \sqrt{s^2} = \sqrt{\frac{\sum_{i=1}^{n}(x_i - \bar{x})^2}{n-1}}$

Eq (14.7)

Another useful quantity is the coefficient of variation that is the ratio of the standard deviation to the mean ( ). This dimensionless value allows a comparison of the relative dispersion of data sets measured in different units.

RELEVANT DATA DISTRIBUTIONS

When a number is used to represent a quantity of interest, a single datum is a representation of a larger population of data. The entire population of data is difficult and perhaps impossible to collect, so a set of samples from the population is taken. This sample set represents the population and shares many of the characteristics of the population from which it was taken. A sample set is typically a manageable collection of data points, or samples, that are reasonable to collect for the data quality required by the LCA being performed. When at least two samples in a set exist, denote this $X_{n=2}$, a range, *max(X)-min(X) = range(X)*, which can be calculated to describe the plausible set of samples that can be assumed given the data. As more and more samples are observed, the range is insufficient to fully describe how the samples are distributed in the set.

A distribution of data is a representation of the frequency at which the data values occur within a certain range. These frequencies, in turn, reflect the probability of a certain datum occurring in the larger set of samples. Characterizing the sample set this way allows one to make inferences about the population. Consider a set of values collected on the mass (kg) of fossil CO_2 emitted to air from the beer life cycle used throughout this book. Table 14.3 shows a portion of the collected data.[3] The table shows the data values, the bins into which the data are grouped, and the frequencies of data occurrences within each bin. In this case, frequencies are used that represent the proportion of the bin count (the number of data points within a bin) to the total number of data points.

A more common way to visualize a data set's distribution is using a histogram. A histogram is a plot of the frequency table shown above. The x-axis represents the bins and the y-axis can be either counts or frequency. A histogram of the data is

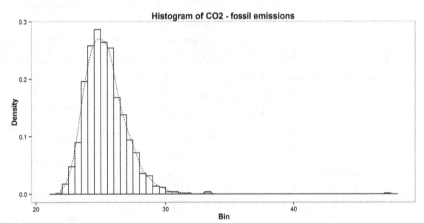

Figure 14.3: Histogram of CO2 Data Shown in Table 14.3

Row	Substance	Compartment	Unit	Ordered Values	Bins	Counts/Bin	Frequency
1	Carbon dioxide, fossil	Air	kg	21.89	21	1	0.001
2	Carbon dioxide, fossil	Air	kg	22.09	23	176	0.176
3	Carbon dioxide, fossil	Air	kg	22.15	25	531	0.531
4	Carbon dioxide, fossil	Air	kg	22.27	27	236	0.236
5	Carbon dioxide, fossil	Air	kg	22.27	29	47	0.047
6	Carbon dioxide, fossil	Air	kg	22.31	31	6	0.006
7	Carbon dioxide, fossil	Air	kg	22.33	33	2	0.002
8	Carbon dioxide, fossil	Air	kg	22.36	35	0	0
9	Carbon dioxide, fossil	Air	kg	22.37	37	0	0
10	Carbon dioxide, fossil	Air	kg	22.48	39	0	0
.	.	.	.	.	41	0	0
.	.	.	.	.	43	0	0
.	.	.	.	.	45	0	0
1000	Carbon dioxide, fossil	Air	kg	47.34	47	1	0.001

Table 14.3: Frequency Table for a Set of CO2, Fossil Emissions Generated from a Reference Flow of 100L of Beer

3 This set of data is simulated data generated from randomly sampling from a 1000 samples generated from a Monte Carlo algorithm.

shown in Figure 14.3.

Distributions can be formed from both discrete and continuous data. As discussed previously, discrete data are those data that take specific values for the observation, while continuous data are data that can take on fractional values. This section will focus on continuous data distributions that are common in LCA. Discrete distributions can be encountered, however, and the interested reader can consult the references at the end of this chapter.

The shapes of distributions help the analyst identify basic characteristics of the data, such as central tendency of the data set (mean and median) and the amount of spread contained within the sample set (variation and standard deviation). For example, after looking at a data distribution, one can decide if the data favors lower or higher values, contains extreme values, or is symmetric. In LCA, knowledge of how the data is distributed helps the analyst assess the uncertainty of the values used as inputs and outputs to the LCA model. This follows from the fact that measured values have error associated with them. The error could be due to measurement technique, instrument error, and uncontrolled sources, like environmental effects, that impart random variation.

The error sources affect the measurement such that upon repeated measurements, a slightly different value is obtained. For example, if a measurement has a value, x_1, and the another measurement is taken, x_2, x_1 and x_2 will differ by $x_2 - x_1 = dx$. As samples are gathered, more evidence supports the true value of the population and uncertainty on that value. The differences between the individual samples define the width of the distribution, while the actual values define the magnitude.

All of these replicate measurements or samples make up a distribution of point estimates that account for not only the quantity of interest, but also the underlying error for all the different error sources (both random and imposed). No matter the source of the error in the data, a value representing this quantity is a point estimate of an underlying distribution. Characterizing the distribution of the data with which you are working is relevant for other reasons, too. It helps us understand the basic structure of the data and how best to work with it. Later sections will discuss how to summarize and analyze data in a statistically sound way. The way in which data is summarized and analyzed can be dependent on the data's underlying distribution. For example, if one uses the

t-test to judge the statistical significance between two means, we should assume that the underlying distribution(s) of the samples is normal or approximately normal.

As mentioned previously, the shape of the distribution is a reflection of the sampled data. Some represent discrete data, like tallying the choice of particular impact method used, while others reflect the nature of continuous data, like the measurement of kg of CO_2 emitted from a production plant. In order to partition the data to compute frequencies, or plot as a distribution, a binning method must be applied to the data. A convenient way to subset, or bin, the data is to use quantiles. Quantiles are used on a set of ordered data in order to divide the data into equally spaced subsets. Mathematically, given an ordered set of data $X = x_1, x_2, x_3, \ldots, x_n$, where the order is defined by the index n, the kth quantile, q_k, is the data value at which the probability of value, x, is less than or equal to k/q. A more concise representation of a quantile can be written as Equation 14.9:

$$p(X < x) \leq \frac{k}{q} \qquad \text{Eq (14.9)}$$

If the boundary value is included, the form is written as in equation 14.10:

$$p(X \leq x) \geq \frac{k}{q} \qquad \text{Eq (14.10)}$$

Equation 14.9 accounts for the fact that the value for the quantile represents the boundary value below which x may be found. Equation 14.10 accounts for the boundary value, which, by Equation 14.9, is greater than the proportion k/q. The ratio k/q can be expressed as a decimal fraction bounded between 0 and 1, or as a percentile bounded between 0% and 100%. Please see the discussion of quantiles in Section 14.1 for additional explanation. Table 14.4 shows the calculation of the CO_2 data into four equal quantiles, which are referred to

Calculation Parameters	Value	Quantile Location	Quantile Value	Interpretation
Data Points (N)	1000	-	-	-
Number of quantiles (q)	4	-	-	-
Quantile (k)	1	250	24.26	25% of the ordered data is less than 24.26
	2	500	25.22	50% of the ordered data is less than 25.22, median
	3	750	26.23	75% of the ordered data is less than 26.23
Expresssion				
Quantile Location = N*(k/q)				

Table 14.4. Quantile Calculation for the Emissions of CO2, Fossil

as quartiles.

Mathematical forms can be found of many data distributions. The probability density function (PDF) is used to calculate the probability of a random variable being drawn from the distribution in question within a particular set of bounds. The PDF is a function, f(x), of a continuous variable x, where f(x) ≥ 0 and the area under the curve is 1. Another common distribution function is the cumulative distribution function

(CDF). The CDF function returns the probability between -∞ and +∞ that a datum, x, is less than a given quantile. As opposed to the general discussion of quantiles earlier, the CDF takes a specific form for each underlying distribution (e.g., normal, binomial, lognormal, and exponential). Many of these functions are parameterized. For instance, the analyst would only need to know the arithmetic mean and standard deviation of a sample set to fully describe an underlying normal distribution.

In LCA, a few distributions represent most of the data collected. The following sub-sections describe these distributions and provide examples of when one might use them in LCA. The PDF will be provided for each distribution, along with the parameters defining them. These formulae are presented for reference and provide some evidence, although not rigorous proof, of the basis for these distributions. Practically, these formulae are implemented as algorithms in many software packages, making them easy to use. It should be cautioned, however, that blind use of these software tools is risky. One should always understand what the calculator produces.

THE NORMAL DISTRIBUTION

The normal distribution is one of the most well known distributions and is the basis for many common statistical hypothesis tests. The probability density function has the form

$$PDF(x; \mu, \sigma) = \frac{1}{\sigma\sqrt{2\pi}} e^{-\frac{1}{2}\left(\frac{x-\mu}{\sigma}\right)^2}, \quad (-\infty < x < \infty) \quad \text{Eq (14.11)}$$

where μ and σ are the population mean and standard deviation. Common measures of location and spread are: the sample mean, $= \frac{1}{n}\sum_{i=1}^{n} x_i$, sample variance $s^2 = \frac{1}{n-1}\sum_{i=1}^{n}(x_i - m)^2$, and the sample standard deviation, $s = \sqrt{\frac{\sum_{i=1}^{n}(x_i-m)^2}{n-1}}$. The sample measures are not necessarily the same as the population parameters. The sample measures are directly calculated from the sample of data drawn from an underlying distribution characterized by the population parameters. The sample measures are only estimates of the true population parameters of the underlying distribution.

Equation 14.11 has the characteristic bell-shaped curve that is fully described with a location and scale parameter. The location parameter defines the peak of the bell and is repre-

sented with the Greek letter μ, which is the population mean. The scale parameter defines the width of the curve and is represented with the Greek letter **σ**, which is the population standard deviation. Equation 14.11 is also known as the Gaussian function. Figure 14.4 illustrates the effects of changing the location and scale parameters. The y axis shows the probability density and the x axis shows the data value. The area under the curve is 1, representing 100% of the population, so that when the standard deviation changes, the probability density

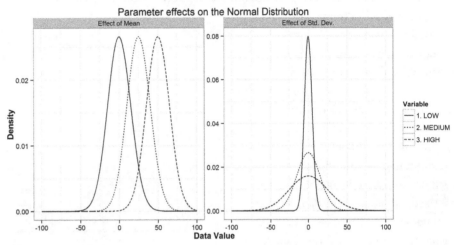

Figure 14.4. Parameter Effects on Normal Distribution
Effects of the different values of mean and standard deviation on the Normal distribution

broadens, but the height decreases.

The normal distribution is one of the most recognizable distributions, given its widespread occurrence in nature, as well as its use in the physical sciences and as the foundation of many statistical methods. It stems from the central limit theorem of additive errors. The concept of additive errors states that a measured quantity is composed of a true value, μ, and error, **ε**.

$$x = \mu + \epsilon \quad \text{Eq (14.12)}$$

The error term arises from all kinds of sources that are assumed independent and random. The sources can be variation in physical and chemical parameters, such as in wind speeds and instrument variations, and variations in humidity and temperature, or they may be variation due to human action, such as process times and energy consumption. Because of the assumption of error independence, the combined effect of all random error sources is additive. The result of additive error is that a set of samples will form a distribution. As more samples are gathered (as N increase), the resolution of the distribution becomes greater and the sample mean approaches the true value of the population.

This limit for observations with additive errors is known

as the central limit theorem (CLT). The CLT states that the mean, or any additive summary, of a set of random, independent, and identically distributed samples will approach a normal distribution as the sample size increases to infinity. The practical significance is that, independent of the underlying form of the sampled distribution, continuous or discrete distribution of the mean will be normal, given enough samples. In LCA, the distribution of shipping weights used in transport scenarios could be considered normal.

In order to illustrate the importance of CLT, consider a set of hypothetical data from a distribution center. The panel labeled "Shipping Distances" shows a population of 50 independent delivery routes defined by their round trip distance. A normal work day consists of a total of 10 independent dispatches drawn randomly from the total population of 50 routes. The average round trip distance was recorded each day, and after 5 days (50 dispatches) the distribution of the daily average is bimodal. As the number of days, N, increases to larger values, the shape of the distribution becomes more normal with a center at the sample mean.

some observed parameters, such as the yield of corn, tend to be the result of the multiplication of many normally distributed parameters, such as sun, rain, and the addition of agricultural chemicals. A set of data is considered lognormal if the log transform of the original data is normal. Lognormal values are all positive and are heavily weighted to the right (i.e., they have right skew), with a few large values defining a long tail.

In Figure 14.6, Panel A shows the histogram of the original data that have the general characteristics of a lognormal distribution. Panel B shows the same data after a log transform

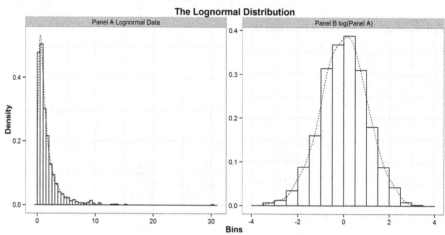

Figure 14.6. The Lognormal Distribution

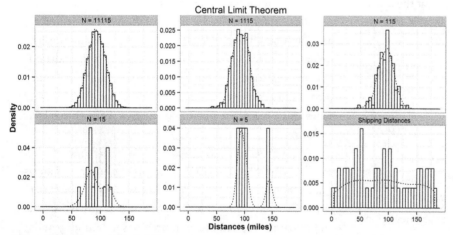

Figure 14.5. Central Limit Theorem
Panel plot illustrating the central limit theorem for additive effects

LOGNORMAL DISTRIBUTION

While the normal distribution tends to be the most common distribution observed in the physical sciences, the lognormal distribution is observed most frequently for environmental data. Multiplying two or more normally-distributed parameters yields a log-normal distribution. For example,

(natural log and log base 10) has been applied. The data in Panel B looks quite similar to a normal distribution. In fact, if the data in Panel B was tested against the normal distribution, the results would indicate that the data in Panel B is not significantly different from a normal distribution. This characteristic of the lognormal distribution allows one to transform the data using a logarithm, as well as perform tests on the data with statistical methods that assume a normal distribution for the results to be robust.

Panel A shows a lognormal distribution drawn from 1000 random samples. Panel B shows the resulting distribution after a logarithmic transformation is applied to the data in Panel A.

Equation 14.13 shows the form of the two parameter lognormal distribution. Like the normal distribution, the lognormal distribution is fully described by a location, $\mu_{\log}$ and shape parameter, $\sigma_{\log}$. The parameters are noted here with a subscript, "log", to emphasize that they represent the mean and standard deviation on the logged scale. More explicitly, the sample mean is $m_{\log} = = \frac{1}{n}\sum_{i=1}^{n}\log x_i$ and the sample standard deviation is, $s_{\log} = \sqrt{\frac{\sum_{i=1}^{n}(\log x_i - m_{\log})^2}{n-1}}$. Unlike the normal distribution, the lognormal distribution is defined only for positive values of x.

$$PDF(x; \mu_{log}, \sigma_{log}) = \frac{1}{x\sigma\sqrt{2\pi}}e^{\frac{-(\log x - \mu_{log})^2}{2\sigma_{log}^2}}, \quad (0<x<\infty) \quad \text{Eq (14.13)}$$

The parameters, $\mu_{\log}$ and $\sigma_{\log}$ represent the arithmetic mean and standard deviation of the variable, x, on the logged scale. These parameters are unitless. The relationships between these parameters to the mean and standard deviation on the measured scale are shown in equations 14.14 and 14.15.

$$\mu = e^{\left(\mu_{log}+\frac{1}{2}\sigma_{log}^2\right)} \quad \text{Eq (14.14)}$$

$$\sigma = e^{\left(\mu_{log}+\frac{1}{2}\sigma_{log}^2\right)}\sqrt{e^{\sigma_{log}^2}-1} \quad \text{Eq (14.15)}$$

While representing the location and shape of the distribution as the arithmetic mean and standard deviation, μ and σ, may be convenient for comparisons on the same measured scale, like kilometers or kilograms, interpreting their effects of the location and spread of the lognormal distribution is more

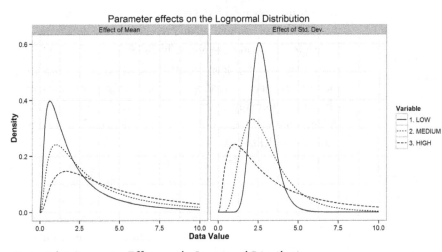

Figure 14.7. Parameter Effects on the Lognormal Distribution

difficult. They represent compound relationships between the mean and variance of the logged data. Panels A and B of Figure 14.7 show the direct effects of changing the mean and standard deviation of the logged data.

Panel A shows the effect of changing the location parameter ($\mu_{\log}$) while keeping the shape parameter ($\sigma_{\log}$) constant. Panel B shows the effect changing shape parameter while the location parameter is kept constant.

Data that follow a lognormal distribution are often summarized by the geometric mean (GM) and geometric standard deviation (GSD). This summarization occurs because a lognormal distribution is a distribution whose values are linked by a proportion, which is a geometric sequence. The geometric mean and geometric standard deviation are different than the more common arithmetic mean and standard deviation.

Much of LCA data can be approximated by either the lognormal or normal distributions. Knowledge of the underlying distribution of the data helps in two key areas of LCA: 1) data analysis and 2) uncertainty analysis. First, data analysis in LCA is a broad category; when analyzing any set of data, the analyst should have some idea of the underlying distribution in order to calculate the right summary measure, use the appropriate significance test, or establish the right error estimate on a final result. As was illustrated briefly for the normal and lognormal distributions, different distributional forms behave differently with respect to changes in location and shape.

Second, one should have a good understanding of the data distribution for estimating the uncertainty. Typically, propagation of error in LCA models is handled by a Monte Carlo simulation that requires the distributional form and the corresponding parameters. The normal and lognormal distributions are used when an adequate amount of supporting data exists for the distributional form. When the data is sparse, or the analyst is uncertain about the data value, other distributions may be more appropriate.

The last few distributions to be discussed in this section are the triangular and uniform distributions. These distributions are seen commonly enough in LCA that they warrant mention, but they are not as pervasive as the normal and lognormal distributions; therefore, these sections are markedly shorter than the first two.

TRIANGULAR DISTRIBUTION

The triangular distribution may be used on occasions when the available data is quite sparse. It may be that after interviews with stakeholders and subject experts that we can define a minimum (a), a maximum (b), and a typical value (c) for the product or process under investigation. In these cases, the triangular distribution may be the best choice to reflect the nature of the data. It can reflect gross assumptions of the data distribution by the definition of the location parameter. The location parameter can be specified to approximate data that we assume to be symmetrically distributed or skewed, as in Figure 14.8. Equation 14.19 shows the generating function for the triangular distribution. The extremes of the distribution are bounded by the minimum (a) and maximum (b) values. The location (c) is given by some value that is considered to be the best estimate, given the sparse data.

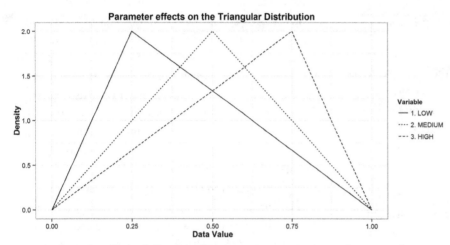

Figure 14.8. The Effects of Changing the Location Parameter, c, on the Triangular Distribution

$$PDF(x; a, b, c) = \begin{cases} \frac{2(x-a)}{(b-a)(c-a)}, & if\ a \leq x \leq c \\ 1 - \frac{2(b-x)}{(b-a)(b-c)}, & if\ c \leq x \leq b \end{cases} \quad Eq\ (14.19)$$

Uniform (rectangular) distribution

Like the triangular distribution, the uniform distribution can be used in cases when the supporting data is not sufficient to adequately define a more precise distribution. In the absence of more detailed evidence, it reflects that the measurement is vague. The probabilities of multiple random draws with values bounded between a minimum, a, and maximum, b, are equal. The expression for the uniform

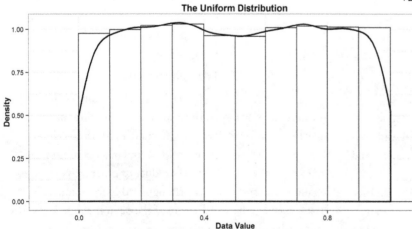

Figure 14.9. The Uniform Distribution

distribution function is given by Equation 14.20 and is shown in Figure 14.9.

$$PDF(a, b) = \frac{1}{(b-a)} \quad Eq\ (14.20)$$

SUMMARIZING VARIATION WITH INTERVAL ESTIMATES

Armed with the knowledge that sampled data can take different distributional forms, understanding the methods of summarizing the underlying distribution is crucial. In Section 14.1, point estimates for central tendency like mean and median were discussed. Standard deviation and the coefficient of variation were discussed as point estimates to convey uncertainty information. In this last section, ground work is laid to discuss confidence intervals and their estimation.

Interval estimation is another way in which information can be conveyed about the variation in the data. In contrast to a point estimate like standard deviation, interval estimation estimates bounds in which the population parameter of interest is most likely found. Different types of interval estimates can be used. One of the most commonly used types is the confidence interval.

The typical confidence interval assumes the data from which it is estimated are normally distributed, although techniques exist for determining the confidence intervals for non-normal data, including mathematical transformations that create normally distributed data (STAT 415 2013) Confidence intervals for an estimated mean derived from a normal data set are defined as:

$$\bar{x} \pm t_{1-\alpha/2, n-1} \frac{s}{\sqrt{n}} \quad Eq\ (14.21)$$

where is the sample mean, is the sample standard deviation, is the number of samples, is the desired significance level, is the confidence coefficient, and is the upper critical value of the t-distribution with n-1 degrees of freedom[4]. A simpler form for a 95% confidence interval derived from a normal distribution is:

$$\bar{x} \pm 1.96 * SE \quad Eq\ (14.22)$$

4 The t-distribution will be discussed in the next section. T-table values can be found in any basic statistics textbook. A full discussion of the use of t-tables is beyond the scope of this book.

In Equation 14.22, the general form is the same as Equation 14.21, but the critical value is replaced with 1.96, which is the critical value from the standard normal distribution for a 95% confidence interval. This replacement assumes a large samples size (N>100). If a large sample size cannot be assumed, the critical value is obtained by the inverse t-distribution using 0.975 as the upper bound to the 95% two-sided confidence interval and sample size as the degrees of freedom. The inverse t-distribution is available in many modern data analysis software packages, including popular spreadsheet packages. The "SE" term in Equation 14.22 is the standard error and is the standard deviation divided by the square root of the sample size.

Equations 14.21 and 14.22 will yield confidence intervals for symmetric distributions whose errors follow a t-distribution (i.e., the normal distribution at large sample sizes). Typically, data used in LCA are represented as coming from a lognormal distribution, as was discussed previously. The lognormal distribution is not symmetric and therefore equations 14.21 and 14.22 cannot be used to estimate a confidence interval. Closed form expressions estimate the confidence interval for a lognormal distribution (Zhou and Gao 1997) The modified Cox expression is shown as Equation 14.23.

$$\bar{x} + \frac{s^2}{2} \pm t_{1-\frac{\alpha}{2},n} * \sqrt{\left(\frac{s^2}{n} + \frac{s^4}{2(n-1)}\right)} \quad \text{Eq (14.23)}$$

In order to use Equation 14.23 with a data set that is thought to follow a lognormal distribution, the user must take the natural logarithm of the original data and calculate the equation's parameters from the log-transformed data. Once all the calculations are made, the upper and lower confidence bounds should be transformed back to the original scale using the exponential function. A question at the end of the chapter explores the process of calculating 95% confidence intervals for the lognormal distribution.

Closed-form expressions for confidence intervals are not available for all distributions. Numerical simulation methods can be used to estimate confidence intervals. The methods are called bootstrap methods. These can be applied as an alternative to a closed-form expression for cases in which no information exists about the type of distribution the sample(s) came from. A few different algorithms have been developed to calculate bootstrap estimates (Barker 2005). Bootstrap methods are available in specialized data analysis software, but are easily implemented using scripts or macros. The list below shows the pseudo code to implement one of the most basic bootstrap algorithms.

FOR i in 1: Sample Size
1. SAMPLE WITH REPLACEMENT

2. CALCULATE PARAMETER OF INTEREST (MEAN, MEDIAN, ETC.)
3. STORE CALCULATED PARAMETER IN A VECTOR, COLUMN, LIST, ETC.
NEXT N
RANK ORDER THE STORED PARAMETERS
FOR A 95% CONFIDENCE INTERVAL, CHOOSE ROWS REPRESENTING 2.5% AND 97.5%.
(i.e., for N = 1000, rows 25 and 975)

The bootstrap approach is quite useful and works according to the CLT. Many samples are drawn from the data available with replacement. The sample size can vary, but often the size is set to the size of the data set. Each sample is then aggregated to some parameter estimate (mean or median), and those estimates are stored to form an approximately normal distribution. For a confidence interval estimate, we should draw at least 1000 sample sets.

Confidence intervals are useful, but are often misinterpreted. They depend on both sample variation and sample size. The confidence interval contains no distributional information. The confidence interval range is defined by endpoints calculated at a chosen significance level within which values are more likely to contain the true mean than the values lying outside the interval. We should not interpret the confidence interval such that the values located toward the middle are more likely than the values closer to one of the bounds.

For example, the set of numbers {-9, -1, 12, 12, -7, 0, 11, 1, -8, 0} has a mean of 1 and standard deviation 8. Its confidence interval is defined from -5 to 7 at the 5% significance level (a 95% confidence interval). The interpretation of this interval is that given the data was drawn from a normal distribution, and the sample size will approach infinity, there is a 95% chance of finding the true value of the sample mean somewhere within the bounds defined by the confidence interval.

14.5 METHODS FOR QUALITATIVE DATA ASSESSMENT

This section describes some common methods for LCA data analysis. First, the importance of viewing and exploring data is stressed through the use of simple summary measures and plotting techniques. Second, formal methods of testing significance are introduced. In this section, some common significance testing methods are presented, such as the family of t-tests, F-tests, and some possible distribution-independent methods. Finally, this section will conclude with regression modeling for analysis and interpretation. As with previous sections, the main goal of this section is inform readers of potential methods to help with a more robust interpretation of the data. The interested reader should consult a statistics textbook for more information on any of these topics (Buthmann 2010)

EXPLORATORY DATA ANALYSIS

Exploratory data analysis is just what it sounds like. The goal is to view data by different methods, with the goal of gaining different perspectives and developing insights based on the data's patterns. John W. Tukey is credited for coining the phrase exploratory data analysis (EDA). He describes it as "detective work – numerical detective work – or counting detective work – or graphical detective work" (Tukey 1976). Taking this mindset as a first step in data analysis allows one to be open to different interpretations of the data, given different perspectives and observed patterns. Once a sufficient inventory of observational trends has been gleaned, the analyst can apply more formal methods to establish the degree of significance of those trends.

While a number of tools have been developed to make analyzing data easier, an analyst must use his or her expertise to decipher the trends and accurately report the potential

consequences of those trends. Falling into the trap of blindly reporting the results of a calculated analysis must be avoided. It should not be forgotten that all aspects of LCA modeling require analyst engagement.

This section will be referring to the data in Table 14.5 for the examples. The data are the calculated impacts from both the substance and process inventories for each beer producer from the LCA of beer production. This data set reflects the flows after the Water Consumption 4.02 impact method was applied at the characterization level. The data are the amount in liters of the water intake from process and substance flows

FLOW ACCOUNTING

Flow accounting is an exploratory method in LCA whose goal is to organize inventory flows based on conditional statements applied to the data. For example, Table 14.5 is a sub group of the total inventory of both substance and processes, after the question, "Which substance and process flows in my total inventory are included in my impact assessment method?" is applied. Prior to imposing this question, there were a total of 1,179 substance flows. After applying the impact

Index	Name	Unit	DataType	Producer A	Producer B	Producer C	Producer D
1	Water, cooling, unspecified natural origin/kg	liters	Inventory	NA	1.36E+02	NA	NA
2	Water, cooling, unspecified natural origin/m3	liters	Inventory	2.46E+03	4.83E+03	1.83E+03	1.83E+03
3	Water, fresh	liters	Inventory	6.13E+02	7.69E+01	1.23E+02	1.50E+02
4	Water, groundwater consumption	liters	Inventory	7.94E+01	3.36E+02	2.13E+02	1.17E+02
5	Water, lake	liters	Inventory	1.60E+03	1.21E+03	5.90E+02	1.36E+03
6	Water, process, unspecified natural origin/kg	liters	Inventory	NA	2.68E+02	NA	NA
7	Water, process, unspecified natural origin/m3	liters	Inventory	NA	NA	2.76E-08	NA
8	Water, process, well, in ground	liters	Inventory	NA	NA	3.12E-03	NA
9	Water, river	liters	Inventory	4.64E+03	3.52E+03	1.81E+03	3.66E+03
10	Water, unspecified natural origin/kg	liters	Inventory	7.69E-02	1.73E-01	5.71E+01	NA
11	Water, unspecified natural origin/m3	liters	Inventory	5.43E+02	1.66E+03	3.62E+02	3.48E+02
12	Water, well, in ground	liters	Inventory	8.57E+04	1.29E+05	7.66E+04	5.06E+04
13	Aluminum can stock body	liters	Process	NA	3.06E+02	NA	NA
14	Aluminum can stock lid	liters	Process	NA	9.83E+01	NA	NA
15	Barley, at farm	liters	Process	7.89E+04	1.11E+05	7.03E+04	4.27E+04
16	Container glass, at plant	liters	Process	-1.10E-02	NA	-8.18E+00	NA
17	Electricity, TPU	liters	Process	NA	NA	3.12E-03	NA
18	Fertilizer - Nitrogen (N) at plant, 2010	liters	Process	2.37E+01	3.35E+01	2.11E+01	1.28E+01
19	Fertilizer - Phosphorous (P) at plant, 2010	liters	Process	1.75E+01	2.47E+01	1.56E+01	9.45E+00
20	Fertilizer - Potassium (K) at plant, 2010	liters	Process	7.00E+00	9.88E+00	6.23E+00	3.78E+00
21	Hop pellets, at plant System	liters	Process	3.68E+03	1.56E+04	5.32E+03	5.44E+03
22	Stainless steel world 304, cold rolled coil, recycled	liters	Process	1.26E+02	1.30E+00	1.77E+01	2.75E+01
23	Stainless steel world 304, cold rolled coil, virgin	liters	Process	1.49E+02	1.53E+00	2.09E+01	3.24E+01
24	Stainless steel world 304, hot rolled coil, recycled	liters	Process	1.25E+02	1.28E+00	1.75E+01	2.71E+01
25	Stainless steel world 304, hot rolled coil, virgin	liters	Process	1.65E+02	1.70E+00	2.32E+01	3.60E+01
26	Tap water, at user/RER WITH US ELECTRICITY U-fiNA	liters	Process	8.75E+03	6.62E+03	3.22E+03	7.46E+03
27	Tap Water, TPU	liters	Process	NA	NA	2.20E+02	NA

Table 14.5. Substance and Process Inventory Contributions to Water Intake per the Reference Flow of 100 L of Beer

assessment, we find 1,167 unused substance flows. A quick glance at the data shows that some flows contribute to some of the producers, but not all. It may also be of interest to know the number of substance and process flows that contributed to each producer impact.

Table 14.6 shows the tally for this query. This type of table is known as a contingency table or cross tabulation table. Contingency tables are summary tables formed by tabulating across variables given some condition. Table 14.7 shows a returned a flow count summarizing flows according to the

	Inventory	Process
Producer A	8	11
Producer B	10	12
Producer C	10	13
Producer D	7	10

Table 14.6. *Count of Contributing Flows to Each Producer by Flow Type*

producer and the type of flow (whether it is a substance or process). The numbers confirm that each producer has a different inventory and process profile. The number of common inventory and process flows could be parsed out to identify the degree of correlation between producers.

Table 14.7 tabulates whether a flow is used for each of the producers. A "1" indicates the flow is used and a "0" indicates it is not used. Table 14.7 provides a convenient table of the used flow for each producer. This not only yields information about the flows used, but potentially some sources of correlation.

Contingency tables allow the analyst to assess different aspects of the data in tabular form. This provides the actual numbers that can be read from the table, as opposed to trying to read them from a plot. Other types of tables that can be formed are proportional tables, frequency tables, and margin tables, to name a few.

	Producer A	Producer B	Producer C	Producer D
Aluminum can stock body	0	1	0	0
Aluminum can stock lid	0	1	0	0
Barley, at farm	1	1	1	1
Container glass, at plant	1	0	1	0
Electricity, TPU	0	0	1	0
Fertilizer - Nitrogen (N) at plant, 2010	1	1	1	1
Fertilizer - Phosphorous (P) at plant, 2010	1	1	1	1
Fertilizer - Potassium (K) at plant, 2010	1	1	1	1
Hop pellets, at plant System	1	1	1	1
Stainless steel world 304, cold rolled coil, recycled	1	1	1	1
Stainless steel world 304, cold rolled coil, virgin	1	1	1	1
Stainless steel world 304, hot rolled coil, recycled	1	1	1	1
Stainless steel world 304, hot rolled coil, virgin	1	1	1	1
Tap water, at user/RER WITH US ELECTRICITY U-fiNA	1	1	1	1
Tap Water, TPU	0	0	1	0
Water, cooling, unspecified natural origin/kg	0	1	0	0
Water, cooling, unspecified natural origin/m3	1	1	1	1
Water, fresh	1	1	1	1
Water, groundwater consumption	1	1	1	1
Water, lake	1	1	1	1
Water, process, unspecified natural origin/kg	0	1	0	0
Water, process, unspecified natural origin/m3	0	0	1	0
Water, process, well, in ground	0	0	1	0
Water, river	1	1	1	1
Water, unspecified natural origin/kg	1	1	1	0
Water, unspecified natural origin/m3	1	1	1	1
Water, well, in ground	1	1	1	1

Table 14.7. *Specific Flows Used by Each Producer*

SUMMARY MEASURES AND EXPLORATORY PLOTS

Beyond flow accounting, aggregating the data set into summary measures can help provide some insight into the structure of the underlying data set. Some of the structural information that can be obtained are location, spread, and potential skew (i.e., how non-symmetric the data is). These summary measures consist of many of the central tendency and variability measures discussed in Section 14.1. Most statistical software packages or add-ons provide a summary table that includes different

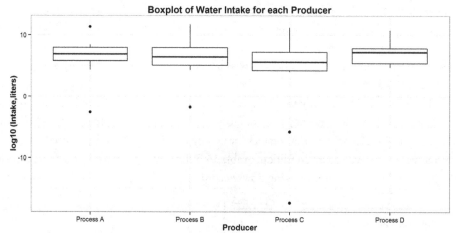

Figure 14.10. Boxplot of Water Intake Data from all Contributing Inventory Flows for Each Beer Producer. The data is plotted on with logged y-scale due to the large range in the data.

Measure	Producer A	Producer B	Producer C	Producer D
NA Count	4	2	2	5
Data Count	8	10	10	7
Min	79	77	123	117
25% Quantile	578	773	287	249
Mean	13664	20081	11648	8301
Median	1597	1661	590	1362
75% Quantile	3546	4172	1820	2745
Max	85724	128936	76604	50640
Sum	95649	140564	81533	58108

Table 14.8. Contributing Inventory Flows for Water Intake

combinations of summary measures. Table 14.8 shows a summary table of the calculated amount of water used during beer production for the substance flows shown in Table 14.5.

As mentioned previously, Table 14.8 was generated from substance flows that were included in the impact method. In this case, the method chosen was Water Consumption 4.02 with a single impact category, water intake in liters. The sample size for each producer is a count of contributing flows. Upon inspection of the table, we see that the sample sizes for the different producers range from 7 to 10 due to the presence of missing flows for some producers, as shown in Table 14.5. Next, we see that the range max-min is quite large for all the producers' ranges from 10^4 to 10^5 orders of magnitude. Next, a jump to the two central tendency measures, mean and median, allows one to judge how symmetric the data is. In this case, the mean is much greater than the median (the 50% quantile). The 25th and 75th quantiles are provided as an estimate of the bounds by which the central most 50% of the data is contained. So in this case it appears that for all the producers, 50% of the data are bounded within an order of magnitude. Finally, the sum across all contributing inventory flows is

shown for this common summary metric. From the summary measures presented, this weighted sum is heavily influenced by a few large values. Figure 14.10 shows a boxplot of the logged data and provides a visual summary of the underlying sample distribution. With the y axis on a log scale, the boxplots appear symmetric and, therefore, the underlying distribution of contributing flows may be lognormal. This conclusion is tentative because the sample set is fairly small with samples size <10 for all producers. The specifics of the boxplot were discussed in Section 16.1.

After exploring the data with just a couple of tables and plots, the following observations can be made. The final sums for each producer are derived from sample sets with less than ten samples. The data appear to follow a lognormal distribution given the values are positive and highly weighted to lower values, with a few large values defining the tail. When the data are plotted on a log scale, they appear more normal. This knowledge will help in the next section when the difference between producers is assessed.

14.6 ASSESSING SIGNIFICANCE

When analyzing LCA data, assessing the significance of the results is necessary. The word significance is rather ambiguous. It has different meanings to different people. In the context of LCA, we can broadly define two classes of significance. The first, and perhaps more familiar, is statistical significance. This type of significance, which will be discussed in the following pages, uses formal statistical tests designed around hypotheses that form the basis of tests to be rejected. The second type of significance is practical significance (Cumming 2012). Practical significance is a generic way to express that significance is subjective and dependent on perspective. Less formal tests and procedures assess practical significance, but communication of both magnitude and spread are essential to convey the certainty contained in the data.

As stated previously, significance testing is most often associated with formal statistical tests to evaluate the model results against some test hypothesis. While this approach is valuable and will be discussed in the next few sub-sections, assessing significance of a result can be made in a more qualitative way as well. The key to both types of assessments is that the results must include both a best estimate and the variability of the estimate. As a fictitious example, imagine that Producers A and B make a similar type of beer.

Typically, their production volumes are similar. However, Producer A embarks on a large advertising campaign that spikes sales. When numbers were tallied at the end of the quarter, Producer A produced 150 ± 25 L, while Producer B produced 110 ± 25L. Given just the magnitude estimates, one might conclude that a significant, beyond-chance increase in production for Producer A. With the addition of the variability information in the form of the standard deviation, the difference between the production volumes is not so apparent.

Monte Carlo is often used for error propagation for the full LCA model, while sample estimates are often made on model inputs. In the case where insufficient data exist to assess the variability of an input, methods do exist to generate error estimates, such as the pedigree approach)

In order to illustrate the concepts of this section more clearly, Monte Carlo simulations were run for each of the four beer producers. One of the crucial questions to answer at the end of an LCA is if the outcome(s) are significantly different for the four producers. The significance question is critical because it directly relates to the use of the LCA in decision-making.

Monte Carlo: Incorporating distributional information

The data used to this point are for a single run of an LCA model. This means that the inventory flows contributing to each producer are point estimates and do not contain distributional information. In order to impose distributional information, model simulations have to be run where some change in the inputs results in a change on the resulting inventory. A common method for generating variation in the model output is a class of simulation algorithms collectively called a Monte Carlo simulation. Monte Carlo simulation derives its name from the geographic location in Monaco that is renowned for its casinos. In this light, it may be easy to infer that Monte Carlo simulations are algorithms designed specifically to execute random draws from some population distribution.

In LCA, input and output data are characterized by a distribution that best reflects the data set. The distribution is a representation of the error associated with the data being used. In order to propagate the error of the data to the results of the model, Monte Carlo simulation is used to sample from all the distributions assigned to the data. After the stop criterion is met, either some convergence measure or predefined number of runs, the model results have a distribution of point estimates from which a best estimate can be made in the presence of the propagated error. Statistically, the model results are now stochastic, meaning the data were generated from random processes that propagate through as error. Until now, the results have been deterministic, meaning that upon successive runs of the model, the results are identical.

The Monte Carlo process is initialized by a seed value. The seed value determines the start of the random chain bounded by the particular definition of the distribution being sampled. Recall that the distribution is defined by its parameters, such as the sample mean and standard deviation. However, for the majority of the data available in LCA software, this information does not exist and must be estimated. In order to generate different sets of random samples for successive Monte Carlo simulations, the seed should be set to a different value each time. If the desire is to generate the same set of random values for independent Monte Carlo runs, then the seed should be set to the same value. Different software packages may have different implementations of a Monte Carlo algorithm, so having a high level of understanding of the version being used is essential. The most common stop criterion is using a predefined number of sample runs. Another common stop criterion is based on the degree of convergence.

Changes in the calculated parameter (mean and median) are measured after each iteration of Monte Carlo simulation. If the change from iteration to iteration is less than a given convergence criterion, the simulation stops. While generally accepted guidelines for setting up and assessing Monte Carlo results is difficult and often machine dependent, it may be advantageous to initially run a model with a small number of runs, <= 100 runs. This shortened run may help with model adjustments prior to running a larger simulation (≥1000) for error propagation. As previously mentioned, the results of Monte Carlo simulations are a set of model results that now form a distribution. The benefits of the additional information are that uncertainty assessments can be made, as well as assessments of comparative significance.

Qualitative significance assessments

The relative magnitudes of impact results can be compared

and analyzed a number of different ways, but how significant are the results once the distribution of those estimates is added to the analysis? The exploratory data analysis section offered

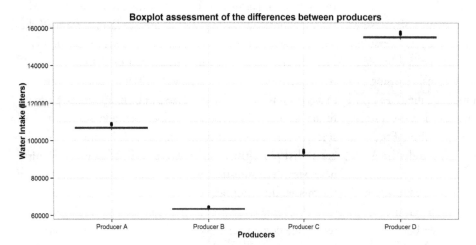

Boxplot assessment of the differences between producers

Figure 14.11. Boxplot Assessment of the Differences between Producers for the Amount of Water Intake. Boxes are compressed due to the fairly large range of total intakes for the four producers. Data were left on the original measurement scale, liters.

some visual techniques, such as the boxplot to assess both magnitude and spread of the data. These visual tools can be used to support arguments of practical significance. Figure 14.11 shows an example boxplot of the simulated results comparing the four producers:

The plotted distributions show that the differences between producers appear significant because no apparent overlap occurs between groups. The closest difference is between Producer A and C, but this difference still appears significant. This may be an adequate representation of the data, but it also may be dubious if the propagated error only accounts for a small percentage of the inventory. This should be investigated and reported because data sources are often of different quality, and uncertainty estimates for inputs and output are left out. In this example, 68% of the flows have error estimates. In some cases, a qualitative analysis may be sufficient for the LCA's stated goals. In other cases, it may be necessary to have more quantitative rigor around assessing significance.

PROBLEM EXERCISES

1. Target: continuous and categorical data. Please refer to Section 14.1 for the subject matter related to this question. For the following examples, indicate whether the data are continuous or categorical.

a. pass/fail outcome of an environmental performance test

b. total organic carbon (TOC) measurement of 25 water samples

c. dry weights of incoming materials

d. time-integrated radiative forcing (Watt.m^2/kg)

e. number of packages shipped (hint: ordinal numbers are the set of whole numbers like count data)

f. transport distance of 3 suppliers that are officially qualified by sourcing

2. Target: translate data units. When dealing with different data providers for an LCA, translating units of the data given to units that are more convenient in an LCA model is often necessary. In many instances, different data types are translated. Categorical and continuous data were introduced in the text. Another type of data that is ordinal data. This count data take on values that are only whole numbers. For example, bills of materials typically have a quantity associated with each line item. The quantity count is usually ordinal because ordering or shipping partial products is unusual. Translate the following data types to categorical, ordinal, or continuous.

a. A quality assurance manager delivers data from a test determining whether to scrap or ship of certain product for a given year. The data states that 1,200 units were shipped and 378 units were scrapped. If the unit is produced in volumes of 1 liter, translate the ship/scrap metrics to volume shipped and scrapped.

b. A process consumes three 55-gallon drums of a chemical. If the process is 87% efficient in using the chemical, calculate the mass equivalent of the chemical's use and waste stream if its density is 0.627 kg/L.

c. An input of 427.6 grams of material A is added to a process. The purchasing department must acquire the material in batches of 10 units per pallet, with each unit

Year 1 (by Month)	Measure	Year 2 (by Month)	Measure
1	13.42	13	16.64
2	12.97	14	14.33
3	2.80	15	8.49
4	10.58	16	16.50
5	19.01	17	21.91
6	9.56	18	15.26
7	5.04	19	16.84
8	9.99	20	22.79
9	16.02	21	32.32
10	12.28	22	50.90
11	9.95	23	28.69
12	11.49	24	35.23

containing 250 grams of material. Calculate the number of pallets, rounding to the nearest whole number, required to run the process 10 times. Assume the efficiency of material use is 100%.

3. Target: initial error assessment, classification, and visualization. Consider the data below:

a. Plot measurement response data as a histogram. How

are the data distributed? Provide qualitative observations.

b. Plot the data as a scatter plot with the measurement response on the y-axis and the date (in months) on the x-axis. Note any observations.

c. From the two visualizations above, classify your observations and list at least three potential errors in the data.

4. Target: precision and accuracy. The table below contains flow measurement data for a waste stream. Four different

Index	System 1	System 2	System 3	System 4
1	5.86	6.82	4.90	5.00
2	8.08	6.62	4.94	5.00
3	5.44	6.59	5.47	5.00
4	6.77	6.68	4.71	5.00
5	8.41	6.59	4.82	5.00
6	6.10	6.69	3.71	5.00
7	6.23	6.64	5.42	5.00
8	6.06	6.48	5.50	5.00
9	6.41	6.72	3.34	5.00
10	6.84	6.67	3.15	5.00

measurement methods were employed: eyeball guesstimate, tribal knowledge/rule of thumb, the bucket test using a 5 gallon bucket, and an ultrasonic flow meter. The flow rates were measured as gallons per hour. The system can tolerate a target flow rate of 6.7 ± 0.1 gal/min. Plot the data in the table below, assess whether the data are precise, accurate, both, or neither, and classify them according to the measurement system used.

5. Target: calculation of confidence intervals. For each system given in the table for Exercise 4, calculate an estimate of the sample mean and the 95% confidence limits.

6. Target: central limit theorem. For this problem, please use the data file, ch16_ex6_data.dat. The data file contains the substance inventory from an LCA. The inventory consists of individual runs from a Monte Carlo simulation and has had the IPCC GWP 100a impact method applied.

a. Plot at least five of the substances as histograms and qualitatively assess if they came from a normal distribution.

b. For each replicate, compute the sums across all the substances. Plot the sums as a histogram. Confirm that the data are approximately normally distributed.

Bias and Uncertainty in LCA

AnnaNicholson

15.1 Bias and Uncertainty

Bias in life cycle assessment (LCA) indicates a systematic error in the assessment process, such that the results do not reflect the actual mean of the samples studied. Uncertainty in LCA, on the other hand, indicates the amount of random variability in the quantitative results of a study, which has been discussed in detail in the previous chapter. Both of these topics are often easier to discuss than they are to identify in most LCAs. This chapter helps clarify these concepts, describe how to evaluate whether they are occurring, and offers approaches to control and reduce them in the LCA process.

15.2 Sources and Types of Bias

Bias can occur in several forms. Sometimes bias is consciously created, such as when impact categories are deliberately excluded from an LCA to focus on a certain damage. Intentionally excluding impact categories that are known to be problematic for a given system being assessed is an unethical practice and should be avoided. LCA practitioners should call out such practices whenever they occur. Likewise, LCA practitioners should support colleagues who refuse to exclude highly affected impact categories at the request of their commissioners.

More commonly, bias occurs unintentionally. It often occurs without the LCA practitioner, or the people who read the results of the LCA, being aware of the bias. For example, a practitioner can select three impact categories for an assessment that seem reasonable: climate change, net energy consumption, and fossil fuel depletion. The practitioner may be unaware that all of these are primarily measuring various aspects of one phenomenon, fossil fuel consumption. This

constitutes a bias because it does not provide a balanced, environmental health assessment. Removing two of these categories, and replacing them with an ecosystem health category and human health impact category, would give a less biased result.

Bias can occur in the LCI data collection process as well. For instance, accidentally neglecting relevant flows can bias the inventory data. This, in turn, can bias the resulting assessed impacts. Likewise, in the LCIA process, inputs and outputs can be modeled in ways that do not represent the actual conditions. For instance, if using European background data for US conditions would create a bias. Also, in the interpretation phase, practitioners can create bias by (often unconsciously) promoting their favored impact categories and downplaying those they favor less. Most kinds of bias are best controlled by careful review of the assumptions and steps taken in the individual steps of the LCA process, and by correcting errors whenever they are uncovered.

Normalization bias

Systematic bias is the inherent tendency of a process to support particular outcomes. External normalization bias can systemically miscalculate the relative scale of a particular impact category's normalized impact results, as compared to other normalized impact category results (Norris 2001). As mentioned in Chapter 12, external normalization can create normalization bias. A practitioner can recognize normalization bias from observing large numbers of characterized and normalized impact results resulting from identical characterization and normalization methods. In these cases, the externally normalized results identify one or a few impact categories as consistently dominant.

The type of normalization is external because it uses an external source of data, such as per capita estimates of impacts in a geographical region per year. Normalization bias is caused by a quantitative disparity between 1) substances accounted for and correlated characterized impacts in the LCIA and 2) the estimated substances emitted in the normalizing region and their correlated impacts values (Heijungs et al. 2007). Normalization bias can create many orders of magnitude difference between normalized impact indicator results, and thus can systematically and inaccurately identify the impact categories with the most significant impacts.

External normalization bias can be identified and reduced. Inventory dataset (ID) normalization eliminates external normalization bias by dividing the impact results by the average normalized results of a large number of processes. It first calculates the percent contribution of each impact category for a large number (preferably many hundreds) of processes assessed with identical characterization and external normalization methods. It then determines the average percent contribution for the entire data set in each impact category. Each impact category is then divided by the value of the average data set contribution. Individual impacts in each impact category are recomputed for each individual process in the study. ID normalization can be used with or without simple weighting. ID normalization calculates each of the individual impact category results of processes that are assessed, while eliminating normalization bias (White and Carty 2010).

15.3 SOURCES AND TYPES OF UNCERTAINTY

Throughout an LCA, estimations and assumptions are made that potentially introduce uncertainty into the final assessment result. All data used in an LCA study possess a certain level of uncertainty. The sources of uncertainty are myriad. It can be caused, for example, by missing primary data during data collection procedures, or by uncertainty in estimated and assumed data.

This inevitable "noise" in any LCA study must be considered when comparing product systems to determine whether differences in environmental impact are real differences or ones caused by this proverbial noise. In the case of the latter, this would mean the environmental impact resulting from two systems is equivalent within the accuracy of the evaluation.

As with any analytical approach, the robustness of LCA is affected by uncertainty in underlying data and the implications of specific methodological decisions. To date, conducting uncertainty analyses is not common practice in LCA. However, the ISO standard states that whenever feasible, uncertainty

analysis should be performed to better explain, and support, life cycle inventory (LCI) conclusions and interpretation of results. Moreover, the standard clearly states that an analysis of results for uncertainty must be conducted for studies intended for public comparative assertions. A proper analysis will be facilitated when we know what types of uncertainties exist and which tools are available to deal with them (Hujibregts 2011). The following framework is proposed to classify three levels of uncertainty in LCA:

1. Statistical uncertainty is any uncertainty that can be characterized in probabilities or by the measure, or estimation of likelihood of occurrence, of an event (Wamelink et al. 2008).

2. Decision rule uncertainty arises whenever ambiguity or controversy exists about how to quantify or compare social objectives (Hertwich et al. 2000) for which a range of outcomes is possible. However, the mechanisms leading to these outcomes do not enable the definition of the probability of any particular outcome. Typically, decision rule uncertainties can be addressed by conducting sensitivity analyses.

3. Model uncertainty is defined as uncertainty about the relationships and mechanisms being studied (Huijbregts 1998). Model uncertainty is introduced through simplifications of reality.

This section summarizes the literature on the sources and implications of uncertainty within LCA. Before moving to that, it first explores previous work that documented the pervasive nature of this issue and the lack of adequate methods to address the effects of uncertainty within LCAs. LCA results are usually presented as point estimates, which strongly overestimate its reliability, and make use of deterministic models, which do not address the variability and uncertainty inherent in input variables. This may lead to decisions that are unnecessarily costly, or mislead public perception about the environmental profile of a product or process (USEPA 1995). These problems have been long-since acknowledged. LCA practitioners lack systematic approaches for determining data quality, and need improved techniques for sensitivity and uncertainty analysis.

Although this matter is often mentioned, real uncertainty assessments, as described in Chapter 6: Data Quality, are rarely performed. In a review of thirty LCA studies, only fourteen (47%) mentioned uncertainty, two (7%) performed qualitative uncertainty analysis, and one (3%) performed quantitative uncertainty analysis (Ross 2001). The fact that no consensus about methodology exists is not surprising. While the recent

ISO standards recommend the use of uncertainty methods (ISO 14040 1997; ISO 14041 1998; ISO 14043 2000), they give little guidance for practical implementation. In recent years, initiatives to understand, incorporate, and reduce uncertainty in LCA have made progress.

For example, a Society of Environmental Toxicology and Chemistry (SETAC) Europe LCA working group on data availability and data quality has developed a framework for modeling data uncertainty, while the Danish Environmental Protection Agency (EPA) has proposed a data collection strategy for reducing uncertainty in LCI (Weidema 2003). Some LCA software platforms, such as SimaPro, provide the ability to analyze uncertainty using Monte Carlo analyses (Goedkoop 2006). Also, the Ecoinvent LCI databases include quantitative uncertainty distributions for parameters in many processes, such as the variability and stochastic error of the figures, which describe the inputs and outputs due to, for example, measurement uncertainties, process specific variations, and temporal variations (Frischknecht 2007).

In the following section, different types of uncertainty appearing in the LCA methodology are presented. Specifically, this section first considers uncertainty around data used in the LCA inventory process and then methodological uncertainty. Within the methodological uncertainty section, uncertainty around framing, allocation, and impact assessment is explored.

15.4 DATA

The robustness of LCA is affected by dependence on data from different countries, operations, and sources, as well as data that are often not specifically collected for LCA purposes, such as enterprise data on energy (DeSmet 1996), and more or less subjective methodological choices. The quality and availability of data for the inventory stage varies widely and is an expressed concern in the impact assessment process (USEPA 1995).

In usual practice, data are collected from a variety of sources: production sites, engineering texts, regulatory reports for industries, or industry literature. In many cases, primary data are not available directly from the most accurate source-production sites, because these data are often considered proprietary. The result is considerable variation in representativeness and accuracy among the many individual data points within a unit-process inventory, and high potential for out-of-date information. In addition, averages, or even point estimates from different sources, may be extrapolated for an entire industry and the data can be out-of-date. Editing old data sets

by projecting future performance and innovation should be considered.

DATA INACCURACY

Data inaccuracy concerns the empirical accuracy of measurements that are used to derive the numerical parameter values (Huijbregts 2000). Measurements can be subject to random error, which can result from imperfections in measuring, sampling instruments and observational techniques, or systematic error from an inherent flaw or bias in the data collection or measurement process (USEPA 1995).

DATA GAPS

The complex data in a comprehensive LCI may have missing parameter values that leave the model with data gaps (Huijbregts 2000). For example, one may know the amount of VOCs (volatile organic compounds) released from one process, while in another one knows the actual compounds emitted. Other issues contributing to gaps or unevenness in data include:

- data confidentiality
- system complexity
- products at an early stage of market development
- context of product unknown
- lack of knowledge around processes and components

MISREPRESENTATIVE DATA

Data gaps may be avoided by using representative data (Huijbregts 2000), typically data from similar processes. However, this data often does not accurately represent the age, geographical origin, or technical performance of the original system or process. These all constitute important factors when determining environmental impact.

15.5 FRAMING

According to the definition of the ISO standard, the system boundary is "the interface between a product system and the environment or other product system" (ISO 14040 1997). Complex systems like industrial and fuel production systems have practically no final limit. One can trace back materials and energy indefinitely, depending on the level of detail used. Therefore, every assessment must limit its analysis at some point. Different studies having different system boundaries may have different results, and this level of detail must be taken into account when comparing them.

The boundary conditions and system definitions in an LCA have great influence on the study results, as do framing issues related to time-scale and spatial considerations. For example, the chosen time horizon to combine potential effects in impact assessment methods is a temporal framing issue, which, for instance, applies to global warming potentials, photochemical ozone creation potentials (Huijbregts 1998), and emissions from landfills (Finnveden 1999). Several LCA studies examine this system boundary setting flexibility. For example, in washing machine assessments, including and excluding the services (i.e., heating, lighting, and compressed air) of the manufacturing plant resulted in different conclusions (Lee, Callaghan, and Allen 1995).

CONCEPTUAL FRAMING/MODEL UNCERTAINTY

Choices are unavoidable in LCA. Because often there is no single choice, making choices implies uncertainty. Uncertainty can derive from choices of functional unit, system boundary, characterization method, weighting method (Huijbregts 1998), marginal or average data (Weidema 1999), and technology level (Lindfors 1995).

IMPACT ASSESSMENT

As discussed in Chapter, impact assessment evaluates the potential environmental consequences of a product system based on the collected inventory data. In other words, this step applies quantitative measures to value the relative severity of different environmental changes. To do so, impact assessment seeks an association or relationship between an inventory data set and particular environmental issues using defined impact categories. For each category endpoint, the environmental and human health sciences have usually identified several compounds or stressors that are related to adverse effects for that endpoint (e.g., sulfur dioxide and nitrogen oxides for acid rain.). These compounds or interventions are then targeted for inventory data collection for particular categories. Finally, a category-specific model is used with simplifying assumptions to convert the inventory data into a numerical category indicator. This category indicator is intended to portray an overall load from the system, a life-cycle stage, or an operation on the specific category.

In principle, impact assessment strives to achieve an accurate model and representative indicator, but impact assessment is limited by the simplifications inherent in LCIA modelling, as well as the limitations in our understanding of environmental mechanisms. The indicator's actual benefits and limitations must be taken into account when making decisions. Disparities between inherent spatial, temporal, and even toxicological characteristics of the categories in LCA all contribute to uncertainty that is made more complex by inventory accounting, calculation, impact assessment, and valuation procedures.

At this level, when subjective judgments are made on scientific evidence, transparency is crucial for communicating how complex modeling approaches generate these indicators. Studies have shown that the selection of different impact assessment methods can lead to different results in terms of preferred product or material. For example, the application of different LCA impact assessment methods can lead to different preferred material in automotive design, showing that environmentally preferred choice of material depends on one's valuing of energy, resource use, and global warming (Saur 2000). Applying more than one impact assessment method in an assessment is a useful exercise to identify the varying results that different methods deliver.

With limited exceptions, current impact assessment methods generally do not provide uncertainty information. The contribution of uncertainty in the characterization factors to the overall impact assessment results is at least as important as the uncertainty in the LCI results.

ALLOCATION

When a process in a product system is related to more than one product, it presents a problem for the LCA practitioner. The allocation of these multiple products, known as co-products, has been one of the most controversial issues in the development of the methodology for LCA, as it may significantly influence, or even determine, the result of the assessments (Ekvall 1997).

WHY QUANTIFY LCA UNCERTAINTY?

In light of the presented sources of variation and uncertainty within the LCA methodology, addressing the sources of methodological and analytical variation of LCA in general is a critical need. Failing to give sufficient consideration to sources of uncertainty in an LCA can lead to erroneous or misleading conclusions that, in the worst case, are ultimately communicated to the public or used for decision-making. As the development and application of Product Category Rules (PCR) and Environmental Product Declarations (EPD) continues to mature globally, LCA practitioners must not dismiss sources of methodological or data uncertainty as too difficult or time-consuming to consider. Although a great deal of research is undergoing in the field of LCA, there is as yet no consensus on how to go about quantifying uncertainty.

Properly characterizing sources of uncertainty becomes crucial when using the LCA standard to make comparative product declarations using the ISO 14040 series. Many current studies fail to comprehensively address uncertainly, merely giving a nod to its existence but not presenting a rigorous uncertainty analysis. As practitioners, this presents an ethical challenge. To confidently state, for example, that Product A is environmentally preferable to Product B, a study must move beyond baseline results for each product system and assess underlying sources of data and methodological uncertainty that may affect the baseline result or, in some cases, change preference across environmental impacts.

If the results for one product system reveals slightly lower environmental impact than another product, yet has much more uncertain model data, this result may switch given a different set of equally acceptable inventory data or practitioner modeling choices. In this case, a definitive product declaration of the environmental preference of the former product is extremely misleading. To this end, consensus-based frameworks are needed to systematically characterize uncertainty.

15.6 Methods to quantify and assess uncertainty

Various methodologies are available to quantify and assess uncertainty when conducting an LCA. This section covers the pedigree matrix approach, Monte Carlo simulations, and sensitivity analysis. It also discusses novel approaches presented in LCA studies and literature. A quantitative uncertainty assessment seeks to quantify all the inherent uncertainties and variations in an LCA. Among others, Hansen and Asbjornsen (1996) used analytical statistics, Ros (2001) used fuzzy logic, and Maurice et al. (2000) used stochastic methods.

Quantitative assessment is particularly problematic because not all types of uncertainty can be analyzed. The methods discussed here are especially applicable for uncertainty analysis of LCIs. However, uncertainties are associated with the nor-

malization factors applied during the life cycle impact assessment of an LCA, as discussed in Huijbregts (2000). Currently, uncertainty data for impact assessment methodologies are not robust. They provide an area for future development. Apart from uncertainty estimation techniques below, many data sources provide uncertainty distribution information that can be incorporated into an inventory. For example, in the case study provided at the end of the chapter, uncertainty information at the dairy farm product system was taken from relevant EPA and IPCC data. Most inventory distributions are log-normally distributed. The lognormal probability distribution function is ideal for modeling skewed distributions when values are low, variances large, and values cannot be negative.

15.6.1 Pedigree matrix for uncertainty estimation

When gathering the data used in an LCA, the mean value is often used to represent data points with no other statistical information that can be used to characterize data uncertainty (e.g., probability distribution function). The pedigree matrix approach was proposed by Weidema and Wesnaes (1996) to estimate uncertainty and is used to assess data quality in LCA. Pedigree analysis, which is discussed in Chapter 6: Data Quality and Chapter 14: Statistics in LCA, outlines the indicator score definitions across six data parameters.

After assigning pedigree matrix values to the underlying unit process data used to model the primary and secondary data in an LCA, an uncertainty factor is calculated, expressed as a contribution to the square of the geometric standard deviation, and attributed to each of the score of the six characteristics.

The square of the geometric standard deviation for the Pedigree matrix values (95% confidence interval – SD_{g95}) can be calculated with the following formula (Frischknecht 2007):

$$SDg_{95} = \sigma_g^2 = \exp\sqrt{[\ln(U_1)]^2 + [\ln(U_2)]^2 + [\ln(U_3)]^2 + [\ln(U_4)] + [\ln(U_5)]^2 + [\ln(U_6)]^2 + [\ln(U_b)]^2}$$

with U_{1-6} representing uncertainty factors assigned from the Pedigree Matrix (see table 6.2) and U_b representing a basic uncertainty factor. The basic uncertainty factor should be calculated based on conventional statistical methods (Jungbluth et al. 2012) and assigned to unit process inventory data used to model a system. Table 15.1 shows uncertainty factors that correlate with the U_{1-6} indicator score (Frischknecht et al. 2005).

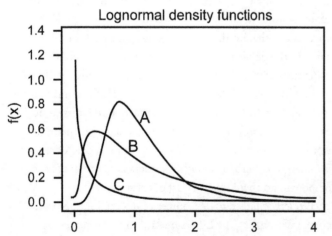

Figure 15.1. Graphs of A: lognormal ($\mu=0$, $\sigma^2=0.25$), B: lognormal ($\mu=0$, $\sigma^2=1.0$), and C: lognormal ($\mu=0$, $\sigma^2=25.0$) density functions

Indicator Score	1	2	3	4	5
Reliability	1.00	1.05	1.10	1.20	1.50
Completeness	1.00	1.02	1.05	1.10	1.20
Temporal Correlation	1.00	1.03	1.10	1.20	1.50
Geographical Correlation	1.00	1.01	1.02	-	1.10
Further technological correlation	1.00	-	1.20	1.50	2.00
Sample size	1.00	1.02	1.05	1.10	1.20

Table 15.1. Default Uncertainty Values (Frishcknecht et al. 2005) These factors contribute to the square of the geometric standard deviation and are applied in conjunction with the pedigree matrix to solve the equation above for the given LCI data. Practitioners should note that the result is a lognormal value hence it can not be used directly as a deviation value but must be interpreted through a lognormal process such as Monte Carlo analysis. This is done prior to applying Monte Carlo Analysis to the entire product system.

Monte Carlo simulations

Once uncertainty data are assigned to underlying inventory data using the Pedigree Matrix approach, Monte Carlo simulations can be run to propagate and determine overall LCA model uncertainty. Using Monte Carlo simulations provides decision makers with far more information than a single estimate of damage. Popular LCA software packages provide Monte Carlo simulation capabilities, although these simulations can also be conducted in a software such as Microsoft Excel.

Monte Carlo simulation involves generating random data inputs over a defined probability distribution and repeating the process over and over. For instance, one can run simulations many times over in order to calculate these same probabilities heuristically. Repeated calculations produce a distribution of the predicted output values, reflecting the combined parameter uncertainties (Huijbregts 1998). When using Monte Carlo to assess uncertainty, running at least 10,000 simulations to fully characterize a system's uncertainty is considered to be good practice.

Each time a Monte Carlo simulation is run, the collective outcome of each calculation forms a distribution, as illustrated in Figure 15.2.

Taylor series expansion method

Another method used to propagate uncertainty is the Taylor series expansion method. Taylor series expansion can be applied to calculate the uncertainty of the output as a function of the uncertainties of the input variables (Hong et al. 2010). This method characterizes model uncertainty without the intensive computing power and time needed to conduct Monte Carlo simulations.

Decision rule uncertainty

Subjective choices are inescapable when conducting life cycle assessments (Huijbregts 1998). Decision rule uncertainty arises whenever ambiguity or debate surrounds how to quantify or compare social objectives (Hertwich et al. 2000). Examples of choices leading to uncertainty in the inventory analysis phase are choice of the functional unit and allocation procedure for multi-output processes, multi-waste processes, and open-loop recycling. Furthermore, different characterization methods can be used for the same impact category and choices within the calculation of characterization factor methods are unavoidable (De Schryver et al. 2010).

Although many weighting methods have been suggested in LCA literature, only a few are operational to date and no general agreement exists as to which one(s) should be preferred. For example, an authorized weighting set could potentially be based on political reduction targets, environmental control and damage costs, and panel preferences in reducing impacts (Huijbregts 1998). Moreover,

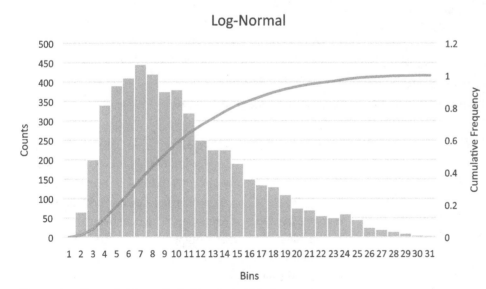

Figure 15.2. Example Monte Carlo Simulation Results

the debate continues about whether to use generic weighting sets or to perform weighting case-by-case in environmental sustainability assessments (Huijbregts 2011).

As such, LCA studies should consider a variety of what-if scenarios designed to investigate parameters, choices, and mechanisms of special interest. Conducting decision rule analyses reveal how sensitive the baseline result is to practitioner decisions made during the goal and scope phase of an LCA. For example, the following parameters may be considered in a sensitivity analysis of an LCA on dairy milk, presented in the case study at the end of this chapter:

- allocation method between beef and dairy milk at a dairy farm
- choice of functional unit (volume versus nutritional content)
- dairy cow feed composition
- allocation between animal feed and human foodstuffs
- refrigeration times at distribution center, retail, and consumer
- consumer transport distances
- product losses at retail and consumer
- container recycling allocation method (e.g., cut-off or system expansion)
- packaging composition and size (e.g., carton versus plastic jug)
- best and worst case scenarios based on a combination of the above parameters for each product system

For example, a sensitivity analysis would involve inputting different data points, such as the maximum and minimum conceivable product loss rates, to test how sensitive the overall model result is to assumptions around a parameter. This exercise can determine which data are driving impact results, and areas where data should be further clarified and assumptions transparently communicated.

Figure 15.3 (page 205) shows sensitivity analysis results from the case study presented at the end of the chapter. The results compare the environmental preference of soy milk to dairy milk across the parameters mentioned above. In this figure, if results fall within a +/- 10% range, the results are considered equivalent for interpretation purposes. Soy milk clearly demonstrates less environmental impact than dairy milk across a majority of impact categories, with the exception of natural land transformation and fossil depletion impacts. In the case of fossil depletion, results are considered by and large equivalent between the soy milk and dairy milk systems with three exceptions: when normalizing results according to nutritive value, when experiencing high consumer product losses,

and when given retail and consumer storage times of thirty or more days.

The equivalence of the soy milk and dairy milk system for fossil depletion impact is consistent with the result from the uncertainty assessment. In the case of natural land transformation, scenarios show that the soy milk system is consistently worse than dairy milk, with the exception of reduced customer transportation. However, interpretation of these results must be coupled with the uncertainty assessment because there is little statistical certainty that the soy milk system land transformation impact is, in fact, greater than the dairy milk system.

MODEL UNCERTAINTY

According to Huijbregts (2011), some aspects cannot be modelled within the present structure of technology assessments. Hertwich et al. (2000) discuss model uncertainty resulting from simplifications that may exclude relevant variables from the analysis. The use of proxy variables may not appropriately represent the variables of interest. For example, most spatial and temporal characteristics are currently lost by the aggregation of emissions in the inventory analysis. Furthermore, in the impact assessment we commonly assume that ecological processes respond in a linear manner to environmental interventions and that intervention thresholds are disregarded. You can refer to Chapter 11:LCIA Methods on impact assessment for more discussion. Further research is required to address these uncertainties. Tools are needed to:

- develop estimation tools for inventory data
- expand system boundaries
- improve impact assessment of metals, land use, and water use, and
- expand geographical coverage of impact indicators.

One way to understand the importance of model improvements is to compare the outcomes of old calculations and new calculations side-by-side.

INTERPRETATION CHALLENGES

When interpreting LCA results, practitioners must be careful of making Type I, II, and III errors when comparing product systems. Conducting a thorough uncertainty analysis should help practitioners have confidence in the interpretation of study results. However, understanding the root cause of each of these types of errors caused by underlying data and methodological variation is important. Statistical tests of uncertainty are based on a null hypothesis. The null hypothesis can take one of two forms: Sample a equals Sample b, or Sample a is greater or less than Sample b.

A Type I error occurs when no meaningful difference exists between overall results, but random sampling caused data to show a statistically significant difference (i.e., association or correlation). In this case, concluding that the two groups are

Legend:
- Equivalent (+/- 10%)
- Soy milk better than milk
- Soy milk worse than milk

	Baseline	Nutrition — Equivalent Calories Consumed	Nutrition — Equivalent Protein Consumed	Processing — Plant 1	Processing — Plant 2	Allocation on Dairy Farm — Protein	Allocation on Dairy Farm — Economic	Allocation on Dairy Farm — Mass	Feed — + silage, - grass	Feed — + concentrate	Feed — - silage, + grass	Feed — - silage, + DG
Climate change	57.1%	47.9%	42.7%	57.8%	55.2%	62.5%	67.0%	70.9%	56.6%	57.6%	57.8%	58.3%
Ozone depletion	34.0%	19.9%	12.0%	33.9%	34.2%	34.2%	34.3%	34.4%	34.0%	34.0%	34.0%	34.1%
Photochemical oxidation	21.5%	4.7%	-4.7%	21.1%	22.4%	28.3%	31.7%	34.7%	20.4%	22.3%	23.2%	23.4%
Acidification	80.0%	75.7%	73.3%	80.1%	79.7%	84.1%	89.6%	94.4%	75.7%	80.1%	85.3%	80.8%
Eutrophication	51.7%	41.4%	35.6%	50.3%	55.1%	57.8%	62.3%	66.3%	51.2%	53.6%	52.3%	55.6%
Agricultural land occupation	75.9%	70.8%	67.9%	75.9%	75.9%	80.2%	85.2%	89.8%	89.4%	97.2%	98.1%	95.6%
Human toxicity, cancer	67.4%	60.4%	56.5%	66.8%	68.8%	72.4%	77.4%	81.7%	68.3%	67.9%	66.8%	75.0%
Human toxicity, non-cancer	90.6%	88.6%	87.5%	90.6%	90.7%	94.0%	100.3%	105.8%	90.6%	91.5%	90.7%	98.0%
Ecotoxicity	86.3%	83.4%	81.8%	86.3%	86.3%	90.4%	96.8%	102.6%	87.1%	86.7%	85.5%	97.6%
Respiratory effects	24.9%	8.8%	-0.2%	25.6%	23.2%	29.9%	32.5%	34.8%	24.6%	26.1%	26.5%	27.2%
Natural land transformation	-31.9%	-60.1%	-75.9%	-35.4%	-23.6%	-32.0%	-32.0%	-32.0%	-27.9%	-26.9%	-27.8%	-27.9%
Fossil depletion	0.7%	-20.5%	-32.4%	0.1%	2.1%	5.4%	7.3%	8.9%	0.4%	1.5%	1.8%	2.7%

	Milk Farm to Processor — 20km (−)	50 km (+)	110 km (+)	140 km (++)	Milk Processor to Retail — 100 km (−)	200 km (+)	400 km (+)	500 km (++)	Warehousing — 2 days (−)	10 days (+)	20 days (++)	30 days (+++)
Climate change	56.5%	56.8%	57.3%	57.6%	55.1%	56.0%	57.9%	58.8%	57.1%	57.0%	56.8%	56.6%
Ozone depletion	33.8%	33.9%	34.0%	34.1%	33.5%	33.7%	34.2%	34.4%	34.0%	34.0%	33.9%	33.9%
Photochemical oxidation	18.6%	20.1%	22.9%	24.3%	10.9%	15.9%	26.0%	31.0%	21.5%	21.4%	21.2%	21.0%
Acidification	79.6%	79.8%	80.2%	80.4%	78.5%	79.2%	80.6%	81.3%	80.0%	79.9%	79.8%	79.7%
Eutrophication	51.6%	51.7%	51.7%	51.8%	51.5%	51.6%	51.8%	51.9%	51.7%	51.7%	51.6%	51.5%
Agricultural land occupation	75.9%	75.9%	75.9%	75.9%	75.9%	75.9%	75.9%	75.9%	75.9%	75.9%	75.9%	75.9%
Human toxicity, cancer	66.9%	67.1%	67.7%	67.9%	65.5%	66.4%	68.2%	69.1%	67.4%	67.3%	67.2%	67.1%
Human toxicity, non-cancer	90.6%	90.6%	90.6%	90.6%	90.6%	90.6%	90.6%	90.6%	90.6%	90.6%	90.6%	90.6%
Ecotoxicity	86.3%	86.3%	86.3%	86.3%	86.3%	86.3%	86.3%	86.3%	86.3%	86.3%	86.3%	86.3%
Respiratory effects	23.8%	24.3%	25.4%	25.9%	20.9%	22.8%	26.6%	28.4%	25.0%	24.7%	24.3%	23.9%
Natural land transformation	-31.9%	-31.9%	-31.9%	-31.9%	-31.9%	-31.9%	-31.9%	-31.9%	-31.8%	-32.1%	-32.4%	-32.8%
Fossil depletion	-1.0%	-0.2%	1.6%	2.4%	-6.7%	-2.6%	3.4%	6.5%	0.8%	0.5%	0.0%	-0.4%

	Retail Storage — 1 day (−)	7 days (+)	12 days (++)	30 days (+++)	Consumer Storage — 5 days (−)	15 days (+)	20 days (++)	30 days (+++)	Consumer Transport — walking (−)	5 km (+)	25 km (+)
Climate change	57.6%	56.0%	54.7%	50.1%	57.5%	56.7%	56.3%	55.5%	62.2%	60.6%	54.2%
Ozone depletion	34.0%	33.9%	33.9%	33.7%	34.0%	34.0%	33.9%	33.9%	35.0%	34.7%	33.4%
Photochemical oxidation	22.1%	20.1%	18.4%	12.1%	22.0%	20.9%	20.4%	19.3%	24.7%	23.7%	19.6%
Acidification	80.4%	79.2%	78.2%	74.8%	80.3%	79.7%	79.4%	78.8%	81.1%	80.8%	79.3%
Eutrophication	52.1%	51.0%	50.0%	46.6%	52.0%	51.4%	51.1%	50.6%	52.3%	52.1%	51.4%
Agricultural land occupation	75.9%	75.9%	75.9%	75.8%	75.9%	75.9%	75.9%	75.9%	75.9%	75.9%	75.9%
Human toxicity, cancer	67.7%	66.8%	66.1%	63.4%	67.6%	67.2%	66.9%	66.5%	69.3%	68.7%	66.3%
Human toxicity, non-cancer	90.6%	90.5%	90.5%	90.2%	90.6%	90.6%	90.6%	90.5%	90.9%	90.8%	90.4%
Ecotoxicity	86.3%	86.3%	86.3%	86.3%	86.3%	86.3%	86.3%	86.3%	86.3%	86.3%	86.3%
Respiratory effects	26.4%	21.9%	18.2%	4.8%	26.0%	23.7%	22.6%	20.3%	31.0%	29.1%	21.4%
Natural land transformation	-31.3%	-33.1%	-34.6%	-39.9%	-31.4%	-32.4%	-32.8%	-33.7%	5.8%	-6.0%	-53.1%
Fossil depletion	1.8%	-1.5%	-4.3%	-14.3%	1.5%	-0.2%	-1.0%	-2.7%	14.3%	10.0%	-6.9%

	Retail Loss — 2% (−)	7% (+)	17% (+)	22% (++)	Consumer Loss — 0% (−)	10% (+)	30% (+)	40% (++)	Best vs. Worst Case — Worst	Best
Climate change	60.5%	58.8%	55.3%	53.6%	63.0%	60.6%	53.5%	49.9%	-2.3%	86.0%
Ozone depletion	39.6%	36.9%	31.1%	26.1%	42.6%	39.5%	28.5%	23.0%	-20.7%	78.0%
Photochemical oxidation	28.8%	25.2%	17.8%	14.1%	32.0%	28.1%	14.9%	8.3%	-73.5%	78.4%
Acidification	81.7%	80.8%	79.1%	78.2%	82.7%	81.7%	78.3%	76.6%	50.8%	91.8%
Eutrophication	56.5%	54.1%	49.3%	46.9%	59.3%	56.3%	47.1%	42.6%	-4.0%	83.7%
Agricultural land occupation	78.1%	77.0%	74.8%	73.8%	79.0%	77.9%	73.9%	71.9%	57.1%	88.9%
Human toxicity, cancer	70.3%	68.8%	65.9%	64.5%	72.0%	70.2%	64.6%	61.8%	28.7%	87.7%
Human toxicity, non-cancer	91.4%	91.0%	90.2%	89.8%	91.8%	91.4%	89.8%	89.0%	82.1%	95.2%
Ecotoxicity	87.6%	86.9%	85.7%	85.1%	88.1%	87.5%	85.2%	84.1%	75.6%	93.1%
Respiratory effects	31.1%	28.0%	21.7%	18.6%	35.1%	31.2%	18.6%	12.3%	-100.4%	81.2%
Natural land transformation	-23.6%	-27.7%	-36.1%	-40.2%	-13.3%	-20.9%	-42.9%	-53.9%	-347.9%	77.4%
Fossil depletion	8.6%	4.7%	-3.3%	-7.3%	14.3%	9.0%	-7.6%	-15.9%	-161.5%	78.2%

Figure 15.3. Sensitivity analysis results of case study comparing soy milk to dairy milk. Baseline results show % preference of soy milk over dairy milk by impact category. Changes in preference are noted across each parameter.

meaningfully different is an error. In the case of a medical test, this would be called a false positive.

A Type II error occurs when a difference (i.e., association or correlation) exists between overall results, but random sampling caused the data to not show a statistically significant difference. Here, concluding that the two system results are not meaningfully different is an error. This would be called a false negative.

To give an example of Type I and II errors, say a company has developed a new auto fuel additive that claims to increase a car's gas mileage. It is tested on a random sample of cars under a random sample of driving conditions and ultimately the cars tested did have somewhat better gas mileage than normal. This result can mean one of two things:

1. The fuel additive does not ultimately make a difference, and the better mileage observed in the sample is due to sampling error (i.e., because of other factors, the mileage tests in the sample just happened to come out higher than average). If we could test all cars under all conditions, we would not see any difference in average mileage in the cars with the additive. This would be the null hypothesis.

2. The difference reflects the fact that the additive does increase gas mileage. If we could test all cars under all conditions, we would see an increase in mileage in the cars with the fuel additive. This would be the alternative hypothesis.

A Type I error occurs when we select the second option (reject the null hypothesis) when it should be the first; we conclude, based on the test, that the additive makes a difference, when it does not.

A Type II error occurs if we decide not to rule out the first option (fail to reject the null hypothesis), even though it is, in fact, true. We can conclude, based on the test, either that it does or does not make a difference, but there was not enough of a difference in the sample to conclude that a difference exists.

Monte Carlo calculations can overestimate uncertainty if products are compared where correlations are not observed. To illustrate, suppose there are two products, Product A and Product B. Product A is made of 10 kg of steel, while Product B is made of 11 kg of the same type of steel. Assume that the uncertainty in the CO_2 output of steel production is extremely high, +/- 100%. If we calculate the Monte Carlo distributions for the CO_2 emissions of product A and B, we cannot say that Product A is better than Product B, as both uncertainty ranges would be overlapping. However, because both products use the same steel, the uncertainty is completely correlated. In order to determine the difference in CO_2 output, the uncertainty is not relevant. We can conclude the obvious fact that Product A will have a 10% lower CO_2 production than Product B because it simply uses 10% less of the same steel.

The term Type III error has two different meanings. One definition essentially states that a Type III error occurs when the "right" answer is obtained by asking the "wrong" question. This is sometimes called a Type 0 error. Another definition is that a Type III error occurs when one correctly concludes two groups are statistically different, but wrong conclusions are made about the direction of the difference. Type III errors are rare. They only happen when random chance leads one to sample low values from a system parameter that is higher, and high data values from the system that is lower.

MEANINGFUL DIFFERENCE

What comprises a meaningful difference? How do we determine if a particular result is significant with a predefined quantitative cut-off (10%, 20%, et al.)? How do we know the point at which the uncertainty is sufficiently large that the entire study is erroneous and a waste of time and money?

When using standard LCA software packages to calculate uncertainty of product systems, such as the probability that the difference in System A is greater or less than System B, the level of significance used is generally 0.05, correlating to a 95% confidence interval in statistics. This level of significance level can be used as a cutoff mark for p-values (each p-value is calculated from the data) or as a desired parameter in the test design.

Popular levels of significance are 10% and 5% and continue through 1%, 0.5%, and 0.1%, although the latter are commonly used in the medical community and not in LCA. If a test of significance gives a p-value lower than or equal to the significance level, the null hypothesis is rejected at that level (Dallal 2012). Such results are informally referred to as ‹statistically significant (at the p = 0.05 level, etc.)[1]. For

1 In statistical significance testing, the p-value is the probability of obtaining a test statistic at least as extreme as the one that was observed, assuming that the null hypothesis is true (Goodman 1999). A researcher will often reject the null hypothesis when the p-value turns out to be less than a certain significance level, often 0.05 or 0.01. Such a result indicates that the observed result would be highly unlikely under the null hypothesis (that is, the observation is highly unlikely to be the result of random chance alone). Many common statistical tests, such as chi-squared tests or Student's t-test, produce test statistics that can be interpreted using p-values.

example, if someone argues that there is only one chance in a thousand this could have happened by coincidence, a 0.001 level of statistical significance is being stated. The lower the significance level chosen, the stronger the evidence required.

The choice of significance level is somewhat arbitrary, but for many applications, a level of 5% is chosen by convention, as illustrated in the case study at the end of this chapter. Generally speaking, an LCA practitioner makes a value judgment based on the underlying data used in the study to define an acceptable quantitative cut-off for the level of uncertainty needed to achieve a meaningful result. The significance level should always be clearly stated when communicating the results of an LCA study.

15.7 CONCLUSION

This chapter provided a practical example of applying an uncertainty assessment to a real-world LCA study that compares multiple product systems. The case illustrated how to isolate sources of uncertainty. Understanding the distributions associated with underlying datasets is crucial when interpreting impact assessment results and drawing conclusions about the environmental performance of one product versus another.

Simply accepting the baseline impact assessment values is not sufficient. A comprehensive evaluation of uncertainty moves beyond accepting these results to understanding the confidence we have in these results and characterizing their significance. Ultimately, the more complete treatment LCA practitioners give to uncertainty when conducting studies, the better equipped we will be in positioning the tool of LCA to be used for practical decision-making in the field.

The LCA community has granted the topic of uncertainty more attention than the topic of bias over the past decade. Both indices are critical to understanding the accuracy and validity of LCA results. The coming decades may provide clearer, less ambiguous approaches to modeling these challenging data characteristics.

EXERCISES

EXERCISE 1. BIAS OR UNCERTAINTY?

Bias and uncertainty can both contribute to drawing incorrect conclusions from LCA. There are more kinds of quantitative methods to approach uncertainty, but sometimes biases can be much more important. It is important to understand the difference. Imagine that you have spent a relaxing Friday afternoon reading many published LCA and noted each attribute listed below in at least one of the papers. Answer if each is an example of potential bias or uncertainty and what might be done to improve this attribute of the LCA.

1. A Monte Carlo approach was used to look at the impact of variability in database model parameters for upstream processes.
2. A food crop LCA includes only global warming potential, land use, and ozone depletion potential as impact categories.
3. A US steel manufactured good LCA uses Ecoinvent datasets based on European operations for process and material inputs.
4. A cleaning product LCA, which is based on a long list of ingredients, uses a surrogate for most of the materials.
5. LCIA results for a system in South America are characterized by a method based on average European conditions.

EXERCISE 2. EXTERNAL NORMALIZATION BIAS

Your commissioner asks you to conduct an LCA on roofing products using a particular characterization method and normalizing with global per capita impacts for the year 2000. You are evaluating a range of roofing products made of asphalt, concrete, terra cotta, wood, and steel. Regardless of which material you use, two impact categories consistently show the largest normalized results by one or two orders of magnitude.

1. What is the likelihood that this constitutes external normalization bias?
2. If so, how could you more thoroughly check if external normalization bias is happening?

EXERCISE 3. USE OF DIFFERENCES IN UNCERTAINTY CALCULATIONS

The case study in this chapter presented the uncertainty assessment results as "dairy milk – soy milk" and other similar differences. Explain the advantages and disadvantages of this reporting method.

EXERCISE 4. MEANINGFUL DIFFERENCE

You work in a team of product developers as the LCA expert in the process of designing a new refrigerator. You modeled all of the inputs and outputs of the refrigerator over its entire life cycle, and documented all assumptions about refrigerator lifetime and energy use. For comparison purposes, you also modeled a previous refrigerator made by the same manufacturer. Your characterized results show that the new refrigerator has 7% lower impacts in four impact categories and 3% lower impacts in six other impact categories.

1. Is this a meaningful difference?

2. How should you communicate this to the rest of the development team?

EXERCISE SOLUTIONS

EXERCISE 1

1. Uncertainty, specifically statistical uncertainty. The kind of analysis described is commonly possible with LCA databases and software tools.

2. Bias, and quite extremely so! LCA of agriculture should include consideration of water and eutrophication—these are both well-known and significant impact categories for agriculture. This is an extreme example of bias and could be worth a letter to the authors or journal editors.

3. Uncertainty, specifically misrepresentative data. In this common practice, the report authors should consider this as part of a sensitivity analysis.

4. Uncertainty, specifically misrepresentative data. In this common in practice, the report authors should consider this as part of a sensitivity analysis.

5. Bias. A sensitivity analysis could be done using a characterization method based on South American conditions or global conditions.

EXERCISE 2

1. The likelihood appears to be high that external normalization bias is occurring, given the very diverse materials that are being assessed.

2. You could select several hundred different materials and processes, characterize and normalize them, and identify if this pattern is evident on the average percent contribution of each impact category.

EXERCISE 3

From a communication point of view, the case study involves comparisons of options, which is a consistent way to compare uncertainty as well. From a statistical point of view, it avoids overstating the influence of shared parameters from the many common upstream processes in the models. A disadvantage is that the difference is not lognormally distributed, whereas the original distributions are.

EXERCISE 4

The average difference of all ten impact categories is 4.6%, so the refrigerator is considered to be below the 5% threshold of meaningful difference.

1. The average difference of all ten impact categories is 4.6%, so the refrigerator is considered to be below the 5% threshold of meaningful difference.

2. You should explain that on the average for the ten impact categories, the new design does not create significantly lower impacts. The threshold is generally considered to be 5%, and stronger evidence of a meaningful difference would be 10% or more.

PARALLEL LIFE CYCLE METHODS

HANNA-LEENA PESONEN AND CATHERINE BENOÎT NORRIS

16.1 LIFE CYCLE COSTING

Life cycle costing (LCC) predates LCA and can be traced back to the 1950s in both the public sector, evaluating the effectiveness of public services such as health care and infrastructure, and systems engineering in the private sector (Swarr et al. 2011). Originally, LCC was not developed in an environmental context. A number of other environmental accounting tools have similarities to LCC, such as full cost accounting, total cost accounting, total cost assessment, and cost-benefit analysis.

Different models have been used to integrate LCA and LCC (e.g., BEES software developed for the building industry by NIST). SETAC published a Code of Practice for Environment LCC, which provides a framework for evaluating decisions from economic perspective as a component of LCSA (Swarr et al. 2011). Because the principles of SETAC Code of Practice come closest to a commonly accepted standard on environmental LCC, we use the SETAC Code of Practice as the framework for presenting LCC.

In general, LCC is conducted in a similar way to environmental LCA. The compilation of a LCC is carried out in four main phases corresponding roughly to the ISO LCA standard (2006a): goal and scope definition, environmental life cycle inventory, interpretation, and reporting and review (Swarr et al. 2011). A separate impact assessment phase does not exist in a LCC (see Figure 16.1), because all inventory data comprise a single monetary unit of measure. Thus, we do not need characterization or weighting of inventory data. Aggregated cost data provide a direct measure of financial impact.

LCC aims to:
- assess the costs and revenues that relate to real money flows along the physical life cycle of a product, and that are to be borne directly by the actors along the life cycle
- compare life cycle costs of alternative products, including associated processes and services
- record the improvements made by a firm with regard to a given product (reporting)
- estimate improvements of the planned product system, including process changes within the life cycle, innovations, supplier replacement, or changes in taxation
- identify win-win situations and trade-offs in the life cycle of a product, once LCC is combined with LCA and, ultimately, also with sLCA (Swarr et al. 2011).

16.2 GOAL AND SCOPE DEFINITION IN LCC

Goal and scope definition in LCC is similar to that of an LCA. Functional unit, as well as the system boundary, should be consistent with that of the parallel LCA. Different parts of the product system may fall below relevant cut-off criteria for the separate LCC and LCA components. For example, research and development as well as marketing may impose significant costs but little environmental impact. The key is that both studies refer to a consistent definition of the product system, and that cut-off criteria do not conflict with the intended goal and scope of the study. Although a LCC is typically conducted to inform decision-making of a particular actor, ideally the data can be presented in a way to fairly inform all actors in the product system (Swarr et al. 2011).

ECONOMIC LIFE CYCLE INVENTORY

The economic life cycle inventory phase is analogous to that of an LCA study, and in most cases will build on a life cycle inventory of physical flows associated with the product system (Swarr et al. 2011). LCC does not require new, revolutionary methods of cost accounting. Instead, it uses traditional methods of cost accounting in a broader context. While traditional cost accounting takes into account the internal costs and revenues of one actor in the product life cycle (e.g., investment, operational, and maintenance costs of the product manufacturer), LCC includes the costs and revenues of all actors, such as raw material production, manufacturing, trade, use, maintenance, and end-of-life, as well as society at large, including so-called external, social, or hidden costs or externalities, such as environmental degradation not covered directly by product manufacturer or user.

The basic idea of LCC is simply defined by UNEP and SETAC (2011) as "an aggregation of all costs that are directly related to a product over its entire life cycle," yet the development of LCC has not been straightforward. At first, attempts were made to include all costs, both internal and external, to LCC. However, monetary valuation of the external costs brought several problems. Examples of the most extreme difficulties include how to evaluate the monetary value of a human or animal life, or the loss of biodiversity. The monetarization is still used in some cost accounting methods (e.g., cost-benefit assessments), but many have questioned its use.

The recent Code of Practice for LCC recommends avoiding monetary valuation of external impacts, and instead proposes presenting both environmental and social impacts as the results of parallel LCA and sLCA (Swarr et al. 2011). This practice will also avoid double-counting of impacts, (externalities) such as impacts of pollution expressed both as physical units in LCA and as monetary units in LCC. In summation, LCC should only include internal costs covered by one or more actors in the product system, which reflect only real monetary flows. In addition, those costs that are likely to be internalized in the decision-relevant time frame can be included. These could be, for example, environmental taxes, the costs of which can already be monetized.

The LCC inventory should be built on a well-defined cost classification system, where cost data are divided to appropriate cost categories (Swarr et al. 2011). Cost categories are product and sector specific; therefore, no general model can be given. Data needs to be restated in a common currency at present value using appropriate exchange and discount rates (Ciroth 2008). Indirect costs and overhead costs need to be properly allocated to specific products. Allocation has been a continuous topic in both LCC and LCA literature (Curran 2007). General cost accounting principles concerning allocation are also useful in LCC. The two most frequently used cost-allocation methods are those based on physical measures such as weight and volume, and those based on market value such as costs allocated in proportion to the estimated value of the products (Drury 2005).

Economic life cycle inventory faces many of the same data access and quality issues faced by LCA. Because everyone is familiar with monetary units, costs data can create a false sense of certainty (Ciroth 2008). In fact, data sources for LCC cost data are diverse, varying from internal accounting and external databases to market prices and expert judgment (Ciroth 2008), and developing a consistent data set for a study can be challenging. Also, cost data are often more volatile than physical units and, therefore, special care has to be paid to sensitivity analyses of the LCC results.

A characteristic problem in any cost accounting is how to treat costs occurring in the future. In traditional cost accounting, this question has been solved through the use of a discount rate. The Code of Practice (Swarr et al. 2011) recommends the use of discounting in LCC as well, even though LCA does not use discounting. Discounting of environmental flows makes little sense, because that would result in future environmental impacts being given a relatively small weight (Gluch and Baumann 2004). In LCC, however, discounting should be used to correctly reflect the dynamics of economy in long-term time horizons. If the product's life cycle is less than two years, the discounting can be neglected. Otherwise, discounting is needed (Swarr et al. 2011). Selection of an appropriate discount rate depends on the goal and scope, and on whether the study is conducted from the perspective of the manufacturer, the user, or another stakeholder.

INTERPRETATION, REPORTING AND REVIEW OF LCC RESULTS

Requirements and procedures for interpretation, communication, and review of LCC results are analogous to the principles of ISO 14040 LCA standard (2006a). Different actors along the product life cycle can view life cycle costs from different perspectives. Correspondingly, the LCC results can be viewed from the perspective of a producer, user, or society, as has been defined in the goal and scope definition of the assessment. For a producer, it might be interesting to identify the total costs for the producer itself. Alternatively, in order to assess competitiveness of the product, identifying consumer costs (costs of ownership) might be crucial. For consumers, total costs of ownership are usually pertinent.

Uncertainty and sensitivity analysis should be conducted for the data in order to identify those that might contain the highest uncertainties in LCC. These typically include project life, discount rates, specific cost assumptions, and raw material and energy prices.

As discussed in Section 16.1, in order to identify environmental and economic (and social) win-win situations or trade-offs, the final results of LCC should ideally be analyzed and presented together with the results of parallel LCA and/or sLCA. One possibility is to plot selected LCA or sLCA results versus LCC results. An example of this is given in Figure 16.1. Yamaguchi et al. (2007) performed LCA and LCC to understand the environmental and economic impacts of various washing machine parts. According to their results, the driver and controller units of the washing machine both have high costs and a high environmental load; housing and basket units that have a high environmental load with small costs; a lid unit shows high cost and low environmental load; and for other units, both environmental and economic impacts are low.

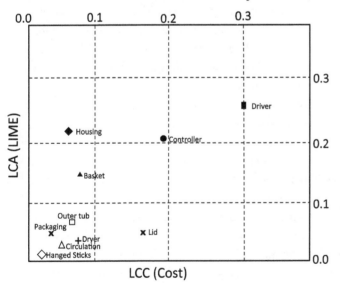

Figure 16.1. Presentation of LCA and LCC Results for a Washing Machine (Yamaguchi et al. 2007)

16.3 SOCIAL LIFE CYCLE ASSESSMENT

Social life cycle assessment (sLCA) is a technique to evaluate positive and negative social impacts along the life cycle of a product (UNEP-SETAC 2009). sLCA research resides at the convergence of social responsibility and industrial ecology. While product system modeling is drawn from the engineering sciences, the issues investigated, the indicators, and the evaluation methods largely stem from the social sciences, particularly the social responsibility field.

With intensified globalization (WEF 2012) and pressure to keep costs low, supply-chain social and ethical issues often becomes under scrutiny. Despite the interest and concerns of multiple stakeholders, and the fact that the division of production in multiple international steps is probably the most significant change in international trade of the past 40 years (Robertson et al. 2009), the question of supply chain social responsibility is still relatively new (Brammer et al. 2011).

In fact, even though large companies have invested substantial efforts in ethical compliance and auditing over the last 20 years, the situation for workers has gotten worse in most places (Harrison 2013). However, increased access to data, methods, and tools help us understand, weigh, and communicate about the social impacts of product supply chains and life cycles. We describe the origins of Social LCA, the assessment framework, and how the technique can be applied.

HISTORY OF SLCA

Several groups developed frameworks for Social LCA and applied their new methods starting from the 1990s (Oeko Institut, BASF, Belgian Ecolabel, and the CIRAIG Simplified LCA tool). However, in the new millennium, efforts have begun to be made by experts, practitioners, and users to reach a global consensus and derive a common framework. An international multi-stakeholder project group was launched in 2004 under the umbrella of the UNEP SETAC Life Cycle Initiative. The project group offered three main contributions: a feasibility study (2006), which confirmed the interest in and the practical feasibility of Social LCA and also pointed to the necessary adaptation of the existing (environmental) LCA standard (ISO 14040 and 14044); the Guidelines for Social LCA (2009), which laid out the framework and described the basis of the technique; and the Methodological Sheets (2010 and 2013), which provided detailed information about each impact subcategory and suggested indicators that can be used.

The first academic article on Social LCA was published in 1996 by O'Brien et al. in the newly created *International Journal of Life Cycle Assessment*. Only a dozen articles were published from 1996 to 2006, and almost all were exclusively published by the journal. About sixty journal articles have been published in total since 1996, with the vast majority of them following the publication of "Guidelines for Social

LCA" in 2009. A wide range of social and economic journals have since published articles including *CIRP Annals – Manufacturing Technology* and the *Journal of Purchasing and Supply Management*.

The rising number of conferences and sessions dedicated to s-LCA mirrors the exponential augmentation in the publication of journal articles, chapters, and books, and testifies of the dynamism of this emerging field.

DIFFERENCES AND SIMILARITIES WITH LCA

Social LCA is an iterative and multidimensional technique (Benoit and Norris 2012) that is applied in a business decision making and reporting environment (Jorgensen 2012). In 2005, a published editorial stated that the ISO standards (ISO 14040 and ISO 14044) applied to social impacts (Weidema 2005). This meant that it was (and is) possible to apply the LCA framework (and the ISO standards) to study the social impacts of product life cycles. It also indicates the necessity of adapting the technique to the social assessment (Hutchins et al. 2008). This statement was leveraged in the UNEP SETAC feasibility study. It constitutes the foundation of the approach. Thus, sLCA follows the same phases as LCA (Goal and Scope, Inventory Analysis, Impact Assessment, and Interpretation).

Social impacts are not generally directly linked to the unit process (Jorgensen et al. 2008). Social impacts are not determined by the elementary flows of the LCA, but rather by the behaviors of the stakeholders involved. That is why identifying and engaging stakeholders in the studies is recommended when possible and relevant (Ramirez and Petti 2011). Some of the differences identified in the sLCA Guidelines and captured in the literature are evident, such as the following examples:

* Characterization models used in sLCA are different than the ones used in LCA (Parent et al. 2010).
* Data sources are different (Jorgensen et al. 2008).
* Steps and methods used to collect data can vary from LCA (Hutchins et al. 2008).
* The significant issues will differ (Lehmann et al. 2011).

Some guidelines clarify new aspects added to the framework:

* Impact subcategories are classified by impact category and stakeholder category (Franze et al. 2011).
* Presenting justifications is recommended when subcategories are not applied in a study (Ciroth et al. 2011).
* The product utility must be expressed in relation to its function in LCA. However, the sLCA Guidelines recommend that the social impacts of the use phase and related product functions also be considered (Moberg et al. 2011).

* Subjective data might be the most appropriate type of data to use in a study (Dreyer et al. 2010).
* Activity variables for life cycle attribute assessment (LCAA) can be used in sLCA to estimate the share attributable to each unit process in a product system (Ekener Petersen et al. 2012).
* A performance reference point was used (Parent et al. 2009).
* The focus on the assessment of positive impacts was accentuated (Hauschild et al. 2008).

This brings nuances, including these examples:

* Impact assessment generally requires geographical location data in sLCA. The demand for the regionalization of assessed environmental impact increases, but the type of geographical information may differ by geographical type and population density versus political boundaries (Zamagni et al. 2012).
* The balance in between quantitative, qualitative, and semi-quantitative data used will generally be different (Hauschild et al. 2008).

Many authors of sLCA journal articles have proposed, supported, or observed the concept that sLCAs make use of data collected at the level of the organization (Jorgensen et al. 2008). However, even when management type data are sought, in most cases they will have to be collected at the level of the production site. Some other authors have proposed that data at the regional or national level are of interest for sLCA (Eckvall 2011). In fact, even if data are collected at the production site, the data are always collected in function of the country specific context.

Some authors propose to combine data collected by several different tools. For example, Lehmann et al. (2011) suggest combining social impact assessment and social life cycle assessment. Dreyer et al. (2010) combine social auditing and sLCA and Benoît et al. (2010) describe how different tools can be mobilized by sLCA and vice-versa.

Figure 16.2 illustrates the scale at which information is collected by different tools, including sLCA, LCA, social audit, country index, and sustainability reports. As we can see, S-LCA is not the only tool to make use of data collected at different scales. However, S-LCA is the tool that sources the most data from different scales. LCA mostly collects data from processes, but it also increasingly collects data at the level of the production site as well as country sectoral data for studies (e.g., using IO models). Social audits mostly use data collected at the production site, but they also make use of data at the scale of the organization, country, and industrial sector. Coun-

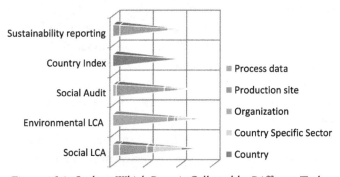

Figure 16.2. Scale at Which Data is Collected by Different Tools

try indices like GINI and SIGI collects data at the scale of a country population. Sustainability reports, collects, and presents, for the most part, information at the scale of the organization, while making some use of production site, country, and industrial sector data.

SOCIAL LCA FRAMEWORK

The preceding section helped us uncover some of the main characteristics of sLCA as compared to LCA. In this section, we will present the framework specific to sLCA. Two figures will help define the topics and process of assessment: Figure 16.3 and Figure 16.4.

The stakeholder classification for sLCA has its origins in stakeholder theory. Stakeholder theory argues that organizations would be more successful, and would gain greater understanding around the complex issues involved in their business, if they engaged with a broader set of stakeholders, or non-shareholders. This theory is widely accepted today, as it forms the basis of social responsibility, and is adopted by virtually all main CSR instruments and framework. Choosing the stakeholder classification as the backbone of the issue matrix by the UNEP SETAC initiative sLCA project group ensures that impact subcategories affecting all relevant groups were taken into consideration when assessing the social and socio-economic impacts of products throughout their

life cycles. The selection of the subcategories was achieved by looking at international agreements, standards, and guidelines that have been developed by multi-stakeholder groups, as they capture consensus of wide audiences.

Figure 16.3 presents the list of impact subcategories recommended by the sLCA Guidelines, as classified by stakeholder group. These subcategories can also be classified by impact category when conducting social life cycle impact assessment.

The methodological sheets provide detailed information on each of the subcategories introduced in the Guidelines, with subcategories organized by stakeholder category. They

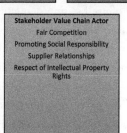

Figure 16.3. List of Stakeholder and Impact Subcategories (Issue Matrix)

Stakeholder categories	Impact categories	Subcategories	Inv. indicators	Inventory data
Workers	Human rights			
Local community	Working conditions			
Society	Health and safety			
Consumers	Cultural heritage			
Value chain actors	Governance			
	Socio-economic repercussions			

Figure 5 – Assessment system from categories to unit of measurement. Adapted from Benoit et al., 2007

Figure 16.4. Social Assessment System. Reproduced from the sLCA Guidelines, this diagram shows the assessment process from the inventory data to the stakeholder and impact categories.

were developed in recognition of the fact that data collection is the most labor-intensive activity when conducting an sLCA. Therefore, different indicators may be used, depending on data availability and the goal and scope of the study. The sheets are meant to inspire sLCA case studies based on the Guidelines, rather than represent a complete set of indicators that must be included and criteria that must be met.

Since 2009, papers have been published that provides input on subcategories that should be added or disregarded in certain contexts, such as informal economy (Ekener-Petersen et al. 2013), or for specific types of assessment, such as technology (Lehmann et al. 2013). Refinement of the methodology is highly probable.

Goal and scope

The sLCA Guidelines state that the ultimate goal of an sLCA is to improve the social conditions of production. Obviously, conducting an sLCA study is not sufficient in and of itself in improving supply chains' social condition of production. Fostering improvements require consistent efforts, dialogue, long-term engagement, and capacity building. Nonetheless, this general goal can serve as a guide when determining the study parameters and follow-ups.

S-LCA is a methodology for providing decision support about the social impacts related to product life cycles (Jørgensen 2012). sLCA is more holistic than comparable tools in the social assessment sphere because of its inclusion of the entire life cycle. It also provides a systematic approach to the evaluation of social impacts.

The variety of goal and scope pursued by practitioners partly explains the diversity of practical approach implemented in the sLCA field (Benoit Norris 2013). One type of sLCA has four dimensions: the determination of the study goals (understand, weigh, educate, and communicate), the selection of social issues to include (from one impact subcategory to all subcategories related to one stakeholder category, or the full spectrum of impact subcategories related to all 5 stakeholder categories), the context of application of the study (from product design to reporting and labeling), and the selection of the life cycle stages to be included in the study boundaries (from one phase to the full life cycle). Detailed guidance on how to select issues to include (impact categories, subcategories, and

indicators) does not yet exist.

One other factor to consider in the goal and scope is how the results might be used. Will they be used as stand-alone sLCA, as a combined environmental and sLCA, or within a life cycle sustainability assessment? Finally, knowing ahead of time can be useful if a generic, desktop-type study will be sufficient, or if the practitioner plans to conduct site-specific assessments, or both. Luckily, sLCA is an iterative technique so one can go back and change the choices made earlier in the process.

Thus, defining the objectives pursued by the study and the context of application are the first tasks. Figure 16.5 summarizes some of the potential objectives of an sLCA. The goal and scope needs to clearly describe which of these goals the study will address.

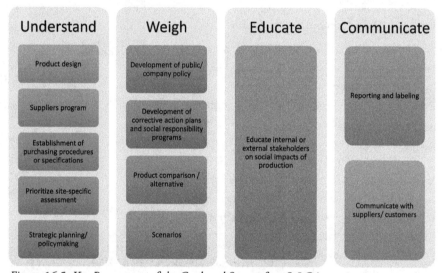

Figure 16.5. Key Parameters of the Goal and Scope of an S-LCA

The scale at which the study will be conducted is also a determining factor that needs to be identified as soon as possible at the onset of the study. Here are some examples of the level at which studies can be conducted:

- product
- product category
- basket of products
- organization (e.g., company)
- company division
- economic sector (e.g., Canadian dairy industry)
- technology
- economy (e.g., trade partner)

Following these first sets of key choices, the stakeholder(s), impact categories and subcategories to be assessed will need to be selected. Conducting a preliminary hotspot assessment can guide the selection process, along with a materiality assessment. A materiality assessment is an exercise in stakeholder engagement designed to gather insight on the relative importance of specific environmental, social and governance (ESG) issues. However, the sLCA guidelines recommend that the exclusion or addition of subcategories be justified and that the justification be included in the report.

In the large majority of cases, a preliminary hotspot or scoping assessment will be recommended (Dreyer 2010). A social hotspot assessment includes the following steps: location information data collection, mapping from existing data and model(s) to the one used for the assessment, entering or linking to the lLCI information, conducting a social life cycle impact assessment, calculating the results and interpreting the results.

Prioritization plays a key role in sLCA and life cycle management. While a product system may involve thousands of processes, the ability to collect site-specific data is limited. Therefore, we should find which impact subcategories and life cycle phases may contribute more significantly to the overall impact or risk.

Once the study parameters are determined, the functional unit will be assigned. The functional unit is a measure of the function of the studied system and it provides a reference to which inputs and outputs can be related. The functional unit required models the product system in sLCA. However, the functional unit might not be used as a way to report about the results (depending of the study goals). The choice of the functional unit will depend of the goal, but will also largely depend on the type of model and data used for product system modeling. If an sLCA input-output database is used, then the functional unit will be in dollars purchased from a sector. If a traditional LCA database will be utilized, then the functional unit may be a quantity such as unit surface protected for a number of years, for weight carried over a distance, and so on.

If the functional unit is not used to communicate about the social impacts, then what can be used to link the social inventory information to the product system? The sLCA Guidelines suggest utilizing Life Cycle Attribute Analysis (LCAA) to resolve this issue. LCAA quantifies the percent of an activity variable that possesses an attribute of interest, such as the percent of worker hours that is at high risk of being child labor. LCAA enables us to speak about results in a way that carries information about the scope of the life cycle. This presents a viable alternative to the use of the functional unit when presenting results, which enables us to relay the results without losing sight of the scope and in a way that may better

reach the audience.

Therefore, S-LCA also requires activity-variable data that will be used in LCAA. The literature (Dreyer et al. 2010) describes a few potential activity variables, including economic value added, acreage, water usage, cost, and worker hours (with worker hours being the most popular). Worker hour data also offers one additional and meaningful parameter to help prioritize further action, such as additional data collection.

SOCIAL LIFE CYCLE INVENTORY

Data collection is also the most daunting, energy-intensive step of a life cycle assessment. A social life cycle inventory includes product modeling data, generic social data, and, when relevant, site-specific data. It also often includes performance reference-point information. Product modeling data are the data necessary to ensure that the assessment planned or underway captures the entire life cycle. It also provides quantitative metrics that can assist when justifying the study boundaries and scoping choices.

As discussed in the preceding section, product modeling in sLCA generally must include some information about the geographical location of the unit process in order to be valid. For this reason, current environmental LCI databases fall short of providing the necessary parameter. The appearance of databases especially designed for sLCA, such as the Social Hotspots Database (Benoît-Norris 2013), fill this gap.

THE SOCIAL HOTSPOTS DATABASE

The Social Hotspots Database (SHDB) is an input/output LCI database. It provides a solution to enable (1) the modeling of product systems and (2) the initial assessment of potential social impacts (i.e., social hotspot or scoping assessment).

The SHDB system (Benoît-Norris et al. 2013) is based on the Global Trade Analysis Project Version 7 (GTAP 2008). A unit process is referred to country-specific sector (CSS). The total database contains data for 57 different sectors in 113 different regions. Most of these regions correspond to individual countries, while others are regions containing multiple countries. Thus, 57*113 = 6441 unit processes are in the database. This input-output model is used to provide estimates of sector- and country-specific activity in product supply chains. The system also calculates worker hours for each activity in the supply chain.

The labor intensity data were developed by converting GTAP data on wage payments into estimates of worker hours for skilled and unskilled labor, in each sector and in each GTAP country/region, by compiling and using wage rate data according to sector and region. These labor hour intensity factors are used together with the social risk level characterizations in order to express social risks and opportunities in terms of work hours, by sector and country, at a given level of risk relative to each of over twenty-two impact subcategories and 100 different indicators. The risk data is addressing five main impact categories: labor rights and decent work, human rights, health and safety, governance, and community.

The SHDB is based upon LCAA. Each unit process has a number of different attributes, or characteristics, relative to a large set of social issues. The activity variable used in the SHDB is worker hours. Thus, the SHDB can be used to identify how many worker-hours are involved for each unit process in the supply chain, for a given final demand (final product or service output from the system). The sociosphere flows are expressed as "worker-hours at a specified level of risk on a given risk indicator, per USD of process output."

When relying on the SHDB is not possible, other strategies can be implemented. When starting with a product system modeled from an LCA process database, practitioners can collect trade data to augment the product system with location information. Lagarde and Macombe (2011) have also demonstrated other ways to model the product system for sLCA consequential assessments.

SOCIAL IMPACT DATA

Social impact data are the data by which we can know how a social impact category or a stakeholder group is affected in the context of a production activity. Researchers commonly make a distinction between causal-chain impacts, performance, and context.

A causal-chain impact is an impact directly attributed to the production activity itself. Social sciences do not currently provide many well-established impact pathways that allow for the assignment of a particular impact to a specific and documented action to be tied to a unit process. The fields of public health and epidemiology are precursors in the search for social impact pathways.

Data on social performance represents the level of realization with respect to a threshold or a best practice. This type of data and assessment using performance reference points is common in the field of corporate social responsibility and is frequently used in sLCA. Contextual data represents the typical social situation in a country and an economic sector/industry. It can be used as background data and for scoping assessments. However, the actual supply chain performance may vary from the average, and so the contextual data may

need to be replaced with site-specific data, depending on the goal and scope of the study.

Different types of studies may necessitate different types of social impact data and often require a mix. Parent et al. (2010) described the main difference between performance reference points and causal-chain impact assessment methods. Jørgensen (2009) showed that the type of data that would need to be collected for evaluating, The impact pathways of child labor might differ from the data that would need to be collected for performance assessments. These distinctions are essential. However, the most critical consideration is to make sure that the planned study will provide the most relevant results for its goal and scope.

GENERIC OR SITE-SPECIFIC

Generic data are data neither collected on site nor specific to a company. Generic data are often collected from large international databases (e.g., ILO or World Bank) or by conducting a literature review (e.g., journal articles, books, or media). Site-specific data are data collected at the level of the production activity. These data can be collected from the company's factory or field where the unit process is (are) located. They may also be collected from the workers, local communities, local NGOs, or other relevant stakeholders for the unit process.

The methodological sheets provide examples of generic and site-specific indicators that may be used, and that list existing sources of data. The sLCA guidelines recommend triangulation to reduce results uncertainty and to improve the validity of findings. Triangulation means to collect data from various sources, using different method to corroborate the findings (or to put them in perspective).

SOCIAL LIFE CYCLE IMPACT ASSESSMENT

Kloepffer (2008) identified several relevant questions about social life cycle impact assessment:

1. How to relate quantitatively existing indicators to the functional unit?
2. How to quantify all impacts?
3. How to interpret indicators results (e.g., low payment)?

These questions highlight numerous challenges about the development of characterization models that are addressed differently by practitioners. In sLCA, the impact categories are logical groupings of sLCA results, related to social issues of interest to stakeholders and decision makers. Midpoints and endpoints exist at different points along a social impact pathway that begins with a social intervention or pressure (e.g., requiring unpaid overtime) and leads to different levels of impacts (e.g., workers' rights and the well-being of workers and their families).

The S-LCA guidelines differentiate between performance assessment (Type 1) and causal chain modeling (Type 2). These two approaches provide different results (Parent et al. 2010).

A Type 1 method utilizes performance reference points in order to assess the relative position and state of a unit process impact subcategory (or indicator) in reference to one or more international instruments or best practice (threshold). This LCIA method helps one understand the magnitude and the significance of the data collected in the inventory phase. Most of the characterization models developed applies performance assessment (Ugaya et al. 2013,). This method requires collecting information specific to these performance reference points (Benoît-Norris et al. 2011).

Performance-based impact assessment methods generally use an ordinal scale that either describes the risk (from extremely high to low), the performance (from non-compliant to best practice) or the degree of management (from uncontrolled to under control). Some methods also compare the results to the context.

sLCA may also include the development of indices. As we have seen, the sLCA process requires multiple aggregation steps. Each subcategory may be assessed by one or more indicators. Impact categories are assessed by multiple subcategories. Ordinal scales may be used at each assessment step, and points can be assigned according to the rank order to render an overall assessment for the impact category under study.

Some of the developed LCIA methods allow calculating the impact in relation to the functional unit, while most of them do not. Using LCAA can allow one to calculate the scope of the life cycle (e.g., labor rights issues are under control), bringing back the crucial life cycle perspective to the assessment.

A Type 2 method assesses social impacts using impact pathways. Each impact pathway makes use of a specific characterization model that translates inventory results in midpoint and endpoint impacts. The Type 2 method attempts to isolate cause-effect chains caused by a specific pressure. For example, requiring excessive working time may cause workers to experience higher stress levels, and high stress levels may cause depression (midpoint), and depression will result in a loss of (psychological) wellbeing and personal health (endpoints).

Because data models and experiences with causal chain modeling are minimal, they will only be discussed briefly. Early works in sLCA have proposed some impact pathways.

Norris (2006) characterized certain health effects resulting from changes in economic activity by reconstructing the Preston curve. Hutchins and Sutherland (2008) examined the relationship between infant mortality and GDP per capita by measuring the effect of changing one supplier (from a different country). Weidema (2006) proposed an approach to sLCA causal chain modeling using quality-adjusted life years (QALY).

More recently, Jørgensen et al. (2010) suggested a pathway linking unemployment and several social impacts. Feschet et al. (2012) published an article about the development of the Preston pathway, whose aim is to measure changes in economic activity generated by the functioning of a product chain and the changes in health status of the population in the country where the economic activity takes place. Most papers conclude highlighting the limitations of the impact pathway method (e.g., local context is not taken into consideration or the necessity of having more than one pathway to analyze) and the necessity for further development.

S-LCA INTERPRETATION

Interpretation is a repeated and systematic process for the identification, description, estimation, and presentation of all the information that have been derived from the other stages. The purpose is to analyze results, to give references, and to lead to conclusions and recommendations that allow taking future decisions. That includes the identification of the most significant social issues, the assessment of data quality, and an evaluation of the exhaustiveness, completeness, and consistency of the study. It represents a systematic evaluation of the needs and opportunities to reduce social risks and impacts, and to enable increased shared benefits. Thus, interpretation should take place continuously during the study. The final output of the analysis can be a set of improvement opportunities or scenarios.

CONCLUSION

While tremendous progress has been made with the development of the methodology and the creation of an international consensus, some challenges remain. Some of these challenges are due to the variety of goal and scope pursued by sLCA studies, some of which lead to the development of impact assessment methods. The availability of tools such as the SHDB makes it easier for academics and practitioners to start to engage and carry sLCAs.

Overall, many of these challenges are also shared with LCA (e.g., uncertainty, data quality and availability, preferred impact assessment method, and verification) and will slowly be overcome as practitioners invest more research efforts. Because of the strong interest in sLCA on the part of companies and academic institutions, there will probably be more progress on methodology development soon. The ultimate goal of sLCA is to serve as an instrument to improve the livelihoods and social conditions of people touched or involved in the life cycle of products. Much work is yet to be done to be done to create caring, fair, and respectful conditions all along global supply chains. Understanding the social impacts is just the first step.

As discussed above in Section 16.1, in order to identify environmental and economic (and social) win-win situations or trade-offs, the final results of LCC should ideally be analyzed and presented together with the results of parallel LCA and/or sLCA. One possibility is to plot selected LCA or sLCA results versus LCC results. Yamaguchi et al. (2007) performed LCA and LCC to understand the environmental and economic impacts of various washing machine parts. These were their results: driver and controller units of the washing machine have both high costs and high environmental load; housing and basket units have high environmental load with small costs; lid unit shows high cost and low environmental load; and for other units, both environmental and economic impacts are low.

16.4 LIFE CYCLE SUSTAINABILITY ASSESSMENT

Since the introduction of the concept of *sustainable development* in the pivotal Brundtland report (WCED 1987), it has become clear that environmental, social, and economic impacts should not be studied in isolation. Rather, in order to reach the ultimate goal of sustainability, all three aspects have to be respected. The three aspects have to be examined hand-in-hand to avoid trade-offs and problem-shifting among them. In a business context, this three-pillar interpretation of sustainability is referred to as corporate (social) responsibility (WBCSD, not dated). Elkington (1997) calls this "the triple bottom line," renaming the three pillars as "people, planet and profit," thereby emphasizing the importance of all three aspects for long-term business results.

Life cycle aspect was included as one of the guiding principles of sustainability by the United Nations (UN) in Johannesburg 2002 (UNEP 2002). As a consequence, the United Nations Environmental Programme (UNEP) and the Society for Environmental Toxicology and Chemistry (SETAC) launched The Life Cycle Initiative to promote an integrated framework of life cycle thinking in the context of products, including all three pillars of sustainability. Klöpffer (2003) has initiated the discussion on the options of how to formulate the procedure to carry out a full sustainability assessment. UNEP and SETAC (2011) published "Towards Life Cycle Sustainability Assessment," in which they argued that while considering the three pillars is possible, a need to develop methods, tools, and capacity still exists. Over the years, the discussion has evolved to a holistic approach in increasing awareness of false optimization and wrong choices, like the burden shifting within or between each domain (environment, economy, and society) or to the future (Pesonen and Horn 2012).

INTEGRATING SOCIAL AND ECONOMIC IMPACTS IN LIFE CYCLE ASSESSMENT

Critics have pointed out, and rightly so, that sustainability assessments of products are still often reduced to a LCA or even mere carbon footprint calculations (Finkbeiner 2009). As discussed above, according to the life cycle community, a life-cycle-based full sustainability assessment should, instead, be pursued. An LCSA of products is grounded in a conceptual framework that uses distinct analyses for each of the three pillars of sustainability--environment, economy, and social equity--with consistent, and ideally identical, system boundaries and the same functional unit (Klöpffer 2003):

LCSA = LCA + S-LCA + LCC

Figure 16.6 presents the framework for an LCSA, based on a common product system with the same functional unit, consistent system boundaries and scope including similar phases (i.e., inventory, impact assessment, and interpretation) for the three parallel assessments (Swarr et al. 2011).

Product System (same functional unit, consistent scope and system boundaries)		
Environmental life cycle inventory	Economic life cycle inventory	Social life cycle inventory
Environmental life cycle impact assessment		Social life cycle impact assessment
Environmental interpretation	Economic interpretation	Social interpretation

Figure 16.6. Framework for an LCSA (Swarr et al., 2011)

Until now, we have had no official standard for LCSA. LCA is the only pillar that has been standardized (ISO 2006a and 2006b). Recently a lot of work has, however, been dedicated within the life cycle society to developing both sLCA and environmental life cycle cost (LCC). UNEP (2009) has published Guidelines for sLCA, and SETAC the Code of Practice for LCC (Swarr et al. 2011). The UNEP and SETAC Life Cycle Initiative has recently published a Guideline for LCSA (UNEP and SETAC 2011) as the first attempt to create common understanding and practices in LCSA through looking at the full sustainability impacts of a product.

According to current understanding, the results of an LCSA cannot be reduced to one quantitative measure, because no general consensus exists about the relative weight of the three aspects. The results of LCA, sLCA, and LCC should rather be presented in parallel to allow the visibility of any trade-offs to the decision maker. The main advantage of this approach is its transparency. As of now, we have no uniform understanding of what the parallel presentation of the LCSA results could look like.

One of the first attempts was published by Traverso et al. (2012) in the form of a Life Cycle Sustainability Dashboard. In this study, a sustainability assessment of the assembly step of photovoltaic (PV) modules production by LCSA was implemented. Figure 16.7 shows the result of the sustainability performance of the three scenarios, one Italian and two German products; the position of the arrow changes according to the sustainability performance of each scenario. The Sustainability Performance Index (SPI) of each scenario is reported by the score and position of the arrow in the dashboard. The dashboard on the bottom right presents the relative distance of sustainability performance of the three scenarios. This allows for an easy-to-understand visualization of the LCSA results and a direct comparison of products.

Experience in the integration of LCA, LCC, and sLCA is limited. However, all the techniques have evolved to a point where we can use all three tools in parallel. The degree to which the integration will be completed in different studies will likely vary.

Some life cycle scientists have argued that the world-shaping potential of LCA is more important than its use as a decision-making tool (Ehrenfeld 1997). Heskanen (2001) illustrated that this worldview is reified and embodied in

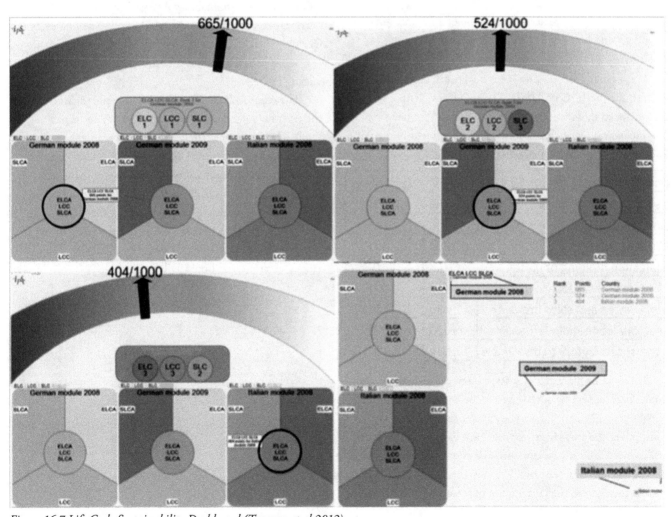

Figure 16.7 Life Cycle Sustainability Dashboard (Traverso et al 2012)

ideas about policy, management, and consumption, and that it has implications for how ordinary people are expected to behave and for how well they can live up to these expectations. LCSA could play a strong role in helping people understand sustainability issues in a systemic and systematic way, assisting institutions and people alike.

PROBLEM EXERCISES

1. Life Cycle Costing

You have been asked by a furniture manufacturer to perform an LCC on one of their chairs. The manager who hired you poses several questions about the types of economic data that included in an LCC. What would be the most accurate way to answer these questions?

a. Does LCC include only the costs to us the manufacturer, or does it also include the costs to the purchaser and the municipality that will landfill the used chair?

b. Does LCC quantify the economic value of environmental damage, human health damage, or resource depletion?

c. The chair is expected to have a life of 24 years. What discount rate should be applied in the LCC study?

2. The manufacturer provides a number of costs associated with the production and distribution of the chair. These are tabulated in dollars per quantities of chairs (or parts for specific quantities of chairs) for each chair element in the table below. What are the total costs to the manufacturer in $/chair?

costs	cost ($)	per # chairs
steel	420	600
nickel plating of steel	408	1200
plastic	1280	800
injection molding of plastic	5.7	30
corrugated cardboard box	40	100
warrantee and sales brochure	290	1000
factory labor and overhead	11020	2000
transport, factory to distributor	192	60

3. The chair manufacturer asks you to estimate the cost of a take-back program for the chair at the end of its useful life in 24 years. The manufacturer estimates that the total cost of shipping the chair from the user back to the manufacturer will be $9.20. Using the discount rate equation below, and assuming a discount rate of 2.9%, how many dollars per chair should they add to the current cost of the chair?

Current cost = future cost / $(1 + r)^T$

where r = discount rate and T = number of years

4. The user will purchase the chair for $127, including tax. Cleaning supplies will be used to maintain the chair every year, amounting to $0.45/year. Assuming a discount rate of 2.9%, what is the current total cost of ownership to the user? The equation for an annuity (the $0.45/year) is:

Current cost = C $((1-(1/1+r)^T)/r)$

where r = discount rate, T = number of years and C = annual payment

5. The chair manufacturer decides to conduct an sLCA on the same chair.

a. Which type of data is necessary to conduct an S-LCA that concerns the inputs (steel, plastic, and corrugated cardboard box) that you will need to request or collect?

b. How will you prioritize site-specific data collection?

c. How can you group the S-LCIA results?

6. Life Cycle Sustainability Assessment

The chair manufacturer would like you to combine the LCC and the sLCA of the chair with an LCA of the chair. Why would it be difficult to interpret the meaning of the resulting LCSA?

7. The commissioner identifies a competitor chair from a different manufacturer and requests you to perform a complete LCSA on that chair in order to compare the chairs for internal communications in the manufacturer of the first chair. After many months of work, you complete the data collection and analysis of the second chair. A considerable amount of data and resulting impacts from the second chair is missing, particularly in the S-LCA and the LCA. What are your options for managing this disparity in data and impact results objectively and impartially?

PROJECT MANAGEMENT

CARINA ALLES, SUSANNE VEITH, IVO MERSIOWSKY

17.1 INTRODUCTION

The process of performing LCA becomes easier if thorough project management practices are applied. As with many other involved tasks, careful project design and planning can help practitioners avoid pitfalls. The job of project management is to assign tasks at a given time using particular resources, then following up to assure that tasks are completed. In the course of designing an LCA project, decisions are made that inform the entire study and project.

Unlike many engineering projects, an LCA project rarely follows a sequential, predefined path; rather, it is an iterative procedure with subsequent steps that loop back to previous ones. For example, the discovery of unforeseen data needs may require a review of earlier steps, such as goal and scope definition. Good project management enables you to anticipate iteration loops and attain the project goal.

Iterations can improve the quality of data needed to satisfy the goal and scope. Conversely, iterations can also adjust a too ambitious goal and scope to the available data. This chapter, is organized around three elements:

1. the goal and scope description
2. key techniques of project management
3. detailed step-by-step instructions on performing an LCA study

17.2 GOAL AND SCOPE DEFINITION

Goal definition is the first phase of any LCA,. It defines the purpose and the decision-making context of the study, along with its intended audience. Goal definition is decisive for all the other phases of the LCA. Therefore, special care must be taken when defining the goal of the LCA study, and sufficient

LCAs rely on input from multiple contributors along the value chain, including individuals from diverse educational backgrounds. A value chain is a series of activities that a company operating in a specific industry performs in order to deliver a valuable product or service to the market. Value chain partners generally include raw material and energy suppliers, customers, and end-of life treatment facilities. A key element of LCA is collaboration along the value chain that should at least ensure a joint interpretation of results, if not for the actual data collection. The practitioner should assemble an interdisciplinary team of individuals who are familiar with the product due to their particular fields (e.g., marketing, R&D, and business functions). This team may also include knowledgeable internal and external experts on the considered technologies and value chains who gather input from customers and other stakeholders to determine the relevant issues early in the process.

time has to be allowed for this step to develop.

The goal and scope discussion seeks to define answers to the questions described in the section below. Decisions taken in the original goal and scope discussion may need to be revisited during the course of the study. Further helpful guidelines on Goal and scope are provided by ISO (2006) and EC JRC (2010).

GOAL DEFINITION

LCA can be used (a) for internal purposes within organizations as part of a wider decision-making context, (b) for external purposes in policy making or marketing, or (c) in standardized communications such as environmental product declarations as per previously established Product Category Rules. As a consequence, both external and internal stakeholders may have to be consulted in the definition of goal and scope. Thinking about this task is crucial in answering a series of questions.

TOPIC OF THE STUDY: WHAT ARE WE STUDYING?

LCA studies may be on a product, process, or system. Being clear about the topic of the study will help avoid confusion later on. Normally, this decision is documented in a flow chart that shows the technosphere flows of the system under study (see Chapter 5:LCI for more information).

Definition of the study goal: Why are we doing this study? Is the study being used to

- Better understand the impacts of the product or service?
- Communicate specific environmental characteristics of a product, process, or service (e.g., carbon footprint, water footprint, or corporate sustainability reports)?
- Support internal decision-making (to show improvement opportunities in the development of a product or process)?
- Support the selection of certain raw materials or vendors by asking questions such as the following: "Are bio-based raw materials the more environmentally sustainable choice compared to the fossil-based alternatives?" and „Is Vendor A's product more environmentally friendly than Vendor B's?"
- Support policy decisions (e.g., tax deductions for products with a lower environmental footprint, or preferential procurement policies)?
- Provide a competitive advantage in the marketplace?
- Support environmental product declarations (EPDs)? In this case, do product category rules (PCRs) exist that provide more detailed guidance on how the sustainability assessment needs to be conducted?
- Support public comparative assertions (e.g., Product A has a lower carbon footprint than product B)?

Note: Public comparative assertions require a critical review of the LCA and the final claims. If comparative assertions are going to be made, an external peer reviewer should participate in the scoping discussions.

Different parties, such as supply chain partners and government authorities, may contribute to shaping the goal and scope. This multi-stakeholder approach means diverse and sometimes conflicting objectives need to be prioritized, consolidated, and balanced.

When discussing the purpose of the study with stakeholders from varying backgrounds (such as marketing and R&D), misunderstandings may arise from the use of different terminology. Even though basic understanding of the concept about LCA is improving, its method and interpretation are still far from commonplace. , Therefore, common terminology and shared understanding about the study goals is essential, especially in an interdisciplinary team.

WHO IS THE AUDIENCE?

To meet the study's objective, the specific needs of key audiences must be addressed, as the questions below illustrate:

- Is the study intended to inform internal stakeholders (e.g. product developers, engineers, and researchers) about hotspots and improvement opportunities of a product or process? If so, the approach will likely be attributional LCA, the number of impact categories may be fewer, the level of quality assurance may be lower, and cutoff rules may be more generous.
- Does the LCA inform policy managers or politicians to make more informed decisions, such as deciding whether investment in a certain process technology would result in a lower carbon footprint on a global scale, or deciding whether to support tax credits for certain biofuels? If so, it may be wise to use a consequential analysis, have a comprehensive set of indicators, and a thorough data quality review process.
- Is the information used to communicate the environmental benefits of a product along the value chain? Communication of LCA results can be made between companies ("business-to-business" or B2B) or between a company/business and the final consumer ("business-to-consumer" or B2C). Normally, such studies are attributional and well documented through a third-party report in accordance with ISO standards. See Chapter 2: International Standards for LCA for more information.

When communicating data to different audiences, be they B2B or B2C, the communication style should be tailored to the individual audience. In LCA, results and conclusions may be communicated beyond the original target audience. These secondary target groups should be anticipated as early as possible so that the conclusions can be "translated" for the target audience. This is discussed in Chapter 18:Reporting LCA Results.

COMMISSIONER

The commissioner of the study needs to be kept informed about the progress of the study at all times. The active involvement of the commissioner is essential during the scoping and final interpretation phase, because the commissioner is likely to be the most familiar with the product, its value chain, and the opinions of critical stakeholders. Commissioners of LCAs often have managing, marketing, and/or research roles, and might be less familiar with the processes and technical details of an LCA. Close cooperation between the commissioner and the LCA practitioner insures that a common understanding is generated and a common goal is pursued.

NEED FOR CRITICAL REVIEW

Decide on the type of the main review:

- single-person review, during which no comparative assertion will be disclosed;
- review by panel, during which comparative assertion will be disclosed;
- stakeholder involvement in addition to the review panel, where varied interests are represented (e.g., policy makers and market actors).

Reviewers, panelists, and stakeholders need to be informed and invited early enough to assure that they will be ready to participate when needed. Aside from the main review by an independent external entity, internal quality checks are recommended and should be planned for in the course of development.

SCOPE DEFINITION

DEFINITION OF FUNCTIONAL UNIT

The functional unit defines the product system being studied and quantifies the service or function delivered by the product. The functional unit provides a reference unit to which the inputs and outputs and final environmental impacts can be related.

In the scoping discussion on a refrigerator LCA, the characteristics of the refrigerator must be defined. These include the size and temperature performance specification of the refrigerator model, the average energy consumption at a certain geographic location to keep a certain amount of product at a defined temperature, and the average life time of the refrigerator.

For LCAs that are targeted towards comparative assertions, the scoping team must assure that the product alternatives are compared on an equivalent functional performance basis (e.g., compare the life cycle of two types of refrigerators that deliver the same cooling amount and performance over a defined life time).

WHAT IS THE SYSTEM BOUNDARY?

Is the LCA analysis carried out on a cradle-to-gate basis or does it include the whole value chain?
Generally, a cradle-to-grave LCA is preferred, because it shows the environmental footprint of a product or service along the whole value chain. However, in certain cases, such as when use phase and end-of-life are the same for two alternative products, practitioners may decide to only focus on a cradle-to-gate assessment.

WHAT IS THE GEOGRAPHIC SCOPE?

Which geographic scope should be covered in the study? Should the LCA only focus on a certain geographic region or should the study be applicable to other regions in the world as well? In the latter case, sensitivity studies can be carried out with varying geographic scope.

The definition of geographic scope is critical for the choice of the appropriate electricity grid, because the energy sources for electricity generation may differ significantly from place to place. Processes that use bio-based feedstocks require a clear definition on where the individual renewable feedstocks are grown, because farming practices and climatic conditions can vary greatly between different locations.

WHAT ARE THE DATA COLLECTION STRATEGY AND DATA REQUIREMENTS?

Many scoping decisions are needed at the beginning of data collection process. Data collection is oftentimes the most time-consuming step and is most likely on the critical path in the project Gantt chart (see the section below).

ENVIRONMENTAL IMPACT CATEGORIES AND MODELS

Data collection is determined by which environmental impact categories will be included in the study (see Chapter 11: LCIA Methods). Responses from stakeholders on communicated issues and concerns should be considered when defining the included impact categories. Furthermore, the selection of impact assessment portfolio methodologies (e.g., CML or TRACI) need to be defined in the scoping discussions.

DATA COLLECTION STRATEGY AND DATA QUALITY REQUIREMENTS

During the course of the goal and scope discussion, the practitioner must decide the data sources that will be used for the individual steps in the value chain. Primary data (collected from employees of the commissioner and suppliers in the value chain) are generally preferred, but secondary data (e.g., generic datasets from databases) are oftentimes used if primary datasets are not available or credible.

CUT-OFF CRITERIA

Data precision depends on the purpose of the study. Less detail is required for a screening LCA. Data requirements are more stringent for a comparative LCA that substantiates public claims. Typical cut-off values for LCAs are 1% of mass and energy, provided that no environmentally significant flows fall under the threshold. If the inventory data quality cannot be met due to lack of access to data or resources, revisions may be made to the scope of the study. Impact categories, where the required data quality cannot be met, can still be considered on a screening level.

Does the study employ attributional LCA or consequential LCA?

This decision will define what kind of LCI data needs to be collected. The attributional life cycle inventory modeling principle is also called „accounting," "book-keeping," "retrospective," or "descriptive" LCA approach. As background data, producer-specific LCI data, or database information referring to average current or historical producer mixes, are generally used. The consequential life cycle inventory modeling principle is also known as "change-oriented," „effect-oriented," „decision-based," or "market-based" LCA methodology. It aims at identifying the consequences that a decision in the foreground system has for other processes and systems of the economy. Hence, the consequential life cycle model does not reflect an actual, forecasted, specific, or average supply-chain. The hypothetical, generic supply chain model is based on expected market-mechanisms, political interactions, and consumer behavior changes. For a consequential LCA, marginal data are needed that reflect those changes. Consequential modeling is described in more detail in Chapter 8: Advanced System Modeling.

Multi Input/Output Processes and Allocation

When working with processes that have multiple inputs and outputs, the environmental burden of the production process needs to be allocated in a way that reflects best physical relationships and/or market realities. Different physical allocation methodologies like mass allocation, heat allocation, or economic allocation methodologies can be applied. Allocation is also described in more detail in Chapter 8: Advanced System Modeling.

Quality Control, Documentation, and Transparency

Chapter 6: Data Quality provides more detail on the requirements for data quality control and reporting. The questions below provide a first set of guidelines for the scope definition.

Quality control

A gap assessment or critical review is recommended for quality assurance purposes. Once comparative assertions are drawn based on the LCA study, ISO 14040 series require a critical review prior to making any public claims.

Documentation

During the course of the study, it is good practice to doc-ument all relevant steps, decisions, assumptions, data sources, and calculation procedures. Those notes are a valuable basis for correct and efficient reporting.

As a final step of the study, a study report is prepared. This report can be used as input for a critical review, or it can be used to document the study framework and context (along with all necessary assumptions and data sources). Documentation can be part of a data set and/or can be presented in the form of a report. Both forms of documentation will build on the extensive notes that were taken and revised during iterations of the LCA work. Confidential and proprietary data and information should be documented in separate confidential reports that are made accessible only to the critical reviewer(s). For LCA studies, a public third-party study report is required, if the target audience is external.

Reproducibility and transparency

Reproducibility and transparency are key principles of a study report. Results and conclusions of the study need to be displayed in a concise and clear fashion that can be understood by the target audience. Enough detail should be provided to stakeholders to enable a reproduction of the study.

17.3 Project Plan Development

Introduction to project management

Careful planning at the onset of an LCA will pay off during the course of the project. Experienced project managers have offered the following reasons why strong project management is essential (Bizmanualz 1999):

1. It provides a process for estimating project resources, time, and costs
2. It delivers project results on time and on budget
3. It communicate project progress, risks, and changes
4. It controls "scope creep" and manages change
5. It defines the critical path to optimally complete the project
6. It prepares for unexpected project issues
7. It documents, transfers, and applies lessons learned from projects

Project management can follow phases:

Phase 1: Planning

In the planning phase, teams will be organized and individual responsibilities will be assigned by means of a RACI chart (see the section below). This phase is connected with the goal and scope definition phase in a LCA.

Phase 2: Scheduling

In the scheduling phase, resources such as financial and

human resource expenditures will be assigned and project tasks will be identified. A Gantt chart can be used to break down the project into smaller tasks, assign a logical sequence to the tasks, and allot sufficient time for the completion of the individual activities. At this point, tasks will be identified that need to happen in sequence, because subsequent tasks may rely on information from a previous activity in order to be completed. Other activities may happen in parallel. Time management is generally less critical for the latter tasks.

PHASE 3: CONTROLLING

The third project phase is the controlling phase, in which the project manager monitors resource allocation, costs, and the progress of the projects. In this phase, the Gantt chart might be adapted based on the feedback from team members about actual progress and resource needs. Costs may be subsequently reallocated to support time-critical tasks.

RACI CHARTS TO SET UP TEAMS

One key aspect of good project planning is to have the right team in place to solve the task at hand, and to assign and communicate the responsibilities in a clear and transparent manner.

In project management, a RACI model is often a useful tool to identify the roles and responsibilities of the team members. RACI stands for:

R = RESPONSIBLE

The person who "owns" the problem and performs the work. In an LCA, the responsible person is often a LCA practitioner or individual business people who help collect input data for a life cycle analysis.

A = APPROVE

The person to whom the team is accountable and who is often the commissioner of the study. In LCAs that are used for research guidance, this role is often taken by a research manager. In LCAs that are used for marketing communications, this person might be the marketing manager of the product.

C = CONSULT

The person to be consulted is often a technical subject matter expert who has the specific knowledge needed in certain parts of the project. Technical experts can either support the data collection or help address specific technical questions that the team may encounter during the course of a project. A technical expert can also help assess competitive technologies and provide input. Internal and external peer reviewers act also as consultants throughout the project. They will validate goal and scope definition, data choices, the methodological approach, and the conclusions of the study. In addition, legal counsels are often referred to for legal advice when formulating life cycle based marketing claims.

I = INFORM

The person who needs to be informed is often a business manager who will use the results of an LCA to make the appropriate business decisions.

Roles are laid out for each task that needs to be accomplished. Table 17.2 is a RACI chart showing typical team responsibilities in the key steps of a LCA

	LCA Practitioner	Business Person	Commissioner	Sunject Matter Expert	Internal/External Reviewer	Business Manager
Goal and Scoping Discussion	R		I,A		C	I
Data Collection	R	R	I,A	C		
Development of LCA Model	R		I,A	C		
Discussion of Results	R		I,A		C	
Interpretation/Business Guidance	R		I,A		C	I

Table 17.1. RACI Chart

Team members can have dual roles: for example, the commissioners of the study and R&D /marketing managers may need to be informed about the goal and scope, as well as approve the final project outline. Responsibilities for the completion of one task may lie in the hands of multiple people. For example, data collection for certain production steps can be conducted by a business person who might be located at the product production site, while other data collection may be done by the LCA practitioner.

When creating a RACI chart, each activity must be assigned to at least one "responsible person" (each activity needs to be followed by an "R") and each member of the team will "own" at least one role (R, A, C and/or I). All action items and project tasks contained in process plan must be present in the RACI chart. The RACI chart can help verify whether all task items in the project plan have a responsible person assigned.

SPECIAL CASE OF EXTERNAL LCA SERVICE PROVIDERS

With the professionalization of LCA and the increasing volume and complexity of LCA projects, enlisting the aid of external service providers is commonplace. The role of a service provider can vary, but here are some typical examples:

• Purchasing LCA databases and modeling software from a

vendor, and possibly using the vendor's software implementation services

- Contracting an LCA service provider with experienced experts to conduct data collection, life cycle modeling, and/or calculation
- Contracting a verification company or a freelancing LCA senior expert to conduct a peer review or critical review of the LCA study
- Involving a facilitator or communication expert who need not be well-versed in LCA, but who acts as a skilled moderator that is open and able to discuss complex technical subjects.

Further reasons for enlisting the assistance of a service provider can include the following:

- The commissioner has insufficient internal capacity available, as industry experts are often booked on a variety of other tasks. This practice is a plain case of outsourcing, coupled with the additional advantage that the solicited expert, due to their experience and routine, may often be more efficient at completing the task.
- There may be demanding requirements of LCA data and modeling expertise; for instance, specific knowledge may be needed in terms of allocation methodologies, recycling systems modeling, and specific impact assessment models such as toxicity assessment. In such cases, the LCA practitioner will want to benefit from the scientific background or long-term experience of senior experts.
- Multi-stakeholder projects (especially in the case of conflicting goals or emotional issues) may need a neutral, ideally renowned, and commonly accepted practitioner. This practice is closely related to involving a facilitator for additional neutrality and arbitration assistance in complex negotiations.

In each case, the practitioner may be in the role of project manager without having to complete the entire project (or even any of it). Instead, one may define the goal and scope of the study, and then use the project management tools discussed here to set up and control the LCA project. Here are ways to effectively involve service providers in your project:

- **Identify suitable service providers**

  This will increasingly rely more on external accreditation rather than claims made by practitioners. Similar to the development of rating agencies, practitioners will need to demonstrate their expertise and professional conduct such as the ACLCA-certified practitioner certification.

- **Select eligible service providers**

  This should be informed by a quality management system,

or at least quality assurance processes, as a minimum requirement. At this time, many service providers are academic experts who freelance and often work in small offices. They can be proficient reviewers and strong partners on methodological issues. Complex projects may require more professionalism in terms of communication and project management. Service providers can be differentiated by projects of strategic importance, staffing commitments, binding deadlines, and internal reviews.

- **Contract with service providers**

  They can take the form of commissioning a (sub-) contract or recruiting a temporary or interim LCA manager. Professional service contracts will include warranties, but LCA project contracts tend to contain vague phrases and ambiguous wording of deliverables. Clear definition of the statement of work, and the method of payment and contingency planning in the contract, will make the LCA process more pleasant and effective for everyone. Once you have contracted an external service provider, you can include them in your RACI chart just as any other team member. Be sure to carefully assign accountabilities (R) based on the service contract in place.

GANTT CHARTS FOR LCA PROJECT SCHEDULING

Considering the complexity and iterative nature of LCAs, a structured approach to project planning is essential to the success of a LCA project. To create a Gantt chart, the LCA project tasks have to be clearly identified and put into a logical sequence. The ILCD handbook (EU JRC 2011) supports LCA project planners with helpful guidelines.

Gantt charts are appropriate tools to support the project scheduling phase. Software tools like Microsoft Project are available to support the planning process and to help build Gantt charts, but they can also be made in spreadsheets.

Gantt charts support managers to make sure that all activities are accounted for, that their order of performance is logical and appropriate, and that completion times for each individual activity (also known as "activity times") and for the whole project are determined. In order to judge whether project deadlines can be met, the longest time path through the network (also known as "critical path") is determined and critical activities are identified. The project manager then pays special attention to the critical tasks. In order to support on-time completion of the activities along the critical path, the project manager may increase resources and financial support in those bottle neck activities.

In order to create a Gantt chart, the project is broken down into individual tasks. The tasks are put into the appropriate sequence and tasks that can be worked on in parallel are identified. Oftentimes a work breakdown structure can be helpful in defining the individual tasks. Figures 17.1 and 17.2 show the work breakdown structure for planning a party, and the network diagram that provides more information on the timing of the individual tasks.

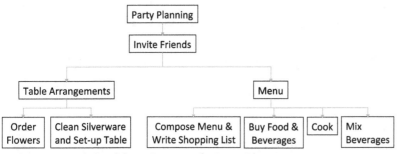

Figure 17.1 Work-breakdown Structure for Party Planning

Based on the work-breakdown structure, a network is prepared. When creating a network diagram, one must:

1. First, calculate the earliest start and the earliest finish from left to right (upper line of numbers). The earliest start of a step is the latest earliest finish of its predecessor.
2. Then calculate the latest finish and the latest start from

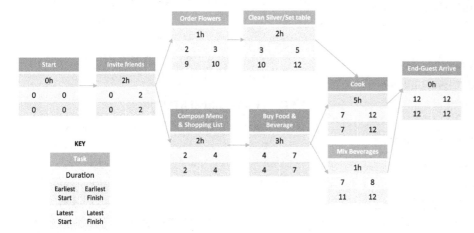

Figure 17.2. Network Diagram for Party Planning

right to left (lower line of numbers). The latest finish of a step is the earliest latest start of its successors

The critical path represents the longest time path through the network. A task is critical if the earliest and latest finish times are equal (see tasks marked in red). The tasks

on the critical path also determine the minimum time needed to finish the project. Any delay in a critical task will cause a delay of the whole project. In the case of the party planning example, a minimum of twelve hours is needed to complete the project (12h = 2h+2h+3h+5h). If only ten hours are available for the preparation of the party, the duration of tasks on the critical path needs to be reduced (e.g., by ordering food through a catering service). In this case, reducing the duration time of activities that are not on the critical path (ordering flowers or setting up the table) is not helpful.

If a task is not critical, it has "slack." Slack time is the free time allotted to each activity. The slack time is the difference between the latest finish and the earliest finish, or the difference between the latest start and earliest start. Slack is the length of time an activity can be delayed without delaying the whole project. In this example, only three tasks have slack. The ordering of flowers and the preparation of the table both have seven hours of slack time. The mixing of the beverages has four hours of slack time. Activities on the critical path have no slack time.

Based on the network diagram, the Gantt chart at the bottom of the page, can be developed. In practice, project management software tools (such as Microsoft Project) will generate the Gantt charts automatically and will keep track of the dependencies. Activities on the critical path are indicated with red rectangles.

HIGH LEVEL VIEW ON THE STAGES OF A LCA PROJECT

In a Gantt chart, the iterations of an LCA study are displayed as a sequence of tasks, framed by the preparatory work at the beginning of

Figure 17.3. Gantt Chart for Party Planning

Task	Timeline in Hours											
	1	2	3	4	5	6	7	8	9	10	11	12
Invitations												
Invite Friends												
Table Preparation												
Order Flowers												
Clean Silver & set table												
Menu Preparation												
Compose Menu 7 Shopping List												
Buy Food & Beverages												
Cook												
Mix Beverages												
End/Guest Arrive												

the project and the final steps required to complete the study at the end (Figure 17.3).

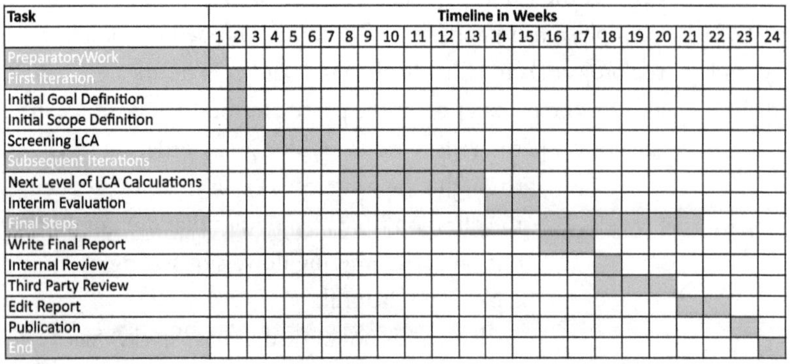

Task	Timeline in Weeks																							
	1	2	3	4	5	6	7	8	9	10	11	12	13	14	15	16	17	18	19	20	21	22	23	24
PreparatoryWork																								
First Iteration																								
Initial Goal Definition																								
Initial Scope Definition																								
Screening LCA																								
Subsequent Iterations																								
Next Level of LCA Calculations																								
Interim Evaluation																								
Final Steps																								
Write Final Report																								
Internal Review																								
Third Party Review																								
Edit Report																								
Publication																								
End																								

Figure 17.4 High Level Gantt Chart of a Typical LCA Project

Figure 17.4 shows a generic Gantt chart for a typical LCA project, based on the tasks outlined in the ILCD handbook (EU JRC 2011). Any LCA will share commonalities with this generic layout, but the project manager should follow the steps outlined in goal and scope definition section to identify specific tasks at the appropriate level of detail. Timelines will depend on the circumstances of the particular LCA project, which in turn will be determined by the scope and complexity of the tasks, as well as by the qualifications and the size of the project team. Detailed task definitions and realistic timelines are prerequisites in the identification of the critical path and thus scheduling for the overall project.

The preparatory steps outlined in Figure 17.5. help establish good management processes and effective communications from the beginning.

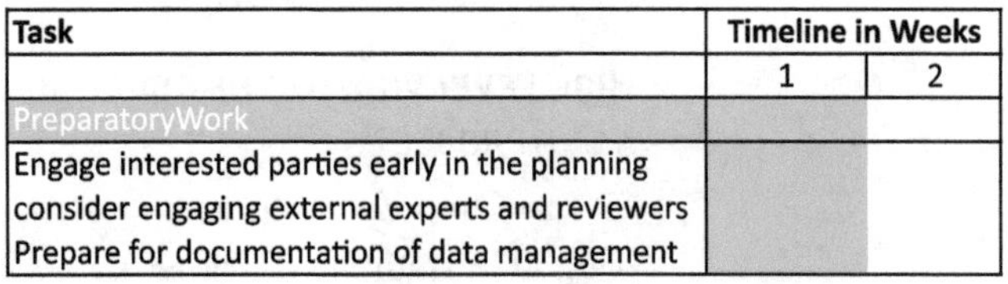

Task	Timeline in Weeks	
	1	2
PreparatoryWork		
Engage interested parties early in the planning consider engaging external experts and reviewers Prepare for documentation of data management		

Figure 17.5. Gantt Chart of the Preparatory Work in a Typical LCA Project

Budgeting of LCA and LCA projects

Gantt charts can help to establish a realistic basis for time and cost estimates of the individual activities within a LCA project, as well as the LCA project as a whole. The iterative nature of LCAs requires that such estimates be revisited in the course of a project. At the outset, determining the costs of an LCA project can be difficult due to many unknown factors, such as data availability, data quality, unexpected results, and increasingly ambitious interpretation. These factors place both the commissioner and the service provider in an awkward position.

Adjustments may be necessary, including revisions to the initial cost estimate as tasks change in duration and intensity, or reductions in the scope of work in order to fit within a capped budget (also known as "target costing").

Always be aware of possible "scope creep" during the course of the project, as new insights will create new desires for additional information. Critical consideration will become necessary when expanding the scope seems attractive, but is not affordable within the given budget. Capped time allotment for low priority items can maintain scope discipline.

Engagement with key players

Reaching out to interested parties and potential contributors at the beginning of the project facilitates seamless communications and optimizes resource planning. The section on RACI teams above provides more guidance on how to set up an effective LCA project team. In particular, it is recommended that critical reviewer(s) be involved from the beginning of the study so that they may contribute in the early stages of goal and scope definition. At this stage, RACI charts identify all essential resources early on in the process so that everybody involved is informed as the project proceeds. Failure to connect with key players early on, or to put the necessary confidentiality agreements in place, can cause long and costly delays.

Set the stage for documentation

Prepare to document all relevant steps taken, decisions and assumptions made, data sources used, and calculations performed. This creates a valuable basis for correct and efficient reporting. Reporting, if foreseen, is the last step of an LCI or LCA study before a critical review, but it begins at the beginning of the process. The preparatory works sets the stage for the first phase of an LCA project, the goal and scope defintion phase. Figure 17.6 shows a high-level Gantt chart of the essential steps.

Goal and scope definition

Because the goal and the scope of an LCA project define the frame of reference for the entire study, they deserve special attention. In many cases, seeking input from different perspectives is helpful, including internal and external stakeholders. If the study is peer reviewed at a later stage, engaging the reviewers in the goal and scope definition can help to ensure alignment and clarity of expectations.

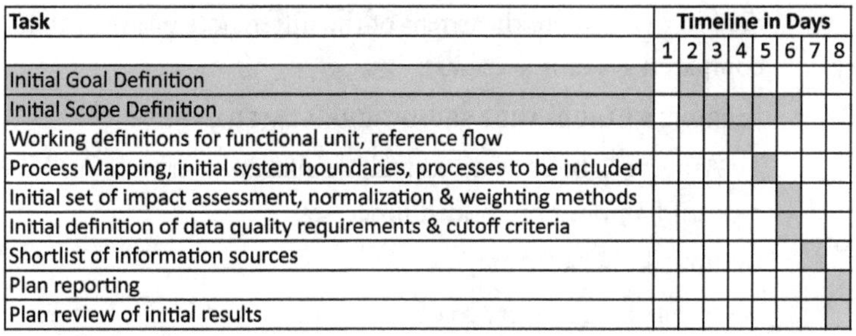

Task	Timeline in Days							
	1	2	3	4	5	6	7	8
Initial Goal Definition	▓	▓						
Initial Scope Definition	▓	▓						
Working definitions for functional unit, reference flow			▓					
Process Mapping, initial system boundaries, processes to be included				▓				
Initial set of impact assessment, normalization & weighting methods					▓			
Initial definition of data quality requirements & cutoff criteria						▓		
Shortlist of information sources							▓	
Plan reporting								▓
Plan review of initial results								▓

Figure 17.6. Gantt Chart of the Goal & Scope Definition Phase in a Typical LCA Project

17.4 FIRST ITERATION: THE SCREENING LCA

A first rough draft of a life cycle inventory system model, along with its impact assessment calculations and analysis, helps to identify key processes, parameters, and assumptions that largely contribute to, or influenc,e the measurement of the environmental impacts of the process or system. In subsequent iterations, the analysis can be refined with special attention to these key processes. Prioritizing the aspects that matter most allows one to deliver the minimum required data quality with minimum effort. A typical screening LCA comprises the steps shown in Figure 17.7:

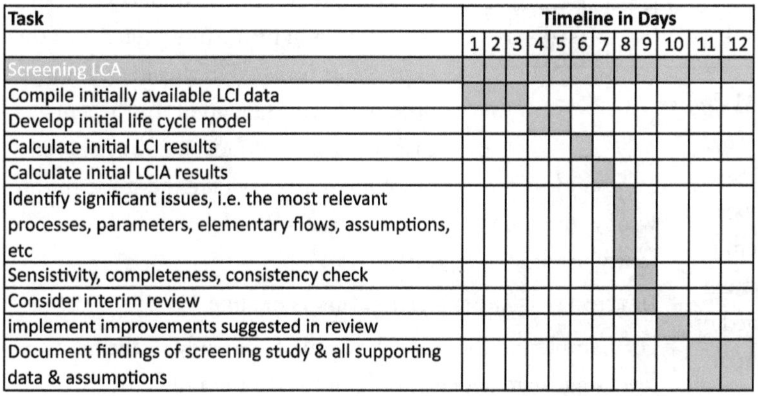

Task	Timeline in Days											
	1	2	3	4	5	6	7	8	9	10	11	12
Screening LCA												
Compile initially available LCI data	▓	▓										
Develop initial life cycle model			▓									
Calculate initial LCI results				▓								
Calculate initial LCIA results					▓							
Identify significant issues, i.e. the most relevant processes, parameters, elementary flows, assumptions, etc							▓					
Sensitivity, completeness, consistency check									▓			
Consider interim review										▓		
implement improvements suggested in review											▓	
Document findings of screening study & all supporting data & assumptions												▓

Figure 17.7 Gantt Chart of a Screening LCA as the First Iteration in a Typical LCA Project

The most time consuming and expensive part of an LCA is generally the inventory collection.

COMPILE INITIALLY AVAILABLE LCI DATA

Any readily available specific foreground data is compiled, be it raw data, unit processes, LCI results, or similar data. These data sets are then supplemented with secondary data, preferably from the suppliers and/or downstream users. Subject matter experts should be consulted to bridge data gaps with reasonable estimates. Confidentiality agreements may be required bring in outside experts. Ample time should be allowed to get the necessary legal documents in place.

Accompanying all the LCI steps with an interim quality control is recommended. For example, having a second set of eyes look over any calculations, as well as check preliminary results for reasonableness, can help identify and correct errors. Once the essential data sets are collected, the data are processed in the following steps:
- Develop an initial, rough life cycle model
- Calculate initial LCI results
- Calculate initial LCIA results
- Where appropriate, calculate normalization and/or weighting.

The next two steps are critical to determine whether the screening LCA is sufficient, to indicate trends and hot spots, and to guide further explorations into the study subject.

STEP 1: IDENTIFY SIGNIFICANT ISSUES

As a first step of the interpretation phase, identify the significant issues, which are the key processes, parameters, elementary flows, and assumptions with the largest contributions and/or relevance for overall environmental impacts, or individually for each impact category.

STEP 2: CHECK THE INITIAL RESULTS

To ensure the validity of the results, the following checks are required:
- sensitivity check
- completeness check
- consistency check

One may also wish to have a third party provide an interim review.

In some instances, the results of the screening LCA may suffice to test a hypothesis or to provide a rough orientation on a topic of concern. The practitioner can then proceed with the closing steps as discussed in the section below. Whenever deeper, more specific insights are desired, more refinement is required. The results of the screening study set the stage for the next iteration.

Regardless whether the screening LCA is the primary deliverable of a LCA project or whether more iterations follow, documenting all relevant results is essential. This includes documenting the supporting data and assumptions with great care as a stand-alone reference document or as a building block in the final report of a bigger study.

17.5 SUBSEQUENT ITERATIONS

Ultimately, the insights gleaned from the interpretation and quality checks in the previous iteration will improve the overall quality of the LCI model. Iterative loops of scope, inventory, impact assessment, and interpretation and quality control are performed until the accuracy, precision, and completeness of the LCI and LCA study meet the requirements posed by the intended application of the results. Insights gained in an iteration may also lead to a necessary revision of the goal of a study if, for example, data limitations cannot be overcome.

Figure 17.8 lists (in a generic fashion) the steps to be processed in subsequent iterations of the LCA project.

Task	Timeline in Days											
	1	2	3	4	5	6	7	8	9	10	11	12
Subsequent Iterations												
Next Level of LCA Calculations												
Review fgoal & scope	▓											
Improve LCA data		▓										
Improve methods & assumptions, characterization factors			▓									
Re-calculate LCIA results				▓								
Repeat sensitivity, completeness, consistency check					▓							
Consider inteim review						▓						
implement improvements suggested in review							▓					
Document findings of this iteration and all supporting data & assumptions										▓	▓	
Interim Evaluation												
Decide if (revised) study objectives met									▓			
If not, repeat steps of second iteration								▓				

Figure 17.8. Gantt Chart of a Typical Iteration Step in a LCA Project

GOAL AND SCOPE REVISION

Check whether the goals are met and whether the scope (e.g., time and geography) still apply fully. If necessary, refine or revise these goals. A key step is reviewing the initial system boundary, especially when co-functions have been excluded or have later been added within the system boundary via system expansion and substitution or allocation.

IMPROVE KEY LCI DATA

Refine key processes, parameters, and elementary flows in the foreground system with directly collected or calculated product- and producer-specific primary and secondary LCI data. Use more accurate, precise, and complete LCI sets for the background system. It may be necessary to collect study-specific LCI data for key processes in the background system if existing third-party data is not of sufficient quality or consistency.

IMPROVE OTHER LCI DATA

Improve the quality of the LCI data for all life cycle stages, activity types, processes, or elementary flows that the sensitivity analysis reveals to be relevant. Using LCI data of sufficient quality is in accordance with the cut-off criteria established in the scope definition. In the case of comparisons, refine LCI

data in cases where the extent of the differences between the compared systems is critical.

IMPROVE METHOD- AND ASSUMPTION-RELATED DATA

Aspects to be considered include allocation criteria, and the type and amount of avoided processes.

IMPROVE LCIA FACTORS

It may be necessary to develop customized LCIA factors or to accept reduced accuracy in the analysis.

CALCULATE AND EVALUATE THE IMPROVED LCIA RESULTS

Check whether the significant issues have been addressed and perform another round of checks for completeness, sensitivity, and consistency in order to capture the improvements generated by the extra efforts invested in this iteration.

ENSURE THAT ALL CHANGES ARE SUFFICIENTLY DOCUMENTED

Pay particular attention to changes in goal and scope as well as underlying assumptions and methodology. If these revisions are not clearly understood, then the results will be misinterpreted and the extra effort invested in this iteration may do more harm than good. In the worst case scenario, misunderstandings based on poor documentation of revisions may take the study in the wrong direction and subsequent iterations can thus deliver faulty results.

DETERMINE IF ANOTHER ITERATION IS NEEDED

Completing the steps above takes the study to the next level stage gate review, in which the decision must be made whether the study meets its (revised) objectives or whether another iteration is needed. As at the end of the first iteration, a third party could be invited to contribute to the review. Carefully check the goal of the study to assure that you do not over-invest in analysis. Doing so is called a "Type IV error," which occurs when a good analysis is provided to answer a question that was not posed by the commissioner of the study. Type IV errors waste resources, including the good will of the commissioner. See Chapter 15 for more discussion.

FINAL STEPS OF A LCA PROJECT

The steps outlined in Figure 17.9 apply generically to any LCA, regardless of the number of iterations involved:

Task	Timeline in Weeks										
	1	2	3	4	5	6	7	8	9	10	11
Final Steps											
Interpretation of results	▓										
Final reporting, building on interim reports		▓	▓								
Final critical review				▓	▓	▓					
Implement improvements suggested in review								▓			
CELEBRATE ACCOMPLISHMENTS											▓

Figure 17.9 Gantt Chart of the Final Steps in a Typical LCA Project

FINAL INTERPRETATION OF RESULTS

If the LCI data and model have reached the intended or required quality, formal results interpretation is the next step. Full LCA studies also include the steps of conclusions and potential recommendations, while also highlighting any limitations that apply.

REPORTING

Prior to a formal critical review, the study report is prepared. It can be part of a data set and/or be a classical third-party report (per ISO 14044). Each will be based on the interim reports created along the iterations of the LCA work. The principles of reporting are reproducibility and transparency. Confidential and proprietary data and information should be documented in separate confidential reports that are made accessible only to the critical reviewer(s).

FINAL REVIEWS

The final review is the last formal step of an LCA study. If the review process has been integrated from the beginning of the project (i.e. from the goal and scope definition all along the iterations), then the final review can build on the previous work and big surprises are unlikely. Even if the review type and reviewer(s) had not been fixed in the related scope chapter, the project team should carefully consider whether a critical review is required for their type of LCI or LCA study and target audience, or for general quality-assurance reasons.

REVISIONS BASED ON THE REVIEW OUTCOME

A final review may lead to corrections in the LCI model. It might even result in fundamental revision of the goal and scope of the study. Such a review performed at the end of a study can result in considerable delays and extra work. Conducting a review at an earlier stage is recommended for this reason.

MISSION ACCOMPLISHED

The revised final deliverable of the LCI or LCA study, potentially together with the study report and review report, is finally available to be distributed to the target audience and in support of the intended applications.

PROBLEM EXERCISES

1. Audience: How should you engage the target audience in the project set-up of an LCA study?

2. Commissioner: What expectations will the commissioner have of the project manager?

3. Scope: Name three examples how the scope definition figures into project management?

4. RACI chart: Which common conflicts of interest becomes obvious from a RACI chart?

5. Gantt chart: How can a practitioner account for the iterative progress of most LCA projects in the Gantt chart?

6. Project controlling: Why is controlling a major part and a common pitfall of project management?

7. Reporting: How do reporting requirements figure into project management?

COMMUNICATING LCA RESULTS

WILLIAM FLANAGAN AND WESLEY INGWERSEN

> *"The single biggest problem*
> *with communication*
> *is the illusion that it has taken place."*
> -George Bernard Shaw

18.1 AN OVERVIEW OF COMMUNICATION IN LCA

The LCA is complete: all data collection, modeling calculations, sensitivity and uncertainty analyses, and interpretation of results. No LCA study is truly complete, however, until results are communicated to stakeholders. Well-planned and careful communication is particularly crucial in the field of LCA, given the complexity of LCA studies and the ISO 14044 requirement that the "results and conclusions of the LCA must be completely and accurately reported without bias to the intended audience" (ISO 14044:2006, 5.1.1).

Communicating about LCA can be challenging. The most effective communication format and the appropriate level of detail depend on the audience and the goal of communication. Who are the audiences, and what is communicated to each audience? When the results of the LCA are to be communicated to a third party (i.e., any interested party other than the commissioner or practitioner of the study), regardless of the format of the communication, a third-party report must be prepared following the ISO 14044 requirements and guidelines (ISO 14044:2006, 5.2). These requirements will be covered in more detail in Section 18.3.

Start preparing for communication of results at the beginning of the study. If the study is commissioned by, or being conducted for internal stakeholders, begin discussion with clients or peers while still planning for the study. LCA practitioners often communicate with a variety of stakeholders during the course of the study, either to provide background information about the goal and scope, communicate data-collection requirements, or provide interim status updates. Once the initial study is complete, beyond the ISO required minimum communication, communication in and beyond the report needs to be tailored to the particular needs of the client. If the study will be submitted for critical review, one may prepare a separate goal and scope document early in the project to solicit feedback from the critical review panel.

What if the LCA study contains confidential or proprietary data? This concern is common within industry, but mechanisms exist to protect sensitive information in the course of conducting and communicating LCA studies. For example, non-disclosure agreements can be negotiated with critical review panel members, and proprietary information can be excluded from external communications (including third-party reports, as defined by ISO 14044).

While this chapter primarily focuses on writing LCA reports, much of the general guidance in sections 18.1 and 18.2 can also apply to other communication formats such as informal updates, presentations, and content prepared for the web. Public marketing claims and environmental product declarations are further discussed in Chapter 19.

RANGE OF AUDIENCES

LCA results often are communicated to a variety of audiences, such as the study commissioner, contributors to the study, various other internal stakeholders (within your own organization in industry, academia, or within government agencies), customers, consumers, the public, or the press. The communication requirements and the most effective communication formats can, and will, differ depending on the intended audience. The study may involve comparative assertions used to support marketing claims, or the study may be aimed at informing public policy. In all cases, care must be taken to ensure that the study results are communicated accurately and transparently, without bias, and at the appropriate level of detail for the intended audience. LCA studies involving comparative assertions intended to be disclosed to the public must adhere to specific reporting and review requirements, as will be further discussed in Section 18.3 (ISO 14044:2006 5.3 and 6).

For all communication efforts, make sure you know your audience and ensure that you are tailoring the message to their needs by using language that they will understand. Various stakeholder groups may have vastly different perspectives and needs based on their roles. Table 18.1 illustrates basic role differences for various functions (Hodder 2011).

In the following sections, common audiences for LCA studies are identified along with key interests related to LCA. Suggestions are provided for the most appropriate types and formats of LCA results to use for these audiences.

CORPORATE BUSINESS TEAM

How can LCA results inform strategic business decisions? For example, LCA approaches can potentially be used to identify hotspots in a value chain that may (1) represent a risk to market vitality or, alternatively, (2) highlight areas where environmental products or technologies could add business value by mitigating environmental issues along the value chain. In this case, a clear presentation of high-level LCA results outlining issues and opportunities, and expressed using business terminology, might be most effective.

CORPORATE ENVIRONMENTAL TEAM

How should product development strategies be focused to achieve optimal environmental benefits? Which business units should be engaged in future LCA studies? LCA that is communicated at a product portfolio level, with a focus on relative impacts in the supply chain, manufacturing, distribution, use, and end-of-life, can be very effective at prioritizing environmental strategies.

MARKETING AND PUBLIC RELATIONS TEAM

What are the environmental features of the products, and how can this information be used to differentiate the products from the competition? What can be communicated about the environmental attributes of the products to shareholders? In this case, the comparative environmental benefits need to be expressed very clearly, along with guidance on what claims can potentially be made for the product, and with clear expression of any trade-offs, complexities, and any additional reporting and review requirements that need to be met before the claims can be used publicly.

TECHNICAL / ENGINEERING TEAM

Are there innovation opportunities to improve the environmental performance of products? What technical changes in materials or manufacturing processes need to be made to reduce impacts? Technical and engineering staffs tend to be rigorous and detail-oriented, and in some cases may demand to see additional levels of technical detail, including results of sensitivity and uncertainty analyses.

DESIGN TEAM

What design choices make the products more environmentally responsible? This audience benefits from hotspot analyses, scenarios, and alternatives recommendations. LCA results incorporated into design tools can be particularly informative to personnel involved in product design.

PURCHASING TEAM

What is the impact of the supply chain, and where should improvement efforts be focused? How much (and what types) of data are needed in order to improve knowledge of impacts related to purchasing? This audience benefits from understanding trends and hotspots within the supply chain, including impacts associated with transport logistics. They also benefit from understanding data needs and prioritizing supply chain data collection strategies.

POLICY MAKERS

LCA studies of potential policy options (e.g., energy policy) may be very broad and of significant consequence to a large number of stakeholders. Entities involved in policy (government policy makers, policy-influencers, and corporate policy teams) need to have a comprehensive understanding of complexities and trade-offs in order to develop policy initiatives that drive the right behavior that results in optimized benefits for society. Communication should include consideration of assumptions and uncertainties inherent in the approach so that policy makers can understand how LCA results may be best utilized.

CUSTOMERS

Business-to-business customers tend to have more time and

Corporate Business Team	Corporate Environmental Team	Marketing	Technical/Engineering Team	Design Team
How can LCA results inform strategic business decisions?	Where can product strategies be focused for optimal environmental benefit?	How can we products be differentiated based on environmental attributes?	Are there innovation opportunities in materials and manufacturing processes?	What design choices make products more environmentally responsible?
Purchasing Team	**Policy Makers**	**Customers (B2B)**	**Consumers**	**Press**
How can supply chain impacts be reduced?	How can policies drive the right behavior to benefit society?	What information is needed to inform purchasing decisions?	Is this product good for the environment (yes/no)?	What's the headline? Is the information fresh, exciting, relevant, and newsworthy?

Figure 18.1. Role-Based Needs for a Range of Audiences (adapted from Hodder 2011)

competence to navigate product-related environmental information, and may or may not be interested in more detailed technical results. LCA results and reports may be relevant to their goals, which may provide an opportunity for direct and ongoing engagement (Hodder 2011).

CONSUMERS

Consumers tend to have little time and technical competence to judge product environmental information, but studies have demonstrated that providing LCA information can increase consumer confidence in green claims (Molina-Murillo and Smith 2009). Consumers are exposed to the information for shorter periods of time, and therefore one should provide simplified information, possibly including a graphical element, to assist them with an "at-a-glance" understanding of the presented information (Hodder 2011). LCA-based environmental product labeling must adhere to ISO standards (ISO 14025:2006), as well as any advertising regulations such as the US FTC Green Guides (US Federal Trade Commission 2012). Consumer product labeling campaigns can also contribute to (or detract from) a company's green brand image. Therefore, the messaging strategy should consider this perspective as well.

MEDIA

Media reporters, publications, and agencies are always looking for new stories related to the environment. One can encourage the press to write articles by issuing a press release. In other cases, members of the press may opportunistically write an article based on a conference presentation, a published journal article, or some other public communication format you may have used. The press will typically be interested in stories that are fresh, exciting, relevant, newsworthy, or perhaps controversial. They will mainly be interested in generating a catchy headline and a concise sound bite, with more or less supporting detail depending on the length of the article. A good strategy here is to prepare succinct, sound bite-sized summaries to increase the odds that a reporter will use the right language to describe the LCA study. If one is in direct contact with a reporter preparing the study, requesting the chance to review the article before going to press is valuable and can ensure that findings are appropriately represented.

18.2 COMMUNICATION BEST PRACTICES IN LCA

We explore reporting requirements in Section 18.3, but first let us discuss some communication fundamentals that may be useful in various forms of LCA communication. The

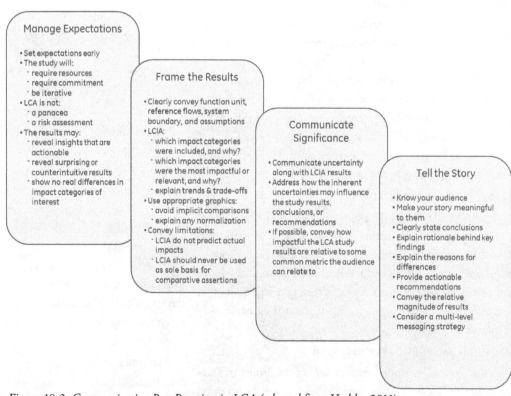

Figure 18.2. Communication Best Practices in LCA (adapted from Hodder 2011)

general outline of this section consists of four main aspects: (1) manage expectations; (2) frame the results; (3) communicate significance; and (4) tell the story (Hodder 2011), as summarized in Figure 18.1.

MANAGE EXPECTATIONS

One of the most important aspects of ensuring good communication is to set expectations from the beginning. Being transparent with participants (particularly the LCA commissioner) about what they should expect during and after the study is good practice when conducting an LCA. Performing the LCA will require resources and commitment, and will be iterative. The results may be useful, supporting product attributes that can be used to differentiate one's product in the marketplace or revealing insights that can be translated into actionable improvements in the product's life cycle. The results may also reveal counterintuitive or surprising trade-offs, or may show no real difference in impact categories of interest. You may wish to communicate to your stakeholders that in many cases the insights gained during an LCA study can be of equal or sometimes greater value than the quantitative results (Hunter 2010).

LCA is not a panacea, nor should the results of an LCA study be communicated as such. One should be careful to express the limitations of LCA to key stakeholders. For example, while LCA is capable of providing detailed insight, it may not be able to tell you which option is greener, better, or more sustainable, and there may be additional requirements that must be met per ISO 14044 before LCA results can be used externally. An LCA study is not a risk assessment, and LCA is not the best tool for assessing chemical hazards, substances of concern, or human health issues. Such expectation setting lays the groundwork for post-study communication efforts that are complete, accurate, and without bias to the intended audience as required by ISO 14044.

FRAME THE RESULTS

After communications have been delivered to their respective audiences, one may not have many opportunities to provide additional context or clarification; thus, one should carefully frame LCA results to ensure proper interpretation.

Be sure to clearly convey the functional unit, reference flows, and system boundaries so that audiences will understand the context and limitations of any comparisons that are being made. For example, if the results of an LCA comparing electric to fossil fuel vehicles are based on a US average electricity grid, make sure to clearly state this and also that these results only apply to the scenario that was studied. The

LCA report author's careful attention to rigor in this regard may help mitigate the proliferation of news stories that make over-simplified and generalized statements such as "A exhibits lower global warming potential than B," when this comparative statement might be valid only for the specific scenario that was studied.

The life cycle impact assessment methodology used in the study should be clearly defined. Which impact categories were included, and why? Within the included set of impact categories, which emerged as the most relevant or impactful? Do certain impact categories trend together? Were there trade-offs between different impact categories, and if so, how should one interpret the trade-offs?

Graphs are often used to convey LCA results. Make sure the appropriate graph types are selected and that they accurately convey the appropriate information. Is the graphic clearly and sufficiently labeled to ensure reader comprehension? Does the graphic representation lead to implicit comparisons that are not intended, such as comparisons between different impact categories? Are you including results for impact categories that show large comparative differences, but are inconsequential relative to other impact categories? If so, is this sufficiently clear to the reader? If anything is inconclusive, explain why. Make sure any use of normalization is thoroughly explained.

Limitations of the study should be clearly expressed. Make sure the reader understands that LCIA results are "relative expressions [that] do not predict impacts on category endpoints, the exceeding of thresholds, safety margins or risks" (ISO 14044 5.2.e.8). Be transparent about other key limitations, such as the following:

- LCA is an evolving science that is reliant on data quality and availability
- The study results only relate to the scenario in question as defined by the functional unit definition, system boundary and assumptions associated with the study
- Value judgments are needed to interpret results and negotiate trade-offs
- "LCIA should never be used as the sole basis for any comparative assertion intended to be disclosed to the public, as additional information is necessary to overcome the inherent limitations of LCIA" (ISO 14044:2006; 4.4.5)

COMMUNICATE SIGNIFICANCE

Some level of data and methodological uncertainty is inherent in all LCA studies. Chapter 13: Interpretation and Chapter 15: Bias and Uncertainty discussed how to apply uncertainty and sensitivity analyses to understand the influence of such factors on the LCIA results. Even when such aspects are included in the study, uncertainty and sensitivity are too often missing from LCA presentations and other communication formats. The ethical and responsible LCA practitioner will strive to navigate uncertainty and communicate it along with the LCIA results, with an explanation of how such uncertainties may influence the study results, conclusions, and recommendations.

Communication of your LCA results may also benefit from conveying the relative magnitude of the impacts of your product, process, or service (or the potential for improvement discussed in your recommendations), as compared to some other benchmark that your audience can relate to (Nissinen et al. 2007). For example, if the study evaluates renewable energy technologies, what is the potential reduction in global warming potential relative to forecasted national or global greenhouse gas emissions? Equivalencies are also often used to convey relative magnitude. For example, you might state that the potential global warming potential of Technology A versus Technology B is comparable to X Average Passenger Cars Driven or Y Average Residential Homes Heated per Year (although one must be sure to have an accurate and reputable basis for such comparisons, and the source of the equivalency should be referenced). Use caution to ensure that your comparisons are accurate and supportable, and that the use of such equivalencies is not in itself misleading. When using equivalencies, the goal should be to make the study results relatable to your audience.

TELL THE STORY

The audience only knows what the audience sees and hears. One must effectively convey the LCA results in an

appropriate manner for the audience and a format consistent with the ISO 14044 requirements. The LCA can be made into more of a story by crafting a narrative exposition of the results, including real-life details about the product supply chain, and using the aid of pictures and other graphical elements.

Consider a multi-level messaging strategy to accommodate audiences with varying technical knowledge of LCA, or varying interest in, and ability to absorb, the information. A three-level messaging strategy might involve:

1. Present high-level qualitative conclusions in non-technical terms
2. Present quantitation of key results, along with interpretive detail
3. Present detailed results with quantification and expository detail

This multi-level messaging approach, if handled properly, should allow any audience to understand the key findings of the study, while also providing the richer layer of detail that a technical audience will need to properly interpret (and assess the credibility of) the results.

ILLUSTRATE COMPLEXITIES SELECTIVELY AND SIMPLY

Important contributions to impacts are often found in parts of a product system well beyond the influence of the manufacturers, or not likely to be associated by the audience with the final product. For example, toxicity impacts in the life cycle of a toy may come from the mining of metal used upstream in the production of raw materials for an internal component of the toy. In the same vein, potential emissions unfamiliar to the clients (e.g., a rare volatile organic compound emitted to air) may be the most significant contributors to some impacts. These results can and should be communicated, but an attempt should be made to illustrate the relationship between the product and source of impact through simplified, but not misleading, language or imagery.

Several other suggestions for telling the story include the following:
- Make it comprehensible to the audience (put it in their terms using their language)
- Convey the relative magnitude of the results (perhaps using equivalencies)
- Explain the rationale behind key findings
- Explain the reasons for differences
- Make the right information stick (and clearly state conclusions)
- Make the results clear and easy to interpret, but also help them appreciate the depth of the analysis by showing them the level of detail and rigor that were used (Hensler 2010)
- Provide actionable recommendations

The concepts in this section are further illustrated in the

sidebar case study titled "A Life Cycle Assessment Comparison of Warp Drive Technologies for 24th Century Multi-Use Applications."

18.3 REPORT REQUIREMENTS PER ISO 14044

The ISO standards for LCA include a section on reporting that specifies requirements that apply to all LCA studies, plus additional detailed requirements for third-party reports and for comparative assertions intended to be disclosed to the public (ISO 14044:2006, 5). LCA practitioners should carefully review these requirements during the scoping phase of the study to ensure that all required report content will be generated and documented during the study. Furthermore, the type and format of the report needs to be defined during the scope phase of the study (ISO 14044:2006, 5.1.1).

plexities and trade-offs inherent in the LCA. The report shall also allow the results and interpretation to be used in a manner consistent with the goals of the study" (ISO 14044:2006, 5.1.1). These requirements apply to all LCA studies regardless of whether the results will be communicated to a third party other than the study commissioner or the study practitioners.

THIRD-PARTY REPORTS

In cases where the "results of the LCA are to be communicated to a third party (i.e., any interested party other than the commissioner or practitioner of the study), regardless of the format of communication, a third-party report needs to be prepared following the ISO 14044 requirements and guidelines" (ISO 14044:2006, 5.2). The third-party report is considered a reference document that shall be made available to any third party to whom the communication is made.

ISO 14044 Requirements and guidance: All LCA reports	ISO 14044 Additional requirements: Third-party reports	ISO 14044 Additional requirements: Comparative assertions
Requirements • The type and format of the report shall be defined during the scope phase of the study • Results and conclusions shall be completely and accurately reported without bias to the intended audience • Results, data, methods, assumptions, and limitations shall be transparent and presented in sufficient detail to allow the reader to comprehend the complexities and trade-offs inherent in the LCA • Report shall allow the results and interpretation to be used in a manner consistent with the goals of the study **Guidance** • Recommended content: a) modifications to the initial scope b) system boundary, including c) description of the unit processes, d) data, including e) choice of impact categories and category indicators *(refer to ISO 14044:2006 for itemized content requirements)*	**Requirements** • When results of the LCA are to be communicated to any third party (i.e., an interested party other than the commissioner or practitioner of the study), regardless of the form of communication, a third-party report shall be prepared • The third-party report is considered a reference document that shall be made available to any third party to whom the communication is made (can include reference to confidential study documentation that is not included in the third-party report) • Required aspects: **a) General** LCA commissioner, practitioner of LCA (internal or external), date of report, and an statement that the study has been conducted according to the requirements of ISO 14044 **b) Goal** Reasons for carrying out the study, its intended applications and target audiences, and a statement as to whether the study intends to support comparative assertions intended to be disclosed to the public **c) Scope** Function, functional unit, system boundary, and cut-off criteria **d) Life cycle inventory analysis** Data collection procedures, description of unit processes, sources of published literature, calculation procedures, data validation, sensitivity analysis, and allocation methods **e) Life cycle impact assessment** 1) LCIA procedures, calculations, results, assumptions, limitations, impact categories and indicators, value choices, normalization, grouping, and weighting 2) A statement that the LCIA results are relative expressions and do not predict impacts on category endpoints, the exceeding of thresholds, safety margins or risks and, when included as part of the LCA, further details on any new impact categories or characterization methods, use of grouping, or transformation methods. **f) Life cycle interpretation** Results, assumptions, limitations, data quality assessment, and transparency with respect to value choices, rationales, and expert judgments **g) Critical review, where applicable** Name and affiliation of reviewers, critical review reports, and responses to recommendations *(refer to ISO 14044:2006 for itemized content requirements)*	**Requirements** • Further reporting requirements for LCA studies in which comparative assertions are intended to be disclosed to the public: a) analysis of material and energy flows to justify their inclusion or exclusion b) assessment of the precision, completeness, and representativeness of data used c) description of the equivalence of the systems being compared d) description of the critical review process e) an evaluation of the completeness of the LCA f) a statement as to whether or not international acceptance exists for the selected category indicators and a justification for their use g) an explanation for the scientific and technical validity, and environmental relevance, of the category indicators used in the study h) the results of the uncertainty and sensitivity analyses i) evaluation of the significance of the differences found • Additional requirements if grouping is used *(refer to ISO 14044:2006 for itemized content requirements)*

Table 18.1 LCA Report Requirements and Guidance per ISO 14044

GENERAL REQUIREMENTS

For all LCA studies, "the results and conclusions of the LCA shall be completely and accurately reported without bias to the intended audience. The results, data, methods, assumptions and limitations shall be transparent and presented in sufficient detail to allow the reader to comprehend the com-

Notice that the third-party report requirement applies regardless of the format of communication. For example, if the LCA results are primarily intended to be communicated via an oral presentation in a forum involving third parties, a strict interpretation of the ISO 14044 requirements and guidelines suggests that a third-party report still needs to be generated as a reference document to ensure that all of the required information about the study is available upon request. If necessary, the third-party report can omit proprietary or confidential data, but should reference any confidential study documentation excluded from the report. This can be particularly important for industry LCA studies.

Figure 18.3 shows a generic LCA report outline that represents one way to organize the required report content. This representative outline is based on several detailed LCA study reports and is provided as a starting point. Of course, the required report elements can be organized in other ways. You may need to adjust the report structure to accommodate specific aspects of your particular LCA study. If you intend to publish the results of your LCA study in a scientific journal, you may want to organize the required report elements in a manner consistent with the format specified by the journal (e.g., introduction, materials and methods, results and discussion, and conclusion). While the third-party LCA report will likely contain much more information than will fit into the length requirements of a typical journal article, the goal and scope, methods, and primary results and discussion can be included in the body of the journal article; also, more technical information from the report, such as tables of data and additional analysis steps taken to aid interpretation, can be included in supplemental material.

The report outline shown in Figure 18.3 will be used to discuss the various required and recommended report elements. The reader should refer to the ISO 14044 requirements and guidelines to ensure that all requirements are fully understood (ISO 14044:2006). These requirements are also summarized at a high level in Table 18.2.

EXECUTIVE SUMMARY

The third-party LCA report should almost always include an executive summary because few readers have time to thoroughly digest the full report. As such, the executive summary is a particularly important part of the report because it must effectively and concisely convey the study results "completely and accurately without bias to the intended audience" (ISO 14044:2006, 5.1.1). If any of the results are misrepresented in the executive summary, the report will be misleading to the majority of readers, regardless of the accuracy of the full report.

INTRODUCTION

As with any report, the introduction must set the stage. What is the context and background of the study? In this section, you may also want to provide a brief overview of LCA, along with references, to ensure that the reader is familiar with

Figure 18.3: Example LCA Report Table of Contents

Executive Summary
1. Introduction
   1.1. Study background and context
   1.2. Introduction to LCA
2. Goal and Scope
   2.1. Goal
   2.2. Function, functional unit, *and reference flow*
   2.3. System description
       2.3.1. General system description
       2.3.2. Description of unit processes (qualitative)
       2.3.3. System boundaries
   2.4. Cut-off Criteria
3. Methodology
   3.1. Assumptions
   3.2. Limitations
       3.2.1. Limitations of LCA methodology
       3.2.2. Limitations of the study
   3.3. Life Cycle Inventory
       3.3.1. Data collection procedures
       3.3.2. Description of unit processes (quantitative)
       3.3.3. Data sources
       3.3.4. Data quality requirements
       3.3.5. Treatment of missing data
       3.3.6. Calculation procedures
       3.3.7. Allocation
       3.3.8. Geographic and temporal relevance
   3.4. Life Cycle Impact Assessment (LCIA)
       3.4.1. Life Cycle Impact Assessment methods
       3.4.2. Impact categories and category indicators
       3.4.3. Normalization (if used)
       3.4.4. Grouping (if used)
       3.4.5. Weighting (if used)
   3.5. Calculation Tool
   3.6. Critical Review (if performed)
4. Life Cycle Inventory (LCI) Results
5. Life Cycle Impact Assessment (LCIA) Results
   5.1. Comparative analysis per functional unit
   5.2. Contribution analysis
6. Interpretation
   6.1. Key findings
   6.2. Evaluation
       6.2.1. Completeness Check
       6.2.2. Consistency Check
       6.2.3. Data Quality Analysis
       6.2.4. Sensitivity Analysis
       6.2.5. Uncertainty Analysis
7. Conclusions and recommendations
8. References
Appendices
       LCI data tables
       LCIA characterization results tables
       Critical Review
           Names and affiliations of reviewers
           Critical review reports
           Responses to critical review recommendations
           Final statement from Critical Review panel chair

Figure 18.3: Example LCA Report Table of Contents

the concept of LCA and has access to additional information if they would like to learn more.

The introduction may also be a good place to include general required elements, such as the date of the report, the LCA commissioner, the LCA practitioner(s), and a statement that the study was conducted according to the requirements of the ISO LCA standards 14040-44.

GOAL AND SCOPE

The goal statement should clearly indicate the reasons for carrying out the study, along with its intended applications and target audiences. If the study contains comparative assertions that we intend to be disclosed to the public, this must also be clearly stated.

The function description should mention any additional functions that will not be included when performing comparisons. The functional unit definition should be related to a tangible performance measurement, and should be consistent with the goal and scope of the study. Although the reference flow is not included in the list of report elements required by ISO 14044, it should be defined as necessary to ensure clarity when subsequently describing the study results.

A system description should be provided, including a general description as well as a detailed description of unit processes (including any allocation related to unit processes). The system boundary should be clearly explained and should mention any omissions of life cycle stages, processes, or data needs, along with associated decision criteria for each omission. Inputs and outputs of the system should consist of (and therefore be described as) elementary flows.

Cut-off criteria for initial inclusion of inputs and outputs should be described, including any assumptions made and the effect of selection criteria on the study results. The ISO requirements and guidelines also stipulate that the study scope should include discussion related to quantification of energy and material inputs and outputs, and any assumptions related to electricity production. If any aspects of the initial scope were modified during the course of the study, the modifications should be described, along with justification for the changes.

For LCA studies that will be submitted for critical review, practitioners often prepare a separate document with the goal and scope to be reviewed by the critical review panel early in the project. This ensures that any panel feedback related to function, functional unit, reference flow, system boundary, and cut-off criteria can be incorporated into the study while the work is in progress. The initial document can subsequently be updated (as necessary) and incorporated into the third-party report.

METHODOLOGY

In many types of technical reports, discussion of methodological aspects often occurs before discussion of the study results. In the representative outline shown in Figure 18.2, many of the methodological aspects have been grouped into a separate section of the report. Depending on your study, the required report elements related to life cycle inventory, life cycle impact assessment, and life cycle interpretation may be incorporated into the methodology section, or they may be incorporated into the respective results sections of the report (or a combination of both, as appropriate).

ISO requirements specify that "results, data, methods, assumptions and limitations should be transparent and presented in sufficient detail to allow the reader to comprehend the complexities and trade-offs inherent in the LCA." Therefore, gather the assumptions and limitations of the study into a designated section of the report. In our representative example, the assumptions and limitations have been included as part of the methodology section, but they could just as easily reside in the Goal and Scope section, or in a separate section altogether. Of course, it may also make sense to further elaborate on specific assumptions and limitations elsewhere in the report, as needed.

Required report elements related to life cycle inventory include:

- data collection procedures
- qualitative and quantitative description of unit processes (in our example outline, we included qualitative unit process descriptions in the Goal and Scope section)
- sources of published literature used as inventory data
- calculation procedures (in our example outline, we include this element in a separate methodology section titled Calculation Tool)
- allocation principles and procedures (including documentation, justification, and uniform application of allocation procedures)
- data validation (including data quality assessment and treatment of missing data)
- sensitivity analysis for refining the system boundary (if appropriate)

Although geographic and temporal relevance of the data is not specifically listed as a reporting requirement by ISO 14044, these elements should also be included in the report. In our example outline, we have included most of these LCI-related elements within the methodology section, although depending on your study some of the elements might best be described in the Life Cycle Inventory Results section of the report.

Methodological aspects related to life cycle impact assessment (LCIA) should be fully described, including the life cycle assessment methods used, along with a description of individual impact categories and category indicators (and a rationale for their inclusion or exclusion). All characterization models, characterization factors, and methods used (including assumptions and limitations) should be described and/or referenced.

If any further procedure is used that transforms the indicator results (grouping, normalization, or weighting), the approaches should be described and justified. Somewhere in the report, there should be a statement indicating that LCIA cannot be the sole basis for any claims intended to be disclosed to the public, and that additional information is necessary to overcome the limitations of LCIA. If the study is subject to critical review, including a description of the critical review process is common practice.

LIFE CYCLE INVENTORY RESULTS

Life cycle inventory studies are life cycle studies that do not include LCIA. All of the LCI study results are described in the LCI Results and Interpretation sections of the report, and there is no LCIA Results section. In other cases, the study conclusions may be based on a combination of LCI and LCIA results.

LIFE CYCLE IMPACT ASSESSMENT RESULTS

The results of the life cycle impact assessment should be fully described and include statements about any limitations of the LCIA results relative to the goal and scope of the study, the relationship of the LCIA results to the goal and scope, and the relationship of the LCIA results to the LCI results. This section should also describe any influence of the life cycle impact assessment method on the results, conclusions, and recommendations of the study.

If normalization, grouping, or weighting are used in the study, the indicator or characterization results reached prior to any transformation shall be made available along with the normalized, grouped, or weighted results. Weighting shall not be used for comparative assertions that we intend to be disclosed to the public (ISO 14044:2006, 4.4.5).

Graphical representation of the LCI and LCIA results may be useful, although care should be taken because this may invite implicit comparisons and conclusions. The LCA practitioner should always strive to carefully describe and qualify any graphical representation of results to ensure accurate interpretation by the reader.

Figures such as bar charts, pie charts, and process tree diagrams are commonly used to convey LCA results, but are often used incorrectly by practitioners. Using the example of an LCA comparing two household cleaning products, Figure 18.4 demonstrates two incorrect uses and an acceptable use. Assume that we would like to present the results from four impact categories together in a single figure to visually compare the two products. Figure 18.4 (a) presents a common mistake – presentation of raw impact characterization scores for the four categories on a common axis. Raw characterization scores use the units of the impact category indicators. In the example here, TRACI 2.1 is being used, so the units are kg N equivalent for eutrophication, kg CO2 equivalent for global warming potential, kg SO2 equivalent for acidification, and kg PM2.5

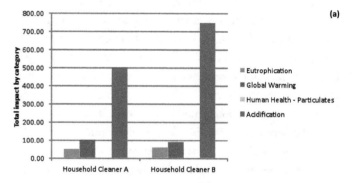

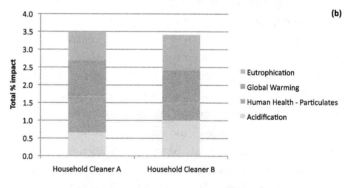

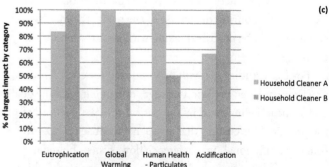

Figure 18.4. Incorrect (a), (b), and Acceptable (c) Example Representations of LCA Results

equivalent for human health effects of airborne particulates.

Figure 18.4 (a) instead falsely conveys that the units are the same and that acidification is by far the most significant contributor, whereas human health particulates are insignificant. In order to present results from multiple impact categories on the same axis, these results either need to be scaled, normalized using published normalization factors (see Chapter 12: Decision Support Calculations and Chapter 13: Interpretation), or weighted and aggregated into areas of protection or a single score. Scaling results relative to the largest impact score in the category among the products being compared is one option. Scaled results are presented in Figure 18.4 (b) and (c). But scaled results should not be combined as in Figure 18.4 (b), as this assumes an equal weighting among all impact categories. Figure 18.4 (c) presents scaled results, but keeps the impact categories separated. Figure 18.4 (c) represents an acceptable way of presenting the results from four different impact categories in a single figure, although some might argue that this form of presentation still leads the reader to assume equal weighting (Laurin et al. 2013). To avoid this type of explanation, a clearly written explanation of figures is essential.

INTERPRETATION

The interpretation of the study describes the meaning of the results of the study, along with any key assumptions and limitations that could potentially affect proper interpretation of results. The report author should fully describe the evaluation of the results, including checks for completeness and consistency, the data quality assessment, and the sensitivity and uncertainty analyses. The interpretation section of the report is also where the LCA practitioner must be transparent with respect to value choices, rationales, and expert judgment used to interpret the study results.

CONCLUSIONS AND RECOMMENDATIONS

Although conclusions and recommendations are part of the interpretation phase of the study, these are often reported in a separate section in keeping with traditional technical report formats. As with the executive summary, description of the study conclusions and recommendations must be accurate and transparent.

COMPARATIVE ASSERTIONS

Specific reporting requirements apply when an LCA study involves comparative assertions that we intend to be disclosed to the public. A comparative assertion is defined as an "environmental claim about the superiority or equivalence of one product versus a competing product that performs the same function" (ISO 14044: 2006, 3.7). The requirements that apply to the critical review as conformity assessment for ISO 14044 are discussed separately in Chapter 2. The review statement, comments of the practitioner, and any response to recommendations made by the reviewer shall be included in the third-party LCA report.

Specific additional elements to be included in the report are as follows (parentheses indicate suggested locations for each element within the representative report structure shown in Figure 18.2):
* an analysis of material and energy flows to justify their inclusion or exclusion (3.3.5; 6.2.4)
* an assessment of the precision, completeness, and representativeness of the data used (6.2)
* a description of the equivalence of the systems being compared (2.1)
* a description of the critical review process (3.6)
* an evaluation of the completeness of the LCIA (6.2.1)
* a statement as to whether international acceptance exists for the selected category indicators and a justification for their use (3.4.1; 3.4.2)
* an explanation for the scientific and technical validity, and environmental relevance, of the category indicators used in the study (3.4.2)
* the results of the sensitivity and uncertainty analyses (6.2.4; 6.2.5)
* an evaluation of the significance of the differences found (6.2; also please see Chapter 14: Statistics)

If grouping is included in the LCA, the following should be included:
* the procedures and results used for grouping
* a statement that conclusions and recommendations derived from grouping are based on value choices
* a justification of the criteria used for normalization and grouping (these can be personal, organizational, or national value choices)
* a statement that ISO 14044 does not specify any specific methodology or support the underlying value choices used to group the impact categories
* a statement that the value choices and judgments within the grouping procedures are the sole responsibilities of the commissioner of the study (e.g., government, community, or organization)

CONCLUSION

LCA is a complex methodology that can provide a rich set of information about the environmental performance of products, but the results of an LCA study are not always straightforward to communicate. While the ISO standards provide minimum requirements for third-party reports and for communication of LCA for product claims such as environmental product declarations, many other forms can be useful for communicating LCA results to a variety of audiences. As important as it is to show the final results of an LCA to users clearly and without bias, understanding the context and limitations of the study are equally important. For internal or commissioned studies, the communication process should begin early to engage clients in the LCA and the final report. When communicating about LCA, the message and level of detail need to be tailored to the audience, so that LCA can be comprehensible and not misinterpreted. Effective communication may be best achieved in a four-step approach of (1) managing expectations, (2) framing the results, (3) communicating significance, and (4) telling the story.

Problem examples

Q1: Following your successful LCA of warp-drive technologies, you are asked to help a colleague prepare some graphs for an LCA of deep-space mining hardware. Using Monte Carlo methods, your colleague has generated the following table of results for the new impact assessment category "Tritium depletion." Your assignment is to use Excel (or other comparable data-analysis software) to prepare a comparative histogram similar to those shown in Table 18.4. (page 243)

A1: Here are instructions for how to generate the required histogram plots in Excel (Excel solution file provided):

1. For the complete set of data (two columns representing results from both Mining applications), determine the maximum and minimum values. Subtract the minimum from the maximum to determine the range.

2. Designate the number of bins (i.e., how many total bars would be in the histogram bar chart); the value is chosen based on visual preference; the solution spreadsheet uses 50.

3. Determine the bin increment by dividing the range (Step 1) by the number of bins (Step 2).

4. Create a column called Bins, starting at 1 and incrementing to the number of bins identified in Step 2 (actually, the number of bins + 1). In the solution spreadsheet, this is a column of incremental values from 1 to 51.

5. Create another column that calculates bin values corresponding to each bin. The first cell in the column contains the minimum value of the data set. The second cell in the column increments the minimum value by one bin increment. Continue incrementing by bin increment until you have reached the maximum value in the data set, corresponding to the largest bin number + 1.

6. Copy the table generated in Step 5 and paste as values into a new area in the spreadsheet.

7. On the Data tab, select Data Analysis (this may need to be enabled as an add-in).

8. Select Histogram and click OK.

9. Select the input range corresponding to the first column of Monte Carlo results (A: Mining App 1).

10. Select the bin range as the second column generated in Step 6.

11. Click the box to enable Chart Output and then select OK. A new sheet will be created containing results from A: Mining App 1, along with a histogram. Label the sheet so you remember which data this histogram corresponds to.

Run	A: Mining App 1 Tritium depletion [MJeq]	B: Mining App 2 Tritium depletion [MJeq]
1	3.033	0.886
2	2.664	0.798
3	3.286	1.121
4	2.795	0.655
5	3.434	0.930
6	3.153	0.838
7	3.935	1.245
8	3.056	0.768
9	2.299	0.549
10	2.692	0.762
11	2.707	0.655
12	2.441	0.521
13	3.012	0.922
14	3.607	1.165
15	2.559	0.694
16	2.663	0.715
17	2.901	0.854
18	2.521	0.737
19	2.789	0.743
20	2.490	0.673
21	3.049	0.926
22	2.466	0.672
23	2.511	0.634
24	2.614	0.802
25	3.150	0.744

Table 18.4

graph with two series corresponding to the second and third columns of data (frequency for each mining app). For both series, the horizontal axis labels should refer to the first column (bin).

Q2: You are tasked with performing an LCA study of a new energy-generating technology that offers the potential for large-scale implementation without reliance on fossil fuel sources. Following the guidance in this chapter, develop a communication plan listing the sequence, timing, and communication formats throughout the LCA project for each of the following audiences: (1) corporate business team; (2) corporate environmental team; (3) marketing and public relations team; (4) customers (B2B); and (5) policy makers.

12. Repeat steps 7-11 for B: Mining App 2. This will create another new sheet containing data and a histogram.

13. Next you will combine the two histogram plots into one. Copy and paste the histogram data for B: Mining App 2 immediately below the data for A: Mining App 1, but with the frequency data shifted one column to the right.

14. Select the cell range created in Step 13 and sort from lowest to highest based on the first (bin) column

15. Select the cell range created in Step 14 and make a bar

LCA-BASED PRODUCT CLAIMS

WESLEY W. INGWERSEN

19.1 INTRODUCTION

Product environmental claims include a wide range of marketing claims, labels, declarations, statements, and reports that are generally intended to distinguish a product as environmentally friendly or "green." They differ from organizational environmental claims in that they refer particularly to an environmental aspect of a specific product and not the organization that manufactures, distributes, or retails a product. Product environmental claims may be considered a subset of product sustainability claims, along with other claims focusing on social aspects of a product and its supply chain (e.g., working conditions and welfare of producers). Ecolabels are the most common form of product environment claims that are directed toward end-consumers and institutional buyers. Ecolabels may provide evidence of validation and certification of a product against a set of criteria, or indicate that a product has an environmental attribute, such as a certain percentage of recycled content. In 2010, over 70 organizations self-reported as managing third-party certified eco-labels (Big Room and WRI 2010). Other unverified claims may appear on or off product packaging. In the North American market, such unverified and non-standardized claims often use such terms as "eco-friendly," "sustainable," or "green," thereby confusing consumers and institutional purchasers.

While the general public sees potential benefits of environmental claims for those interested in greener consumption, institutional buyers also benefit from the claims by using them to make purchases that meet green procurement mandates or collectively reduce the organization's environmental footprint.

The growing demand for greener products (and, consequently, green claims) not only provides opportunities for financial growth, such as improving market share or creating new markets, but also has the potential to drive industry to innovate towards reducing the environmental impact of their practices. If environmental claims work to create significant shifts toward cleaner production in the market, the general public may benefit from a cleaner and healthier environment.

The variety and number of product environmental claims grow in usage and importance in the global marketplace. They continue to penetrate into new regions and market sectors. Increasing use of environmental claims has been met with deserved skepticism about their meaning and validity. Greenwashing is the practice of making unjustified environmental claims for marketing purposes, thereby misleading prospective purchasers (TerraChoice 2007). Greenwashing erodes consumer trust and confidence in green products, and investor confidence in manufacturers that produce environmentally friendly products.

Regulatory bodies such as the US Federal Trade Commission (FTC) and US Environmental Protection Agency (EPA) have acted to curb greenwashing. The FTC maintains guidelines for the communication of product environmental information, which are commonly referred to as the Green Guides (FTC 2010). The US EPA has developed a basic set of tools and materials to help in identification of greener products, and organizations like TerraChoice and the Green Product Roundtable have worked to educate consumers and companies to better understand and interpret environmental claims (The Keystone Center, 2012).

The application, accuracy, and information conveyed in product claims vary depending on the type of claim and the specific program. The ISO 14000 series of standards provide requirements for businesses to identify and control their environmental performance. They also provide requirements for environmental claims in the 14020 series. The three types of claims described in 14020 series include Type I eco-labels based on third-party standards and verification, Type II self-declarations, and Type III environmental product declarations based on LCA. Table 19.1 presents a comparison of the three label types. Type III claims are neutral, quantitative, full life cycle, and they permit differentiation from other products in the category. Type III declarations require verification, but do not mandate specific practices or prohibit specific chemicals and can be used for any category of products.

While these international standards establish a taxonomy for product environmental claims, many claims do not fall clearly under any of the aforementioned types of claims. For example, Type I-like ecolabels have the verification and certification processes associated with a Type I label, but they focus on a single issue (United Nations 2009). But the ISO types are useful for making some distinctions. LCA can contribute to all three types of environmental claims, but in particular it is required to make Type III and Type III-like declarations.

19.2 EPDs, Carbon Footprints, and Other Types of LCA-based Claims

LCA-based product claims use LCA to quantify and report one or more environmental impacts for a product, and may also provide additional information beyond typical LCA results. Claims that focus on more than one environmental impact are generally referred to as multi-attribute claims. Claims that use LCA to quantify just one environmental impact are referred to as single-attribute claims.

Single-attribute claims

Product carbon footprints (PCFs) are the most common

	Type I (eco-labels)	Type II (self-declarations)	Type III (LCA-based declarations)
Neutral	No	No	Yes
Valid for any product in a category	No	Yes	Yes
Appears with product on shelf	Yes	Yes	No (a symbol or link to the full declaration may be present)
May require specific practices	Yes	No	No
May prohibit use of certain inputs	Yes	No	No
Considers the entire life cycle	Maybe (recommended)	No	Yes
Requires independent verification	Yes	No	Yes
Ease of understanding by consumers	Yes	Maybe (depends on topic)	No (depends on presentation)
Permits differentiation from other products with the same label	No	No	Yes
Quantitative	No	No	Yes

Table 19.1. Comparison of Three ISO Product Claim Types

form of single attribute LCA-based claims. PCF labels and reports provide a single value for the carbon footprint, or the product's global warming potential, in CO_2-equivalents. This single value is obtained by summing up the greenhouse gases emitted over the full or a partial life cycle of the product using a common unit of measurement. Several standards have been developed, or are in development, for estimating the carbon footprint of products. These standards include PAS 2050 (BSI 2011), the GHG Protocol Product Life Cycle Accounting and Reporting Standard (GHG Protocol 2011), the Japanese carbon footprint of products (JISC 2009), and the ISO TR 14067.

The drivers for the strong recent interest in PCFs are as follows: (1) the increasing perception of climate change as the key environmental problem of the 21st century and associated market demand for information on the causes of this problem; (2) the ease of comprehending a single value which represents the results of a LCA; and (3) and the general consensus on the methodology for estimating global warming potential based on the work of the International Panel on Climate Change (IPCC).

Single-attribute labels may also cover other topics, such as water depletion, in the form of a product water footprint. A standard for product water footprint is currently under development by the ISO (ISO/CD 14046 2012). While promising easier general acceptance, methodological issues related to the life cycle inventory as well as impact assessment are still unresolved with these single-attribute approaches, such as how biogenic carbon should be accounted for in carbon footprints or how water footprint can be improved by taking into account water scarcity (Finkbeiner 2009). Furthermore, the limitation to a single impact category comes at the expense of one of the primary strengths of the LCA approach: preventing burden shifting to other impact categories by quantifying many environmental impacts of products (if not an exhaustive number of them).

Multi-attribute claims

Environmental product declarations (EPDs) are Type III declarations that are often described as nutrition labels with environmental information. EPDs are the most standardized form of multi-attribute LCA-based claims. EPDs and their variants may also be referred to as eco-profiles (PlasticsEurope 2011) or product footprints. The first standard developed for multi-attribute claims, or life cycle based product claims for that matter, was ISO 14025, which was first published in 2000 and revised in 2006[1]. The format of an EPD can vary based on the intended audience and program goals, but convention-

ally take the form of reports and do not appear on or with a product on the shelf, although a symbol or link to the full declaration will often appear on the product. An exception to this is the French EPD program, which appears on the product and contains three to five impact indicator results, always including the carbon footprint.

EPDs are similar to PCFs in that they lead to quantitative LCA results associated with a single product. They differ from PCFs in that EPDs are typically multi-attribute, and are less amenable to presentation to the final consumer in the form of a label. ISO 14025 specifies that EPDs can include life cycle inventory and life cycle impact results, and other ancillary data or results estimated beyond the life cycle assessment (e.g., lists of known toxins used in manufacturing and estimated impacts on biodiversity). EPDs also include general descriptive data about a product, dates of publication and period of validity, and qualifying statements about accuracy limitations and lack of comparability with EPDs from other programs.

EPDs have a history that pre-dates the current version of the ISO 14025:2006 standard, with long-standing programs in Europe, Japan, and South Korea. In these markets, EPDs are present for products representing a wide variety of industrial sectors. In North America, EPDs are just beginning to take hold (Schenck 2009) and are most prevalent in the building and furnishing sectors, where an additional ISO standard defines EPDs for building products – ISO 21930 (2007). Presently, we have no reliable estimate for the number of EPDs that have been published globally, but hundreds have been published in government-sponsored programs in Europe and Asia alone (Ingwersen and Stevenson 2012).

EPDs intended to inform consumers are generally scoped to include the full life cycle, while those for intermediate products (e.g., unfinished metals and chemical precursors) are based on a cradle-to-gate scope. While manufacturers may wish to provide EPDs for intermediate products, they may also just provide the life cycle inventory or the LCA results of the product of concern, which is otherwise known as information modules. They can be used as data inputs and thus assembled, along with other sources of data, to make EPDs for downstream products.

EPDs stipulated by ISO 14025 were originally intended to represent a product from one manufacturer. More recently, another variant of EPD has emerged for representing an average product in a sector, called a Sector EPD (Strazza et al. 2010). Sector EPDs may provide a useful benchmark for the industri-

---

1 The standard entered another cycle of revision in 2013.

al sector, or a product category for consumers and manufacturers. In the IBU program (Germany) for construction products, group EPDs are published to represent worst-cases scenarios.

Recent developments in France and in the European Union to create national and regional programs for multi-attribute claims have emerged in the last few years. In France, omnibus national legislation, also known as the *Grenelle environnement laws*, included a pilot program for multi-attribute claims (Cros, Fourdrin, and Réthoré 2010) that provided end-consumers with quantitative environmental information on common consumer products[2]. The goal of the pilot program was to educate consumers and to support more sustainable purchasing. The pilot involved voluntary participation from 168 national and transnational manufacturers of consumer products.

Preliminary results of the pilot indicate that the program was useful for participants, but recognized the need for better data and methods, more efficient tools, and wider institutional communication (Bortzmeyer 2012). In 2013, the European Commission announced the launch of a pilot Product Environmental Footprint program (EC 2013), which is another Type-III like multi-attribute declaration program. The program will be open to all companies operating in the European Union, including multi-nationals, and will be implemented for any product category where sufficient interest exists.

ENVIRONMENTAL PRODUCT DECLARATION FOR BEER

ISO 14025 and other LCA-based product claim standards offer some flexibility in the structure and format of claims so that they can be tailored to the intended audience. This flexibility permits creative ways of displaying LCA results and additional information, while assuring the results and associated language are compliant. Figure 19.1 demonstrates an example EPD for a craft beer. The format of the EPD is a one-page summary with reference to the full EPD via a (QR) code that can be quickly scanned. The brewery (Harmon Brewing Company) and the product (Mt.

Takhoma Blonde Ale) are clearly indicated in the images and text. Results for four impact categories (climate change, water use, eutrophication, and land use) are presented with reference to a 12 oz. bottle of the beer. At the bottom, the EPD features non-expert explanations of the four impact categories. In smaller text, the EPD provides other relevant information such as the validator, registration number, expiration date, and the program logo. No comparisons to any other products are made. EPDs are considered neutral as they do not contain assertions of superiority over other products, but permit the reader to make these comparisons and use this information to make decisions. Therefore, information presented in an EPD should not allow for misinterpretation, and should sufficiently convey modeling assumptions and uncertainties that may affect the interpretation of the LCA results.

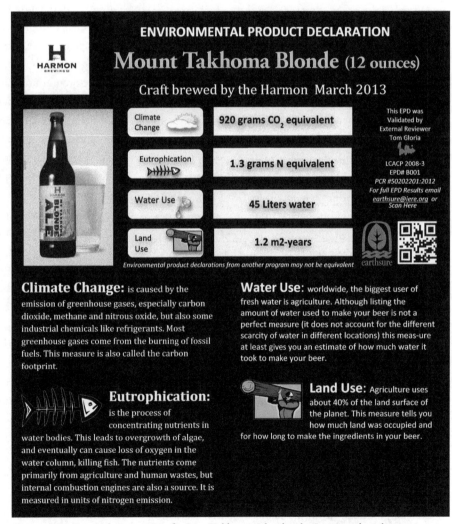

Figure 19.1 Example Beer EPD for Mt. Takhoma Blonde Ale. Reprinted with permission from Earthsure.

2 www.developpement-durable.gouv.fr/National-experimentation-for-the.html

19.3 The Role of Product Category Rules and Program Operators

As emphasized in earlier chapters, LCA involves a large number of choices in each of its stages. These choices include, but are not limited to, the definition of the product system and functional unit, selection of sources of inventory data, development of unit processes, selection and implementation of impact assessment methodologies, and interpretation and communication of results. If LCA is to be performed in a standardized manner for all products, these choices need to be predetermined based on relevance and scientific soundness to enable comparison. The complexity in performing an LCA, the different processes to be included depending on the type of product, and the ability of an LCA to allow comparison only across equivalent functional units—dictates that the rules are defined at the product category level. Product categories are groups of products that fulfill a common function (ISO 2006). Product categories should be specific enough to have a shared functional unit and broad enough to allow for various alternatives to be compared.

To this end, ISO 14025 sets out the need for product category rules (PCRs). Other product claim standards call for equivalent rules that are known by names defined in the various standards – product rules (GHG Protocol Product Standard: 2011), supplementary guidance (PAS 2050: 2011), carbon footprint product category rules (ISO/DIS 14067: 2012) and product environmental footprint category rules (EU PEF: 2012). The Guidance for Product Category Rule Development describe what PCRs should contain and who should be involved in the development process (Ingwersen and Subramanian 2013).

PCRs should define clearly the types of products covered by the PCR and period of validity, the types of claims permitted, the LCA rules, the rules for inclusion of additional information and reporting, references to existing PCRs, and author and history information. ISO recommends that PCRs be based on an underlying LCA, as well as other potentially relevant non-LCA studies of a product in the category. PCRs have a uniquely critical quality. PCRs are developed with inputs from manufacturers and third-party stakeholders, as well as LCA experts in an open-committee process. Therefore, they usually embody a deep and accurate knowledge of the life cycle of products within the category, as well as use best practices for development.

Product categories may be defined by product function, but are often classified using codes in one or more product classification systems. Classification systems provide a hierarchical structure to organize product categories in a systematic manner and provide discrete definitions. These systems are useful to organize PCRs and EPDs such that they can be easily searched for, and provide other insights such as identification of product categories where PCRs and EPDs are widely used. Presently, two product classification systems are being widely used by different programs for mapping PCRs: United Nations Central Product Classification (UN CPC) and United Nations Standard Products and Services Code (UNSPSC).

ISO 14025 does not set up one universal program to facilitate and maintain PCRs, or oversee the publication of EPDs. Instead, it provides basic guidelines for the creation of program operators to fulfill these functions. A variety of types of organizations can fulfill the role of program operators. Depending on the origin of the program (public or private), more than one program operator may be active in a market.[3] Other program claim standards specify the program to manage the PCR process (e.g., PAS 2050) or may not specify a program at all (e.g., GHG Protocol Product Standard). But where program operators are present, they develop a set of program rules that describe procedures for the publication of PCRs and LCA-based claims through their programs. These program rules generally supplement one of the life cycle-based claim standards. Once a PCR is developed for a product category, manufacturers make the claims by working with the program operator. The program operator assures standard and PCR compliance, manages third-party verification of the claims, and may publish the claims (in the case of EPDs) on its website.

The EPD for Mt. Takhoma Blond Ale in Figure 19.1 is based on a PCR for beer published by the Earthsure[4] program operator (IERE 2012). Like other PCRs for EPD programs, the PCR was developed to be compliant with ISO 14025 and the program's general program instructions (IERE 2010). The first section of the PCR defines the document information and the basic goal and scope of the PCR. The Earthsure program uses the UNSPSC product classification system for identifying the product category. Also specified are the functional unit (12 oz. of beer consumed by the customer) and the scope (cradle-to-grave). The PCR includes substantial instructions on the inventory analysis and the LCIA.

3 An informal list of program operators can be found at www.pcrguidance.org/programoperators/. For more information on national EPD programs in different regions of the world, see the supplemental information in Ingwersen and Stevenson (2012)
4 http://iere.org/earthsure.aspx. The PCR for beer is freely available for download.

The inventory analysis includes a system diagram indicating all the processes included in the scope, accompanied with text detailing how the inventory collection for the stages should be handled. The LCIA instructions define a list of the impact categories to be reported and the methods for use in reporting those impacts. Although impacts in only four categories are presented in the EPD (see Figure 19.1), the PCR specifies nine impact categories be reported in the full EPD. The only additional information that needed to be reported included a list of the ingredients composing 95% of the dry weight of the product. Other sections specify background information, definitions and acronyms used, and standards that referred to the format of the EPD. The appendices include a mock EPD, a list of parties invited to participate in the PCR development, a data-gathering spreadsheet, and the report from the review of the PCR by a formal review panel.

The beer PCR was adapted for the North American context from a PCR first published by the International EPD system. Because no universal program operator exists, PCRs for the same or an overlapping product category may be published in different programs. For the Earthsure beer PCR, justifications for creation of a new PCR are explicitly stated with reference to the original PCR. But more often, PCRs have been duplicated without an acknowledgement of an existing PCR. In a detailed comparison of duplicate PCRs in four product categories, Subramanian et al. (2012) show that this duplication generally resulted in rules for LCA and claims that are not compatible.

19.4 USE AND COMMUNICATION OF LCA-BASED CLAIMS AND PCRS

The following sections describe the potential benefits of using LCA-based claims in comparisons with other types of claims, and the benefits of the collaboratively developed PCR.

THE BENEFITS OF LCA-BASED CLAIMS

In comparison to other types of claims, LCA-based claims have numerous benefits (Costello and Schenck 2009). LCA based claims are quantitative, comprehensive, transparent, flexible, modular, neutral, and comparable. Unlike claims that just provide poorly-defined descriptors like "sustainable," or even those based on more elaborate criteria, LCA-based claims provide quantitative results in reference to a defined unit of a product. Although these claims may not be considered complete in encompassing of all environmental or other aspects of a product's provenance, the claims are comprehensive in scope and, except for the single-attribute claims, present a selection of relevant environmental impacts. While these results may not be intuitive or easily understood without a basic background in environmental science and life cycle thinking, the results are based on a transparent rule set (the PCR), which are, ideally, created by a balanced group of experts and stakeholders in the relevant product category. LCA-based claims are flexible in format and use because multiple standards and programs facilitate claims in different regions and markets. As LCA-based claims become more prevalent, these claims are likely to be used in a modular fashion. EPDs for intermediate products will be used as information modules to make EPDs for consumer products, offering benefits in efficiency, and consistency in the claims. As mentioned above, LCA-based claims are neutral because they present information without explicitly comparing the product to an existing performance standard or other products in the category. However, one of their advantages for use in production and purchasing is that the information in claims based on the same PCR is comparable.

THE BENEFITS OF PCRS BEYOND LIFE CYCLE-BASED CLAIMS

PCRs may benefit efforts for product sustainability beyond serving as common rules for claims for a particular category of products. Some of the benefits of PCRs derive from the process as much as the outcome. Bringing industry, as well as outside stakeholders, to the table to create a set of transparent rules and parameters for quantitatively assessing products may be educational and encourage productive cooperation toward reducing product environmental impacts. Industry representatives are often unaware of LCA, and the application of life cycle concepts and the PCR process can provide a learning opportunity.

On the other hand, PCR development may also be informative for third parties unaware of the complexities of measuring sustainability for a certain category of products. If the process of PCR development is managed openly and effectively publicized, it may also draw the attention of industry members who have yet to engage in an environmental assessment of their products. An industry-academic collaboration called The Carbon Leadership Forum has successfully used the process of PCR development to engage a variety of public and private partners in the building industry.[5] Once a PCR has been registered, its benefits extended beyond the product category of concern. Because of the modular nature of product systems and LCA studies, a PCR for a product input can be used as the foundation of a related PCR. For example,

5 www.carbonleadershipforum.org/

if a PCR already exists for a basic dairy product such as milk, it can be adopted and extended to develop PCRs for product categories for which milk is a constituent (such as ice cream).

19.5 OVERCOMING OBSTACLES IN THE CREATION AND USE OF LCA-BASED PRODUCT CLAIMS

The creation of comparable LCA-based claims faces numerous challenges (Ingwersen and Stevenson 2012) and the conveyance of life cycle based information in a standardized fashion (Jungbluth et al. 2012). These challenges are associated with the following: the development of LCA data and methodology; the use of additional information to address impacts of concern beyond the purview of LCA; the existence of multiple standards and programs for LCA-based claims; the challenge of industry-wide acceptance of PCRs; conveying LCA-based information to the end-user; and, lastly, claims as barriers to trade and commerce. These obstacles represent threats to the reliability, legitimacy, and spread of LCA-based claims, but initiatives are currently working to help overcome these obstacles.

AVAILABILITY OF LCA DATA AND METHODOLOGY

The robustness of an LCA and, thus, of LCA-based claims is dependent on various factors covered in previous chapters, including availability and quality of inventory data (Chapters 4-6), modeling approaches (Chapters 6-7), impact assessment (Chapters 10-12), and handling of uncertainty (Chapter 15). While inventory for industrial and agricultural processes and infrastructure is more extensive in Europe, the absence of appropriate data to represent processes in most regions of the world constrains the growth of LCA-based claims (Ingwersen and Stevenson 2012). When data are available, differences in inventory methodology, documentation, nomenclature, transparency of data, and data formats create incompatibilities that further limit the usefulness of data and methods. Like with any process of standardization or rule development, until we have adequate data and accepted methods available, there will likely not be support to develop a consensus standard or rule.

SUPPLEMENTING CLAIMS WITH NON-LCA INFORMATION

Although LCA is a comprehensive tool for assessing the environmental impacts of goods of services, it does not address all environmental impact of products. Unless LCA is coupled with life cycle costing or social life cycle assessment (Chapter 16), it does not capture information on economic impacts or human rights concerns related to product supply chains. For example, LCA does not address all impacts, such as direct site-specific impacts of chemicals or groundwater contamination and work-

er exposure, nor does it capture direct impacts on ecosystem goods and services. This has not only led to the evolution of LCA, but also to the recognition that other tools must be used alongside LCA to inform product sustainability assessment.

Thus far, issues beyond LCA are not being addressed in LCA-based claims. The International EPD program provides methods and recommendation for inclusion of other environmental indicators, such as ecological footprint and virtual water content, in some of its PCRs (Environdec 2010) . But other environmental, economic and social indicators have, to date, not been included in PCRs. The failure to address other environmental and social issues (e.g., direct impacts on biodiversity or impacts on worker wellbeing) has been criticized by those in the traditional environmental NGO community (Grant 2011) who historically have been heavily involved in product-certification schemes. Expanding the use of additional information to capture other product sustainability information acknowledged to be of critical importance to other stakeholders will be critical to the wider acceptance of these claims. Conversely, expanding the scope of LCA and derived claims based on unsupported and unscientific information erodes confidence in EPDs.

CONFLICTS AMONG LCA-BASED CLAIM STANDARDS

As described in Section 19.2, multiple standards exist for LCA-based claims; even in the case of EPD programs compliant with the ISO 14025 standard, program operators provide their own distinct program instructions. While the existence of many standards may have helped promote and expand the use of these claims among a variety of audiences in different regions of the world, it has also resulted in inconsistencies in the claims and, in particular, the PCRs underlying claims. Three notable initiatives to overcome inconsistencies among standards and programs are mutual recognition agreements between programs, common guidance for PCR development, and common registries for maintaining PCRs and claims from various programs.

A number of mutual agreements have been recently been established between various program operators for joint recognition of PCRs and/or claims published by either program. Mutual agreements to recognize PCRs and claims developed in the partnering program have recently been signed between the following groups: IBU (Germany) and the International EPD system, Earthsure and the International EPD system, and Earthsure and JEMAI (Japan). These agreements between program operators represent steps toward cooperation and harmonization (Del Borghi 2012).

One obstacle that deters collaboration and sharing of PCRs between programs, as well as recognition of programs, is the lack of a centralized location for publication and/or notification of the availability of PCRs and claims. A concept to overcome this obstacle is the creation of one or more central Internet repositories. Two proposals for repositories that have garnered attention are the global PCR database by the Global Environmental Declarations Network (GEDnet) and a quality regulated PCR registry for the Americas by the American Center for Life Cycle Assessment (ACLCA). Currently, both organizations are exploring business models that do not act as a barrier to accessibility. While there have been discussions around a claims registry, no concrete steps have been taken yet. These efforts represent the need to optimize time and cost in the creation and management of PCRs and product claims.

In attempt to overcome unnecessary inconsistencies in PCRs, a group of experts from over 40 organizations representing roughly 14 countries and regions called the Guidance for Product Category Rule Development Initiative[6] has recently developed a guidance document for the development of PCRs (W. Ingwersen & Subramanian, 2013). The Guidance aims to provide additional instructions for developing PCRs for LCA-based product claim standards with the goal of making PCRs more consistent and robust and reducing requirements to duplicate PCRs for compliance with multiple standards. The Guidance clarifies the role of parties to PCR development, specifies required and recommended elements, and provides a set of best practices for PCR development.

Industry collaboration and PCR acceptance

The field of sustainability, due to its innate complexity, has brought together competing businesses to improve their common understanding of sustainability and to establish industry benchmarks for improvement (Golden, Subramanian, and Zimmerman 2011). Consensus-based PCRs for product claims are best created with broad industry representation, often through industry associations and the participation of outside experts from NGOs, government, and academia.

The development of consensus-based PCR development has recently occurred in the building product, office furniture, and apparel categories. A PCR for building thermal insulation products was developed under the auspices of the American Chemistry Council, and included participation from six industry associations representing manufacturers of various types of insulation products and insulation materials (Levy et al. 2011). The Business + Institutional Furniture Manufactur-

ers Association[7] (BIFMA) is a not-for-profit trade association that has been actively pushing boundaries by developing PCRs for product categories that fall under its coverage with the collective support and participation of its members. The Sustainable Apparel Coalition[8] (SAC) is another not-for-profit organization that has brought together manufactures and suppliers from around the world to create a guidance document that streamlines the creation of PCRs for product categories that fall under its coverage. These efforts show collaboration and harmonization among competitors who seek to move their industries forward in a collective manner.

Conveying LCA-based information to the end-user

Due to the innate complexity in LCAs, EPDs evolved primarily in the context of business-to-business transactions in which buyers with LCA expertise could use EPDs to aid in purchasing decisions. While EPDs for business-to-consumer transactions are not forbidden, the traditional, full EPD report format is not suitable for most consumers who do not have the background to interpret results from an LCA.

Carbon footprint labels have been one consumer-oriented application of LCA-based claims. Carbon Trust has been a leading certifier of carbon-based labels in the European market, working with the major supermarket Tesco to label more than 900 products with carbon footprints estimated according to the PAS 2050 standard (Tesco 2012). In Japan, carbon footprints also appear on hundreds of consumer products.[9]

Multi-attribute claims present more of a challenge. Condensing the presentation of EPD results into a shorter format is a critical step. The one page EPD summary in Figure 19.1 is an example of a format more appropriate for a final consumer.

Presenting more intuitive graphical elements with values as relative percentage scores (0-100%) is another recommendation for consumer-facing EPDs (Christiansen, Wesnæs, and Weidema 2006). Limiting the number of impacts presented to no more than three was recommended in the French national pilot program, and manufacturers were allowed to explore creative ways of presenting this impact information (Bortzmeyer 2012). Some industry consortia and manufacturers have moved in the direction of aggregating LCA-based information, along with other information into a single-score index, such as the Outdoor Industry Association's Eco Index and Walmart's Sustainable Product Index (O'Shea, Golden, and

---

6 www.pcrguidance.org
7 www.bifma.org
8 www.apparelcoalition.org
9 www.cfp-japan.jp/english

Olander 2012), but these approaches begin to deviate from ISO 14025 and the other product standards.

The potential effect on consumer decision making of LCA-based claims is a factor of the format of the claims, the context, the awareness of consumers, and the product category (Jungbluth et al. 2012). The formula for successful use of LCA-based claims in the consumer market is still largely unknown.

AVOIDING BARRIERS TO TRADE AND COMMERCE

Foreign products are often present in the marketplace where environmental claims are associated with products. For all forms of environmental product claims, concerns have been raised that these claims could potentially be non-tariff technical barriers to trade under the rules governing international trade (Staffin 1996). With regard to LCA-based claims, a further concern is the potential expense and challenges that might be associated with collecting data for claims, acquiring expertise, certifying, and publishing these claims, all of which could disadvantage smaller and medium size enterprises (SMEs) as well as foreign producers.

To avoid potential problems and allay fears, claim standards should be voluntary and participation should be possible for any entity. To keep costs down and make LCA-based claims possible for small businesses, program rules and PCRs should be designed with small businesses in mind, and additional support and tools may be needed to provide capacity to SMEs and foreign entities (Scarinci et al. 2012).

19.6 CONCLUSION

LCA-based claims offer a standardized means of providing product-specific LCA results to the public. LCA-based claims may take various forms and present results of one or more environmental impacts, as well as convey additional environmental information. These claims are a more recent development and are still scarce in the marketplace. Effective and legitimate LCA-based claims face many obstacles, including the lack of sufficient data and consensus on methodologies, the need to acknowledge and include other forms of product sustainability information, the existence of numerous and sometimes conflicting standards and product category rules, a lack of an identified formula for successful presentation to final consumers, and fears of creating barriers to trade and commerce. Many of these obstacles are being addressed, and others require more effort and investment. However, these claims offer the promise of quantified, transparent, and verified environmental information on products, a means of sector-wide collaboration toward improvement of environ-

mental performance, and a potential means of comparing the environmental performance of products.

PROBLEM EXERCISES

1. Compare the two EPDs for X and Y beers. Which beer appears to be a superior choice from an environmental perspective? Referring to the PCR, describe the process for estimating climate change impact. What life cycle stages does this impact cover, and what emissions might be related to that impact? What is the contribution of the bottling stage to this impact?

2. Your company would like to make environmental claims for a new line of bottled juice products. Particular interest has been devoted to developing a product carbon footprint and marketing the product with a carbon footprint label in North American markets. Delineate the steps necessary to make this claim in a manner compliant with the GHG Protocol Product Life Cycle Accounting and Reporting Standard (www.ghgprotocol.org/standards/product-standard). Include consideration of the need for program operators, PCRs, and LCA experts. Discuss the advantages and disadvantages of this approach to a environmental claim in reference to at least one alternative.

3. Perform an Internet search for PCRs for refrigerators at the PCR Library maintained by the Environment and Development Foundation (EDF) (http://pcr-library.edf.org.tw). Use the PCR Comparison Template (www.lcacenter.org/product-category-rule.aspx) to compare all PCRs found. Determine where inconsistencies exist, what potential effect these inconsistencies might have on the final LCA results, and what changes would be made to align the PCRs.

AUTHORS

Datu Buyung Agusdinata is an assistant professor at the department of Industrial & Systems Engineering at the Northern Illinois University (NIU). He is also a faculty associate with the Environment, Sustainability, and Energy Institute at NIU

Carina Alles is the Global Sustainability Offering Manager for DuPont Titanium Technologies. She earned her PhD in Chemical Engineering at the University of Karlsruhe

Mikhail Chester is an Assistant Professor in the department of Civil, Environmental, and Sustainable Engineering at Arizona State University. His research focuses on the energy and environmental assessment of infrastructure systems and urban activities.

Andreas Ciroth is founder and director of GreenDelta (before Oct 15 2012: GreenDeltaTC), a consulting and software company with focus on sustainability assessment and life cycle analyses. Environmental engineer by education, he finished his PhD (error calculation in LCA) in 2001 at TU Berlin and is working since then in sustainability consulting in research, industry, and policy contexts.

Matthew Eckelman is an Assistant Professor of Environmental Engineering at Northeastern University in Boston, MA. His life cycle assessment work covers a range of topics, with particular emphasis on metals, nanomaterials, and benefits/impacts to public health.

Matthias Finkbeiner is Professor and Chair of Sustainable Engineering at the Technical University Berlin. He is Chair of the ISO Committee for Life Cycle Assessment (ISO/TC207/SC5).

Bill Flanagan leads the Ecoassessment Center of Excellence for the General Electric Company. He graduated from Virginia Tech in 1985 and received a PhD in Chemical Engineering from the University of Connecticut in 1991, is an ACLCA-certified LCA Professional and a member of the ACLCA Advisory Council, and serves on the External Advisory Board for the University of Michigan's Center for Sustainable Systems.

Roland Geyer is an Assistant Professor at UCSB's Bren School of Environmental Science and Management. He has graduate degrees in physics and engineering and studies pollution prevention strategies based on recycling, reuse, and material and technology substitution

Tom Gloria is the founder and CEO of Industrial Ecology Consultants. In addition to an active consulting practice in LCA and sustainability, he teaches Industrial Ecology at Harvard Extension School's Sustainability and Environmental Management Graduate Program

Troy R. Hawkins is the Director of Life Cycle Assessment for Enviance. He has been working on the development of input output models for life cycle assessment for roughly 10 years, including developing a mixed unit input output model for tracking metal flows through the U.S. economy and the EXIOPOL multi-regional input output model for understanding the environmental outcomes associated with international supply chains.

Dr. Rich Helling is a director of sustainability and life cycle assessment at The Dow Chemical Company. He has degrees from Harvey Mudd College and MIT, and over 25 years of industrial experience in R&D, manufacturing, management and sustainability.

Andrew Henderson is an assistant professor at the School of Public Health, University of Texas Health Science Center at Houston, where his focus is on bridging sustainability, public health, and environmental engineering

Shawn Hunter, PhD, LCACP is a Sustainability and Life Cycle Assessment Leader at The Dow Chemical Company, where he serves as an expert in sustainability and LCA and works to integrate sustainability and life-cycle thinking into Dow's business units.

Wesley Ingwersen, Ph.D., leads research and application of LCA with the U.S. Environmental Protection Agency's National Risk Management Research Laboratory in Cincinnati, Ohio.

Dr. Greg Keoleian is the Peter M. Wege Endowed Professor of Sustainable Systems at the University of Michigan and serves as Director of the Center for Sustainable Systems. He has appointments as Professor in the School of Natural Resources and Environment and Professor in the Department of Civil and Environmental Engineering.

Chris Koffler is the Technical Director of PE INTERNATIONAL, Inc. He is responsible for the quality of all North American LCA and EPD consulting projects, methodological development, and in key selected areas, such as Automotive, as primary lead.

Dr.-Ing. Sébastien Lasvaux works as a Research Engineer at the University of Paris-East in the Environment and Life Cycle Engineering Division of the Scientific and Technical Centre of Buildings (CSTB). He has 7 years experiences in LCA applied to the construction sector and contributes to the development of data, methods and tools for building practitioners willing to assess the environmental impacts of their projects.

Christoph J. Meinrenken is an instructor of Industrial Ecology and Life Cycle Assessment in the Department of Earth and Environmental Engineering and associate research scientist at Columbia University's Earth Institute. His research focusses on computer modeling to elucidate and improve the technological and economic performance of low carbon energy systems.

Dr. Ivo Mersiowsky is Business Line Manager, Sustainable Products & Strategy with DEKRA Consulting GmbH. Holding a PhD in Environmental Engineering, Ivo is an expert in product lifecycle and sustainability with a focus on communication strategies in the chemical and plastics industries.

Anna Nicholson earned two Masters degrees from MIT. She works at UL Environment in their ecolabelling program.

Catherine Benoît Norris is the Social Assessments and Strategy director at New Earth, a non-profit organization. She is the co-creator and executive director of New Earth Social Hotspots Database project She was the lead editor of the Guidelines for Social-LCA that were published by the UNEP Life Cycle Initiative.

Hanna-Leena Pesonen is Professor in Corporate Environmental Management at the School of Business and Economics of the University of Jyväskylä, Finland. She is co-author of SETAC's code of practice for environmental life cycle costing (LCC), and subject editor concerning LCC topics in Journal of Life Cycle Assessment

Matt Pietrzykowski works at General Electric Company and enjoys working with data in all its forms. He especially likes turning data into action by providing robust analytics to support decisions, identify key drivers for stakeholders, and provide predictive analytics to mitigate risk or identify opportunities

Tom Redick has an international environmental law practice in St. Louis, Missouri, with clients in the high-technology and agricultural biotechnology industry sectors. He handles issues relating to regulatory approval, liability avoidance and compliance with industry standards addressing socioeconomic and environmental impacts. He earned his B.A. in Ethical Philosophy and his J.D. from University of Michigan

Beverly Sauer is a senior LCA manager at Franklin Associates, A Division of ERG. She has been conducting and managing LCAs for public and private clients on a wide range of products and materials since 1990.

Rita Schenck is the founder and executive director of the Institute for Environmental Research and Education, the parent of the American Center for Life Cycle Assessment, which she co-founded. She represented the U.S. in negotiating the LCA standards in the 1990's. She earned her doctorate in Oceanography from the University of Rhode Island in 1984

Bengt Steen is an adjunct professor at Chalmers University. He represented Sweden in developing the ISO standards in the 1990s and he continues his practice in valuation in LCA.

Dr. Susanne Veith is a consultant for sustainability and economic evaluations at E. I. du Pont de Nemours and Company (DuPont). She holds a PhD in chemical engineering from the Swiss Federal Institute of Technology (ETH) in Zurich, Switzerland and a MBA in International Management from Drexel University in Philadelphia, USA

Fu Zhao is an Associate Professor in the School of Mechanical Engineering and Division of Environmental and Ecological Engineering at Purdue University

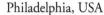

Thorsten Volz has 20 years of experience in the field of LCA. His PhD thesis in LCA was the foundation of the GaBi System. he was responsible for the GaBi development development team at PE International.

Dr. Christopher L. Weber is a LCA expert and Principal Analyst at Enviance, Inc. Environmental Business Intelligence.

Philip White is an assistant professor at Arizona State University, where his work focuses on LCA in design. He is the developer of the highly respected OKALA method.

Ron Wroczynski is a Senior Scientist in the Ecoassessment Center of Excellence at GE Global Research, Niskayuna, NY. Since 2008, his research activities have been centered on environmental life cycle assessment. Ron's current interests focus on the application of statistical and data analysis tools to improve the analysis and understanding of life cycle assessment results

REFERENCES

CHAPTER 1

Argonne National Laboratory. 2012. Greenhouse Gas, Regulated Emissions, and Energy Use in Transportation Model, USA.

Banerjee, A., B. Solomon. 2003. Eco-labeling for energy efficiency and sustainability: a meta-evaluation of US programs, *Energy Policy*, Volume 31, Issue 2, January 2003, Pages 109–123.

Baumann, H., A. Tillman. 2004. *The Hitchhiker's Guide to LCA: an orientation in life cycle assessment methodology and application*, Studentlitteratur AB, Lund.

Boustead, I, G. Hancock. 1979. LCA – How it came About, *Int. J. of LCA* 1(3) 147-150.

BUS. 1984. Bundesamt für Umeltschutz, *Oekobilanzen von Packstoffen*. Schriftenriehe Umweltschutz, Nr. 24, Bern.

CA. 2007. www.arb.ca.gov/fuels/lcfs/eos0107.pdf

CA. 2008. www.dtsc.ca.gov/PollutionPrevention/GreenChemistryInitiative/upload/ab_1879_GCI.pdf

Cooper, J.; Fava, J. 2006. Life Cycle Assessment Practitioner Survey: Summary of Results, *J. of Ind. Ecology*.

Fiala, N. 2008. Measuring sustainability, *Ecological Economics* 67 (4): 519–525. doi:10.1016/j.ecolecon.2008.07.023.

Franke, M. 1984. *Umweltauswerkungen duch Getränkeverpackungen. Systematik zur Ermittlung der Umweltauswirkungen von komplexen Prozessen am Beispiel von Einweg- und Mehrweg-Getränkegehältern*. EF-Verlag für Energie- und Umwelt-technik, Berlin.

Franklin, W., R. Hunt. 1972. *Environmental Impacts of Polystyrene and Molded Pulp Meat Trays, a Summary*, Mobile Chemical Company, Mecedon, NY.

Frischknecht R. et.al. 2003. *Implementation of Life Cycle Impact Assessment Methods*, Ecoinvent Report #3, Swiss Centre for Life Cycle Inventories, Dubendorf, Switzerland.

Graedel, T. 1996. Weighted Matrices as Product Life Cycle Assessment Tools. *Int. J. of LCA* 1 (2) 85–89.

Graedel, T., B. Allenby 2003. *Industrial Ecology*, Prentice Hall Publishing, Upper Saddle River, NJ, USA.

Guinée, G. et al. 2002. *Handbook on Life Cycle Assessment: Operational Guide to ISO Standards*, Kluwer Academic Pub., Dordrecht.

Heijungs, R., U. de Haes, U., P. White, J. Golden. 2008. *Life Cycle Assessment Training Kit*. www.unep.fr/scp/lcinitiative/publications/training/index.htm

Heine, L. 2008. *Green Screen for Safer Chemicals Version 1.0*, Clean Production Action, Medford, MA, USA. www.cleanproduction.org.

Hendrickson, C., L. Lave, S. Matthews. 2006. *Environmental Life Cycle Assessment of Goods and Services: An Input-Output Approach*, RFF Press.

Horvath, A. and C. Hendrickson. 1998. Comparison of Environmental Implications of Asphalt and Steel-Reinforced Concrete Pavements, *Transp. Res. Record* 1626 105-113. http://dx.doi.org/10.3141/1626-13

Houe, B., B. Grabot. 2009. Assessing the compliance of a product with an eco-label: From standards to constraints, *Int. J. of Production Economics*, Volume 121, Issue 1, Sep. 2009, pp 21–38

Hunkeler, D., K. Lichtenvort, G. Rebitzer. 2008. *Environmental Life Cycle Costing*, SETAC, Pensacola, FL.

International Programme on Chemical Safety. 2000. *General Scientific Principles of Chemical Safety*, UNEP/IPCS Training Module No. 3, Section B.

ISO, ANSI & NSF. 1997. *ISO 14040: Environmental Management – Life Cycle Assessment – Principles and Framework.*

ISO, ANSI & NSF. 2006. *ISO 14044: Environmental Management – Life Cycle Assessment – Principles and Framework.*

Lave, L., E. Cobas, C. Hendrickson, F. McMichael. 1995. Using Input-Output Analysis to Estimate Economy-Wide Discharges, *Env. Science and Tech.* 29(9), pp. 153-161.

OECD. 1991. *Estimation of greenhouse gas emissions and sinks: Final report from the OECD experts meeting*, Feb. 1991, Paris.

Pehnt, M. 2006. Dynamic Life Cycle Assessment (LCA) of Renewable Energy Technologies, *Renewable Energy* 31(1) 55-71. http://dx.doi.org/10.1016/j.renene.2005.03.002

Sathre, M., M. Chester, J. Cain, E. Masinet, 2012. A framework for environmental assessment of CO2 capture and storage systems, *Energy* 37(1) 540-548. http://dx.doi.org/10.1016/j.energy.2011.10.050

Schenck, R. 2010. *LCA for Mere Mortals: a primer on life cycle assessment*, IERE, Vashon Island, WA.

Schmidt-Bleek, F. 1993. *Das MIPS Konzept- Faktor 10*, Droemer Knaur, Munich.

Searchinger, T., et al. 2008. Use of U.S. Croplands for Biofuels Increases Greenhouse Gases through Emissions from Land-Use Change, *Science* 319(5867) 1238-1240. http://dx.doi.org/10.1126/science.1151861

U. S. Environmental Protection Agency. 2010. *Defining Life Cycle Assessment (LCA).*

Wackernagel, M., W. Rees. 1996. *Our ecological footprint*, New Society Publishers, Gabriola Islands, Canada.

CHAPTER 2

Berger, M., M. Finkbeiner. 2010. Water footprinting—How to address water use in life cycle assessment? *Sustainability.* 2(4): 919–944

Berger, M., M. Finkbeiner. 2012. Methodological challenges in volumetric and impact-oriented water footprints. *J. of Ind Ecol.* DOI: 10.1111/j.1530-9290.2012.00495.x

Finkbeiner, M., A. Inaba, R. Tan, K. Christiansen, and H-J Klüppel. 2006. The new international standards for life cycle assessment: ISO 14040 and ISO 14044. *Int. J. of LCA* 11(2): 80–85

Finkbeiner, M. 2009. Carbon footprinting – opportunities and threats, *Int. J. of LCA* 14(2) 91–94

Finkbeiner, M., E. Schau, A. Lehmann, and M. Traverso. 2010. Towards Life Cycle Sustainability Assessment, *Sustainability* 2010, 2, 3309-3322, doi:10.3390/su2103309

Finkbeiner, M. 2011. *Towards life cycle sustainability management.* Springer Berlin/Heidelberg. ISBN: 978-94-007-1898-2

Finkbeiner, M. 2012. From the 40s to the 70s—the future of LCA in the ISO 14000 family. *Int. J. of LCA* DOI: 10.1007/s11367-012-0492-x

ISO (2010) *Environmental management – The ISO 14000 family of International Standards. International Organization for Standardization.* Geneva, Switzerland. ISBN 978-92-67-10500-0

ISO (2012a) Who develops ISO-Standards. www.iso.org/iso/home/standards_development/who-develops-iso-standards.htm. Accessed July 2012

ISO (2012b) Standards Development. www.iso.org/iso/home/standards_development.htm. Accessed July 2012

ISO 14025 (2006) *Environmental labels and declarations, Type III—Environmental Declarations – Principles and Procedures.* Geneva, Switzerland

ISO 14040 (1997) *Environmental Management – Life Cycle Assessment – Principles and Framework.* Geneva, Switzerland

ISO 14040 (2006) *Environmental Management – Life Cycle Assessment – Principles and Framework.* Geneva, Switzerland

This is excerpted from ISO 14040:2006, Section 3.18 on page 4, with the permission of ANSI on behalf of ISO. (c) ISO 2014 - All rights reserved

This is excerpted from ISO 14040:2006, Section 3.19 on page 4, with the permission of ANSI on behalf of ISO. (c) ISO 2014 - All rights reserved

This is excerpted from ISO 14040:2006, Section 3.2 on page 2, with the permission of ANSI on behalf of ISO. (c) ISO 2014 - All rights reserved

This is excerpted from ISO 14040:2006, Section 3.29 on page 5, with the permission of ANSI on behalf of ISO. (c) ISO 2014 - All rights reserved

This is excerpted from ISO 14040:2006, Section 3.36 on page 5, with the permission of ANSI on behalf of ISO. (c) ISO 2014 - All rights reserved

This is excerpted from ISO 14040:2006, Section 3.39 on page 5, with the permission of ANSI on behalf of ISO. (c) ISO 2014 - All rights reserved

This is excerpted from ISO 14040:2006, Section 3.4 on page 2, with the permission of ANSI on behalf of ISO. (c) ISO 2014 - All rights reserved

This is excerpted from ISO 14040:2006, Section 3.45 on page 6, with the permission of ANSI on behalf of ISO. (c) ISO 2014 - All rights reserved

This is excerpted from ISO 14040:2006, Section 3.5 on page 2, with the permission of ANSI on behalf of ISO. (c) ISO 2014 - All rights reserved

This is excerpted from ISO 14040:2006, Section 3.6 on page 2, with the permission of ANSI on behalf of ISO. (c) ISO 2014 - All rights reserved

This is excerpted from ISO 14040:2006, Section 3.9 on page 2, with the permission of ANSI on behalf of ISO. (c) ISO 2014 - All rights reserved

This is excerpted from ISO 14040:2006, Section 4.1.2 on pages 6 - 7, with the permission of ANSI on behalf of ISO. (c) ISO 2014 - All rights reserved

This is excerpted from ISO 14040:2006, Section 4.1.3 on page 7, with the permission of ANSI on behalf of ISO. (c) ISO 2014 - All rights reserved

This is excerpted from ISO 14040:2006, Section 4.1.4 on page 7, with the permission of ANSI on behalf of ISO. (c) ISO 2014 - All rights reserved

This is excerpted from ISO 14040:2006, Section 4.1.5 on page 7, with the permission of ANSI on behalf of ISO. (c) ISO 2014 - All rights reserved

This is excerpted from ISO 14040:2006, Section 4.1.6 on page 7, with the permission of ANSI on behalf of ISO.

(c) ISO 2014 - All rights reserved

This is excerpted from ISO 14040:2006, Section 4.1.7 on page 7, with the permission of ANSI on behalf of ISO. (c) ISO 2014 - All rights reserved

This is excerpted from ISO 14040:2006, Section 4.1.8 on page 7, with the permission of ANSI on behalf of ISO. (c) ISO 2014 - All rights reserved

Excerpted from ISO 14040:2006, Section 6 on pages 16 -17, with the permission of ANSI on behalf of ISO. (c) ISO 2014 - All rights reserved

Excerpted from ISO 14040:2006, Section 7.1 on page 17, with the permission of ANSI on behalf of ISO. (c) ISO 2014 - All rights reserved

ISO 14041 (1998) *Environmental Management – Life Cycle Assessment – Goal and Scope Definition and Inventory Analysis.* Geneva, Switzerland

ISO 14042 (2000) *Environmental Management – Life Cycle Assessment – Life Cycle Impact Assessment.* Geneva, Switzerland

ISO 14043 (2000) *Environmental Management – Life Cycle Assessment – Life Cycle Interpretation.* Geneva, Switzerland

ISO 14044 (2006) *Environmental Management – Life Cycle Assessment – Requirements and Guidelines.* Geneva, Switzerland

This is excerpted from ISO 14044:2006, Section 6.1 on page 31, with the permission of ANSI on behalf of ISO. (c) ISO 2014 - All rights reserved

This is an adapted excerpt from ISO 14044:2006, Section 4.1 on page 6, with the permission of ANSI on behalf of ISO. (c) ISO 2014 - All rights reserved

This is excerpted from ISO 14044:2006, Section 4.3.4.2 on page 13, with the permission of ANSI on behalf of ISO. (c) ISO 2014 - All rights reserved

This is excerpted from ISO 14044:2006, Section 4.4.5 on page 23, with the permission of ANSI on behalf of ISO. (c) ISO 2014 - All rights reserved

This is excerpted from ISO 14044:2006, Section 4.5.1 on page 23, with the permission of ANSI on behalf of ISO. (c) ISO 2014 - All rights reserved

This is excerpted from ISO 14044:2006, Section 4.5.1.1 on page 23, with the permission of ANSI on behalf of ISO. (c) ISO 2014 - All rights reserved

This is excerpted from ISO 14044:2006, Section 4.5.4 on page 27, with the permission of ANSI on behalf of ISO. (c) ISO 2014 - All rights reserved

This is excerpted from ISO 14044:2006, Section 6.3 on pages 31 -32, with the permission of ANSI on behalf of ISO.

(c) ISO 2014 - All rights reserved

This is excerpted from ISO 14044:2006, Section 4.4.2.1 on page 16, with the permission of ANSI on behalf of ISO. (c) ISO 2014 - All rights reserved

This is excerpted from ISO 14044:2006, Section 4.4.3.1 on pages 20 - 21 , with the permission of ANSI on behalf of ISO. (c) ISO 2014 - All rights reserved

ISO 14045 (2012) *Environmental Management – Eco-efficiency Assessment of Product Systems – Principles, Requirements and Guidelines.* Geneva, Switzerland

This is excerpted from ISO 14045:2012, Introduction on page v, with the permission of ANSI on behalf of ISO. (c) ISO 2014 - All rights reserved

ISO 14046.CD.1 (2012) *Water Footprint – Requirements and Guidelines.* Geneva, Switzerland

This is an adapted excerpt from ISO 14046:2012, Scope on page 10, with the permission of ANSI on behalf of ISO. (c) ISO 2014 - All rights reserved

This is an adapted excerpt from ISO 14046:2012, Section 3.3.2 on page 13, with the permission of ANSI on behalf of ISO. (c) ISO 2014 - All rights reserved

ISO/TR 14047 (2012) *Environmental Management – Life Cycle Assessment – Illustrative examples on how to apply ISO 14044 to impact assessment situations.* Geneva, Switzerland

SO/TS 14048 (2002) *Environmental Management – Life Cycle Assessment – Data Documentation Format.* Geneva, Switzerland

ISO/TR 14049 (2012) *Environmental Management – Life Cycle Assessment – Illustrative examples on how to apply ISO 14044 to goal and scope definition and inventory analysis.* Geneva, Switzerland

ISO/TS 14067 (2013) *Carbon Footprint of Products –Requirements and Guidelines for Quantification and Communication.* Geneva, Switzerland

This is an adapted excerpt from ISO/TS 14067:2013, Introduction on pages v and vi, with the permission of ANSI on behalf of ISO. (c) ISO 2014 - All rights reserved

Kloepffer, W. 2008. Life Cycle Sustainability Assessment of Products. *Int. J. of LCA* 13: 89-95

Marsmann, M., H-J. Klüppel H-J, and K. Saur. 1997. Foreword – Development of Life Cycle Thinking. Special Issue: Current LCA-ISO Activities. *Int. J. of LCA* 2(1): 2–4

Marsmann, M., 2000. The ISO 14040 Family. *Int. J. of LCA* 5(6): 317–318

SETAC. 1993. Society of Environmental Toxicology and Chemistry (SETAC): Guidelines for Life-Cycle Assessment, A "Code of Practice"; SETAC Workshop in Sesimbra 31.03.-03.04.1993, Brussels

CHAPTER 3

Aristotle. *The Ethics.* English Translation T. Taylor, J.A. Smith, ed. 1908-1931.

Cox, S. 2008. *Sick Planet: Corporate Food and Medicine.* London: Pluto Press.

Esty, D., & A. Winston. 2006. *GREEN TO GOLD: HOW SMART COMPANIES USE ENVIRONMENTAL STRATEGY TO INNOVATE, CREATE VALUE, AND BUILD COMPETITIVE ADVANTAGE.* New Haven, CT: Yale University Press.

Federal Trade Commission. 2012, 10 01. *FTC Issues Revised "Green Guides".* Retrieved January 2013: www.ftc.gov/opa/2012/10/greenguides.shtm

Federal Trade Commission. 1992. Green Guides. *16 Code of Federal Regulation Part 260 .* Washington, DC: U.S. Printing Office.

Leopold, A. 1949. *A Sand County Almanac.* London: Oxford University Press.

CHAPTER 4

European Commission - Joint Research Centre - Institute for Environment and Sustainability. 2010. *International Reference Life Cycle Data System (ILCD) Handbook - Nomenclature and other conventions.* First edition. EUR 24384 EN. Luxembourg. Publications Office of the European Union

Franklin Associates. 2011. *Cradle-to-Gate Life Cycle Inventory of Nine Plastic Resins and Four Polyurethane Precursors.* Published by the Plastics Division of the American Chemistry Council. Franklin Associates, A Division of ERG.

Frischknecht R. et al. 2007. The environmental relevance of capital goods in life cycle assessments of products and services. *Int. J. of LCA.*

ISO 14040. 2006. *Environmental Management – Life Cycle Assessment –Principles and Framework.* International Organization of Standardization

ISO 14044. 2006. *Environmental Management – Life Cycle Assessment – Requirements and Guidelines.* International Organization of Standardization

ISOPA. 2012. *Eco-profiles and Environmental Product Declarations of the European Plastics Manufacturers. Toluene Diisocyanate (TDI) & Methylenediphenyl Diisocyanate (MDI).*

Spath, P.L., M. Mann, and D. Kerr. 1999. *Life Cycle Assessment of Coal-fired Power Production*. Golden, CO. NREL. Report No.: NREL/TP-570-25119.

United Nations Environment Programme: 2009. *Guidelines for Social Life Cycle Assessment of Products*. Benoit, Catherine, editor.

US EPA AP-42: *Compilation of Air Pollutant Emission Factors*. Accessible at www.epa.gov/otaq/ap42.htm

World Resources Institute Greenhouse Gas Protocol. *Sector-specific and cross-sector guidance tools for estimating greenhouse gas emissions*. www.ghgprotocol.org/calculation-tools/all-tools

CHAPTER 5

Bayliss, C. 2013. Personal Communication, Deputy Secretary General, International Aluminium Institute. (IAI), London, UK.

Bhander, G.S., M. Hauschild, and T. McAloone. 2003. Implementing Life Cycle Assessment in Product Development. *Env. Progress* 22(4): 255-267.

CEN. 2012. Sustainability of Construction Works - Environmental Product Declarations - Core Rules for the Product Category of Construction Products. European Committee for Standardization, Brussels.

Ciroth, A., S. Mueller, B. Weidema, and P. Lesage. 2012. Refining the Pedigree Matrix Approach in Ecoinvent: Towards Empirical Uncertainty Factors. Presented at the LCA XII, Tacoma, FL, http://lcacenter.org/lcaxii/final-presentations/693.pdf.

Hagelucken, M., K. Schischke, J. Muller, and H. Griese. 2003. *Welcome to the Jungle – Survival of the Fittest Environmental Indicator*. Proceedings of Electronics Goes Green 2004+ Congress.

Hischier, R., and I. Reichart. 2003. Multifunctional Electronic Media – Traditional Media. The Problem of an Adequate Functional Unit. *Int J LCA* 8(4): 201-208.

Huijbregts, M.A.J., et al., 2006. Is Cumulative Fossil Energy Demand a Useful Indicator for the Environmental Performance of Products? *Environ Sci Technol* 40(3): 641-648.

IAI. Life Cycle Assessment of Aluminium. Inventory Data for the Primary Aluminium Industry, 2005 Update, September 2007, International Aluminium Institute, (IAI), London, UK.

ISO. 2000. ISO/TR 14049: 2000. *Environmental Management — Life Cycle Assessment — Examples of Application of ISO 14044 to Goal and Scope Definition and Inventory Analysis*. International Organization for Standardization. Geneva, Switzerland.

ISO. 2006a. ISO 14044: 2006. *Environmental Management — Life Cycle Assessment — Requirements and Guidelines*. International Organization for Standardization. Geneva, Switzerland.

ISO. 2006b. ISO 14040: 2006. *Environmental Labels and Declarations — Life Cycle Assessment — Principles and Framework*. International Organization for Standardization. Geneva, Switzerland.

ISO. 2006c. ISO 14025: 2006. *Environmental Labels and Declarations -- Type III Environmental Declarations -- Principles and Procedures*. International Organization for Standardization. Geneva, Switzerland.

Koffler, C., M. Baitz, and A. Koehler. 2012. *Addressing Uncertainty in LCI Data with Particular Emphasis on Variability in Upstream Supply Chains*. Whitepaper. PE International. www.pe-international.com/ resources/whitepapers/.

Kuczenski, B., and R. Geyer. 2013. PET Bottle Reverse Logistics – Environmental Performance of California's CRV Program. *Int J. of LCA* 18(2): 456 – 471.

Matheys, J., W., et al. 2007. Influence of Functional Unit on the Life Cycle Assessment of Traction Batteries. *Int J. of LCA* 12(3): 191 – 196.

McDowell, M.A., C. D. Fryar, C. L. Ogden, and K. M. Flegal. 2008. Anthropometric Reference Data for Children and Adults: United States, 2003–2006. *National Health Statistics Report No. 10*. US Department of Health and Human Services.

Meinrenken, C.J., S. M. Kaufman, S. Ramesh, and K. S. Lackner. 2012. Fast Carbon Footprinting for Large Product Portfolios. *J. of Industrial Ecology* 16: 669–679. doi: 10.1111/j.1530-9290.2012.00463.x.

Rebitzer, G. 2005. Enhancing the Application Efficiency of Life Cycle Assessment for Industrial Uses. Dissertation. École Polytechnique Fédérale de Lausanne (EPFL).

Simon, R., E. Rice, T. Kingsbury, and D. Dornfeld. 2012. A Comparison of Life Cycle Assessment (LCA) Software in Packaging Applications. *Laboratory for Manufacturing and Sustainability*. University of California, Berkeley.

Smith Cooper, J. 2003. Specifying Functional Units and Reference Flows for Comparable Alternatives. *Int J. of LCA* 8(6): 337-349.

Sun, M., C. J. Rydh, and H. Kaebernick. 2003. Material Grouping for Simplified Product Life Cycle Assessment. Int J Sustain Design 3 (1-2): 45-48.

Todd, J.A., and M. A. Curran, eds. 1999. *Streamlined Life Cycle Assessment: A Final Report from the SETAC North America Streamlined LCA Working Group.* SETAC. Pensacola, FL.

USGS. 1986. *World Bauxite Resources.* Professional Paper 1076-B, United States Geological Survey (USGS), Washington, D.C.

CHAPTER 6

Ciroth, A. 2009. *Mathematical Analysis of Ecoinvent Data, Final Report.* Commissioned by the Ecoinvent Centre, June 2009.

Ciroth, A. 2010. The UNEP/SETAC Database Registry. Presentation, SETAC Annual Meeting, May 2010, Seville, Spain.

Ciroth, A., S. Lundie and G. Huppes. 2008. *Data Quality, Validation, Uncertainty in LCA.* UNEP/SETAC Life Cycle Initiative, Life Cycle Inventory (LCI), Task Force 3, Methodological Consistency: Inventory Methods in LCA: Towards Consistency and Improvement. VDM-Verlag, 2008, 41-66.

Ciroth and Schebek. 2011. Perspectives in Life Cycle Inventory Modeling. Presentation at LCA XI Conf., accessed Nov. 2012, www.greendelta.com/uploads/media/LCAXI_perspectives.pdf.

Ciroth, A. B. Weidema, and P. Lesage. 2012. Refining the Pedigree Matrix Approach in ecoinvent. Ecoinvent Project report, May 2012.

Ciroth, A. Refining the Pedigree Matrix Approach in Ecoinvent: Towards Empirical Uncertainty Factors. Presentation, Swiss Discussion Forum, Sept. 2013

Draucker, L., S. M. Kaufmann, R. ter Kuile, and C. J. Meinrenken. 2011. Moving Forward on Product Carbon Footprinting. *J. of Industrial Ecology* 15(2).

ecoinvent. 2007. Frischknecht, R. and N. Jungbluth, eds. Overview and Methodology, Data v2.0. *Ecoinvent Report No. 1*, Dübendorf, 2007.

ecoinvent. 2012. Weidema, et al. Overview and Methodology, Data Quality Guideline for the Ecoinvent Database Version 3 (Revision 2). *Ecoinvent Report No. 1*(3), St. Gallen, 2012.

ecoinvent. 2012a. Access Nov. 2012, www.ecoinvent.org/ecoinvent-v3/ecoeditor-for-version-3/.

Frischknecht, R., and N. Jungbluth, eds. 2003. Overview and Methodology Data v1.01 (2003). *Ecoinvent Report No. 1*, Dübendorf, December 2003.

GHG-Protocol. 2011 Product Life Cycle Accounting and Reporting Standard.

Heijungs, R. 1996. Identification of Key Issues for Further Investigation in Improving the Reliability of Life-Cycle Assessments. *J. of Cleaner Production* 4(3-4): 159-166.

JRC. 2012. *ELCD Core Database Version II.* Access Nov. 2012, http://lca.jrc.ec.europa.eu/lcainfohub/datasetArea.vm.

JRC. 2010. *ILCD Handbook: Specific guide for Life Cycle Inventory (LCI) Data Sets* (First ed.). http://lct.jrc.ec.europa.eu/pdf-directory/ILCD-Handbook-Specific-guide-for-LCI-online-12March2010.pdf

Klöpffer, W. 2012. The Critical Review of Life Cycle Assessment Studies According To ISO 14040 and 14044. *Int. J. of LCA* 17: 1087-1093

McCarthy, S., and J. Cooper 2012. USDA's Digital Commons: Agricultural LCI Data. Presentation at the LCA XII conference, access Nov. 2012, http://lcacenter.org/lcaxii/final-presentations/650.pdf

Meinrenken, C.J., S. M. Kaufmann, Siddharth Ramesh, and K. S. Lackner. 2012. Fast Carbon Footprinting for Large Product Portfolios. *J. of Industrial Ecology* 16(5).

Meinrenken, C. J., A.N. Garvan, B. Sauerhaft, and K. S. Lackner. 2014. Combining LCA with Data Science to Inform Portfolio-Level Value Chain Engineering: A Case Study At Pepsico Inc. *J. of Industrial Ecology*.

NREL. 2012. U.S. Life Cycle Inventory Database." Accessed November 2012, www.nrel.gov/lci/.

Ripley, B.D. 2004. Robust Statistics, M.Sc. in Applied Statistics MT2004. Accessed Oct. 2013, www.stats.ox.ac.uk/pub/StatMeth/Robust.pdf.

SETAC. 1992. *Life-Cycle Assessment Data Quality: A Conceptual Framework.* J. Fava, et al., eds. Published by SETAC, 1994.

Wolf, M.A. 2011. Personal Communication after Stakeholder Workshop. SETAC Meeting, May 2009, Seville, Spain.

CHAPTER 7

Bullard, C., and R. Herendeen. 1975. The Energy Cost of Goods and Services. *Energy Policy* 268.

Deng, L., C. Babbitt, and E. Williams. 2011. Economic-balance Hybrid LCA Extended with Uncertainty Analysis: Case Study of a Laptop Computer. *J. of Cleaner Production* 19(11): 1198-1206.

EPA. *Sustainable Materials Management: The Road Ahead.* Washington, D.C.: US EPA, 2009. EPA. *Analysis of the Life Cycle Impacts and Potential for Avoided Impacts Associated with Single-Family Homes.* Washington, D.C.:

U.S. Environmental Protection Agency, 2013.

Ewing, B., et al. 2012. Integrating Ecological and Water Footprint Accounting in a Multi-Regional Input-Output Framework. *Ecological Indicators* 23: 1-8.

Herendeen, R. A. 1978. Total Energy Cost of Household Consumption in Norway, 1973. *Energy* 3: 615-630.

Horowitz, K., and M. Planting. *Concepts and Methods of the U.S. Input-Output Accounts.* Washington, D.C.: US Bureau of Economic Analysis, 2006.

Horvath, A., and C. Hendrickson. 1998. Steel vs. Steel-Reinforced Concrete Bridges: Environmental Assessment. *ASCE J. of Infrastructure Systems* 4(3): 111-117.

Lankey, R., and F. McMichael. 2000. Life-Cycle Methods for Comparing Primary and Rechargeable Batteries. *Env. Science & Tech.* 34: 2299-2304.

Lave, L., H. Maclean, C. Hendrickson, and R. Lankey. 2000. Life-Cycle Analysis of Alternative Automobile Fuel/Propulsion Technologies. *Env. Science & Tech.* 34(17): 3598-3605.

Leontief, W. 1970a and 1970b. Environmental Repercussions and the Economic Structure: An Input-Output Approach. *The Review of Economics and Statistics* 52(3): 262-271.

Miller, R., and P. Blair. 1985. *Input-Output Analysis: Foundations and Extensions* (1st ed.). Englewood Cliffs, NJ: Prentice-Hall.

Minx, J., T. Wiedmann, J. Barrett, and S. Suh. 2008. *Methods Review to Support the PAS Process for the Calculation of the Greenhouse Gas Emissions Embodied in Goods and Services.* London: UK DEFRA.

Murray, J., and R. Wood, eds. *The Sustainability Practitioner's Guide to Input-Output Analysis.* Common Ground Publishing

Nakamura, S., and Y. Kondo. 2002. Input-Output Analysis of Waste Management. *J. of Industrial Ecology* 6: 39-64.

Norris, G., F. Croce, and O. Jolliet. 2002. Energy Burdens of Conventional Wholesale and Retail Portions of Product Life Cycles. *J. of Industrial Ecology* 6(2): 59-69.

Peters, G., and E. Hertwich. Opportunities and Challenges for Environmental MRIO Modeling. Presented at 16th Int. Input-Output Conference, Instanbul, 2007.

Stewart, R. L., B. Stone, and M. Streitwieser. *U.S. Benchmark Input-Output Accounts, 2002.* Washington, D.C.: U.S. Bureau of Economic Analysis, 2007.

Suh, S. 2006. Are Services Better for Climate Change? *Env. Science and Tech.* 40(21): 6555-6560.

Suh, S., ed. *Handbook of Input-Output Economics in Industrial Ecology.* Springer, 2009.

Suh, S., et al.. 2004. System Boundary Selection in Life-Cycle Inventories Using Hybrid Approaches. *Env. Science & Tech., 38*(3), 657-664.

Tukker, A., and B. Jansen. 2006. Environment Impacts of Products - A Detailed Review of Studies. *J. of Industrial Ecology* 10(3): 159-182.

Tukker, A., et al. 2009. Towards a Global Multi-Regional Environmentally Extended Input-Output Database. *Ecological Economics* 68(7): 1928-1937.

Wiedmann, T. 2009. A Review of Recent Multi-region Input-Output Models Used for Consumption-based Emission and Resource Accounting." *Ecological Economics* 69(2): 211-222.

Williams, E. 2004. Energy Intensity of Computer Manufacturing: Hybrid Assessment Combining Process and Economic Input-Output Methods. *Env. Science & Tech.* 38(22): 6166-6174.

Williams, E., C. L. Weber, and T. Hawkins. 2009. Hybrid Framework for Characterizing and Managing Uncertainty in Life Cycle Inventories. *J. of Industrial Ecology* 13(6): 928-944.

CHAPTER 8

Agusdinata, D., F. Zhao, K. Ileleji, and D. DeLaurentis. 2011. Life Cycle Assessment of Potential Biojet Fuel Production in the United States. *Env. Science & Tech.* 45: 9133-9143.

Brandt, A. R. 2012. Variability and Uncertainty in Life Cycle Assessment Models for Greenhouse Gas Emissions from Canadian Oil Sands Production. *Env. Science & Tech.* 46: 1253-1261.

Dalgaard, R., J. Schmidt, N. Halberg, P. Christensen, M. Thrane, and W. A. Pengue. 2008. LCA of Soybean Meal. *Int. J. of LCA* 13: 240-254.

de Wit, et al.. 2010. Competition Between Biofuels: Modeling Technological Learning and Cost Reductions Over Time. *Biomass & Bioenergy* 34: 203-217.

Earles, J. M., and A. Halog. 2011. Consequential Life Cycle Assessment: A Review. *Int. J. of LCA* 16: 445-453.

Ekvall, T., and A. Tillman. 1997. Open-Loop Recycling: Criteria for Allocation Procedures. *Int. J. of LCA* 2(3): 155-162.

Ekvall, T., and B. P. Weidema. 2004. System Boundaries and Input Data in Consequential Life Cycle Inventory Analysis. *Int. J. of LCA* 9: 161-171.

European Commission - Joint Research Centre - Institute

for Environment and Sustainability. International Reference Life Cycle Data System (ILCD) Handbook - General Guide for Life Cycle Assessment - Detailed Guidance. 2010, EUR 24708 EN, Luxembourg 2010.

Hofstetter, P., and G. A. Norris. 2003. Why and How Should We Assess Occupational Health Impacts in Integrated Product Policy? *Env. Science & Tech.* 37: 2025-2035.

Maclean, H. and L. Lave. 2003. Life Cycle Assessment of Automobile/Fuel Options. *Env. Science & Tech.*, 37: 5445-5452.

Majeau-Bettez, G., T. Hawkins, and A. Stromman. 2011. Life Cycle Environmental Assessment of Lithium-Ion and Nickel Metal Hydride Batteries for Plug-In Hybrid and Battery Electric Vehicles. *Env. Science & Tech.* 45: 4548-4554.

Marshall, A. Principles of Economics. Library of Economics and Liberty. Retrieved December, 2013, www.econlib.org/library/Marshall/marP33.html

Mueller, S. Detailed Report: 2008 National Dry Mill Corn Ethanol Survey. U. of Il. at Chicago, Energy Resources Center, May, 2010.

Nicholson, A.L., et al. 2009. End-Of-Life LCA Allocation Methods: Open Loop Recycling Impacts on Robustness of Material Selection Decisions. *Sustainable Systems and Tech.* 1-6.

Patzek, T. W. 2009. A First Law Thermodynamic Analysis of Biodiesel Production from Soybean. *Bulletin of Science Technology Society* 29(3): 194-204.

Pradhan, A., D. S. Shrestha, A. McAloon, W. Yee, M. Haas, and J. A. Duffield. 2011. Energy Life-Cycle Assessment of Soybean Biodiesel Revisited. Transactions of the Asabe 54: 1031-1039.

Samaras, C. and K. Meisterling. 2008. Life Cycle Assessment of Greenhouse Gas Emissions from Plug-In Hybrid Vehicles: Implications for Policy. *Env. Science & Tech.* 42: 3170-3176.

Sheehan, J., V. Camobreco, J. Duffield, M. Graboski, and H. Shapouri. Life Cycle Inventory of Biodiesel and Petroleum Diesel for Use in an Urban Bus. NREL/SR-580-24089,1998.

Small, K.A. and K. Van Dender. 2007. Fuel Efficiency and Motor Vehicle Travel: The Declining Rebound Effect. *Energy Journal* 28: 25-51.

US EPA. Documentation for Greenhouse Gas Emission and Energy Factors Used in the Waste Reduction Model (WARM). www.epa.gov/climatechange/waste/SWMGHGreport.html

Wang, M., et al. 2011. Energy and Greenhouse Gas Emission Effects of Corn and Cellulosic Ethanol with Technology Improvements and Land Use Changes. *Biomass and Bioenergy* 35: 1885-1896.

Weidema, B., N. Frees, and A. Nielsen. 1999. Marginal Production Technologies for Life Cycle Inventories. *Int. J. of LCA.* 4: 48-56.

CHAPTER 9

Bao, Y. Y. 2003. Nutrient distributions and their limitation on phytoplankton in the Yellow Sea and the East China Sea. *J. of Applied Ecology* , 14(7):1122-6.

Bolt, B. A. 2006. *Earthquakes: 5th Edition 2006 Centennial Update THe 1906 Big One.* New York: W.H. Freeman.

Cox, S. 2008. *Sick Planet: Corporate Food and Medicine.* London: Pluto Press.

Esty, D., & A. Winston. 2006. *GREEN TO GOLD.* New Haven, CT, USA: Yale University Press.

Federal Trade Commission. 2012, 10 01. *FTC Issues Revised "Green Guides".* Retrieved January 18, 2013, from Federal Trade Ccmmission: www.ftc.gov/opa/2012/10/greenguides.shtm

Federal Trade Commission. 1992. Green Guides. *16 Code of Federal Regulation Part 260* . Washington, DC: U.S. Printing Office.

Holzer, T., R. Gabrysch, and E. Verbeek. 1983. Faulting arrested by control of ground-water withdrawal in Houston, TX. *Earthquake Information Bulletin* , 15: 204 - 209.

Leopold, A. 1949. *A Sand County Almanac.* London: Oxford U. Press.

Mora, C., D. Tittensor, A. Simpson, and B. Worm. 2011. How Many Species Are There on Earth and In the Ocean. *PLoS Biol* , 9(8): e1001127.

United Nations Water. (n.d.). *Statistics: Graphs and Maps Water Resources.* Retrieved April 1, 2013, from United Nations : http://www.unwater.org/statistics_res.html

Zierenberg, R., M. Adams, & A. Arp. 2000. Life in extreme environments: hydrothermal vents. *Proceedings of the National Academy of Science* , 12961-12962.

CHAPTER 10

Butterfield, P. 2002. Upstream Reflections on Environmental Health: An Abbreviated History and Framework for Action. *Advances in Nursing Science* 25(1): 32–49.

Crettaz, P., et al. 2002. "Assessing Human Health Response in Life Cycle Assessment Using ED10s and DALYs: Part

1—Cancer Effects." *Risk Analysis* 22(5): 931–946.

CUAHSI. 2012. CUAHSI Portal. *Consortium of Universities for the Advancement of Hydrologic Science, Inc.* www.cuahsi.org/.

EC/COLIPA. 2013. SEURAT-1 - Towards the Replacement of in vivo Repeated Dose Systemic Toxicity Testing. www.seurat-1.eu/.

Egeghy, P., R. Judson, S. Gangwal, S. Mosher, D. Smith, J. Vail, and E. A. Cohen-Hubal. 2012. The Exposure Data Landscape for Manufactured Chemicals. *Science of the Total Env.* .414: 159–166.

Ernstoff, A., A. Henderson, P. Fantke, S. Chung, and O. Jolliet, in prep. Direct Consumer Product Exposure in a Life Cycle Assessment Framework: Defining the Product Intake Fraction.

Friis, R. *Essentials of Environmental Health.* Sudbury: Jones and Bartlett, 2007.

Friis, R., and T. Sellers. *Epidemiology for Public Health Practice.* Sudbury: Jones and Bartlett, 2009.

Goedkoop, M., et al. *ReCiPe 2008: Report 1: Characterisation.* The Netherlands: Ministry of Housing, Spatial Planning, and Env. (VROM), 2009.

Henderson, A.D., et al. 2011. USEtox Fate and Ecotoxicity Factors for Comparative Assessment of Toxic Emissions in LCA: Sensitivity to Key Chemical Properties. *Int. J. of LCA* 16(8): 701–709.

Huijbregts, et al. 2005. Human-Toxicological Effect and Damage Factors of Carcinogenic and Noncarcinogenic Chemicals. *Int. Env. Assessment and Mgmt.* 1(3): 181–244.

Humbert, S., V. Rossi, M. Margni, O. Jolliet, and Y. Loerincik. 2009. Life cycle Assessment of Two Baby Food Packaging Alternatives: Glass Jars Vs. Plastic Pots. *Int. J. of LCA.* 14(2): 95–106.

Johnson, S. *The Ghost Map: The Story of London's Most Terrifying Epidemic-And How it Changed Science, Cities, and the Modern World.* New York: Riverhead Books, 2006.

Krewitt, W., Det al. Indicators for Human Toxicity in Life-Cycle Impact Assessment. In *Life Cycle Impact Assessment,* ed. by H.de Haes, Pensacola: SETAC Press, 2002.

Lefor, A. 2005. Scientific Misconduct and Unethical Human Experimentation: Historic Parallels and Moral Implications. *Nutrition (Burbank, Calif.)* 21(7-8): 878–882.

McKone, T. 1993. *CalTOX, A Multimedia Total Exposure Model for Hazardous-Waste Sites.* Livermore: Lawrence Livermore National Laboratory.

Mehaffey, M., M. Nash, T. Wade, D. Ebert, K. Jones, and A. Rager. 2005. Linking Land Cover and Water Quality in New York City'S Water Supply Watersheds. *Env. Monit. and Assess.* 107(1-3): 29–44.

Milà i Canals, L., et al. 2007. Key Elements in a Framework for Land Use Impact Assessment within LCA. *Int. J. of LCA.* 12(1): 5–15.

Murray, C., and A. Lopez, eds. 1996. *Global Burden of Disease: A Comprehensive Assessment of Mortality and Disability from Diseases, Injuries, and Risk Factors.* 1st ed. Harvard Sch of Public Health.

NIEHS. 1999. Killer Environment: National Institute of Environmental Health Sciences Forum. *Env. Health Perspectives* 107(2): A62–A63.

Pekkanen, J. and N. Pearce. 2001. Environmental Epidemiology: Challenges and Opportunities. *Env. Health Perspectives* 109(1): 1.

Perkel, J. 2012. Life Science Technologies: Animal-Free Toxicology: Sometimes, In Vitro Is Better. *Science.*

Pfister, S., A. Koehler, and S. Hellweg. 2009. Assessing the Environmental Impacts of Freshwater Consumption in LCA. *Env. Science & Tech.* 43(11): 4098–4104.

Rebitzer, G., et al. 2004. Life Cycle Assessment: Part 1: Framework, Goal and Scope Definition, Inventory Analysis, and Applications. *Env. Int.* 30(5): 701–720.

Rosenbaum, R., T. Bachmann, L. Gold, et al. 2008. USEtox - The UNEP-SETAC Toxicity Model: Recommended Characterisation Factors. *Int. J. of LCA* .13(7): 532–546.

Rosenbaum, R., et al. 2011. USEtox Human Exposure and Toxicity Factors for Comparative Assessment of Toxic Emissions in LCA: Sensitivity to Key Chemical Properties. *Int. J. of LCA.* 16(8): 710–727.

Rosling, H. 2006. Stats That Reshape Your Worldview. Ted Talks: www.ted.com. www.ted.com/talks/hans_rosling_shows_the_best_stats_you_ve_ever_seen.html.

Roy, P., L. Deschênes, and M. Margni. 2012. Life Cycle Impact Assessment of Terrestrial Acidification: Modeling Spatially Explicit Soil Sensitivity. *Env. Science & Tech.* 46(15): 8270–8278.

Scavia, D. and S. B. Bricker. 2006. Coastal Eutrophication Assessment in the United States. *Biogeochemistry* 79(1-2): 187–208.

Smith, V.H. and D. W. Schindler. 2009. Eutrophication Science: Where Do We Go From Here?. *Trends in Ecology & Evolution* 24 (4): 201–207.

Tufte, E.R. *The Visual Display of Quantitative Information.* 2nd ed. Graphics Press, 2001.

Udo de Haes, H.A. 2006. How to Approach Land Use in LCIA Or, How to Avoid the Cinderella Effect? *Int. J. of LCA.* 11(4): 219–221.

Udo de Haes, Helias A., G. Finnveden, Mark Goedkoop, et al., eds. *Life Cycle Impact Assessment: Striving Towards Best Practice.* Pensacola: SETAC Press, 2002.

US Congress. 1991. Public Welfare: Department of Health and Human Services: Protection of Human Subjects. *Code of Federal Regulations.*

US EPA. 2013a. Integrated Risk Information System (IRIS). *Integrated Risk Information System (IRIS).* www.epa.gov/IRIS/.

US EPA. 2013b. Computational Toxicology Research Program. www.epa.gov/ncct/.

CHAPTER 11

Bare, J. 2011. TRACI 2.0: The Tool for the Reduction and Assessment of Chemical and Other Environmental Impacts 2.0. *Clean Technologies and Env. Policy* 13(5):

Bare, J. C., G. A. Norris, D. W. Pennington, and T. McKone. 2003. TRACI – The Tool for the Reduction and Assessment of Chemical and Other Environmental Impacts." *J. of Ind. Ecology* 6(3): 49-78.

Bare, J. C. 2012. Tool for the Reduction and Assessment of Chemical and other Environmental Impacts TRACI), US EPA, www.epa.gov/nrmrl/std/traci/traci.html

Carter, W.P.L. 1994. Development of Ozone Reactivity Scales for Volatile Organic Compounds. *J. of Air and Waste Mngt. Assn.* 44: 881-899.

Carter, W.P.L. 2012. SAPRC Atmospheric Chemical Mechanisms and VOC Reactivity Scales, University of California, Riverside, accessed March 2012, www.engr.ucr.edu/~carter/SAPRC/#mec12210

Derwent, R., M. Jenkin, S. Saunders, and M. Pilling. 1998. Photochemical Ozone Creation Potentials for Organic Compounds in Northwest Europe Calculated. *Atmospheric Env.* 32: 2429-2441

Dreicer, M., V. Tort, and P. Manen. 1995. ExternE, Externalities of Energy, Vol. 5 Nuclear, Centre d'étude sur l'Evaluation de la Protection dans le domaine nucléaire (CEPN), edit by JOULE, Luxembourg.

European Commission (2013) Annex II: Product Environmental Footprint (PEF) Guide to the Commission Recommendation on the use of common methods to measure and communicate the life cycle environmental performance of products and organisations DRAFT, Brussels, Belgium.

Frischknecht, R., A. Braunschweig, P. Hofstetter, and P. Suter. 2000. Human Health Damages Due to Ionizing Radiation in Life Cycle Impact Assessment. *Env. Impact Assess. Review* 20: 159–189.

Frischknecht, R., R. Steiner, and N. Jungbluth. 2009. The Ecological Scarcity Method: Eco-Factors 2006: A Method for Impact Assessment. Environmental Studies 0906.

Goedkoop, M., and R. Spriensma. 2001. The Eco-Indicator 99: A Damage Oriented Method for Life Cycle Impact Assessment. *Annex report Eco-indicator 99*: 1-81.

Goedkoop M. et al.. ReCiPe 2008: A Life Cycle Impact Assessment Method Which Comprises Harmonised Category Indicators at the Midpoint and the Endpoint Level.

Guinée, Jeroen, ed. 2002. *Handbook on Life Cycle Assessment. Operational Guide to the ISO Standards.* Dordrecht: Kluwer Academic Publishers

Hauschild, M., et al. 2008. Building a Model Based on Scientific Consensus for Life Cycle Impact: Assessment of Chemicals. *Env. Science and Tech.* 42(19): 7032-7036

Henderson A., et al. 2011. USEtox Fate and Ecotoxicity Factors for Comparative Assessment of Toxic Emissions in Life Cycle Assessment. *Int. J. of LCA.* 16(8): 701-709

Hettelingh, J.P., et al. 2007. Critical Loads and Dynamic Modelling to Assess European Areas at Risk of Acidification and Eutrophication. *Water Air Soil Pollution Focus* 7: 379–384.

Hofstetter P. 1998. *Perspectives in Life Cycle Impact Assessment. A Structured Approach to Combine Models of the Technosphere, Ecosphere and Valuesphere.* Boston: Kluwer Academic Publishers,

Huijbregts, M., M. Hauschild, O. Jolliet, M. Margni, T. McKone, R. Rosenbaum, and D. van de Meent. 2010. USEtox User Manual, accessed June 2013, www.usetox.org (unpublished manual).

Humbert S., et al. 2009. Assessing Regional Intake Fractions and Human Damage Factors in North America. *Science of the Total Env.* 407: 4812-4820.

Humbert, S., et al.. Intake Fraction for Particulate Matter: Recommendations for Life Cycle Impact Assessment. *Env. Science and Tech.* 45: 4808-4816.

ISO 14040. 2006. Environmental Management - Life Cycle Assessment -Principles and Framework." International Organisation for Standardisation (ISO), Geneva

ISO 14044. 2006. Environmental management - Life Cycle Assessment -Requirements and Guidelines. International Organisation for Standardisation (ISO), Geneva

Jolliet, O. et al. 2004. The LCIA Midpoint-Damage Framework of the UNEP/SETAC Life Cycle Initiative. *Int. J. of LCA.* 9(6): 394-404.

Jolliet, O., M. Margni, R. Charles, S. Humbert, J. Payet, G. Rebitzer, and R. Rosenbaum. 2003. "IMPACT 2002+: A New Life Cycle Impact Assessment Methodology. *Int. J. of LCA.* 8(6): 324-330.

Milà i Canals, et al. 2007. "Key Elements in a Framework for Land Use Impact Assessment within LCA." *International Journal of LCA* 12: 5-15

Milà i Canals, L., J. Romanyà, and S. Cowell. 2007. Method for Assessing Impacts on Life Support Functions (LSF) Related to the Use of 'Fertile Land' in LCA). *J Clean Prod* 15: 1426-1440

Milà i Canals, L., I. Muñoz, and S.J. McLaren. 2007. LCA Methodology and Modeling Considerations for Vegetable Production and Consumption. *CES Working Papers*

Müller-Wenk, R. 1998. *Depletion of Abiotic Resources Weighted on the Base of "Virtual" Impacts of Lower Grade Deposits in Future.* IWÖ Diskussionsbeitrag Nr. 57, Universität St. Gallen.

National Aeronautics and Space Administration (NASA). "Ozone," Earth Observatory – NASA, accessed June 16, 2013, http://earthobservatory.nasa.gov/Features/Ozone/

Norris, G. A. 2003. Impact Characterization in the Tool for the Reduction and Assessment of Chemical and Other Environmental Impacts. *J. of Ind. Ecology* 6(3-4): 79-101

OECD Environment Directorate. OECD Key Environmental Indicators 2004. Access May 2007, Posch, M., et al. 2008. Role of Atmospheric Dispersion Models and Ecosystem Sensitivity in the Determination of Characterization Factors. *Int. J. of LCA.* (13): 477–486.

Redfield, A. , B. Ketchum, and F. Richards. The influence of organisms on the composition of sea water." *Proceedings of the Second Int. Water Pollution Research Conf..* Oxford: Pergamon Press, 1963.

Rosenbaum, R., et al. 2008. USEtox-The UNEP-SETAC Toxicity Model: Recommended Characterization Factors for Human Toxicity and Freshwater Ecotoxicity. *Int. J. of LCA.* 13(7): 532-546.

Rosenbaum R., T. McKone, and O. Jolliet. 2009. CKow - A Dynamic Model for Chemical Transfer to Meat and Milk. *Env. Science and Tech.* 43(21): 8191-8198.

Rosenbaum, R., et al.. 2011. USEtox Human Exposure and Toxicity Factors for Comparative Assessment of Toxic Emissions in Life Cycle Analysis. *Int. J. of LCA.* 16(8), 710-727.

Rosenbaum, R., M. Margni, and O. Jolliet. 2007. A Flexible Matrix Algebra Framework for the Multimedia Multipathway Modeling of Emission to Impacts. *Environment Int.*33(5): 624-634

Seppälä, J., M. Posch, M. Johansson, and J. Hettelingh. 2006. Country-Dependent Characterization Factors for Acidification and Terrestrial Eutrophication. *Int. J. of LCA.* 11(6): 403-416.

Solomon et al. 2011. Technical Summary. Climate Change 2007: The Physical Science Basis. Contribution of Working Group I to the Fourth Assessment Report of the Intergovernmental Panel on Climate Change. Updated Nov. 2011. *In* IPCC. Cambridge U. Press, Cambridge, UK and NY, USA.

Struijs, J., H.. van Wijnen, A. van Dijk, and M. Huijbregts. 2009. Ozone layer depletion. *ReCiPe 2008*, accessed January, 2009, www.lcia-recipe.net

Struijs, J., A. Beusen, H. van Jaarsveld, and M. A. J. Huijbregts. 2009. Aquatic Eutrophication. *ReCiPe 2008*, accessed Jan., 2009, www.lcia-recipe.net

Tarrason, L., et al. 2006. Transboundary Acidification, Eutrophication and Ground Level Ozone in Europe from 1990 to 2004 in Support for the Review of the Gothenburg Protocol. Oslo, Norway.

United Nationals Framework Convention on Climate Change (UNFCCC). 2008. Kyoto Protocol Reference Manual: access Jun 2014, http://unfccc.int/resource/docs/publications/08_unfccc_kp_ref_manual.pdf.

US Environmental Protection Agency. 2008. Ozone Layer Depletion – Science, www.epa.gov/ozone/science/sc_fact.html

US Environmental Protection Agency. 2008. Ozone Layer Depletion - Science, Class I Ozone-depleting Substances,_ww.epa.gov/ozone/science/ods/classone.html

US Environmental Protection Agency. 2008. Ozone Layer Depletion - Science, Class II Ozone-depleting Substances.,www.epa.gov/ozone/science/ods/classtwo.html

US Environmental Protection Agency. 2013. Six Common Pollutants: Particulate Matter. www.epa.gov/pm/health.html

World Meteorological Organization (WMO). 2003. Scientific Assessment of Ozone Depletion: 2002, Global Ozone Research and Monitoring Project—Report No. 47. Geneva, Switzerland.

Wenzel, H., and M. Hauschild.1997. *Environmental Assessment of Products, Volume 2: Scientific Backgrounds.* London, Springer Science & Bus. Media.

Wenzel, H., M. Z. Hauschild, and L. Alting. 1997. *Environmental Assessment of Products, Volume 1: Methodology, Tools and Case Studies in Product Development.* Springer Science & Bus. Media.

CHAPTER 12

Ahbe, S., A. Braunschweig, and R. Müller-Wenk. Methodik für Oekobilanzen auf der Basis ökologischer Optimierung. Schriftenreihe Umwelt Nr 133, Bundesamt für Umwelt, Wald und landschaft (BUWAL), Bern, 1990 Reviewed in Nordic Guidelines on LCA, 1995:20, Nordic Council of Ministers, Copenhagen 1995.

Alroth, S., M. Nilsson, G. Finnveden, O. Hjelm, and E. Hochschorner. 2011. Weighting and Evaluation in Environmental System Analysis Tools. *J. of Cleaner Prod.* 19(2-3): 145-156.

Bare, J., and T. Gloria. 2006. Critical Analysis of the Mathematical Relationships and Comprehensiveness of Life Cycle Impact Assessment Approaches. Environmental Science and Technology 40(4): 1104-1113.

Brundtland, G., et al. 1987. Report of the World Commission on Environment and Development: *Our Common Future,* Cambridge, UK, and New York: Cambridge University Press.

Figueira, J., S. Greco, and M. Ehrgott, eds. *Multiple Criteria Decision Analysis: State of the Art Surveys.* New York: Springer Science & Business Media, 2005.

Finnveden, G. 1996. Valuation Methods within the Framework of Life Cycle Assessment, Swedish Environmental Research Institute, IVL-Report B 1231, Sweden

Frischknecht, R., et al.. 2007. Implementation of Life Cycle Impact Assessment Methods. *Ecoinvent report No. 3:* v2.0. Swiss Centre for Life Cycle Inventories, Dübendorf.

Gloria, T., B. Lippiatt, and J. Cooper. 2007. Life Cycle Impact Assessment Weights to Support Environmentally Preferable Purchasing in the United States. Environmental Science and Technology 41(21): 7551-7557.

Goedkoop, M., and R. Spriensma. 1999. The Eco-indicator 99: A Damage Oriented Method for Life Cycle Impact Assessment., Pré Consultants, Amersfoort, The Netherlands *Annex report EI 99*: 1-81.

Goodkoop, M., et al.. 2012. ReCiPe 2008: A Life Cycle Impact Assessment Method Which Comprises Harmonised Category Indicators, accessed December 2013, www.lcia-recipe.net

Hediger, W. 2000. Sustainable Development and Social Welfare. *Ecological Economics* 32: 481-492.

ILCD Handbook, (2010) JRC, European Commission, Institute for Environment and Sustainability

ISO (2006a) ISO 14040: Environmental Management–Life Cycle Assessment–Principles and Framework. ISO 14040:2006(E), International Standards Organization

ISO (2006b) ISO 14044: Environmental Management–Life Cycle Assessment–Requirements and Guidelines. ISO 14044:2006(E), International Standards Organization

Itsubo, N., and A. Inaba. 2003. A New LCA Method: LIME Has Been Completed. *Int. J. of LCA* 8(5): 305.

Johnsen, F., F. Moltu, and S. Løkke. 2013. Review of Criteria for Evaluating LCA Weighting Methods. *Int. J. of LCA* 18(4); 840-849.

Lippiatt, B. 2007. Building for Environmental and Economic Sustainability Technical Manual and User Guide 4.0. National Institute of Standards and Technology, Gaithersburg, Maryland

Majeau-Bettez, G., T. Hawkins, and A. Hammer Strømman. 2011. Life Cycle Environmental Assessment of Lithium-Ion and Nickel Metal Hydride Batteries. *Environ. Sci. and Technol.* 45: 4548–4554.

Munda, G. 2005. Multiple Criteria Decision Analysis and Sustainable Development. *Multiple Criteria Decision Analysis.* Figueira, ed. Springer Int. Series in Operations Res.and Mngt. Science, 953-986.

Norris, G. 2001. The Requirement for Congruence in Normalization. *Int. J. of LCA*, 6(2): 85-88.

Reap, J., F. Roman, S. Duncan, and B. Bras. 2008. A Survey of Unresolved Problems in Life Cycle Assessment: Part 2: Impact Assessment and Interpretation. *Int. J. of LCA* 13:374–388.

Rowley, H., G. Peters, S. Lundie, and S. Moore. 2012. Aggregating Sustainability Indicators: Beyond the Weighted Sum. *J. of Environmental Management* 111:24e33.

Steen, B. 1997. On Uncertainty Priority Setting and Sensitivity of LCA-based Priority Setting. *J. Cleaner Production.* 5(4): 255-262.

Steen, B. 1999. A Systematic Approach to Environmental Priority Strategies in Product Development (EPS), Version 2000. – Models and Data, Chalmers U.of Tech., Centre for Env. Assess. (CPM) Report 1999:5, Gothenburg 1999

Steen, B. 2006. Describing Values in Relation to Choices in

LCA." *Int. J. of LCA* 11(3): 277-283.

Vogtländer, J., H. Brezet, and C. Hendriks. 2000. "The Virtual Eco-Costs '99, A Single LCA-Based Indicator for Sustainability and the Eco-Costs – Value Ratio (EVR) Model for Economic Allocation." *Int. J. of LCA* 5(6).

White, P., and M. Carty. 2010. Reducing Normalization Bias through Inventory Dataset Normalization. *Int. J LCA* 15(9):994-1013.

CHAPTER 13

Frischknecht, R., N. Jungbluth, H. J. Althaus, G. Doka, T. Heck, S. Hellweg, R. Hischier, T. Nemecek, G. Rebitzer, M. Spielmann, and G. Wernet. 2007. Overview and Methodology. *Ecoinvent report No. 1.*

Goedkoop, M., and R. Spriensma. 2001. The Eco-indictator 99: A Damage Oriented Method for Life Cycle Impact Assessment Methodology Report, accessed June, 2011, www.pre.nl

Guinée, J. B., et al. 2001. *Life Cycle Assessment: An Operational Guide to the ISO Standards.*

Heijungs, R., Goedkoop, M., Struijs, J., Effting, S., Sevenster, M., Huppes, G. 2003. Towards a Life Cycle Impact Assessment Method which Comprises Category Indicators at the Midpoint and the Endpoint Level. Retrieved from www.leidenuniv.nl/cml/ssp/publications/recipe_phase1.pdf.

Helling, R., R. Paludetto, A. Benvenuti, and G. Lista. 2012. The Impact of Bio-Feedstock Choice on the LCA of Renewable Shoe Soles. Proceedings LCA XI conf., American Center for LCA, Tacoma, WA.

Hoefnagels, R., E. Smeets, and A. Faaij. 2010. Greenhouse Gas Footprints of Different Biofuel Production Systems. *Renewable and Sustainable Energy Reviews* 14: 1661-1694.

ISO (2006). ISO 14044: Environmental management – Life cycle assessment – Requirements and guidelines, International Standards Organisation.

Keiji, K. 2004. Beer and Health: Preventive Effects of Beer Components on Lifestyle-Related Diseases. *BioFactors* 22(1-4): 303-310.

Norris, G. 2001. The Requirement for Congruence in Normalization. *Int. J. of LCA* 6(2), 85-88.

Schmidt, Mario. 2006. Der Einsatz von Sankey-Diagrammen im Stoffstrommanagement. *Beitraege der Hochschule Pforzheim*, 124.

Weber, C. L., P. Jaramillo, J. Marriott, and C. Samaras. 2010. Life Cycle Assessment and Grid Electricity: What Do We Know and What Can We Know? *Env. Science & Tech.* 44 6), 1895-1901.

CHAPTER 14

Aitchison, J., and J. Brown. *The Lognormal Distribution.* Cambridge: Cambridge U. Press, 1957.

Barker, N. A Practical Introduction to the Bootstrap Using the SAS System. Presented at SAS Conf. Heidelberg, Germany, Oct. 2005.

Blackwood, L. G. 1992. The Lognormal Distribution, Environmental Data, and Radiological Monitoring. *Env. Monitoring and Assessment* 21(3): 193-210.

Central Tendency. Retrieved Jan 2013, http://en.wikipedia.org/wiki/Central_tendency

Cumming, G. *Understanding the New Statistics: Effect Sizes, Confidence Intervals, and Meta-Analysis.* New York: Routledge, 2012.

Dalgaard, P. *Introductory Statistics with R.* Springer, 2008.

Easton, V. J., and J. H. McColl. Statistics Glossary (Vol. 1.1.) Retrieved Jan. 2013. www.stats.gla.ac.uk/steps/glossary/presenting_data.html#catdat

Forbes, C., M. Evans, N. Hastings, and B. Peacock. *Statistical Distributions.* Wiley Online Library, 2011.

Gelman, A., J. Carlin, H. Stern, and D. Rubin. *Bayesian Data Analysis.* Chapman & Hall/CRC, 2003.

Gelman, A., & Hill, J. *Data Analysis Using Regression and Multilevel/Hierarchical Models.* Cambridge; Cambridge U. Press, 2006.

ISO14040. 2006. Environmental Management - Life Cycle Assessment - Principles and Framework: Int. Standards Organization.

ISO14044. 2006. Environmental Management–Life Cycle Assessment–Requirements and Guidelines: Int. Standards Organization.JMP Statistical Discovery Software. Retrieved Jan., 2013, www.jmp.com/software/jmp10/

Kiemele, M. , S. Schmidt, and R. Berdine. *Basic Statistics: Tools for Continuous Improvement* (Fourth Edition). Air Academy Press & Associates, 1999.

Kruschke, J. *Doing Bayesian Data Analysis: A Tutorial Introduction with R and BUGS.* Academic Press, 2010.

Levine, D., R. Smidt, and P. Ramsey. *Applied Statistics for Engineers and Scientists.* Upper Saddle River, NJ. Prentice Hall, 2001.

Mendenhall, W., and T. Sincich. *Statistics for Engineering and the Sciences.* Prentice-Hall, Inc, 2006.

Minitab 16. 2013. www.minitab.com/en-US/products/minitab/

Olsson, U. 2005. Confidence Intervals for the Mean of a Log-Normal Distribution. *J. of Statistics Edu.* 13(1): n1.

Ott, W. R. 1990. A Physical Explanation of the Lognormality

of Pollutant Concentrations. *J. of the Air & Waste Mngt. Assn* 40(10): 1378-1383.

Ott, W. R. 1994. *Environmental Statistics and Data Analysis.* CRC.

Reimann, C., and P. Filzmoser. 2000. Normal And Lognormal Data Distribution In Geochemistry. Consequences For The Statistical Treatment Of Geochemical And Environmental Data.*Env. Geology* 39(9): 1001-1014.

Sauro, J. Usable Stats. Retrieved Jan, 2013, www.usablestats.com/lessons/datatypes2

STAT 415 Intro to Mathematical Statistics. Retrieved Jan 2013, https://onlinecourses.science.psu.edu/stat414/node/261

Statgraphics Centurion XVI.I. www.statgraphics.com/

Weidema, B. P. 1998. Multi-User Test of the Data Quality Matrix for Product Life Cycle Inventory Data. *Int. J. of LCA.* 3(5): 259-265.

Zhou, X. H., and S. Gao. 1997. Confidence Intervals for the Log Normal Mean. *Statistics in Medicine* 16(7): 783-790.

CHAPTER 15

Bengtsson, M., and B. Steen. 2000. Weighting in LCA- Approaches and Applications. *Env. Progress* 19(2): 101-109.

Dallal, Gerard E. 2012. *The Little Handbook of Statistical Practice.*

De Schryver, A. (2010). Value choices in life cycle impact assessment. PhD Thesis. Radboud University, Nijmegen. The Netherlands

DeSmet, B. and S. M. 1996. LCI Data and Data Quality. *Int. J. of LCA* 1(2): 96-104.

Ekvall, T., and A. Tillman. 1997. Open-Loop Recycling: Criteria for Allocation Procedures. *Int. J. of LCA* 2(3): 155-162.

Finnveden, G. 1999. A Critical Review of Weighting Methods for Life Cycle Assessment, in FMS, Forskningsgruppen för miljöstrategiska studier. Stockholms Universitet/Systemekologioch FOA.

Finnveden, G. 1999. Methodological Aspects of Life Cycle Assessment of Integrated Solid Waste Management Systems. *Resources, Conservation, and Recycling* 26: 173–187.

Frischknecht, R. and N. Jungbluth 2007. EcoInvent Data v2.0: Overview and Methodology, in EcoInvent Report No. 1. Swiss Centre for Life Cycle Inventories: Dübendorf.

Goedkoop, M., A.D. Schryver, and M. Oele. 2006. SimaPro 7: Introduction to LCA. Pré Consultants.

Goodman, S.N. 1999. Toward Evidence-Based Medical Statistics. 1: The P Value Fallacy. *Annals of Internal Medicine* 130: 995–1004.

Hanssen, O.J., and A. A. Asbjornsen. 1996. Statistical Properties of Emission Data in Life Cycle Assessments. *J. Cleaner Prod.* 4(3-7094): 149–57. 710

Hertwich, E., T. McKone, and W. Pease. 2000. A Systematic Uncertainty Analysis of an Evaluative Fate and Exposure Model. *Risk Analysis* 20(4): 439–454.

Huijbregts, M. 1998. Application of Uncertainty and Variability in LCA. Part I: A General Framework for the Analysis of Uncertainty and Variability in Life Cycle Assessment. *Int. J. of LCA* 3(5): 273-280.

Huijbregts, M., U. Thisssen, T. Jager, D. Van de Meent, and A. Ragas. 2000. Priority Assessment of Toxic Substances in Life Cycle Assessment (Part II). *Chemosphere* 41(4): 575–88.

Huijbregts, M. Modeling Data Uncertainty in Life Cycle Inventories. Draft paper prepared for the SETAC Data Quality Workgroup, 2000.

Huijbregts, M. 2011. A Proposal for Uncertainty Classification and Application in PROSUITE. PROSUITE.

ISO 14040. 1997. Environmental Management - Life Cycle Assessment - Principles and Framework. International Organization for Standardization: Geneva.

ISO 14041. 1998. Environmental management – Life Cycle Assessment – Goal and Scope Definition and Inventory Analysis. in International Organisation for Standardisation (ISO). Geneva.

ISO 14043. 2000. Environmental Management – Life Cycle Assessment – Life Cycle Interpretation. in International Organisation for Standardisation (ISO). Geneva.

Jungbluth, N. et al. 2012. *Pedigree Matrix for Estimation of Uncertainties, Annex to the Quality Guidelines.* ecoivent 2000 Team, ESU services, Ulster Switzerland

Lee, J., P. Callaghan, and D. Allen. 1995. Critical Review of Life Cycle Analysis and Assessment Techniques and their Application to Commercial Activities. *Resources, Conservation and Recycling* 13: 37-56.

Lindfors, L.G., et al. 1995. Data Quality. In Tech. Report No. 5, TemaNord 1995:502, Nordic Council of Min. Copenhagen.

Maurice, B., T. Frischknecht, V. Coehlo-Schwirtz., and K. Ungerbuhler. 2000. Uncertainty Analysis in Life Cycle Inventory –Production of Electricity in Coal Power Plants. *J. Cleaner Product* 8(2): 95-108.

Pré Consultants. 2000. *Eco-Indicator 99 Manual for Designers: A Damage Oriented Method for Life Cycle Impact Assessment.*

Ros M. Unsicherheit. 1998. und Fuzziness in o¨kologischen Bewer-tungen—Orientierungen zu einer robusten Praxis der Okobilanzierung. ETH Zurich. Ph.D. Thesis.

Ross, S., et al. 2001. "How LCA Studies Deal with Uncertainty." *Int. J. of LCA* 7(1): 47-52.

Saur, K., J. Fava, and S. Spatari 2000. Life Cycle Case Study: Automobile Fenders. *Env. Progress* 19(2): 72-82.

Steen, B. 1997. On Uncertainty and Sensitivity of LCA-Based Priority Setting. *J. of Clean. Prod.* 5(4): 255-262.

Steen, B. 1999. A Systematic Approach to Environmental Priority Strategies in Product Development (EPS). Version 2000 General System Characteristics, CPM Report 1999: 4., Centre for Env. Asmt. of Pro. and Mat. Sys., Chalmers U.

USEPA. *Guidelines for Assessing the Quality of Life Cycle Inventory Analysis.* Office of Solid Waste, U.E.P.A. Office of Solid Waste, Editor. Washington, DC, 1995.

Vigon, B. and M.A. Curran. 1993. Life Cycle Improvements Analysis: Procedure Development and Demonstration" In IEEE Int. Symp. of Electron. and the Env.. Arlington, Virginia: Inst. of Electr. and Electron. Engineers,, Piscataway, NJ.

Weidema, B., and M. Wesnæs. 1996.Data Quality Management for Life Cycle Inventories — An Example of Using Data Quality Indicators. *J. of Clean. Prod.* 4: 167–174.

Weidema, B., et al. 1999. Marginal Production Technologies for Life Cycle Inventories. *Int. J. of LCA* 4(1): 48-56.

Weidema, B., et al. 2003.Reducing Uncertainty in LCI: Developing a Data Collection Strategy. *Env. Project* 862. Danish EPA: Copenhagen.

WhiteWave Foods Company. 2012. A Comparative Life Cycle Assessment of Plant Based Beverages and Conventional Dairy Milk.

Chapter 16

Benoît, C., G. Norris, S. Valdivia, A. Ciroth, A. Moberg, U. Bos, and S. Prakash, S. 2010. The Guidelines for Social Life Cycle Assessment of Products: Just In Time. *Int. J. of LCA.* 15(2): 156–163.

Benoît-Norris, C. 2012. Social Life Cycle Assessment: A Technique Providing a New Wealth of Information to Inform Sustainability-Related Decision Making. In *Life Cycle Assessment Handbook.* M. Curran, ed. 433–452.

Scrivener Publishing LLC.

Benoît-Norris, C. 2013. Data for Social LCA. *Int. J. of LCA.* accessed Sep 2013, http://link.springer.com/article/10.1007/s11367-013-0644-7.

Brammer, S., S. Hoejmose, and A. Millington. 2011. Managing Sustainable Global Supply Chains. Network for Business Sus. accessed Aug 2013, www.nbs.net/knowledge/supply-chains.

Ciroth, A. 2008. Cost Data Quality Considerations for Eco-Efficiency Measures. *Ecological Economics* 68(16): 1583-1590.

Ciroth, A., and J. Franze. *LCA of an Ecolabeled Notebook - Consideration of Social and Environmental Impacts Along the Entire Life Cycle.* Berlin: 2011.

Drury, C. *Management and Cost Accounting.* London: Thomson, 2005.

Ehrenfeld, J. 1997. The Importance of LCA—Warts and All. *J. of Ind. Ecology* 1(2): 41–49.

Elkington, J. *Cannibals with Forks: The Triple Bottom Line of the 21ˢᵗ Century Business.* Oxford: Capstone Publishing, 1997.

Finkbeiner, M. 2009. Carbon Footprinting - Opportunities and Threats. *Int. J. of LCA.* 14: 91-94.

Gluch, P. and H. Baumann. 2004. The Life Cycle Costing (LCC) Approach: A Conceptual Discussion of its Usefulness for Environmental Decision-Making. *Building and Environment* 39(5): 571-580.

Grießhammer, R., et al. 2006. Feasibility Study: Integration of Social Aspects into LCA. Freiburg, retrieved from http://lcinitiative.unep.fr/.

Hunkeler D., K. Lichtenvort, and G. Rebitzer. eds. *Environmental Life Cycle Costing.* Pensacola: SETAC, 2008.

[ISO] International Standards Organization. 2006a. *International Standard ISO 14040: Environmental Management – Life Cycle Assessment – Principles and Framework.* Geneva (CH): ISO.

[ISO] International Standards Organization. 2006b. *International Standard ISO 14044: Environmental Management – Life Cycle Assessment – Requirements and Guidelines.* Geneva (CH): ISO.

Jørgensen, A., L. Dreyer, and A. Wangel. 2012. Addressing the Effect of Social Life Cycle Assessments. *Int. J. of LCA.* 17(6): 828–839.

Jørgensen, A., M. Hauschild, M. Jørgensen, and A. Wangel. 2009. Relevance and Feasibility of Social Life Cycle Assessment from a Company Perspective. *Int. J. LCA.* 14(3): 204–214.

Klöpffer, W. 2003. Life-Cycle Based Methods for Sustainable Product Development. *Int. J. of LCA*. 8(3): 157-159.. 2006. Social Sustainability Social Indicators. *Int. J. of LCA*. 11(1): 3–15.

Moberg, A., et al. 2009. Using a Life-Cycle Perspective to Assess Potential Social Impacts of ICT Services - A Pre-Study. KTH Centre for Sustainable Communications, TRITA-SUS ISSN: 1654-479X; 2009:1.

Norris, G.A. 2006. Social Impacts in Product Life Cycles Towards Life Cycle Attribute Assessment. *Int. J. of LCA*. 1(1): 97–104.

O'Brien, M., A. Doig, and R. Clift. 1996. Social and Environmental Life Cycle Assessment (SELCA) *Int. J. of LCA*. 1(4): 231-237.

Parent, J., C. Cucuzzella, and J. Revéret. 2010. Impact Assessment in SLCA: Sorting the SLCIA Methods According to Their Outcomes. *Int. J. of LCA*. 15(2): 164–171.

Pesonen, H. L. 1999. From Material Flows to Cash Flows, An Extension to Traditional Material Flow Modelling. Jyväskylä (FI): University of Jyväskylä.

Pesonen, H. L., and S. Horn. 2012. Evaluating the Sustainability SWOT as a Streamlined Tool for Life Cycle Sustainability Assessment. *Int. J. of LCA*, DOI 10.1007/s11367-012-0456-1.

Robertson, R., et al., eds. 2009. Globalization, Wages and the Quality of Jobs. The World Bank: DC.

Swarr T., D. Hunkeler, W. Klöpffer, H. Pesonen, A. Ciroth, A. Brent, and R. Pagan. *Environmental Life Cycle Costing: A Code of Practice*. Pensacola: SETAC, 2011.

Traverso, M., F. Asdrubali, A. Francia, and M. Finkbeiner. 2012. Towards Life Cycle Sustainability Assessment: An Implementation to Photovoltaic Modules. *Int. J. of LCA*. 17: 1068-1079.

[UNEP] United Nations Environmental Programme. 2010. Methodological Sheets of Sub-Categories of Impact for a Social LCA. Accessed Nov 2011, http://lcinitiative.unep.fr.

[UNEP and SETAC] United Nations Environmental Programme and Society of Environmental Toxicology and Chemistry. *Towards a Life Cycle Sustainability Assessment*. Valdivia, S., C. Ugaya, G. Sonnemann, and J. Hildenbrand, eds. Paris: UNEP/SETAC Life Cycle Initiative, 2011.

Weidema, B. P. 2005. "ISO 14044 Also Applies to Social LCA." *Int. J. of LCA*. 10(6): 381.

Weidema, B. P. 2006. The Integration of Economic and Social Aspects in Life Cycle Impact Assessment. *Int. J. of LCA*.
11(1): 89–96.

World Economic Forum. 2012. *The Shifting Geography of Global Value Chains*. Access Aug. 2013, www.voxeu.org/article/shifting-geography-global-value-chains-implications-developing-countries-and-trade-policy.

Yamaguchi, H., N. Itsubo, S. Lee, M. Motoshita, A. Inaba, N. Yamamoto, and Y. Miyano. 2007. *Lifecycle Management Methodology Using Lifecycle Cost Benefit Analysis for Washing Machines*. Zürich (CH): 3rd Int. Conference on LCManagement.

Zamagni, A., O. Amerighi, and P. Buttol. 2011. Strengths or Bias in Social LCA? *Int. J. of LCA*. 16(7): 596–598.

CHAPTER 17

Bizmanualz. 2013. The Top Ten Reasons Why You Need Project Management. Retrieved from www.bizmanualz.com/blog/the-top-ten-reasons-why-you-need-project-management.html

EU JRC. 2010. International Reference Life Cycle Data System (ILCD) Handbook: General Guidance Document for Life Cycle Assessment (LCA). www.bookshop.europe.eu, LBNA24708ENC-002.pdf)

EU JRC. 2011. International Reference Life Cycle Data System (ILCD) Handbook: Recommendations for Life cycle Impact Assessment in the European Context. http://publications.jrc.ec.europa.eu/repository/handle/111111111/26229?mode=full

GHG Protocol. 2011. Product Life Cycle Accounting and Reporting Standard. www.ghgprotocol.org/standards

Heizer, J. H., and B. Render. *Principles of Operations Management*. New Jersey: Pearson Prentice Hall, 2008.

ISO14040:2006. Principles and Framework.

ISO14044:2006. Requirements and Guidelines.

ISO/TS 14048:2002. Data Documentation Format.

ISO/TR 14047:2003. Examples of Application of ISO 14042.

ISO/TR 14049:2000. Examples of Application of ISO 14041 to Goal and Scope Definition and Inventory Analysis.

CHAPTER 18

Caroselli, M. *Leadership Skills for Managers*. New York: McGraw-Hill, 2000.

Hensler, C. 2010. Effectively Communicating LCA Results in Industry, Internally and Externally. LCA X, Nov. 2010, Portland, OR, available at http://lcacenter.org/lcax/presentations-final/27.pdf

Hodder, C. Best Practices in LCA Results Communication. LCA XI Short-Course, October 3, 2011, Chicago, IL

Hunter, S., and M. McCaffrey. 2010. The Anatomy of Communicating LCA. LCA X, Nov. 2010, Portland, OR, http://lcacenter.org/lcax/presentations-final/216.pdf

ISO 14025:2006. Environmental Labels and Declarations – Type III Environmental Declarations – Principles and Procedures.

ISO 14044:2006. Environmental Management – Life Cycle Assessment – Requirements and Guidelines

Laurin, L., V. Prado-Lope, and T. Seager. Using Only Characterization Results in LCA—Is This the Best We Can Do? Paper presented at the LCA XIII, October 1-3, Orlando, FL., http://lcacenter.org/lcaxiii/abstracts/abstract-dynamic.php?id=891.

Molina-Murillo, S., and T. Smith. 2009. Exploring the Use and Impact of LCA-Based Information in Corporate Communications. *Int. J. of LCA. 14*(2): 184-194. doi: 10.1007/s11367-008-0042-8

Nissinen, et al. 2007. Developing Benchmarks for Consumer-Oriented Life Cycle Assessment-Based Environmental Information on Products, Services and Consumption Patterns. *J. of Cleaner Production* 15(6): 538-549.

Schenck, R. 2000. *LCA for Mere Mortals.* The Institute for Environmental Research and Education. www.lcacenter.org/lca-for-mere-mortals.aspx

U.S. Federal Trade Commission. 2012. Environmental Claims: Summary of the Green Guides. http://business.ftc.gov/documents/environmental-claims-summary-green-guides

Chapter 19

Big Room, and WRI. Global Ecolabel Monitor 2010.Washington, DC: Big Room and the World Resources Insitute, 2010.

Bortzmeyer, Martin. *Preliminary Outcomes of the French Experiment.* Paper presented at the PCF World Forum, 2012, Berlin, Germany, www.youtube.com/watch?v=IQGv-DiIoGhE

BSI. PAS 2050: 2011 Specification for the Assessment of the Life Cycle Greenhouse Gas Emissions of Goods and Services. British Standards Institute, 2011.

Christiansen, K., Wesnæs, and B. Weidema. 2006. Consumer Demands on Type III Environmental Declarations Report Commissioned by ANEC–The Consumer Voice in Standardisation. *www.anec.org/attachments/ANEC-R&T.*

Costello, A., and R. Schenck. 2009. Environmental Product Declarations and Product Category Rules. webinar for the US EPA: American Center for LCA.

Cros, C., E. Fourdrin, and O. Réthoré. 2010. The French Initiative on Environmental Information of Mass Market Products. *Int. J. of LCA.* 15: 537-539.

Del Borghi, Adriana. 2012. LCA and Communication: Environmental Product Declaration. *Int. J. of LCA.* 1-3. doi: 10.1007/s11367-012-0513-9

EC. 2013. Communication from the Commission to the European Parliament and the Council: Building the Single Market for Green Products Facilitating Better Information on the Environmental Performance of Products and Organisations. COM 2013/0916/final. Brussels.

Environdec. 2010. Product Category Rules CPC Class 2211 Processed Liquid Milk. PCR 2010:12. Version 1.0: The International EPD System.

Finkbeiner, Matthias. 2009. Carbon Footprinting—Opportunities and Threats. *Int. J. of LCA.* 14(2): 91-94. doi: 10.1007/s11367-009-0064-x

FTC. Guides for the Use of Environmental Marketing Claims. Proposed Rule *Federal Register.* Washington, DC: Federal Trade Commission, 2010.

Golden, J. , V. Subramanian, and J. Zimmerman. 2011. Sustainability and Commerce Trends. *J. of Industrial Ecology,* 15(6): 821-824. doi: 10.1111/j.1530-9290.2011.00381.x

Grant, Jason. 2011. [Personal communication].

IERE. Earthsure Environmental Product Declarations General Program. Tacoma: Institute for Env. Research and Education, 2010.

IERE. Beer Product Category Rule. Earthsure 50202201. Tacoma: Institute for Envl Research and Education, 2012.

Ingwersen, W., and V. Subramanian. 2013. Guidance for Product Category Rule Development, v1.0. The Guidance for Product Category Rule Development Initiative, www.pcrguidance.org.

Ingwersen, W., S. Clare, D. Acuña, M. Charles, C. Koshal, and A. Quiros. *Environmental Product Declarations: An Introduction and Recommendations for Their Use in Costa Rica.* Gainesville, FL: U. of Florida Levin College of Law Conservation Clinic, 2009.

Ingwersen, W. and M. Stevenson. 2012. Can We Compare the Environmental Performance of This Product to That One? *J. of Cleaner Production* 24: 102-108. doi: 10.1016/j.jclepro.2011.10.040

ISO 21930. ISO 21930 Sustainability in Building Construction — Environmental Declaration of Building Products. Geneva: International Standards Organization, 2007.

ISO. 14025: Environmental Labelling and Declarations – Type III Environmental Declarations – Principles And Procedures *International Standard*. Geneva, Switzerland: International Organization for Standardization, 2006.

ISO/CD 14046. Environmental Management -- Water Footprint -- Principles, Requirements and Guidelines. International Standards Organization, 2012.

Jungbluth, N., S. Büsser, R. Frischknecht, K. Flury, and M. Stucki. 2012. Feasibility of Environmental Product Information Based on Life Cycle Thinking. *Journal of Cleaner Production* 28(0): 187-197.

Levy, Mi., D. Wisner, J. Phelan, and G. Pavlovich. Development of a Product Category Rule for Building Envelope Thermal Insulation in the U.S. LCA XI Conference, Oct. 2011, Chicago, IL.

Molina-Murillo, S., and T. Smith. 2009. Exploring the Use and Impact of LCA-based information in Corporate Communications. *Int. J. of LCA.t* 14(2): 184-194. doi: 10.1007/s11367-008-0042-8

O'Shea, T., J. Golden, and L. Olander. 2012. Sustainability and Earth Resources: Life Cycle Assessment Modeling." *Business Strategy and the Env.*, n/a-n/a. doi: 10.1002/bse.1745

PlasticsEurope. 2011. Eco-profiles: PlasticsEurope and Eco-Profiles" Retrieved December, 2012, www.plasticseurope.org/plastics-sustainability/life-cycle-thinking/more-on-lct/eco-profiles.aspx

Scarinci, C., C. Peña, W. Ingwersen, and V. Subramanian. 2012. PCR Guidance Development Process and its Importance to the Latin American Region. Presented at the CILCA 2013, Mendoza.

Schenck, R. The Outlook and Opportunity for Type III Environmental Product Declarations in the United States of America *White Paper*. Vashon: Institute for Env. Research and Education, 2009.

Staffin, E. 1996. Trade Barrier or Trade Boon-A Critical Evaluation of Environmental Labeling and Its Role in the Greening of World Trade. *Colum. J. Envtl L.* 21: 205.

Strazza, C., A. Del Borghi, G. Blengini, and M. Gallo. 2010. Definition of the Methodology for a Sector EPD: Case Study Of Italian Cement. *Int. J. of LCA.* 15(6), 540-548. doi: 10.1007/s11367-010-0198-x

Subramanian, V., W. Ingwersen, C. Hensler, and H. Collie. 2012. Comparing Product Category Rules From Different Programs. *Int. J. of LCA.* 17(7): 892-903. doi: 10.1007/s11367-012-0419-6

TerraChoice. The 'Six Sins of GreenwashingTM' A Study of Environmental Claims in North American Consumer Markets. Ottawa, Canada: TerraChoice Environmental Marketing Inc., 2007.

Tesco. 2012. Product Carbon Footprint Summary.

The Keystone Center. 2012. Green Products Roundtable. Retrieved Dec.2012, www.keystone.org/policy-initiatives-center-for-science-a-public-policy/environment/green-products-roundtable.html

United Nations. 2009. A Guide To Environmental Labels - For Procurement Practitioners.

US EPA. 2012. Greener Products. Retrieved Dec. 2012, www.epa.gov/greenerproducts

SOLUTIONS

PROBLEM EXERCISE SOLUTIONS

CHAPTER 1

1.

Clothing user:	Transports to store
	Purchases shirt and detergent
	Transports home
	Disposes of packaging for shirt and detergent
	Undergoes many cycles of use, including washing and drying
	Eventual disposal or gifting for a second life
Clothing manufacturer:	Extracts materials from nature
	Processes raw materials into refined materials
	Fabricates shirt
	Packages shirt
	Transports shirt to distribution system
Machine manufacturer:	Extracts materials from nature
	Processes raw materials into refined materials
	Manufactures machines
	Distributes machines
	Use of machines for washing and drying
	Repair of machines
	End-of-life treatment: recycling or landfilling
Detergent manufacturer:	Extracts materials from nature
	Processes raw materials into refined materials
	Manufactures detergent
	Distributes through retail network

Disposal of detergent packaging

Use of water and energy in washing and drying processes

Disposal of used detergent in wastewater

Municipality: Collects clothing and detergent packaging

Transports to landfill or recycling center

Impacts in landfill or recycling center

Cleaning and distribution of freshwater to user

Collection and treatment of wastewater

The washing machine manufacturer has no control over the impacts associated with the production and formulation of detergent used in the washing machine, or the quantity of detergent used. Therefore it is possible for them to leave that input out of their assessment.

2. This exercise has thousands of correct possible answers, as long as each of the services described is delivered in three clearly different physical systems.

3. The appropriate mix of methods should include a site-specific environmental risk assessment (ERA) for the factory, and could also include any of the following methods for the environmental quality control or evaluation of the asphalt tiles (other methods could also be applied):

EMS   To monitor annual environmental performance of the factory

Toxicity Screening   To remove the most toxic substances in tile formulation

Environmental Procurement   To help suppliers deliver more environmentally benign materials

Checklists   To follow rules of thumb to improve environmental performance

LCA   To understand in detail the tile impacts over their entire life cycle

4. If a decision maker wants to use LCA to decide between multiple alternatives, then the decision would change if LCA was used over an assessment that focuses on a single phase in the life cycle. LCA does not always need to be performed. For example, if a LCA is performed, and the practitioner determines that 95% of impacts occur during the use phase, then an analysis of strictly the use phase may have led the practitioner to the same conclusion. A review of existing research on a topic can sometimes guide the practitioner towards knowing whether or not a LCA should be performed.

5. Because the system is not vertically integrated (i.e., the chemical company does not own and operate the natural gas extraction process), the chemical company will need to consider internal and supply chain strategies separately. For internal processes, the chemical company would consider the purpose of the toxic chemical, how its use affects the final product, and how an alternative may change the final product. If an alternative were to be identified, but compromised the quality of the final product, then the chemical company might need to add another chemical that may pose new risks. For the Oil and Gas Company's natural gas extraction processes, the chemical company could either work directly with their supplier to understand why the toxics are used or could survey other providers to determine if their processes are less harmful. If the chemical company decides to work with the Oil and Gas Company, then it would need to determine if it is possible to remove the toxic chemical and produce the same quality natural gas or use the toxic chemical, but institute protective measures to reduce exposure.

6. We should consider the lifetime of the life cycle processes and the number of passenger kilometers traveled that each process facilitates. The impacts associated with the car should be limited to the fuel combusted and some assumption should be made about the average number of passengers in a car. Similarly, because buses use the existing roadway infrastructure, fuel combustion and average occupancy are the two primary considerations. For new rail service, new infrastructure must be constructed before the trains begin operation. The components of this infrastructure will have a finite lifetime (such as ten years) and in that lifetime will be able to support the trains that will be providing a certain number of passenger kilometers. A defined number of passenger-kilometers will be travelled on the track per ten-year period. Furthermore, if the trains run on electricity which may or may not have a large potential impact, then it may be managed differently in the LCA than if electricity had a small impact. Assumptions about the average occupancy of the trains should also be made.

Chapter 2

1. Although no right or wrong answers to these questions exist, you should clearly and thoroughly describe the process you went through to locate the complete standards. If you were successful in finding them, you should clearly describe your impression of the content in the standards in at least three sentences.

2. A) No. This is the broad, non-specific claim that the ISO standards were created to prevent. It would also be a violation (in the U.S.) of "greenwashing" guidelines established by the Federal Trade Commission (FTC).

    B) Yes, in the sense that this is not specifically an environmental claim; it is simply a statement of a product attribute that can be directly measured by the purchaser. Therefore, so the ISO LCA standards do not apply.

    C) Yes, or at least probably so. The phrasing of the claim strongly suggests that the key elements of LCA are there, including a functional unit, a benchmark, and consideration of multiple potential impacts.

    D) Yes.

    E) Yes, in the sense that this is a claim of broad environmental superiority. It is a statement of a product attribute that can be directly measured by the purchaser, so the ISO LCA standards do not apply.

    F) No. This is the broad, non-specific claim that the ISO standards were created to prevent.

    G) No. This is an inappropriate use of LCA, as LCA does not predict absolute or precise environmental impacts.

3 The type of LCA described here is intended for "public disclosure of comparative assertions." ISO requires a panel of at least three independent qualified reviewers for this type of review, and you have found three presumably independent people who know LCA. But knowledge of LCA is not the sole factor that determines who is qualified. Knowledge of the subject area (in this case, recycled materials, tires production, and tire performance) should be considered in selecting the panel. You, as an expert in food and packaging, likely do not have such knowledge, and colleagues from previous projects may not either. Including experts in the subject area, such as those from industry groups or consultancies, is an excellent practice, even if they are not expert in LCA.

4. Although this is a comparative LCA, the goal is clearly for an internal decision at the company, not for any external communications or marketing claims. A review by a single expert who is a part of the company but also independent of the project, should be sufficient.

5.

    1. No. Calculation of category indicator results is a mandatory element.

    2. No. Characterization models should be internationally recognized. It would be acceptable (and encouraged) to present new methods to calculate potential impacts, but they should not be used for claims until they have been more widely tested and published.

    3. Yes. Normalization and weighting are optional elements.

    4. Yes, or at least probably. Although LCA is meant to include multiple impact categories, many of the single-attribute methods are based on the ISO 14040/14044 framework.

    5. No. LCAs for comparative assertions must employ a sufficiently comprehensive set of category indicators.

    6. No, or at least probably not. For attributional LCA, it is likely that data exist to either avoid allocation or make allocation on a physical basis, which are preferred in ISO 14044.

Chapter 3

1. This exploratory exercise has no single predetermined correct answer. The more you explore, the more you learn.

2. The client should be cautioned against making this claim. It may be useful to reference the case law on this topic and make reference to the requirement of a comparative assertion.

3. The particular aspects should include a discussion of the importance of environmental outcomes, the lack of technical understanding on the part of the users of the study, the potential long delay between the decision and the environmental effects, and the complexity of connecting observed environmental outcomes with their stressors.

4. Deep ecology is the belief that natural system hold value in and of themselves, and that causing damage to them is wrong.

5. The arguments should include one or more of the following: 1) the LCA can uncover unknown risks in the product supply chain,

which can help manage the risk; 2) the company has an ethical imperative to understand its impacts; and 3) the company needs a license to operate as part of its corporate social responsibility program.

CHAPTER 4
1.

	Elementary	Intermediate	If intermediate, why?
Bauxite ore, in ground, at mine site	●		
Iron ore, at blast furnace		●	Ore has been mined and transported
Trees, standing, in forest	●		
Corn plant, in field	●		
Wood chips, at sawmill		●	Chips are from trees that have been cut, transported to mill, and processed
Corn grain, at wet mill		●	Corn has been harvested, separated from ear, and transported
Limestone, crushed, at mine site		●	Stone has been extracted and crushed
Coal, in ground	●		
Crushed coal, at utility plant		●	Coal has been mined, crushed, and transported
Electricity, from coal, at generating facility		●	Electricity is not a flow from nature but a product of a utility, in this case generated from combustion of mined and processed coal
Grid electricity, at manufacturing plant		●	Similar to previous electricity example, but this case also includes transmission & distribution
Tap water input to washing operation		●	Tap water has been treated and pumped to user
River water, at plant inlet	●		
Stack emissions to scrubber		●	Emissions are being routed to control equipment before being released to environment
Vehicle tailpipe emissions	●		
Fugitive methane from landfill	●		
Methane emissions to flare		●	Emissions are being routed to flare for combustion instead of released to environment
CO2 from methane flare	●		
Spent plating solution to treatment plant		●	Solution will be treated before released to environment

1. Two confidential process data sets are shown in the figures below. Process A uses input of raw materials (RM) X and Y to produce Product A, while Process B uses inputs of RM X, RM Z, and Product A to produce Product B. Use the data sets for the two proprietary processes to prepare a combined data set for 100 kg output of Product B that protects the details of the individual processes.

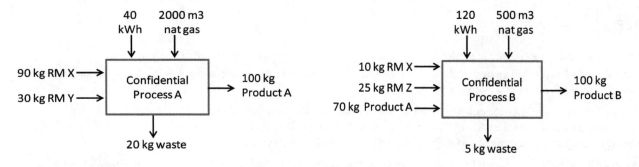

	per 100 kg Product A	per 100 kg Product B	combined inputs for A+B, for 100 kg output from Product B
Inputs (RM = raw material)			
kg RM X	90	10	73
kg RM Y	30		21
kg RM Z		25	25
kg Product A		70	
kWh	40	120	148
cu m nat gas	2000	500	1900
check total kg inputs	120	105	119
Outputs			
kg Product A	100		
kg Product B		100	100
kg solid waste to landfill	20	5	19
check total kg outputs	120	105	119

1. This open-ended problem has many possible solutions.

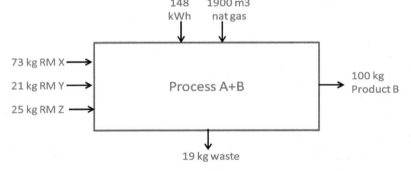

CHAPTER 5

1. Primary packaging reference flows:

2. Electricity consumption for cooling:

	Amstel Light™	Samuel Adams™ Boston Lager	Bud Light™ Platinum
reference flow [L]	202.4	145.2	118.8
no. of 12oz bottles	570.3	409.1	334.8
bottle reference flow [kg]	114.1	81.8	67.0
bottle cap reference flow [kg]	1.1	0.8	0.7

1 liter = 33.8 fl oz; 12 oz bottle ≈ 200 g; bottle cap ≈ 2 g

3. Reference flows per calorific value

	Amstel Light™	Samuel Adams™ Boston Lager	Bud Light™ Platinum
reference flow [kg]	213	152	125
drinking temperature [F]	45	46	43
drinking temperature [K]	280.4	280.9	279.3
room temperature [K]	298.2		
electricity reference flow [kWh]	15.6	10.8	9.7

Electricity reference flow [kWh] = (room temp [K] − drink temp [K]) * heat capacity [kJ/kg*K] * beer ref flow [kg] * COP * 0.00028 kWh/kJ

Chosen functional unit: 2,500 cal/day for age group 21-25 (50% men, 50% women) according to www.cnpp.usda.gov/publications/usdafoodpatterns/estimatedcalorieneedsperdaytable.pdf

FOR DEMONSTRATION PURPOSES ONLY! BEER IS NOT A BALANCED DIET!

	Amstel Light™	Samuel Adams™ Boston Lager	Bud Light™ Platinum
calorific value per 12 fl oz [cal]	95	170	137
beer reference flow [kg]	9.9	5.6	6.9
bottle reference flow [kg]	5.3	2.9	3.6
cap reference flow [kg]	0.05	0.03	0.04
electricity reference flow [kWh]	0.7	0.4	0.5

beer reference flow = (2,500 cal / cal per 12 oz) * 12 * 0.03 l/oz * 1.05 kg/l

bottle reference flow = (2,500 cal / cal per 12 oz) * 0.2 kg

cap reference flow = (2,500 cal / cal per 12 oz) * 0.002 kg

electricity reference flow [kWh] = (room temp [K] − drink temp [K]) * heat capacity [kJ/kg*K] * beer ref flow [kg] * COP * 0.00028 kWh/kJ

CHAPTER 6

1. Data quality is a relative concept, depending on LCA's scope/objective. If data quality is insufficient, the key objectives of LCA, namely impact reduction and/or comparison, cannot be achieved.

2. (i) It depends on scope (ii) concept of uncertainty and accuracy.

3. The sources/types include the following: measurement error, parameter error, argument error, allocation error, physical model error, methodological error, system boundary error, timespan error, location error, and technological quality error.

4. A systemic error is an error in the underlying structure of the LCA that will push the results of the LCA in a particular direction. For instance, if a person models impact categories that do not include land use or ecological damage for an agricultural product, he or she will produce a result that does not consider the impacts on land use and ecological damage.

   A random error is more likely an error made in the measurement process, where incorrect values were accidentally recorded. For instance, a person who meant to measure grams measures kilograms instead and reports it as grams.

5. Students should demonstrate understanding that representative (i.e. primary) data is usually, but not always, of lower uncertainty than secondary data, and be able to give a concrete example.

6. Carefully read all documentation provided for the LCI.

   Review the data quality indicators for the LCI.

   Perform a pedigree matrix evaluation of the LCI.

   Conduct a critical review of the LCI.

7. Students should demonstrate awareness that pedigree matrix is essentially a tool to infer quantitative measures of uncertainty (e.g., low, medium, or high) if direct statistical measures such as standard deviation are unavailable.

8. How the pedigree matrix is applied is situational. It depends on the product system that is being assessed.

   Assigning a pedigree value can be a subjective process by which different people can assign different values.

9. Demonstrate understanding and inter-dependence of three quantities: (i) an input data item's relative accuracy, (ii) an input data item's contribution to a total LCA impact, and (iii) the relative accuracy of the total LCA impact.

10. The goal of LCI critical review is to ensure an overall consistent, comprehensive, peer-reviewed, and up-to date data set.

    The steps in performing a critical review are 1) creating a quality guideline that defines nomenclature, units, a set of default flows, modeling and measurement procedures, and possibly testing routines, and 2) identifying a local admini strator to perform the review.

CHAPTER 7

1

Note: Many possible numerical answers are possible, depending on the choices made about scaling factors and industry sectors. It is important to explain sources and assumptions as part of the solution.

To use the Eco-LCA tool from Carnegie-Mellon[1], we need to convert the prices to a 2002 basis. There are several ways to do this. The simplest is to use an online inflation calculator of the consumer price index, choosing 2002 as the benchmark year. Another way is to use a producer price index for a sector, such as "new industrial building construction," but this free source does not go back to 2002 or 1995 in most categories. The Bureau of Labor Statistics has a wide range of data available, including labor costs. A good source would be to use an industry-specific price index, such as those by *Chemical Engineering* magazine, Marshall & Swift, or IHS. These price indexes typically are not free, so they are more difficult to access. The table able below shows index values for consumer prices, a producer price, and chemical equipment prices (hypothetical):

Index type	Source	1995 value	2002 value	2010 value
Consumer price	www.bls.gov/data/inflation_calculator.htm	$(2002) 1.18	$(2002) 1.00	$(2002) 0.83
Producer price	http://data.bls.gov/pdq/querytool.jsp?survey=pc for New Industrial Building Construction	X	X	$(2007) 1.12
Labor cost	http://data.bls.gov/cgi-bin/dsrv?pr for manufacturing labor	95.3	92.6	96.3
Chemical Equipment Costs	Hypothetical, but representative of *Chemical Engineering* or IHS	500	570	690

An equipment cost from 1995 would be converted to a 2002 value by using the chemical equipment cost index and multiplying the cost by 570/500; a labor cost from 2010 can be scaled by using the labor cost data and multiplying the 2010 value by 92.6/96.3. Scale costs are reported in the table on page 282:

1 www.eiolca.net/> [Accessed May, 2014]

Item	Cost, $	Year	Scaling method	2002 cost
Tanks and related equipment, installed	35,000,000	1995	Chem. Equip. index	39,900,000
Valves and pumps	300,000	1995	Chem. Equip. index	342,000
Analysis building	450,000	1995	Chem. Equip. index	456,000
Trucks	100,000	1995	CPI	118,000
Operating labor	750,000	2010	Labor costs	721,000
Chemical analysis	50,000	2010	CPI	41,500

The next step is to convert these costs into greenhouse gas emissions, using the EIO-LCA tool (or a similar one). As these are all purchased items by the manufacturing company, a purchase price data set is used. Labor, unless supplied as a service, is not included as a burdened input – this fact would need to be communicated back to the project sponsor!

Item	Cost, 2002 $	Industry category name	Category number	Tons, CO_2eq
Tanks and related equipment, installed	39,900,000	Nonresidential manufacturing structures	230102	17,400
Valves and pumps	342,000	Pump and pumping equipment manufacturing	333911	181
Analysis building	456,000	Nonresidential commercial and health care structures	230101	269
Trucks	118,000	Light Truck and Utility Vehicle Manufacturing	336112	66
Total, from capital				17,916
Operating labor (annual)	721,000	NA	NA	NA
Chemical analysis (annual)	41,500	Environmental and other technical consulting services	5416A0	6
Total, from operations				6

The final step is to relate these numbers to the production. The greenhouse gases from capital can be allocated over the life of the plant, which can be assumed to be 20 years (although other assumptions are possible!), giving 896 tons CO_2eq per year. Adding the 6 tons of emissions from the chemical analysis services and dividing by the production gives 0.18 kg CO_2eq/kg of production.

Answer for Exercise 2:

Impact category	value	Unit
Global Warming	570	kg CO2 eq
Acidification Air	0.771	kg SO2 eq
HH Criteria Air	0.165	kg PM10 eq
Eutrophication Air	0.023	kg N eq
Eutrophication Water	0	kg N eq
Ozone Depletion	0.002	kg CFC-11 eq
Smog Air	13.5	kg O3 eq
EcoToxicity (low)	0.001	kg 2,4D eq
HH Cancer (low)	0.025	kg benzene eq
HH NonCancer (low)	8.44	kg toluene eq
EcoToxicity (high)	0.002	kg 2,4D eq
HH Cancer (high)	0.238	kg benzene eq
HH NonCancer (high)	10.9	kg toluene eq

3

The cost information can be converted to environmental metrics using the EIO-LCA tool (or a similar one). As these are all purchased items by the manufacturing company, a purchase price data set is used. Note that many numerical answers are possible, depending on the choices made about industry sectors. The detailed descriptions of the sectors are essential to refer to in making these assignments. It is important to explain sources and assumptions as part of the solution.

Input	Cost per year, millions US$ (2002)	Industry detailed sector name	Industry detailed sector number
Soy oil	8	Soybean and other oilseed processing	31122A
Sodium hydroxide (caustic)	1	Alkalies and chlorine manufacturing	325181
MDI (methylene diphenyl diisocyanate)	15	Other basic organic chemical manufacturing	325190
Flame retardant	0.5	All other chemical product and preparation manufacturing	
Pentane	0.3	Petrochemical manufacturing	325310
Electricity	0.2	Power generation and supply	221100
Heat (from gas boiler)	1	Natural gas distribution	221200

Using this information and the ECO-LCA tool, the footprint characterized using TRACI is at it appears below. The final column shows the total results divided by the annual production of 10,000,000 kg:

TRACI impact category	TRACI impact metric	Total for annual production	Impact per kg of product
Glob Warm	kg CO2e	72000000	7.2
Acidif Air	kg SO2e	347000	0.0347
HH Crit Air	kg PM10e	284000	0.0284
Eutro Air	kg Ne	12700	0.00127
Eutro Water	kg Ne	95.2	9.52E-06
OzoneDep	kg CFC-11e	179	1.79E-05
Smog Air	kg O3e	3950000	0.395
EcoTox (low)	kg 2,4D	1590	0.000159
HH Cancer (low)	kg benzene eq	21600	0.00216
HH NonCancer (low)	kg toluene eq	10000000	1
EcoTox (high)	kg 2,4D	2130	0.000213
HH Cancer (high)	kg benzene eq	46600	0.00466
HH NonCancer (high)	kg toluene eq	41200000	4.12

Taken in isolation, these results do not indicate the "goodness" or "badness" of the product, but could be used to compare to a benchmark, to another product, or as an internal estimate for EPD (most Product Category Rules for EPD would not allow solely EIO-LCA for the calculation of potential impacts).

4.

The first step is to select a tool and database. This answer was done with the ECO-LCA tool, using US purchase cost data from 2002. Other tools can be used, although as the numbers are for sales, a purchase cost is likely more appropriate than a producer cost. The next step is to select appropriate industry sectors for the listed market segments. This can be aided by getting additional information from the company's annual report about the specific products sold into each market, but one can get reasonably close with the information given. There is often more than one reasonable assignment of market segment to an industry sector, so the specific numerical answers can be different from the ones that follow. The table below gives one set of possible assignments.

Market segment	Sales, 2010 (in millions of 2002 US $)	Industry detailed sector name	Industry sector code
Plastics	$9,587	Plastics material and resin manufacturing	325211
Chemicals and Energy	$2,893	Other basic organic chemical manufacturing	325190
Hydrocarbons	$4,517	Petrochemical manufacturing	325110
Corporate (ventures, insurance etc.)	$285	Securities, commodity contracts, investments	523000
Electronics and Specialty	$4,183	All other chemical product and preparation manufacturing	325940
Coatings and Infrastructure	$4,453	Paint and coating manufacturing	325510
Health & Ag Sciences	$4,041	Pesticide and other agricultural chemical manufacturing	325320
Performance Systems	$5,541	Urethane and other foam product (except polystyrene)	326150
Performance Products	$9,049	Adhesive manufacturing	325520

Using this data, the total greenhouse gas emissions related to these sales is 75.6 million tons CO_2eq. The company's reported greenhouse gas emissions were 39.2 million tons of CO_2eq, which represents 52% of the total. It is very common for the amount of emissions occurring upstream of a company to be of the same magnitude, or even much greater, than those of the company itself.

5.

Because process LCI and EIO LCI are created in different ways, you should clearly state the following:

a. In the Inventory Analysis, the two different data types are being used in the assessment.

b. In the LCA Interpretation, describe the fact that while process LCI data is based entirely on measured emissions and resource use associated with a process, EIO LCI data is based on correlations between financial transactions and between market sectors in a country and emissions data for those sectors. Because the nature of the two types of data is different, there is a higher degree of uncertainty about the accuracy of the assessment results than if the entire assessment had been conducted either with all process LCI data or all EIO LCI data.

6.

1. IO-LCA is the best choice, given the time constraints.

2. Process (or hybrid) LCA is the best choice, as IO-LCA is unlikely to identify differences related to the specific choices of similar materials.

3. Although either might be allowed, process LCA is the most likely choice. Most Product Category Rules on which an environmental product declaration is made do not allow using only IO-LCA.

4. IO-LCA is the best choice. It is better able to account for the different kinds of inputs and burdens between capital goods and operating costs.

5. Process LCA is likely the best, since the client is back-integrated and, therefore, has primary data available for most or all of the production inputs.

6. Process LCA is the best choice, since the focus is on end-of-life options that are not typically in IO-LCA databases, but do occur in regions with presumably much different cost structures.

CHAPTER 8

1.

Hydrogen fuel cells produce electricity to power automobiles. Electricity is one product of the chemical reaction between hydrogen and oxygen, with water and heat as by-products. For the purpose of alternative fuels, hydrogen can be produced in several ways, which include:

- an electrolysis process that passes electricity through water between two electrodes (electricity is generated by renewables such as wind or solar is the cleanest way to produce hydrogen)

- a gasification process resources known as pyrolysis of biomass (e.g., agricultural residues like peanut shells)

- steam reforming of hydrocarbon fuels (e.g., natural gas, methanol, ethanol, petroleum distillates, liquid propane, and gasified coal)

- extraction from renewable biogases emitted from landfills, and anaerobic digester gas generated at wastewater treatment plants and breweries

2.

The PHEVs problem can be addressed by going through and considering the following elements:
- production and end of life of the conventional vehicle (base vehicle)

- production, recycling, and disposal of storage batteries

- production of liquid fuel to power the IC engine

- generation of electricity used to replenish or re-charge the batteries

a. The best PHEV configuration from an environmental impacts perspective involves the following:

	Configuration Choice
Production and end of life of the conventional vehicle (base vehicle)	The conventional vehicle with a high proportion of the components used (e.g., plastics, metals, rubber and glass parts) comes from reused, recycled, or remanufactured materials.
Production, recycling, and disposal of storage batteries	The major contributor of environmental impacts for batteries is the supply of the materials (e.g., copper and aluminum) for the production of the anode and cathode, as well as cable. Therefore, the preferred choice will be storage batteries with high proportion of reused, recycled, or remanufactured materials. Also crucial is the avoidance of landfills, particularly for hazardous and non-degradable materials.
Production of liquid fuel to power the IC engine	1.1.1.1 This includes next-generation liquid biofuels such as those derived from algae that use waste water, marginal land, and supply from power plant.
Generation of electricity used to replenish or recharge the batteries	Most environmental impacts of PHEVs occurs during the operation of the vehicle. The choice of electricity source will therefore play a key role. The best electricity is from a hydropower source, since it produces the least carbon footprints: 15 g CO_2 e/ kWh e. An alternative is geothermal energy when the carbon footprints are zero.

a. The worst PHEV configuration from environmental impacts perspective involves the following:

	Configuration Choice
Production and end of life of the conventional vehicle (base vehicle)	The conventional vehicle with a low proportion of the components used (e.g., plastics, metals, rubber and glass parts) comes from reused, recycled, or remanufactured materials.
Production, recycling, and disposal the storage batteries	The least preferred choice will be storage batteries with a low proportion of reused, recycled, or remanufactured materials.
Production of liquid fuel to power the IC engine	The liquid fuel is derived from fossil fuels with significant environment burdens (e.g., oil sands). The worst option may include first-generation biofuels whose production competes with food production and causes deforestation (e.g., palm oil, soybean, and corn)
Generation of electricity used to replenish or recharge the batteries	The electricity is generated from *coal* since it produces the largest carbon footprints: 863 g CO_2 e/ kWh e

3.

The five questions below can be used to identify the technologies that will be affected by the introduction of solar and wind energy:

What time horizon does the study apply to?

A typical service life of solar panels and wind turbine is about 20 to 30 years.

Does the change only affect specific processes or a market?

The change would affect the electricity generation process. There are several markets involved, including electricity for home residents and electricity generated by utility companies. Due to the intermittent nature of solar and wind energy, and without a reliable energy storage system, the technologies may not be the baseload power generation (especially for utility companies).

What is the trend in the volume of the affected market?

The share of power generation from conventional fuels such as coal and oil has decreased over the years (see Table 8.1).

a. Is there potential to provide an increase or reduction in production capacity?

There are a lot of potentials to increase the market share of wind and solar energy production capacity. In fact, the potential production capacity for wind energy, for example, has been estimated to be 40 times of the current electricity demand.

b. Is the technology the most or least preferred?

With government subsidies in the form of guaranteed feed-in tariffs or tax rebates, wind and solar technology can become the most preferred technology. The competitiveness of the technologies will also depend on factors such as the price of fossil fuels that power conventional energy generation.

4.

Increased orange juice production will affect the animal feed market. Currently, the market is served mostly by corn and soybean (major sources), and by oats, wheat, and barleys (minor sources). These crops are grown by farmers. Food processing industries (e.g., milling and brewing) also supply the animal feed market with their byproducts. Biofuel industries generate protein-rich byproducts (e.g., camelina meal and algae residue) that can be used as animal feed. As described in the main text, the impact of additional orange peel supply to the other animal feed products will depend on the price-demand elasticity characteristic of each product.

5.

The reaction: $2NaCl + CaCO_3 \longrightarrow Na_2CO_3 + CaCl_2$

Here we have two products coming out the process. If mass-based allocation, the allocation coefficients corresponding to soda ash and calcium chloride are 0.5 and 0.5, respectively, while the same amount of soda ash and calcium chloride are produced.

On the emission side, assuming perfect combustion of natural gas, we have as follows:

$$CH_4 + 2O_2 \longrightarrow CO_2 + 2H_2O$$

That is, per kg natural gas burnt, $44/16=2.75$ kg CO_2 will be released.

Based on mass balance, the process will also produce $1.5+1.2-1.0-1.0=0.7$ kg waste residue.

After mass based allocation, we have the following inputs/outputs for the soda ash/calcium chloride production process are produced (since they have same allocation coefficients, the numbers are the same for both processes):

Using price based allocation:

Total revenue per ton soda ash produced:

$240/ton x 1 ton + $160/ton x 1 ton = $400

Allocation coefficient of soda ash: $240/400=0.6$

Allocation coefficient of calcium chloride: $160/400=0.4$

	Substance	Quantity	Unit
Inputs	Sodium chloride	1.5*0.5=0.75	kg
	limestone	1.2*0.5=0.6	kg
	Natural gas	0.2*0.5=0.1	kg
Products	Soda ash or calcium chloride	1	kg
Emissions	Carbon dioxide	0.2*2.75*0.5=0.275	Kg
	Residue	0.7*0.5=0.35	Kg

After allocation, the inventory for making soda ash:

	Substance	Quantity	Unit
Inputs	Sodium chloride	1.5*0.6=0.9	Kg
	limestone	1.2*0.6=0.72	Kg
	Natural gas	0.2*0.6=0.12	Kg
Products	Soda ash	1	Kg
Emissions	Carbon dioxide	0.2*2.75*0.6=0.33	kg
	Residue	0.7*0.6=0.42	kg

The inventory for making calcium chloride:

	Substance	Quantity	Unit
Inputs	Sodium chloride	1.5*0.4=0.6	kg
	limestone	1.2*0.4=0.48	Kg
	Natural gas	0.2*0.4=0.08	Kg
Products	Soda ash or calcium chloride	1	Kg
Emissions	Carbon dioxide	0.2*2.75*0.4=0.22	Kg
	Residue	0.7*0.4=0.28	Kg

6.

For the CHP plant, the rate of coal combustion is:

72 GJ/hour ÷ 25 MJ/kg =2880 kg/hour

Rate of CO_2 emission: 2880 kg/hour *2.5 kg/kg = 7200 kg/hour

Total energy content in products:

6000 kWh*3.6MJ/kW-hr*0.0001GJ/MJ + 32.4 GJ = 54 GJ

Using energy based allocation, the allocation coefficient for electricity is:

6000 kW-hr*3.6MJ/kW-hr*0.001 GJ/MJ ÷ 64 GJ = 21.6÷54=0.4

CO_2 emission rate: 7200 kg*0.4÷6000 kW-hr = 0.48 kg/kW-hr

For the traditional coal fired power plant, coal consumption per kW-hr electricity produced:

1 kW-hr*3.6 MJ ÷ 35% ÷ 25 MJ/kg = 0.4114 kg

CO_2 emission rate: 0.4114 kg *2.5kg/kg=1.03 kg.

So electricity from the CHP plant has lower CO_2 emission rate.

7.

32.4 GJ heat from CHP plant will substitute the same amount of heat from coal fired boiler, which requires combustion of coal in the amount of:

32.4 GJ*1000MJ/GJ ÷ 90% ÷ 25 MJ/kg = 1440 kg

This corresponds to CO_2 emission of:

1440 kg * 2.5 kg/kg = 3600 kg

Emission due to electricity generation: 7200 kg – 3600 kg = 3600 kg

CO_2 emission rate: 3600 kg / 6000 kW-hr = 0.6 kg/kW-hr

8.

We could use different approaches to deal with this open-loop recycling case. Burning 1 kg waste packaging will generate 20MJ*30%÷3.6MJ/kW-hr = 1.67 kW-hr electricity. For simplicity, we assume a 100% collection rate.
Option 1: Cut-off method
Plastic packaging has a cumulative energy demand of 90 MJ/kg. Electricity generated from recycled waste packaging has a negligible cumulative-energy demand.
Option 2: Lost-of-quality method
Here we one should treat embedded energy as the quality. Thus, the plastic packaging has a cumulative energy demand of 70 MJ/kg. The electricity generated has a cumulative energy demand of 20MJ/1.67kW-hr = 12 MJ/kW-hr.
Option 3. Closed-loop method
Plastic packaging has a cumulative energy demand of 45 MJ/kg. The electricity generated has a cumulative energy demand of 45MJ/1.67kW-hr = 27 MJ/kW-hr.
Option 4. 50/50 method
Same as in Option 3. It is also possible to use substitution, assuming electricity from waste packaging represents marginal supply.

CHAPTER 9

1.

How big is the potential sea level rise? A study of glacier losses, excluding Greenland and Antarctica, showed that about 1,000 km³ per year was being lost from glaciers (Dyurgerov and Meier 2000). Arctic losses were smaller. The total volume of the ocean is about 1.3 billion km³. The average depth of the ocean is about 3,700 meters. This places the increase per year at about 2.7 mm per year. But remember, quite large errors occur in this measurement, perhaps as much as a factor of two.
The sea level can rise for other reasons, too. As seawater heats, it expands slightly. Unfortunately, how much it expands depends on its salinity and its current temperature and pressure, and parts of the ocean are as much as 10,000 meters deep. Furthermore, only about the top 500 meters of the ocean above the permanent thermocline comes in contact with the air temperature with any great regularity. Interpolating from published data[1] we can estimate that the coefficient of expansion is about 0.0125% per degree centigrade. The average annual global air temperature is increasing about 0.02 degrees per year[2], and this means we could expect

about 1.2 mm of sea level rise per year, based on heating the top 500 meters of the ocean. Again, this is an estimate with significant potential error.

Nevertheless, the predicted sea level rise from heating seawater above the permanent thermocline and from melting glaciers comes to about 3.9 mm of sea level rise per year, not far from the observed 3.1 mm/year. This is fairly accurate for a back-of-the envelope calculation. It looks like melting glaciers are a bigger source of sea level rise than is ocean heating, at least for now. But the errors in these crude calculations are quite large, and the estimates from the two sources are of the same order of magnitude. It would be premature to come down on one or the other mechanism as the major source just yet.

2.

$$CH_4 + 2O_2 \longrightarrow CO_2 + 2H_2O$$

Methane combustion uses more oxygen than carbon combustion and creates water, which carbon combustion does not.

3.

Ecosystem services are processes provided by natural systems that would otherwise have to be provided by human activities. Examples of ecosystems services include cleaning water and air, production of oxygen, flood control, provision of building materials.

4.

The four types of chemical bonds are (in order of strongest to weakest) ionic, covalent, hydrogen and van der Waals.

5.

The seasonal thermocline is shallow and occurs in the summer. It is broken up by fall and winter storms and wind/wave action. The permanent thermocline is always there and is deeper than the seasonal thermocline.

6.

A contaminant can degrade through microbial or chemical action, and it can be removed through sedimentation.

1 http://publishing.cdlib.org/ucpressebooks/view?docId=kt167nb66r&chunk.id=d3_4_ch03&toc.id=ch03&brand=Ecosystem services are eschol

2 http://data.giss.nasa.gov/gistemp/graphs_v3/

CHAPTER 10

1.

In active sampling, a human (or other agent) physically takes a hands-on sample. A specific location (or locations at a site) are targeted for sample collection, and samples are taken from the environment. In passive sampling, equipment is usually left at the location of interest; as analytes move by the equipment, they diffuse into the sampler for later analysis.

Active sampling is good for measuring something at a precise location and a precise moment in time. Passive samplers tend to integrate measurements over time, giving us a picture of the overall, or average, conditions at the location of interest. Sampler capabilities, remoteness of location, and especially cost and availability of resources are major considerations in choosing a sampling method.

2.

Samples that are measured in the field, immediately after collection, provide a more accurate representation of the environment from which they were sampled. During transport from the field to the lab, contamination by the sample container, off-gassing, and biological or chemical changes can all change the sample. However, sophisticated analytical equipment does not travel well, meaning that the precision of laboratory analyses are generally higher. So if the analyte of concern occurs at trace levels that require highly sensitive equipment, field analyses may be impractical.

3.

Environmental measurements often reflect concentrations of chemical substances in the environment, but in a broader sense, environmental measurements can describe any quantifiable aspect of environment (e.g., species diversity). Emissions of a substance in a life cycle inventory will lead to a marginal increase in the concentration of that substance in an environmental compartment, with the baseline determined through direct environmental measurements. Environmental measurements are also used to help modelers calibrate their fate and transport models. For example, if a model predicts that a certain lake should have a concentration of 10 mg per liter of an analyte, but the measured concentrations is 10 µg per liter, there is a problem either with the measurements or with the model. In a broader sense, measurements of biodiversity loss help LCIA modelers connect concentrations in the environment to broader ecosystem impacts.

4.

a. Car trip

i. Driving a gasoline powered car may emit carbon dioxide, nitrogen oxides, carbon monoxide, and particulate matter. Carbon dioxide may contribute to climate change, which has a global scale of impact. Nitrogen oxides may contribute to acidification or eutrophication, which could be transported regional-scale distances. Nitrogen oxides may also interact with volatile organic compounds (VOCs) to form ozone, which has local human health impacts. Carbon monoxide and particulate matter tend to be local as well, although the latter may be transported regional distances.

ii. While driving an electric car, there are no tailpipe emissions associated with the actual driving; however, the production of electricity that charges the batteries has impacts that depend on the type of electricity. Coal-powered generation of electricity produces carbon dioxide (see above) as well as sulfur oxides (which can contribute to regional acidification), mercury (which can be transported globally and has direct human toxicity impacts), and other metals and particulate matter (see above). Finally, production and disposal of the battery for the electric car may have associated impacts. For example, if the battery is disposed of improperly, toxic metals may wind up in the local groundwater supply.

b. Depending on climatic conditions and the population distribution around the plant, benzene emissions may have either regional or local impacts to human health. In addition to climate, as well as the distance that benzene is transported depends on the chemical's half-life, a descriptor of how long lived a substance is in the environment.

c. Fertilizer can contribute to environmental problems by providing terrestrial or aquatic ecosystems with excess nutrients, specifically nitrogen, phosphorus, and, to a lesser extent, potassium. Excess nutrients can cause algal blooms that subsequently deplete dissolved oxygen in waters, an effect is known as eutrophication. Some of the effects may be local, but is also possible that fertilizers be transported hundreds of miles and cause impacts far downriver (e.g., eutrophication of the Gulf of Mexico). Therefore, the effect can also be regional.

d. The fate of pesticides depends on meteorological conditions and the half-life of the pesticide in question, as well as the method of application. Of critical importance is the health of the workers who apply this pesticide, which is certainly a local impact. In general, pesticides would be expected to have local impacts, rather than regional or global impacts. Pesticides may be transported across long distances (we can find trace concentrations of many pesticides all over the world, even in Antarctica), but the health impacts that current epidemiological methods are able to identify tend to be in local communities.

e. Assuming that we are not using a new hybrid tractor, the emissions from the tailpipe of the tractor may be similar to those from a gasoline-powered car (part 1a), though with different proportions. This is particularly true if the tractor runs on diesel fuel, which has higher particulate matter emissions than gasoline.

f. Processing electronics takes many forms, with varying levels of worker exposure and emission controls. Potential emissions inside the smelter that may have local health implications for workers. In a worst-case scenario, toxic metals can be amended to local soils or waterways, and organic compounds may be emitted to the air. Depending on their transport properties, they may travel regional distances, but largely the effects of these substances would be felt by the local population and ecosystem.

5.

Bringing a spatial component to LCIA, GIS allows researchers and practitioners to add a level of sophistication to analyses by differentiating the susceptibility of localized ecosystems and populations to impacts. For example, acid rain causes different effects depending on the buffering capacity of local water bodies, which is highly location-specific. Emissions of benzene on an unpopulated island do not have health impacts, but those same emissions would have health impacts in an urban area. Incorporating GIS enables an analyst to identify specific locations and communities that may be affected, which can lead to more actionable policy than using national-scale LCIA methods.

6.

Morbidity refers to being sick or injured, while mortality refers to the death of a person or animal.

7.

One large challenge faced by any regulatory body is gathering sound data to set standards. In the absence of human epidemiological data, animal studies are often used. This requires extrapolating results from one animal species to another. The EPA uses a factor of 10 for this. It is also a challenge to adequately protect the entire human population, because some people are more susceptible than others to certain diseases. The question of disease, whether it is the effect of measuring whether test rats have cancer or another outcome, is an important consideration as well. Finally, although the study of the rats was two years, one must take care and consider whether exposure is chronic (i.e., occurring over a long time) or acute (i.e., occurring over a short time).

8.

The dose-response curve shown has a sigmoidal, or S-shaped, form. At low doses, doubling the dose may have a negligible effect on the response. In the middle of the curve, a small increase in the dose may result in a large increase in response. Therefore, the effect of doubling the dose depends on where the starting dose is on the dose-response curve, which indicates that this is a nonlinear respond to this particular toxicant.

9.

The fate factor describes a substance's persistence in the environment. This persistence is a function of several parameters, including how quickly it biodegrades, the extent to which it undergoes photodegradation, the extent to which it adsorbs organic material, and the extent to which it volatilizes into the air. One way to interpret the fate factor is in terms of mass of the receiving compartment. Therefore, if a substance A has double the fate factor of B, and if both are emitted at equal rates, then one would expect to find A in the environment at double the concentration of B.

10.

The effect factor describes the inherent hazard of a substance. For humans, this could be toxicity; for biodiversity, it could be effect on species density. Therefore, if substance A has double the effect factor of substance B, and if they are both present in the environment at equal concentrations, one would expect the effect of substance A to be double the effect of substance B.

Table – comparing FF, EF, CF

	Substance A	Substance B	Substance C
Fate factor (in water)	FF_A	$FF_B = 0.5\ FF_A$	$FF_C = FF_A$
Effect factor (ingestion)	EF_A	$EF_B = 2\ EF_A$	$EF_C = 0.5 * EF_A$
Characterization factor	CF_A		

11.

The relevant formulas are:

a. $CF_B = FF_B * EF_B = 0.5\ FF_A * 2\ EF_A = FF_A * EF_A = CF_A$
$CF_C = FF_C * EF_C = FF_A * 0.5\ EF_A = 0.5 * FF_A * EF_A = 0.5\ CF_A$

b. $EF_B > EF_A > EF_C$
At equal concentrations, substance C would have the lowest effect.

c. Note that $FF_B > FF_A$
Therefore, you would expect to find substance B at higher concentrations in the water (double that of A). Perhaps substance B is harder for microorganisms to biodegrade, or it does not tend to absorb organic matter and settle to sediment at the bottom of the lake. It may also be that substance A is more volatile (so most of A's mass is in the air).

CHAPTER 11

1.

Maximum Incremental Radiation; Photochemical smog

2.

Distribution into environmental compartments; uptake through breathing, drinking, consumption of above and below-ground plant consumption.

3.

An endpoint method calculates the impacts at the level of the impact, e.g. human health, ecological health, while a midpoint method calculates the impacts closer to the point of environmental intervention (the inventory results)

4.

A regional method is likely to be a better predictor of the impacts of an environmental intervention, but it does not permit aggregations with impacts from other areas, and is thus applicable only to unit processes within the relevant region.

CHAPTER 12

1.

Global impacts		NiMeH	NMC	LFP
Global warming potential, 100 years	kg CO_2-eq	3.5	1.9	1.4
Fossil resource depletion potential	kg oil-eq	0.99	0.45	0.37
Metal depletion potential	kg Fe-eq	1.1	0.85	0.30
Ozone depletion potential, infinity	kg CFC-11-eq	1.0×10^{-5}	1.1×10^{-5}	7.5×10^{-6}
Regional impacts				
Particulate matter formation potential	kg PM10-eq	2.3×10^{-2}	3.6×10^{-3}	2.1×10^{-3}
Photooxidant formation potential	kg NMVOC	1.7×10^{-2}	4.5×10^{-3}	3.0×10^{-3}
Terrestial acidification potential, 100 years	kg SO_2-eq	9.8×10^{-2}	1.2×10^{-2}	6.5×10^{-3}
Regional or local impacts				
Marine ecotoxicity potential, infinity	kg 1,4-DCB-eq	0.13	5.6×10^{-3}	3.7×10^{-2}
Marine eutrophication potential	kg N-eq	4.0×10^{-3}	2.5×10^{-3}	1.9×10^{-3}
Local impacts				
Terrestrial ecotoxicity	kg 1,4-DCB-eq	1.2×10^{-3}	3.1×10^{-4}	1.7×10^{-4}
Freshwater ecotoxicity potential, infinity	kg 1,4-DCB-eq	0.13	5.1×10^{-2}	3.4×10^{-2}
Freshwater eutrophication potential	kg P-eq	4.5×10^{-3}	2.7×10^{-3}	2.0×10^{-3}
Human toxicity potential, infinity	kg 1,4-DCB-eq	5.6	4.1	2.7

2.A)

Steel Foundry (S)

inputs		outputs	
Scrap iron	0.12 lb./lb. sheet steel	Carbon dioxide (CO_2)	1.88 lb./lb sheet steel
Pig iron (from ore)	0.92 lb./ lb. sheet steel	Nitrous oxides (NOx)	0.0045 lb./lb. sheet steel
Electricity	0.1 kW-hr/ lb. sheet steel	Particulates	0.013 lb./lb. sheet steel
Natural gas	0.017 MJ/lb. sheet steel	Sulfur dioxide (SO_2)	0.0049 lb./lb. sheet steel Scrap
iron	0.04 lb. /lb. sheet steel	Primary product	lb. sheet steel

Refrigerator Manufacturing Plant (R)

inputs		outputs	
Sheet metal	98 lb./refrigerator	Carbon dioxide (CO_2)	44.4 lb./ refrigerator
Other components	111 lb./ refrigerator	Nitrous oxides (NOx)	0.089 lb./ refrigerator
Electricity	61 kW-hr/refrigerator	Sulfur dioxide (SO_2)	0.24lb./ refrigerator
Fuel oil	0.38 lb./ refrigerator	Scrap (iron) steel	7 lb./ refrigerator
Primary product		refrigerator manufactured	

Refrigerators sold in Mexico (M)

inputs		outputs	
Refrigerators	202 lb./ refrigerator	Carbon dioxide (CO_2)	6.4 lb./ refrigerator
Transport, 6 ton truck	200 miles/ref.	Nitrous oxides (NOx)	0.03 lb/ refrigerator
Sulfur dioxide (SO_2)	0.007 lb/ refrigerator		
Particulates	0.004 lb/ refrigerator		
Primary product	refrigerator sold		

2B)

Refrigerator plant sheet steel consumption / annual steel foundry sheet steel production
= 5.5 million lb. / 90,000 million lb. = 0.000061 = 0.0061%

2C)

Refrigerator manufacturer annual production/ annual refrigerators sold in Mexico =

56000 refrigerators / 2.3 million refrigerators = 0.024 = 2.4%

3A)

TRACI 2.1 Impact category	Internally normalized unit	Refrig. 1	Refrig. 2	Refrig. 3	Refrig. 4
Ozone depletion	-	1.00	0.42	1.42	0.67
Global warming	-	1.00	0.39	1.40	0.72
Smog	-	1.00	0.45	1.38	0.77
Acidification	-	1.00	0.43	1.36	0.73
Eutrophication	-	1.00	0.46	1.37	0.76
Carcinogens	-	1.00	0.52	1.56	2.17
Non carcinogens	-	1.00	0.48	1.37	0.71
Respiratory effects	-	1.00	0.44	1.44	0.72
Ecotoxicity	-	1.00	0.23	0.74	0.36
Fossil fuel depletion	-	1.00	0.41	1.36	0.70

3B)

Regardless of which refrigerator is used for internal normalization, the proportional relationship among the each of the refrigerators within each impact category remains constant.

4A)

TRACI 2.1 Impact category	Externally normalized unit	Refrig. 1	Refrig. 2	Refrig. 3	Refrig. 4
Ozone depletion	year-capita	2.80E-04	1.20E-04	4.00E-04	1.90E-04
Global warming	year-capita	4.20E-02	1.60E-02	5.80E-02	3.00E-02
Smog	year-capita	3.40E-02	1.50E-02	4.60E-02	2.60E-02
Acidification	year-capita	7.40E-02	3.20E-02	1.00E-01	5.40E-02
Eutrophication	year-capita	3.00E-02	1.40E-02	4.20E-02	2.30E-02
Carcinogens	year-capita	3.50E-01	1.80E-01	5.50E-01	7.60E-01
Non carcinogens	year-capita	6.60E-02	3.20E-02	5.50E-01	4.70E-02
Respiratory effects	year-capita	1.80E-02	7.90E-03	2.60E-02	1.30E-02
Ecotoxicity	year-capita	7.40E-02	1.70E-02	5.50E-02	2.60E-02
Fossil fuel depletion	year-capita	8.20E-04	3.40E-04	1.10E-03	5.80E-04

4B)

The externally normalized value compares the scale of the refrigerator impact to the scale of the average impact created by one person in the US in 2008

TRACI 2.1 Impact category	Externally normalized unit	Refrig. 1	Refrig. 2	Refrig. 3	Refrig. 4
Ozone depletion	year-capita	2.80E-04	1.20E-04	4.00E-04	1.90E-04
Global warming	year-capita	4.20E-02	1.60E-02	5.80E-02	3.00E-02
Smog	year-capita	3.40E-02	1.50E-02	4.60E-02	2.60E-02
Acidification	year-capita	7.40E-02	3.20E-02	1.00E-01	5.40E-02
Eutrophication	year-capita	3.00E-02	1.40E-02	4.20E-02	2.30E-02
Carcinogens	year-capita	3.50E-01	1.80E-01	5.50E-01	7.60E-01
Non carcinogens	year-capita	6.60E-02	3.20E-02	5.50E-01	4.70E-02
Respiratory effects	year-capita	1.80E-02	7.90E-03	2.60E-02	1.30E-02
Ecotoxicity	year-capita	7.40E-02	1.70E-02	5.50E-02	2.60E-02
Fossil fuel depletion	year-capita	8.20E-04	3.40E-04	1.10E-03	5.80E-04

5A)

Equally weighted internally normalized impacts

TRACI 2.1 Impact category	Internally normalized unit	Refrig. 1	Refrig. 2	Refrig. 3	Refrig. 4
Ozone depletion	-	0.10	0.042	0.142	0.067
Global warming	-	0.10	0.039	0.140	0.072
Smog	-	0.10	0.045	0.138	0.077
Acidification	-	0.10	0.043	0.136	0.073
Eutrophication	-	0.10	0.046	0.137	0.076
Carcinogens	-	0.10	0.052	0.156	0.217
Non carcinogens	-	0.10	0.048	0.137	0.071
Respiratory effects	-	0.10	0.044	0.144	0.072
Ecotoxicity	-	0.10	0.023	0.074	0.036
Fossil fuel depletion	-	0.10	0.041	0.136	0.070
total	-	1.0	0.43	1.34	0.83

Refrigerator 3 has the largest total normalized and weighted impacts and refrigerator 2 has the lowest total impacts.

5B)

Equally weighted externally normalized impacts

TRACI 2.1 Impact category	Externally normalized unit	Refrig. 1	Refrig. 2	Refrig. 3	Refrig. 4
Ozone depletion	year-capita	2.80E-05	1.20E-05	4.00E-05	1.90E-05
Global warming	year-capita	4.20E-03	1.60E-03	5.80E-03	3.00E-03
Smog	year-capita	3.40E-03	1.50E-03	4.60E-03	2.60E-03
Acidification	year-capita	7.40E-03	3.20E-03	1.00E-02	5.40E-03
Eutrophication	year-capita	3.00E-03	1.40E-03	4.20E-03	2.30E-03
Carcinogens	year-capita	3.50E-02	1.80E-02	5.50E-02	7.60E-02
Non carcinogens	year-capita	6.60E-03	3.20E-03	9.10E-03	4.70E-03
Respiratory effects	year-capita	1.80E-03	7.90E-04	2.60E-03	1.30E-03
Ecotoxicity	year-capita	7.40E-03	1.70E-03	5.50E-03	2.60E-03
Fossil fuel depletion	year-capita	8.20E-05	3.40E-05	1.10E-04	5.80E-05
total	year-capita	6.90E-02	3.20E-02	9.70E-02	9.80E-02

Refrigerator 4 now has slightly larger total normalized and weighted impacts than refrigerator 3and refrigerator 2 still has the lowest total impacts.

5C) AND 5D)

The answers to these exercises depend entirely on the values that your group create and the reasons for those values. This exercise demonstrates the subjectivity and group dynamics that are part of creating weight values.

6A)

LCI data	kg	EI99, HA Ecopoints/kg	EPS 2000d ELU/kg	Recipe units/kg	EI99, HA Ecopoints	EPS 2000d ELU	Recipe Recipe
CO2 to air	1500	0,0297	0,108	0,197	4,46E+01	1,62E+02	2,96E+02
NOx to air	6	1,39	2,13	1,703	8,34E+00	1,28E+01	1,02E+01
SOx to air	-200	0,66	3,27	1,551	-1,32E+02	-6,54E+02	-3,10E+02
Oil from ground	480	0,14	0,5	0,206	6,72E+01	2,40E+02	9,89E+01
Cr from ground	0,125	0,0218	84,9	0	2,73E-03	1,06E+01	0,00E+00
Cu from ground	0,0025	0,873	208	0,027	2,18E-03	5,20E-01	6,75E-05
Sum					-1,19E+01	-2,28E+02	9,44E+01

6B)

The difference is mainly from different weightings on SO_2 versus CO_2 and oil resource. This, in turn, depends on the different time perspectives and weighting principles applied.

6C)

The result obtained by the Recipe method would call for a more thorough analysis before the decision is made. Maybe a solution can be found that satisfies all weighing methods.

7.

Because normalized results are desired and because an internationally relevant result is desired, TRACI is skipped because it only offers North American normalization, and Impact 2002+ is skipped because it only offers European normalization. CML offers world normalization from 1995, while ReCiPe offers world normalization values from 2008 and end-point characterization. After looking more carefully at the impact categories and LCIA methods used in each method, you select the most appropriate method of the two for this client.

CHAPTER 13

1.1. False	1.3. True	1.5. True	1.7. True	1.9. True	1.11. True
1.2. True	1.4. False	1.6. False	1.8. False	1.10. False	1.12. True

2.

a. The single dominant contribution "electricity, natural gas" corresponds to the use-phase electricity that is consumed over the three-year life span of the module. Similar to many products with a relatively long and energy-consuming useful life (e.g., appliances and automobiles), we expect that the use-phase energy consumption will be a significant contributor to the life-cycle burdens.

b. The "electricity, natural gas" corresponds to the specific electricity consumed during the use phase of the module because this electricity the largest contribution to the GWP results. The much smaller contribution "electricity, from grid" accounts for the electricity used during production of the module. Thus, natural gas was assumed as the source of all electricity during operation of the wastewater treatment plant in Singapore. This selection is reasonable as a first pass estimate because the International Energy Agency (IEA) reports that 77% of electricity in Singapore is produced from natural gas, with the balance produced from oil, wastes, and biofuels; information on energy production by country is readily available from this and other sources.

c. As use-phase electricity is the most significant GWP burden over the life cycle of the module, it represents an opportunity for life-cycle improvement. For the owners of the wastewater treatment plant in Singapore, investing in a renewable or other low-carbon form of electricity to power the wastewater treatment plant could be one way to significantly lower the GWP burden of the process.

d. For the RO module producers, running their own manufacturing processes more efficiently could have some impact on the life cycle performance of the RO modules. However, the larger life-cycle opportunity lies in designing and producing a module that is more efficient in the use phase. This would mean designing an RO module that requires less electricity consumption and/or can operate for a longer period of time before requiring cleaning. A key life-cycle focus by the manufacturer should thus be to reduce the use-phase inputs required per m^3 of water purified. Designing a module that lasts for more than 3 years would also help, but it would likely be less significant than addressing the use phase burdens because the production and disposal burdens represent only a small fraction of the life-cycle GHG emissions shown in the figure.

e. They need to consider the option, but also look at the cost to have the same impact by other means. Using a visual estimation from the graph, EDTA contributes ~5% of the GWP, so a 20% reduction in its GWP could reduce the overall GWP by ~1%. They now know the cost to get this ~1% benefit, and they can compare it to their other options to get the same, or larger, benefits and their available funds.

f. The single dominant contribution "electricity, natural gas" corresponds to the use-phase electricity that is consumed over the three-year life span of the module. Similar to many products with a relatively long and energy-consuming useful life (e.g., appliances and automobiles), we expect that the use-phase energy consumption will be a significant contributor to the life-cycle burdens.

g. The "electricity, natural gas" corresponds to the specific electricity consumed during the use phase of the module because this electricity the largest contribution to the GWP results. The much smaller contribution "electricity, from grid" accounts for the electricity used during production of the module. Thus, natural gas was assumed as the source of all electricity during operation of the wastewater treatment plant in Singapore. This selection is reasonable as a first pass estimate because the International Energy Agency (IEA) reports that 77% of electricity in Singapore is produced from natural gas, with the balance produced from oil, wastes, and biofuels; information on energy production by country is readily available from this and other sources.

h. As use-phase electricity is the most significant GWP burden over the life cycle of the module, it represents an opportunity for life-cycle improvement. For the owners of the wastewater treatment plant in Singapore, investing in a renewable or other low-carbon form of electricity to power the wastewater treatment plant could be one way to significantly lower the GWP burden of the process.

i. For the RO module producers, running their own manufacturing processes more efficiently could have some impact on the life cycle performance of the RO modules. However, the larger life-cycle opportunity lies in designing and producing a module that is more efficient in the use phase. This would mean designing an RO module that requires less electricity consumption and/or can operate for a longer period of time before requiring cleaning. A key life-cycle focus by the manufacturer should thus be to reduce the use-phase inputs required per m^3 of water purified. Designing a module that lasts for more than 3 years would also help, but it would likely be less significant than addressing the use phase burdens because the production and disposal burdens represent only a small fraction of the life-cycle GHG emissions shown in the figure.

j. They need to consider the option, but also look at the cost to have the same impact by other means. Using a visual estimation from the graph, EDTA contributes ~5% of the GWP, so a 20% reduction in its GWP could reduce the overall GWP by ~1%. They now know the cost to get this ~1% benefit, and they can compare it to their other options to get the same, or larger, benefits and their available funds.

3.

a. The conclusion reached is not supported by the data. The indicator USEtox carries orders-of-magnitude differences in its characterization factors, due to the extreme complexity in modeling the fate, transport, and effect of chemical substances released into the environment. The three indicator results shown have differences of only 10 – 13%. The results in each category are hence on the same order of magnitude, and from a USEtox perspective nearly identical. Based on these results, we cannot conclude that one technology has a human toxicity or ecotoxicity advantage over the other.

b. The results are presented poorly, in what could be interpreted as a greenwashing attempt to emphasize the small difference between the results and, therefore, make the results look more significant (~50% difference in the indicator results) than they actually are (10 to 13%). The authors should present their results instead on a y-axis that begins at 0, rather than at 0.88.

c. No scientific methodology, including LCA, can be used to show that one product is truly "better for the planet" than another, as this idea is related more to the values of a particular audience than to scientific results. Even if the USEtox results indicated a toxicity advantage for the new technology, toxicity is only one dimension of sustainability and several others exist related to the overall sustainability performance of a product. Hence, answering whether a product is "better for the planet" is a completely subjective question. Moreover, no universally correct definition exists of what it means to be "better for the planet." In fact, making this type of claim is ambiguous and potentially misleading. Claims of this type should be avoided, particularly when using LCA results.

4. No. The analyst may be right, but he/she has reached a conclusion too hastily. If the energy use is largely electricity, then it may be appropriate to use a state average electricity model, which will have much lower GWP than the US average as shown in Figure 13.11. This could lower the calculated GWP for PacNorWest even more. Also, using data for only one month of operation could lead to significant errors. If a large part of the energy input relates to the weather conditions, then data from August (e.g., a warmer month in Seattle) might be much different than data from January (e.g., a colder month in Seattle.). If the waste disposal is based on bills or accounting information, it may lag the production information by a month or two, so it could represent production from another month. The analyst has found some exciting early indications, but needs to do some more work.

5. The production of fundamental raw materials BDO, HCHO, and C2H2 account for 85% of the CFED, so the production of these materials, or their efficient use in the process, would likely be more fruitful to focus on to have an impact in CFED. The impact of acids and bases are <1%, so an increase in steam or electricity could easily overcome the benefit of eliminating the acids and bases. The analysis could certainly change by looking at other impact categories. Not all potential impacts correlate CFED. Human and ecotoxicities, for example, can be highly influenced by emissions related to inputs that might otherwise be considered small.

6. The only conclusion that can be drawn from the figures is that the results depend on the choice of allocation method. In this case, we see no obvious advantage for either the plastic or the glass application. What is clear is, however, is that the plastic product moves from being GHG-advantaged to GHG-disadvantaged as the ethylene takes on more of the steam cracker burdens. This scenario can be difficult to interpret because no scientific basis exists for selecting a single allocation method over the others. LCA practitioner need to recognize that alternative allocation methods can lead to different results; when this happens, there may be no conclusion that can be drawn from the study. In selecting allocation methods, choosing LCI data, and choosing LCIA methods, ISO guidance is that the choices should be consistent with the goal and scope, and should enable the goal and scope to be met. Expert judgment and previous experience is required to truly put this guidance into practice when confronting an example similar to the steam cracker allocation example presented in this problem. As each LCA is unique, we have no prescriptive rule for how to handle a scenario such as that shown in this problem. However, with the proper expert input, which might be obtained from LCA expert colleagues or from a critical review, we can develop an interpretation that is fair and consistent with the scientific results of the study.

7. Given the pressure that your CEO may be exerting to make a claim at the conference, this response can be a tricky one. However, many factors make such a claim of "15.4% GHG savings" difficult to support from a scientific and LCA perspective.

• The geographical and market scope of the two studies are completely different. The EPD focuses on water produced for the market in Italy, which is a country with a relatively high standard of living. The focus of your water is on Sub-Saharan Africa, which is a completely different situation from Italy. From an LCA perspective, the functional unit is not the same, as the two products are focused on meeting the needs of different markets. From a functional unit perspective, in order to compare the Italy water results with your results, the transport of the water from Italy to the market in Africa would need to be added to the study. However, even this would be hypothetical, because that the water producer in Italy would probably not target the Sub-Saharan Africa market for production. In any case, as the Sub-Saharan Africa market is not the one addressed by the EPD, the results cannot be compared with your LCA.

• In general, studies that have been conducted independently are not comparable. Even if they were conducted with the same functional unit (delivery of 2L drinking water to the Sub-Saharan Africa market), no conclusion can be drawn by comparing the two GWP results. The studies could have been conducted with a different goal and scope, used different system boundaries, made different assumptions, and/or used different GWP indicators. In order to truly understand how your drinking water compares with the Italian study, you, as the LCA expert, would need to review the details of the Italian study and potentially conduct your own study according to a new goal defined specifically to understand how the two drinking water products compare.

• As an additional point, even if a comparison was done in a single study that compared the environmental impacts of two products serving the same market, a claim of 15.4% is a difficult one to make without some supporting uncertainty analysis. Given the general uncertainty associated with LCI data and LCIA indicators, a claim made to a tenth of a percent is difficult to substantiate in LCA. Given comparable GWP results of 95.8 g CO_2-eq/L and 81 g CO_2-eq/L, it would be more transparent to say that there was a difference of "about 15%" between the two options.

Given the issues discussed above, you should gently inform your CEO that the studies cannot be compared and that such a claim cannot be made. Instead, you may want to encourage your CEO to focus his/her sustainability discussions on other points related to your product, such as the fact that the product seeks to help meet the drinking water needs of an extremely water-challenged region. The fact that your product is focused on solving one of the world's main challenges is likely the more important sustainability message than comparison to some other drinking water solution, even if it is in the same market.

CHAPTER 14
SOLUTION 1.

     a. categorical

     b. continuous

     c. categorical

     c. continuous

     e. continuous

     f. Either, depending in your perspective, it can be continuous if the expected number of suppliers will change and the distances are treated as a single population. But it can also be discrete. Given that these are sourced suppliers, the distance between supplier and receiver should not change and thy have ben qualified, which means a constraint on the population.

SOLUTION 2.

     a. mL / unit * if (shipped, 1200, 378)
     b. 3*gal/drum * 3.78541 L / gal * 0.627 kg / L, if (waste, 0.13, 0.87)
     c. Ceil (10 *427.6 gram A/10*250 grams)

SOLUTION 3.

a. The data appear to have a tail extending to the right. This tail has single dominant value as the extreme. Most of the data is within the range 0-20. This may be a lognormal distribution.

b. The data appear to be nonlinear. There may be a linear trend from 0-15, but then the data grow faster. There seems to be some perodicity with the time index. This is harder to identify given the noise in the data. There also appears to be an offset from zero that could be due to positive bias.

c. Data offset perhaps due to positive bias   2. Periodic structure
   with time   3. Outlier at ~22 months.

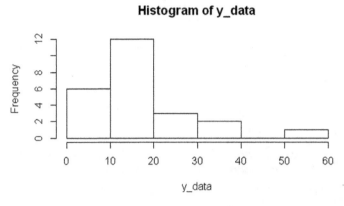

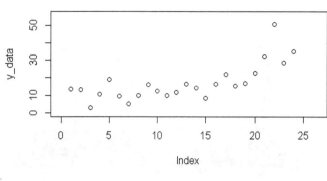

Solution 4.

		m	n	s	95% confidence level		
					level	lower	upper
System 1	Bucket	6.62	10.00	0.95	0.98	6.03	7.21
System 2	Meter	6.65	10.00	0.09	0.98	6.60	6.71
System 3	Guess	4.60	10.00	0.88	0.98	4.05	5.14
System 4	Thumb	5.00	10.00	0.00	0.98	5.00	5.00

Solution 5.

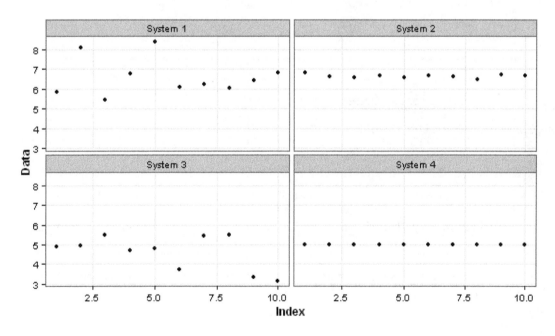

Solution 6.

Five randomly selected substances:

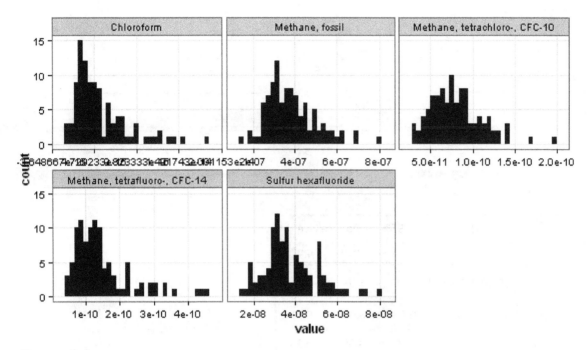

Histogram of sums:

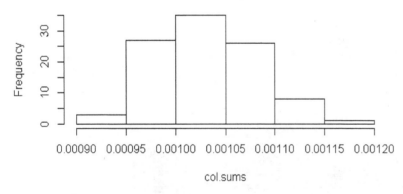

CHAPTER 15

1

1. Uncertainty, specifically statistical uncertainty. The kind of analysis described is commonly possible with LCA databases and software tools.

2. Bias, and quite extremely so! LCA of agriculture should include consideration of water and eutrophication—these are both well-known and significant impact categories for agriculture. This is an extreme example of bias and could be worth a letter to the authors or journal editors.

3. Uncertainty, specifically misrepresentative data. In this common practice, the report authors should consider this as part of a sensitivity analysis.

4. Uncertainty, specifically misrepresentative data. In this common in practice, the report authors should consider this as part of a sensitivity analysis.

5. Bias. A sensitivity analysis could be done using a characterization method based on South American conditions or global conditions.

2

1. The likelihood appears to be high that external normalization bias is occurring, given the very diverse materials that are being assessed.

2. You could select several hundred different materials and processes, characterize and normalize them, and identify if this pattern is evident on the average percent contribution of each impact category.

3.

From a communication point of view, the case study involves comparisons of options, which is a consistent way to compare uncertainty as well. From a statistical point of view, it avoids overstating the influence of shared parameters from the many common upstream processes in the models. A disadvantage is that the difference is not lognormally distributed, whereas the original distributions are.

4.

a. The average difference of all ten impact categories is 4.6%, so the refrigerator is considered to be below the 5% threshold of meaningful difference.

b. You should explain that on the average for the ten impact categories, the new design does not create significantly lower impacts. The threshold is generally considered to be 5%, and stronger evidence of a meaningful difference would be 10% or more.

CHAPTER 16

1.

a. LCC can include costs to one or more actors in the life cycle of the chair, as is decided by the commissioner of the study. All the costs should be internal and directly tangible to the actors in the study (e.g., manufacturing and distribution costs to the manufacturer, purchase and maintenance costs to the user, and landfill costs to the municipality).

b. LCC does not quantify the economic value of these external costs. These impacts can be assessed in a parallel LCA study. If these external costs were included in a LCC and that LCC were coupled with an LCA, the external costs would constitute double counting, which should always be avoided.

c. The discount rate can be any rate that the commissioner requests, possibly 1% – 5%.

2.

costs	cost ($)	per # chairs	
steel	420	600	0.70
nickel plating of steel	408	1200	0.34
plastic	1280	800	1.60
injection molding of plastic	5.7	30	0.19
corrugated cardboard box	40	100	0.40
warrantee and sales brochure	290	1000	0.29
factory labor and overhead	11020	2000	5.51
transport, factory to distributor	192	60	3.20
		Total	12.23

3.

Current cost to manufacturer = future cost / $(1 + r)^T$ = $\$9.40/(1.029)^{24}$ = \$4.63

4.

Current cost of cleaning = C $((1-(1/1+r)^T)/r)$ = $\$0.45((1-(1/1.029)^{24})/0.029)$ = \$7.70

\$127 (purchase price) + \$7.70 (cleaning over lifetime) = \$134.70 (total current cost to user)

5

a. Country of origin data or estimates of country of origin from trade data (GTAP, UN ITC)

b. By conducting a hotspot assessment.

c. By impact categories, by impact subcategories and by stakeholder categories.
The five stakeholder categories that should be represented in a sLCA include: Worker, Consumer, Local Community, Society, and Value Chain Actor

6.

LCSA of the chair assesses in parallel the LCC, the sLCA and the LCA of the chair. It will be difficult to interpret the results of the LCSA of the chair because we have nothing with which to compare. The values will stand alone and it will not be evident what attributes (or impact categories) represent strong or weak performance.

7.

You could include the impact categories (in the LCC, sLCA or LCA) for which the second chair has no results and explain the disparities in the study report. Alternately, you could remove those impact categories for which there is no data in the second chair from the report altogether, and explain in the report that those impact categories that lacked data. The latter approach is less likely to be misinterpreted by the recipients of the study. Perhaps a future iteration of the study could complete the missing data and provide a more complete LCSA.

CHAPTER 17

1.
Conduct a stakeholder analysis to find out how each person or group should be involved (RACI), as well as what they expect to learn from the study (content and format).

2.
Certainty, he or she will expect on-time results, transparency on progress through milestones, budget development, and feedback on cost implications of any changes.

3.
These are the three examples:

    1. Defines effort and budget imposed by scalable parts of study (e.g., the number of plants to be included in data collection)

    2. Defines effort arising from necessary level of detail (e.g., the number and type of impact indicators)

    3. Imposes need of critical review if study is comparative and meant for external communication

4.
Commissioner (A) may request detail or outcomes that are beyond the project's scope or budget, as perceived by practitioner (R); interested parties (I) may have business interest in study and through organizational power evolve to become secondary decision makers (A).

5.
A practitioner can show separate work streams for each iteration that clearly indicate what deliverables are required at what time so that well-informed decisions can be made for the transitions from one iteration to the next.

6.
Tasks are usually underestimated in terms of effort (resources, time, and costs), which leads to delays and budget overruns. This can be mitigated by allowing for safety margins (estimated time multiplied by factor 1.2—2). During project performance, the practitioner should give an early warning of things not going as planned

7.

Project reporting (such as time tracking and scorecards) show progress and if things develop in unforeseen ways; reporting of study contents (goal, scope, methods, and outcomes) is defined by goal and scope and thus determines the required effort to a large extent. Remember to always allow sufficient time for the documentation phase.

CHAPTER 18

1.

This exercise guides you through the process to help you make and understand histograms

2.

These answers can be somewhat subjective, as there could be a number of suitable communication approaches. The goal of the problem is to ensure that the students carefully read and apply the chapter guidance, and have them think through possible communication strategies. One possible set of communication plans is as follows:

Corporate business team

Begin communicating with the corporate business team before the project starts and keep them informed throughout the project (through emails and presentations). Advise them of any issues that arise during the project, including any delays or need for additional resources (funding, access to data or personnel), or any surprises in preliminary results (red flags and unexpected trade-offs or benefits).
Communication formats will typically be focused on informal presentations.

Corporate environmental team

Begin communicating with the corporate environmental team before the project starts and keep them informed through the project. The approach here might be very similar to the communication approach for the corporate business team, but more focused on environmental issues, such as regulatory or liability aspects.
Communication formats will typically be focused on informal presentations.

Marketing and public relations team

Depending on the project, you might begin communicating with the marketing and public relations team prior to the start of the project, or sometime during the project once it becomes clear that there are marketing or public relations aspects to consider once the project is complete.
You will likely work with this team in developing (or supporting the development of) marketing collateral. Your communications with the team itself regarding project progress and outcomes may be through informal presentations. Later in the project, you may be tasked with submitting drafts of marketing collateral, or with providing content supporting the development of marketing collateral.

Customers (B2B)

Business-to-business customer engagement very often occurs through direct communication, but can also involve web-based media. The LCA study should be subjected to third-party critical review per ISO 14040-44 prior to development of communication materials for customer engagement. Likely communication formats may include:

- white papers

- brochures

- in person presentations

- conference presentations

- webinars

- videos

- interactive media (e.g., iPad)

- web-based media (e.g., reports, videos, graphical content, and full detailed report)

Policy makers

Communication with policy makers likely does not begin until the LCA study has been completed and subjected to third-party critical review per ISO 14040-44. Likely communication formats may include:

- submission of the full detailed report (in instances where a policy body is soliciting input)

- submission of a white paper with reference to the full detailed report

- publication in peer-reviewed journal to engage iterative scientific debate (i.e., opportunity for replication and verification, or to solicit alternative viewpoints) and establish additional technical credibility

- in-person presentations at public forums

CHAPTER 19

1. The impact covers all stages of the product life cycle. Emissions might be N2) emissions and CO2 emissions from agriculture, CO2 and CH4 emissions from the se of electricity and fossil fuels, and methane emissions from end-of-life wastewater treatment. Bottling contributes energy-use impacts in creating the bottle and in the bottling step.

2. You should identify that the GHG protocol does not require a program operator. There are many detailed auditing requirements. An advantage is that the analysis can be simpler than an EPD analysis. A disadvantage is that a carbon footprint only looks at climate change, while ignoring many other impacts, and thus looks at only one slice of environmental sustainability. Therefore, it cannot be used as a sustainability claim without extensive further work.

3. This is a complex exercise that will have quite different results, depending on the people doing the analysis and the current PCRs available. Some of the expected differences uncovered could include the following: functional units, rules about end of life, allocation methods, rules about background data, rules about the electric grid, and rules about exclusion of capital goods.

GLOSSARY

ACLCA – American Center for Life Cycle Assessment

affected party - any party affected by the results of a water footprint assessment or a life cycle assessment [ISO 14046:2012]

allocation - partitioning the input or output flows of a process or a product system between the product system under study and one or more other product systems [ISO 14040:2006]

ancillary input - material input that is used by the unit process producing the product, but which does not constitute part of the product [ISO 14040:2006]

ancillary product - building product that enables another building product to fulfill its purpose in the intended application [ISO 21930:2007]

ASTM - American Society for Testing and Materials

attributional modeling – LCI modeling framework that inventories the inputs and output flows of all processes of a system as they occur [JRC 2010]

biogenic carbon – carbon derived from biomass [ISO 14067:2011]

biogenic CO2 – CO2 obtained by the oxidation of biogenic carbon [ISO 14067:2011]

biomass - material of biological origin excluding material embedded in geological formations and material transformed to fossilized material. This includes organic material (both living and dead), e.g. trees, crops, grasses, tree litter, algae, animals and waste of biological origin, e.g. manure. [ISO 14067:2011]

brackish water - water containing salts at a concentration less than that of seawater, but in amounts that exceed normally acceptable standards for municipal, domestic and irrigation uses [ISO 14046:2012]

building product - goods or services used during the life cycle of a building or other construction works [ISO 21930:2007]

carbon dioxide equivalent - calculated mass for comparing the radiative forcing of a greenhouse gas to that of carbon dioxide. The carbon dioxide equivalent is calculated by multiplying the mass of a given greenhouse gas by its global warming potential. [ISO 14064-1:2006]

carbon footprint of a product – sum of greenhouse gas emissions and removals in a product system, expressed as CO2 equivalent and based on a life cycle assessment. The CO2 equivalent of a specific amount of a greenhouse gas is calculated as the mass of a given greenhouse gas multiplied by its global warming potential [ISO 14067:2011]

carbon footprint of a product claim - claim pertaining to the CFP made by the producer, manufacturer or duly authorized supplier or distributor. CFP claims may take the form of statements alone or in conjunction with symbols or graphics on product or package labels, or in product literature, technical bulletins, advertising, publicity, telemarketing, as well as digital or electronic media, such as the internet [ISO 14067:2011]

carbon footprint of a product communication program – program for the development and use of CFP communication based on a set of operating rules. The program may be voluntary or mandatory, international, national or sub-national. [ISO 14067:2011]

carbon footprint of a product declaration - declaration of the CFP made according to the CFP-PCR or appropriate Type III environmental declaration according to the PCR [ISO 14067:2011]

carbon footprint of a product external communication report - report based on the CFP study report intended to be publically available [ISO 14067:2011]

carbon footprint of a product label – means of marking products with their CFP within a particular product category according to the CFP communication program requirements [ISO 14067:2011]

carbon footprint of a product performance tracking report - report comparing the CFP of the same product over time [ISO 14067:2011]

carbon footprint of a product-product category rules - set of specific rules, requirements and guidelines for quantification and communication on the CFP for one or more product categories [ISO 14067:2011]

carbon footprint of a product program operator - body or bodies that conduct a CFP communication program. A CFP program operator can be a company or a group of companies, industrial sector or trade association, public authorities or agencies, or an independent scientific body or other organization [ISO 14067:2011]

carbon footprint of a product study - study which includes the quantification and reporting of the CFP or the partial CFP [ISO 14067:2011]

carbon footprint of a product study report – report on a CFP study [ISO 14067:2011]

carbon footprint of a product verification - confirmation of the validity of an environmental claim using specific predetermined criteria and procedures with assurance of data reliability [ISO 14021:1999]

carbon footprint of a product verifier - competent person, body or team that carries out a CFP verification [ISO 14067:2011]

carbon storage in a product - carbon removed from the atmosphere and stored as carbon in a product [ISO 14067:2011]

CAS number - Chemical Registration numbers provided by the Chemical Abstracts Service

category endpoint - attribute or aspect of natural environment, human health, or resources, identifying an environmental issue giving cause for concern [ISO 14040:2006]

category indicator – see impact category indicator

certification body - body operating a product certification system [ISO/IEC Guide 65:1996]

CFP – see carbon footprint of a product

characterization - calculation of category indicator results [ISO 14044:2006]

characterization factor - factor derived from a characterization model which is applied to convert an assigned life cycle inventory analysis result to the common unit of the category indicator. The common unit allows calculation of the category indicator result. [ISO 14040:2006]

classification – assignment of LCI results to impact categories [ISO 14044:2006]

closed-loop product systems - material is recycled in the same product system [ISO 14044:2006]

comparative assertion - environmental claim regarding the superiority or equivalence of one product versus a competing product that performs the same function [ISO 14040:2006]

competence - demonstrated personal attributes and demonstrated ability to apply knowledge and skills [ISO 19011:2002]

complementary product – see ancillary product

completeness – percentage of locations reporting primary data from the potential number in existence for each data category in a unit process [ISO 14041:1998]

completeness check - process of verifying whether information from the phases of a life cycle assessment is sufficient for reaching conclusions in accordance with the goal and scope definition [ISO 14040:2006]

consequential modeling - LCI modeling principle that identifies and models all processes in the background system of a system in consequence of decisions made in the foreground system. [JRC 2010]

consistency check - process of verifying that the assumptions, methods and data are consistently applied throughout the study and are in accordance with the goal and scope definition performed before conclusions are reached [ISO 14040:2006]

consumer - individual member of the general public purchasing or using goods, property or services for private purposes [ISO 14025:2006]

contribution analysis – contribution of life cycle stages or groups of processes to the total result are examined by, expressing the contribution as a percent of the total [ISO 14044:2006]

co-product - any of two or more products coming from the same unit process or product system [ISO 14040:2006]

CO2e, CO2 equivalent - see carbon dioxide equivalent

CPC – Central Product Classification

cradle-to-gate – Scope of an LCA study covering all unit processed from raw material production to the point at which the product leaves a facility, usually a manufacturing facility. Cradle to gate studies are often performed on raw materials such as plastic resins or metal rods or sheets.

cradle-to-grave - assessment that considers impacts at each stage of a product's life cycle, from the time natural resources are extracted from the ground and processed through each subsequent stage of manufacturing, transportation, product use, recycling, and ultimately, disposal. [JRC 2012]

critical review - process intended to ensure consistency between a life cycle assessment and the principles and the requirements of the International Standards on life cycle assessment described in ISO 14044 [ISO 14040:2006]

cut-off criteria - specification of the amount of material or energy flow or the level of significance associated with unit processes or product system to be excluded from a study [ISO 14040:2006]

damage approach – see end-point method

data commissioner – person(s) or organization(s) which commissions the data collection and documentation [ISO 14048:2002]

data documentation format – structure of documentation of data. This includes data fields, sets of data fields and their relationship [ISO 14048:2002]

data documentor - person(s) or organization(s) responsible for entering the data into the data documentation format in use [ISO 14048:2002]

data field - container for specified data with a specified data type [ISO 14048:2002]

data quality - characteristics of data that relate to their ability to satisfy stated requirements [ISO 14040:2006]

data quality analysis - understanding the reliability of the collection of indicator results, the LCIA profile [ISO 14044:2006]

data source – origin of data [ISO 14048:2002]

data type – nature of the data e.g. units, quantitative, short string, free text, numerical, logical [ISO 14048:2002]

declared unit - quantity of a building product for use as a reference unit in an EPD, based on LCA, for the expression of environmental information needed in information modules. The declared unit is used where the function and the reference scenario for the whole life cycle, on the building level, cannot be stated [ISO 21930:2007]

design for environment - methods supporting product developers in reducing the total environmental impact of a product early in the product development process. This includes reducing resource consumption as well as emissions and waste [JRC 2012]

direct land use change - change in human use or management of land at the location of the production, use or disposal of raw materials, intermediate products and final products or wastes in the product system being assessed [ISO 14067:2011]

dLUC - see direct land use change

ecodesign – see design for environment

eco-efficiency – joint analysis of the environmental and economic implications of a product or technology, aiming to support choosing the method for production, service, disposal or recovery that makes most ecological and economic sense, ensuring optimum conservation of resources, minimum emissions and waste generation at a low overall cost. [JRC 2012]

ecosphere - raw materials taken from nature or returned to nature

elementary flow - material or energy entering the system being studied that has been drawn from the environment without previous human transformation or material or energy leaving the system being studied that is released into the environment without subsequent human transformation . [ISO 14040:2006]

elementary water flow - water entering the system being studied and that has been drawn from the environment, or water leaving the system being studied that is released into the environment [ISO 14046:2012]

end-of-life product - product at the end of its useful life that will potentially undergo reuse, recycling, or recovery [JRC 2010]

end-point method - methods for modeling the effects of emissions directly for the protection targets (natural environment's ecosystems, human health, resource availability). Endpoint methods typically follow the midpoint modeling considering the severity and reversibility of effects and the models' uncertainties [JRC 2012]

energy flow - input to or output from a unit process or product system, quantified in energy units. Energy flow that is an input can be called an energy input; energy flow that is an output can be called an energy output. [ISO 14040:2006]

environmental aspect - element of an organization's activities, products or services which can interact with the environment [ISO 14020:2000]

environmental declaration - claim which indicates the environmental aspects of a product or service. An environmental label or declaration may take the form of a statement, symbol or graphic on a product or package label, in product literature, in technical bulletins, in advertising or in publicity, amongst other things. [ISO 14020:2000]

environmental impact - any change to the environment, whether adverse or beneficial, wholly or partially resulting from an organization's environmental aspects [ISO 14001:2004]

environmental label – see environmental declaration

environmental mechanism - system of physical, chemical and biological processes for a given impact category, linking the life cycle inventory analysis results to category indicators and to category endpoints [ISO 14040:2006]

environmental product declaration – see type III environmental declaration

EPD – see type III environmental declaration

evaluation - element within the life cycle interpretation phase intended to establish confidence in the results of the life cycle assessment. Evaluation includes completeness check, sensitivity check, consistency check, and any other validation that may be required according to the goal and scope definition of the study [ISO 14040:2006]

evapotranspiration - transfer of water to atmosphere as a result of evaporation and plant transpiration [ISO 14046:2012]

feedstock energy - heat of combustion of a raw material input that is not used as an energy source to a product system, expressed in terms of higher heating value or lower heating value [ISO 14040:2006]

final product – product which requires no additional transformation prior to its use [ISO 14041:1998]

fossil carbon - carbon which is contained in fossilized material. Examples of fossilized material are coal, oil and natural gas. [ISO 14067:2011]

fossil water - groundwater body (deep aquifer) that is a non-renewable water resource [ISO 14046:2012]

freshwater - water having a low concentration of dissolved solids. Freshwater typically contains less than 1000 milligrams per liter of dissolved solids and is generally accepted as suitable for abstraction and treatment to produce potable water. The concentration of total dissolved solids can vary considerably over space and/or time. [ISO 14046:2012]

fugitive emission – uncontrolled emission to air, water or land [ISO 14041:1998]

functional unit - quantified performance of a product system, process, or organization for use as a reference unit. As the CFP treats information on a product, the functional unit can be a product unit, sales unit or service unit. [ISO 14040:2006]

gate - point at which the building product or material leaves the factory before it becomes an input into another manufacturing process or before it goes to the distributor, a factory or building site [ISO 21930:2007]

gate-to-gate - scope of a partial life cycle study that covers the unit processes within a facility, but neither upstream nor downstream of the facility. Often this represents the lowest process level at hich data are available and is therefore also a unit process.

geographic coverage - geographical area from which data for unit processes should be collected to satisfy the goal of the study [ISO 14044:2006]

GHG – see greenhouse gas

global warming potential - characterization factor describing the mass of carbon dioxide that has the same accumulated radiative forcing over a given period of time as one mass unit of a given greenhouse gas [ISO 14064-1:2006]

greenhouse gas - gaseous constituent of the atmosphere, both natural and anthropogenic, that absorbs and emits radiation at specific wavelengths within the spectrum of infrared radiation emitted by the earth's surface, the atmosphere, and clouds. Water vapor and ozone are anthropogenic as well as natural greenhouse gases but are not included as recognized greenhouse gases due to difficulties, in most cases, in isolating the human-induced component of global warming attributable to their presence in the atmosphere. [ISO 14064-1:2006]

greenhouse gas emission - mass of a greenhouse gas released to the atmosphere [ISO 14064-1:2006]

greenhouse gas emission factor - mass of a greenhouse gas emitted relative to an input or an output of a unit process or a combination of unit processes [ISO 14067:2011]

greenhouse gas removal - mass of a greenhouse gas removed from the atmosphere [ISO 14064-1:2006]

greenhouse gas sink - process that removes a greenhouse gas from the atmosphere. The process can be natural or anthropogenic. [ISO 14067:2011]

greenhouse gas source - process that releases a greenhouse gas into the atmosphere. The process can be natural or anthropogenic. [ISO 14067:2011]

gravity analysis - a statistical procedure that identifies those data having the greatest contribution to the indicator result. These items may then be investigated with increased priority to ensure that sound decisions are made [ISO 14044:2006]

groundwater - water within the saturated zone in which interstices of rock or other materials are filled with water [ISO 14046:2012]

grouping - the assignment of impact categories into one or more sets as predefined in the goal and scope definition, and it may involve sorting and/or ranking [ISO 14044:2006]

GWP - see global warming potential

hybrid LCA – An LCA analysis that includes both process LCI data derived from measures or calculations of emissions and resource use based on a physical reference flow and economic input-output data derived from sectoral estimates of emissions and resource use derived from economic input-output data using currency as the reference flow.

iLUC - see indirect land use change

impact category - class representing environmental issues of concern to which life cycle inventory analysis results may be assigned [ISO 14040:2006]

impact category indicator - quantifiable representation of an impact category [ISO 14040:2006]

indirect land use change – change in the use or management of land which is a consequence of the production, use or disposal of raw materials, intermediate products and final products or wastes in the product system, but which is not taking place at the location of the activities that cause the change [ISO 14067:2011]

information module - compilation of data covering a unit process or a combination of unit processes that are part of the life cycle of a product. One or more information modules can be the basis of a type III environmental product declaration or partial CFP, and several information modules can be the basis of a CFP. [ISO 14067:2011]

input - product, material or energy flow that enters a unit process. Products and materials include raw materials, intermediate products and co-products. [ISO 14040:2006]

interested party - person or group of people that holds a view that can affect the organization [ISO/DIS 20121:2011]; individual or group concerned with or affected by the environmental performance of a product system, or by the results of the life cycle assessment [ISO 14040:2006]; person or body interested in or affected by the development and use of a Type III environmental declaration [ISO 14025-1:2006]

intermediate flow - product, material or energy flow occurring between unit processes of the product system being studied [ISO 14040:2006]

intermediate product - output from a unit process that is input to other unit processes that require further transformation within the system [ISO 14040:2006]

International Organization for Standardization - a worldwide federation of national standards bodies. The work of preparing International Standards is normally carried out through ISO technical committees. Each member body interested in a subject for which a technical committee has been established has the right to be represented on that committee. International organizations, governmental and non-governmental, in liaison with ISO, also take part in the work.

ISO – see International Organization for Standardization

LCA – see life cycle assessment

LCACP – see life cycle assessment certified professional

LCC – see life cycle cost

LCI – see life cycle inventory analysis

LCIA – see life cycle impact assessment

life cycle - consecutive and interlinked stages of a product system, from raw material acquisition or generation from natural resources to the final disposal. Product includes any goods or services [ISO 14040:2006]

life cycle assessment – compilation and evaluation of the inputs, outputs and the potential environmental impacts of a product system throughout its life cycle [ISO 14040:2006]

life cycle assessment certified professional - an individual in good standing under the American Center for Life Cycle Assessment certification program, or other comparable program developed for LCA professionals under the ISO 17024 standard

life cycle cost - total cost linked to the purchase, operation, and disposal of a product OR the cost of a product or service over its entire life cycle including external costs [JRC 2012]

life cycle impact assessment - phase of life cycle assessment aimed at understanding and evaluating the magnitude and significance of the potential environmental impacts for a product system throughout the life cycle of the product [ISO 14040:2006]

life cycle interpretation - phase of life cycle assessment in which the findings of either the life cycle inventory analysis or the life cycle impact assessment, or both, are evaluated in relation to the defined goal and scope in order to reach conclusions and recommendations [ISO 14040:2006]

life cycle inventory analysis - phase of life cycle assessment involving the compilation and quantification of inputs and outputs for a product throughout its life cycle [ISO 14040:2006]

life cycle inventory analysis result - outcome of a life cycle inventory analysis that catalogues the flows crossing the system boundary and provides the starting point for life cycle impact assessment [ISO 14040:2006]

life cycle stage – activities associated with the production and delivery of raw materials or generation of natural resources to the final disposal

life cycle thinking – a decision-support approach that integrates existing consumption and production strategies towards a more coherent policy making and in industry, employing a bundle of life cycle based approaches and tools. By considering the whole life cycle, the shifting of problems from one life cycle stage to another, from one geographic area to another and from one environmental medium or protection target to another is avoided. [JRC 2012]

mass balance - a requirement of the first law of thermodynamics that mass is neither created nor destroyed. In the context of LCA, all inputs and outputs of a unit process, system process or product system should have the mass of the inputs equal the mas of the outputs.

mid-point method - a term that specifies the results of traditional LCIA characterization and normalization methods as indicators located between emission and endpoint damages in the impact pathway at the point where it is judged that further modeling involves too much uncertainty. [JRC 2012]

nomenclature – sets of rules to name and classify data in a consistent and unique way [ISO 14048:2002]

non-renewable resource - resource that exists in a fixed amount that cannot be replenished on a human time scale [ISO 21930:2007]

non-renewable water resource - a water body that has a negligible rate of natural recharge on the human time-scale [ISO 14046:2012]

normalization - calculating the magnitude of category indicator results relative to reference information. The aim of the normalization is to understand better the relative magnitude for each indicator result of the product system under study. [ISO 14044:2006]

offsetting – mechanism for compensating for all or for a part of the CFP through the prevention of the release of, reduction in, or removal of an amount of greenhouse gas emissions in a process outside the boundary of the product system. e.g. external investment in renewable energy technologies; energy efficiency measures; afforestation/reforestation. Offsetting is not allowed in the CFP quantification and thus is not reflected in any CFP communication [ISO 14021:1999/FDAM 1:2011]

open-loop product systems – material is recycled into other product systems and the material undergoes a change to its inherent properties [ISO 14044:2006]

organization – company, corporation, firm, enterprise, authority or institution, or part or combination thereof, whether incorporated or not, public or private, that has its own functions and administration [ISO 14001:2004]

output – product, material or energy flow that leaves a unit process. Products and materials include raw materials, intermediate products, co-products and releases. [ISO 14040:2006]

partial carbon footprint of a product – sum of greenhouse gas emissions and removals of one or more selected process(es) of a product system, expressed as CO_2 equivalent and based on a life cycle assessment. A partial CFP often covers processes that model specific stages of the life cycle. The partial CFP is based on or compiled from specific processes or information modules which are part of a product system and may form the basis for quantification of a CFP [ISO 14067:2011]

PCR – see product category rules

precision – measure of the variability of the data values for each data category expressed [ISO 14041:1998]

primary data – quantified value of a unit process or an activity within the product system obtained from a direct measurement or a calculation based on direct measurements at its original source. Primary data need not necessarily originate from the product system under study. Primary data may include GHG emission factors and/or GHG activity data. [ISO 14067:2011]

process - set of interrelated or interacting activities that transforms inputs into outputs [ISO 14040:2006], [ISO 9000:2005]

process energy - energy input required for operating the process or equipment within a unit process, excluding energy inputs for production and delivery of the energy itself [ISO 14040:2006]

product – any goods or service. The product can be categorized as follows: service (e.g. transport, implementation of events, electricity); software (e.g. computer program); hardware (e.g. engine mechanical part); processed material (e.g. lubricant, ore, fuel); unprocessed material (e.g. agricultural produce). Services have tangible and intangible elements. Provision of a service can involve, for example, the following: an activity performed on a customer-supplied tangible product (e.g. automobile to be repaired); an activity performed on a customer-supplied intangible product (e.g. the income statement needed to prepare a tax return); the delivery of an intangible product (e.g. the delivery of information in the context of knowledge transmission); the creation of ambience for the customer (e.g. in hotels and restaurants).Software consists of information and is generally intangible and can be in the form of approaches, transactions or procedures. Hardware is generally tangible and its amount is a countable characteristic. Processed materials are generally tangible and their amount is a continuous characteristic. [ISO 14040:2006]

product category – group of products that can fulfill equivalent functions [ISO 14025:2006]

product category rules - set of specific rules, requirements and guidelines for developing Type III environmental declarations for one or more product categories. PCR include quantification rules compliant with ISO 14044. [ISO 14025:2006]

product category rules review - process whereby a third party panel verifies the product category rules [ISO 14025:2006]

product chain – see supply chain

product flow - products entering from or leaving to another product system [ISO 14040:2006]

product system - collection of unit processes with elementary flows and product flows, performing one or more defined functions and which models the life cycle of a product [ISO 14040:2006]

program operator - body or bodies that conduct a Type III environmental declaration program. A program operator can be a company or a group of companies, industrial sector or trade association, public authorities or agencies, or an independent scientific body or other organization. [ISO 14025:2006]

raw material - primary or secondary material that is used to produce a product. Secondary material includes recycled material. [ISO 14040:2006]

reference flow - measure of the outputs from processes in a given product system required to fulfill the function expressed by the functional unit [ISO 14040:2006]

reference service life - service life of a building product that is known or expected under a particular set, i.e., a reference set, of in-use conditions and that may form the basis of estimating the service life under other in-use conditions. The reference service life is applied in the functional unit/declared unit. [ISO 21930:2007]

releases - emissions to air and discharges to water and soil [ISO 14040:2006]

renewable resource - resource that is grown, naturally replenished or cleansed on a human time scale. A renewable resource is capable of being exhausted but can last indefinitely with proper stewardship. [ISO 21930:2007]

representativeness – qualitative assessment of degree to which the data set reflects the true population of interest. Considerations could include geographical, time period and technology coverages. [ISO 14041:1998]

seawater - water from a sea or ocean or coastal area. Seawater has a concentration of dissolved solids greater than or equal to 30,000 milligrams per liter. [ISO 14046:2012]

secondary data - data obtained from sources other than a direct measurement or a calculation based on direct measurements at the original source within the product system. Such sources can include databases, published literature, national inventories and other generic sources [ISO 14067:2011]

secondary fuels - fuels or fuel products that are derived from primary fuels [ISO 21930:2007]

sensitivity analysis - systematic procedures for estimating the effects of the choices made regarding methods and data on the outcome of a study [ISO 14040:2006]

sensitivity check - process of verifying that the information obtained from a sensitivity analysis is relevant for reaching the conclusions and for giving recommendations [ISO 14040:2006]

service life - period of time during which a product in use meets or exceeds the performance requirements [ISO 15686-1:2011]

site-specific data - data obtained from a direct measurement or a calculation based on direct measurement at its original source within the product system. All site-specific data are "primary data" but not all primary data are site-specific data because they may also relate to a different product system [ISO 14067:2011]

SME - small and medium sized enterprises

supply chain - parties involved, through upstream and downstream linkages, in processes and activities delivering value in the form of products to the end user. In practice, the expression "interlinked chain" applies from suppliers to those involved in end-of-life processing which may include vendors, manufacturing facilities, logistics providers, internal distribution centers, distributors, wholesalers and other entities that lead to the end user. [ISO/TR 14062:2002]

supplier – party that is responsible for ensuring that products meet and, if applicable, continue to meet, the requirements on which the certification is based [ISO/IEC Guide 65:1996]

surface water - all water in overland flow and storage, for example rivers and lakes, excluding seawater [ISO 14046:2012]

system boundary - set of criteria specifying which unit processes are part of a product system [ISO 14040:2006]

system expansion - Adding specific processes or products and the related life cycle inventories to the analyzed system. Used to make several multifunctional systems with an only partly equivalent set of functions comparable within LCA [JRC 2010]

technology coverage – technology mix (e.g. weighted average of the actual process mix, best available technology or worst operating unit) [ISO 14041:1998]

technosphere flow – all modified products and services. Technosphere flows are always accompanied by a financial exchange

third-party - person or body that is recognized as being independent of the parties involved, as concerns the issues in question. "Parties involved" are usually supplier ("first party") and purchaser ("second party") interests. [ISO 14024:1999]

time-related coverage – the desired age of data and the minimum length of time over which data should be collected [ISO 14041:1998]

transparency - open, comprehensive and understandable presentation of information [ISO 14040:2006]

type I environmental declaration - a voluntary, multiple-criteria based, third party program that awards a license which authorizes the use of environmental labels on products indicating overall environmental preferability of a product within a product category based on life cycle considerations [ISO 14024:1999]

type II environmental declaration - informative environmental self-declaration claims [ISO 14021:1999]

type III environmental declaration - environmental declaration providing quantified environmental data using predetermined parameters and, where relevant, additional environmental information. The predetermined parameters are based on the ISO 14040 series of standards, which is made up of ISO 14040 and ISO 14044. The additional environmental information may be quantitative or qualitative. [ISO 14025:2006]

type III environmental declaration program - voluntary program for the development and use of Type III environmental declarations, based on a set of operating rules [ISO 14025:2006]

uncertainty - parameter associated with the result of quantification which characterizes the dispersion of the values that could be reasonably attributed to the quantified amount. Uncertainty information typically specifies quantitative estimates of the likely dispersion of values and a qualitative description of the likely causes of the dispersion [ISO 14064-1: 2006].

uncertainty analysis - systematic procedure to quantify the uncertainty introduced in the results of a life cycle inventory analysis due to the cumulative effects of model imprecision, input uncertainty and data variability. Either ranges or probability distributions are used to determine uncertainty in the results. [ISO 14040:2006]

unit process - smallest element considered in the life cycle inventory analysis (3.5.6) for which input and output data are quantified [ISO 14040:2006]

United Nations Standard Products and Services Code - an open, global multi-sector standard for efficient, accurate classification of products and services.

UNSPSC – see United Nations Standard Products and Services Code

value chain – see supply chain

verification - confirmation, through the provision of objective evidence, that specified requirements have been fulfilled [ISO 9000:2005]

verification criteria - policy, procedure or requirement used as a reference against which evidence is compared. Verification criteria may be established by governments, GHG programs voluntary reporting initiatives, standards or good practice guidance [ISO 14064-1:2006]

verifier - person or body that carries out verification [ISO 14025:2006]

waste - substances or objects which the holder intends or is required to dispose of [ISO 14040:2006]

water abstraction – see water withdrawal

water availability - metric describing to which extent humans and ecosystems have sufficient water resources for their needs. Water availability depends on the location. The level of temporal and geographical coverage and resolution for evaluating water availability depends on the goal and scope. Water quality can also influence availability, e.g. if the quality is not sufficient to meet the user's needs. [ISO 14040:2006]

water body - any accumulation of water which has definite hydrological, hydrogeomorphological, physical, chemical and biological characteristics in a given geographical area. Examples of water body includes: lakes, rivers, aquifers, sea, icebergs and glaciers. The geographical resolution of a water body should be determined at the goal and scope stage: it may regroup different small water bodies. [ISO 14046:2012]

water consumption - water withdrawal where release back to the same water body does not occur, e.g. because of evaporation, evapotranspiration, product integration or discharge into a different drainage basin or the sea [ISO 14046:2012]

water footprint - metric(s) that quantify(ies) the potential environmental impacts related to water. If not all comprehensive water related impacts have not been assessed, then the term water footprint shall only be applied with qualification [ISO 14046:2012]

water footprint assessment - compilation and evaluation of the inputs, outputs and the potential environmental impacts related to water of a product, process or organization [ISO 14046:2012]

water footprint inventory analysis - phase of water footprint assessment involving compilation and quantification of inputs and outputs related to water for products, processes or organizations as defined in the goal and scope [ISO 14046:2012]

water footprint profile - compilation of impact category indicator results addressing potential environmental impacts related to water [ISO 14046:2012]

water impact assessment - phase of a water footprint assessment aimed at understanding and evaluating the magnitude and significance of the potential environmental impact(s) related to water of a product, process or organization [ISO 14046:2012]

water quality - chemical, physical (e.g. thermal) and biological characteristics of water with respect to its suitability for an intended use by human or ecosystems [ISO 14046:2012]

water use - any use of water by human activity. Use includes, but is not limited to, any water withdrawal, water discharge or other human activity within the water body including in-stream or in situ uses such as fishing, recreation, transportation [ISO 14046:2012]

water withdrawal - anthropogenic removal of water from any water body, either permanently or temporarily [ISO 14046:2012]

weighting – process of converting and possibly aggregating indicator results across impact categories using numerical factors based on value-choices; data prior to weighting should remain available [ISO 14044:2006]

well-to-wheel – A scope of an LCA study that studies the fuel cycle of a wheeled vehicle. It includes fossil fuel extraction through combustion.

WTO – World Trade Organization

REFERENCES

ISO/IEC, 1996. IEC Guide 65: General requirements for bodies operating product certification systems

ISO, 1999a. 14021: Environmental labels and declarations — Self-declared environmental claims (Type II environmental labelling)

ISO, 1999b. 14024: Environmental labels and declarations — Type I environmental labelling — Principles and Procedures

ISO, 2000. 14020: Environmental labels and declarations General principles

ISO, 2002. 14048: Environmental management — Life cycle assessment — Data documentation format

ISO/TR, 2002. 14062: Environmental management -- Integrating environmental aspects into product design and development

ISO, 2004. 14001: Environmental management systems Requirements with guidance for use

ISO, 2005. 9000: Quality management systems -- Fundamentals and vocabulary

ISO, 2006a. 14025: Environmental labels and declarations — Type III environmental declarations — Principles and procedures

ISO, 2006b. 14040: Environmental management — Life cycle assessment — Principles and framework

ISO, 2006c. 14064-1: Greenhouse gases Part 1: Specification with guidance at the organization level for quantification and reporting of greenhouse gas emissions and removals

ISO, 2007. 21930: Sustainability in building construction — Environmental declaration of building products

ISO, 2011a. 14067: Carbon footprint of products Requirements and guidelines for quantification and communication

ISO, 2011b. 15686-1: Buildings and constructed assets -- Service life planning -- Part 1: General principles and framework

ISO, 2012. 14046-1:2012, Water footprin Principles, requirements, and guidelines

EC JRC, 2010. International Reference Life Cycle Data System Handbook. European Comission Joint Reseach Centre and the Institute for Environment and Sustainability, available at: http://lct.jrc.ec.europa.eu/assessment/publications.

EC JRC, 2012. LCA Info Hub Glossary, available at: http://lca.jrc.ec.europa.eu/lcainfohub/glossary.vm

PERIODIC TABLE

Legend:
- element name → **HYDROGEN**
- atomic number → **1**
- chemical Symbol → **H**
- atomic weight (u) → **1,008**

- Alkali metal
- Alkaline earth metal
- Lanthanide
- Actinide
- Transition metal
- Post-transition metal
- Metalloid
- Nonmetal
- Halogen
- Noble gas
- ● Solid
- ● Liquid
- ● Gas
- ● Unknown

Group	IA	IIA	IIIB	IVB	VB	VIB	VIIB	VIIIB	VIIIB	VIIIB	IB	IIB	IIIA	IVA	VA	VIA	VIIA	VIIIA

Period 1
- HYDROGEN 1 H 1,008
- HELIUM 2 He 4,003

Period 2
- LITHIUM 3 Li 6,941
- BERYLLIUM 4 Be 9,012
- BORON 5 B 10,81
- CARBON 6 C 12,01
- NITROGEN 7 N 14,01
- OXYGEN 8 O 16,00
- FLUORINE 9 F 19,00
- NEON 10 Ne 20,18

Period 3
- SODIUM 11 Na 22,99
- MAGNESIUM 12 Mg 24,31
- ALUMINIUM 13 Al 26,98
- SILICON 14 Si 28,09
- PHOSPHORUS 15 P 30,97
- SULFUR 16 S 32,07
- CHLORINE 17 Cl 35,45
- ARGON 18 Ar 39,95

Period 4
- POTASSIUM 19 K 39,10
- CALCIUM 20 Ca 40,08
- SCANDIUM 21 Sc 44,96
- TITANIUM 22 Ti 47,87
- VANADIUM 23 V 50,94
- CHROMIUM 24 Cr 52,00
- MANGANESE 25 Mn 54,94
- IRON 26 Fe 55,85
- COBALT 27 Co 58,93
- NICKEL 28 Ni 58,69
- COPPER 29 Cu 63,55
- ZINC 30 Zn 65,39
- GALLIUM 31 Ga 69,72
- GERMANIUM 32 Ge 72,59
- ARSENIC 33 As 74,92
- SELENIUM 34 Se 78,96
- BROMINE 35 Br 79,90
- KRYPTON 36 Kr 83,80

Period 5
- RUBIDIUM 37 Rb 85,47
- STRONTIUM 38 Sr 87,62
- YTTRIUM 39 Y 88,91
- ZIRCONIUM 40 Zr 91,22
- NIOBIUM 41 Nb 92,91
- MOLYBDENUM 42 Mo 95,94
- TECHNETIUM 43 Tc (98,91)
- RUTHENIUM 44 Ru 101,1
- RHODIUM 45 Rh 102,9
- PALLADIUM 46 Pd 106,4
- SILVER 47 Ag 107,9
- CADMIUM 48 Cd 112,4
- INDIUM 49 In 114,8
- TIN 50 Sn 118,7
- ANTIMONY 51 Sb 121,8
- TELLURIUM 52 Te 127,6
- IODINE 53 I 126,9
- XENON 54 Xe 131,3

Period 6
- CAESIUM 55 Cs 132,9
- BARIUM 56 Ba 137,3
- LANTHANUM 57 La 138,9
- HAFNIUM 72 Hf 178,5
- TANTALUM 73 Ta 180,9
- TUNGSTEN 74 W 183,9
- RHENIUM 75 Re 186,2
- OSMIUM 76 Os 190,2
- IRIDIUM 77 Ir 192,2
- PLATINUM 78 Pt 195,1
- GOLD 79 Au 197,0
- MERCURY 80 Hg 200,6
- THALLIUM 81 Tl 204,4
- LEAD 82 Pb 207,2
- BISMUTH 83 Bi 209,0
- POLONIUM 84 Po (210,0)
- ASTATINE 85 At (210,0)
- RADON 86 Rn (222,0)

Period 7
- FRANCIUM 87 Fr (223,0)
- RADIUM 88 Ra (226,0)
- ACTINIUM 89 Ac (227,0)
- RUTHERFORDIUM 104 Rf ——
- DUBNIUM 105 Db ——
- SEABORGIUM 106 Sg ——
- BOHRIUM 107 Bh ——
- HASSIUM 108 Hs ——
- COPERNICIUM 112 Cn ——

Lanthanides
- CERIUM 58 Ce 140,1
- PRASEODYMIUM 59 Pr 140,9
- NEODYMIUM 60 Nd 144,2
- PROMETHIUM 61 Pm (144,9)
- SAMARIUM 62 Sm 150,4
- EUROPIUM 63 Eu 152,0
- GADOLINIUM 64 Gd 157,3
- TERBIUM 65 Tb 158,9
- DYSPROSIUM 66 Dy 162,5
- HOLMIUM 67 Ho 164,9
- ERBIUM 68 Er 167,3
- THULIUM 69 Tm 168,9
- YTTERBIUM 70 Yb 173,0
- LUTETIUM 71 Lu 175,0

Actinides
- THORIUM 90 Th (232,0)
- PROTACTINIUM 91 Pa (231,0)
- URANIUM 92 U (238,0)
- NEPTUNIUM 93 Np (237,0)
- PLUTONIUM 94 Pu (239,1)
- AMERICIUM 95 Am (243,1)
- CURIUM 96 Cm (247,1)
- BERKELIUM 97 Bk (247,1)
- CALIFORNIUM 98 Cf (252,1)
- EINSTEINIUM 99 Es (252,1)
- FERMIUM 100 Fm (257,1)
- MENDELEVIUM 101 Md (256,1)
- NOBELIUM 102 No (259,1)
- LAWRENCIUM 103 Lr (260,1)

period

CPSIA information can be obtained
at www.ICGtesting.com
Printed in the USA
BVHW051221300121
598879BV00002B/19